# Recent Progress in Parasitology

# Recent Progress in Parasitology

Editor: Henry Evans

R CALLISTO REFERENCE

www.callistoreference.com

**Callisto Reference,**
118-35 Queens Blvd., Suite 400,
Forest Hills, NY 11375, USA

Visit us on the World Wide Web at:
www.callistoreference.com

ISBN: 978-1-63239-923-6 (Hardback)

**Cataloging-in-Publication Data**

Recent progress in parasitology / edited by Henry Evans.
    p. cm.
Includes bibliographical references and index.
ISBN 978-1-63239-923-6
1. Parasitology. 2. Molecular parasitology. 3. Biology. I. Evans, Henry.
QL757 .R43 2018
577.857--dc23

# Table of Contents

# Preface

Parasitology studies parasites, their interaction with each other and with hosts. This book on parasitology presents researches and studies performed by experts from across the globe. Parasites are unwanted organisms that cause and spread diseases and infections. Parasitology can be further divided into sub-disciplines such as medical parasitology, parasite ecology, veterinary parasitology, etc. This book, with its detailed analyses and data, will prove immensely beneficial to professionals and students involved in this area at various levels. Coherent flow of topics, student-friendly language and extensive use of examples make this book an invaluable source of knowledge. This book will be of great help to students and academicians in the fields of immunology, vector biology and biochemistry.

This book has been the outcome of endless efforts put in by authors and researchers on various issues and topics within the field. The book is a comprehensive collection of significant researches that are addressed in a variety of chapters. It will surely enhance the knowledge of the field among readers across the globe.

It gives us an immense pleasure to thank our researchers and authors for their efforts to submit their piece of writing before the deadlines. Finally in the end, I would like to thank my family and colleagues who have been a great source of inspiration and support.

Editor

# Partial Protective Effect of Intranasal Immunization with Recombinant *Toxoplasma gondii* Rhoptry Protein 17 against Toxoplasmosis in Mice

**Hai-Long Wang**[1]*, **Tie-E Zhang**[1], **Li-Tian Yin**[2], **Min Pang**[3], **Li Guan**[1], **Hong-Li Liu**[1], **Jian-Hong Zhang**[1], **Xiao-Li Meng**[1], **Ji-Zhong Bai**[4], **Guo-Ping Zheng**[5], **Guo-Rong Yin**[1]*

1 Research Institute of Medical Parasitology, Shanxi Medical University, Taiyuan, Shanxi, PR China, 2 Department of Physiology, Key Laboratory of Cellular Physiology Co-constructed by Province and Ministry of Education, Shanxi Medical University, Taiyuan, Shanxi, PR China, 3 Department of Respiratory, the First Affiliated Hospital, Shanxi Medical University, Taiyuan, Shanxi, PR China, 4 Department of Physiology, Faculty of Medical and Health Sciences, University of Auckland, Auckland, New Zealand, 5 Department of Biochemistry and Molecular Biology, Shanxi Medical University, Taiyuan, Shanxi, PR China

## Abstract

*Toxoplasma gondii* (*T. gondii*) is an obligate intracellular protozoan parasite that infects a variety of mammals, including humans. An effective vaccine for this parasite is therefore needed. In this study, RH strain *T. gondii* rhoptry protein 17 was expressed in bacteria as a fusion with glutathione S-transferase (GST) and the recombinant proteins (rTgROP17) were purified via GST-affinity chromatography. BALB/c mice were nasally immunised with rTgROP17, and induction of immune responses and protection against chronic and lethal *T. gondii* infections were investigated. The results revealed that mice immunised with rTgROP17 produced high levels of specific anti-rTgROP17 IgGs and a mixed IgG1/IgG2a response of IgG2a predominance. The systemic immune response was associated with increased production of Th1 (IFN-γand IL-2) and Th2 (IL-4) cytokines, and enhanced lymphoproliferation (stimulation index, SI) in the mice immunised with rTgROP17. Strong mucosal immune responses with increased secretion of TgROP17-specific secretory IgA (SIgA) in nasal, vaginal and intestinal washes were also observed in these mice. The vaccinated mice displayed apparent protection against chronic RH strain infection as evidenced by their lower liver and brain parasite burdens (59.17% and 49.08%, respectively) than those of the controls. The vaccinated mice also exhibited significant protection against lethal infection of the virulent RH strain (survival increased by 50%) compared to the controls. Our data demonstrate that rTgROP17 can trigger strong systemic and mucosal immune responses against *T. gondii* and that ROP17 is a promising candidate vaccine for toxoplasmosis.

**Editor:** Mauricio M. Rodrigues, Federal University of São Paulo, Brazil

**Funding:** This work was supported by the National Natural Science Foundation of China (No. 81071374), the Natural Science Found of Shanxi Province (No. 2012011036-2), the Technique Innovation of Shanxi Medical University (No. 01201103), the 331 Early Career Researcher Grant of Shanxi Medical University (No. 201202), the Research Fund for the Doctoral Program of Shanxi Medical University (No. 03201307), the College Students Innovation and Entrepreneurship Training Programs of Shanxi Province (No. 2011120) and the Advanced Program for Returned Scholars of Shanxi Province Science and Technology Activities. This work was also supported by the Biology Postdoctoral Mobile Research Station of Shanxi Medical University. The funders had no role in study design, data collection and analysis, decision to publish, or preparation of the manuscript.

**Competing Interests:** The authors have declared that no competing interests exist.

* Email: guorongyin@163.com (GRY); longwty@163.com (HLW)

## Introduction

*Toxoplasma gondii* is an obligate intracellular parasite of the phylum *Apicomplexa* [1] that is found worldwide. *T. gondii* has a gigantic intermediate host range that comprises nearly any bird and mammal, including humans. Symptoms associated with *Toxoplasma* infection in humans range from none to severe and can be fatal. For example, *T. gondii* infection is usually asymptomatic but occasionally triggers mild symptoms when *T. gondii* infects immunocompetent hosts. When *T. gondii* infects immunocompromised hosts, such as AIDS patients or malignancy patients, it can lead to severe or even lethal damage [2,3]. In addition, *T. gondii* infection of livestocks can also result in serious economic losses due to abortion, stillbirth, and neonatal death. Moreover, infected livestock are a major route of *T. gondii* transmission to humans [4]. Therefore, the development of

effective and safe methods, such as vaccines [5], to control *T. gondii* infection is crucial for human health and animal husbandry.

Currently, candidate vaccines that have been testing in mice are in the focus of protective antigen selections from membrane-associated surface antigens, excreted-secreted dense granule proteins, rhoptry proteins and micronemal proteins [6–8]. Of interests, rhoptry proteins (ROPs) excreted by rhoptries of the apical secretory organelles are involved in parasitic invasion [9]. Some of the ROPs, such as ROP16 and ROP18, are serine-threonine kinases named as ROP kinases (ROPK) and play the role of virulence factors [10–13]. Recently, ROP16 and ROP18 have been used as immunogens to vaccinate mice, and enabled the mice to produce considerable cellular and humoral immune responses that partly protected the mice against *T. gondii* infection [14–17].

Rhoptry protein 17 (ROP17), which belongs to the ROP2 superfamily, is predicted to be a ROPK [18] and possesses a key ATP-binding domain and conserved residues in its catalytic triad region (KDD) [18,19] as ROP16 and ROP18 [11,20]. Our previous study showed that ROP17 has kinase activity because it can phosphorylate c-Jun in HEK 293T cells [21]. Given that ROP16 and ROP18 can induce mice to produce protective immune responses against *Toxoplasma*, we hypothesize that ROP17 is a candidate vaccine for *T. gondii*.

Recombinant *T. gondii* ROP17 (rTgROP17) protein was produced in *Escherichia coli* (*E. coli*) and shown to exhibit specific antigenicity in our previous study [22]. The objective of the present study was to examine the systemic and mucosal immune responses elicited in BALB/c mice after nasal immunization with rTgROP17 and to evaluate the protective efficacy of rTgROP17 against both chronic and lethal challenges with *T. gondii* in mice.

## Materials and Methods

### Mice, parasites and ethics statement

Female BALB/c mice at the age of 6 weeks were purchased from the Institute of Laboratory Animals of the Chinese Academy of Medical Science (Beijing, China) and used for immunization. All of the mice were maintained under standard, pathogen-free conditions. The tachyzoites of the virulent *T. gondii* RH strain were used as a challenge for the immunized mice, and preparations of *T. gondii* genomic DNA were kindly provided by the Health Science Centre of Peking University (Beijing, China). The parasites were maintained and collected from the peritoneal cavity of infected BALB/c mice in our laboratory according a previously described method [23]. All experimental animal procedures were approved by the Laboratory Animal Use and Care Committee of Shanxi Medical University (permit Number: SXMU-2011-16) and the Ethics Committee of Animal Experiments of Shanxi Medical University (permit Number: 20110320-1). All surgeries were performed under sodium pentobarbital anaesthesia, and all possible efforts were made to minimize the suffering of the experimental mice.

### Expression and purification of rTgROP17

Recombinant *T. gondii* ROP17 protein (rTgROP17) was expressed in *E. coli* Rosetta (DE3) strain [22]. Briefly, the open reading frame of TgROP17 gene (GenBank Accession No. AM075203.1) was amplified with a pair of specific primers, and the RT-PCR product was cloned into the prokaryotic expression pGEX-6P-1 vector (Merck Biosciences, Germany). The recombinant pGEX-6P-1/TgROP17 plasmid was transferred into *E. coli* DH5a, and positive clones were selected via double restriction enzyme digestion and DNA sequencing. The successful pGEX-6P-1/TgROP17 construct was transformed into *E. coli* Rosetta (DE3), and its expression was induced with IPTG and analysed with SDS-PAGE. The antigenicity of the resultant rTgROP17 was analysed with rabbit antiserum for *T. gondii* using Western blotting assays.

All of the subsequent purification steps were performed at 4°C. The clear supernatants were applied onto a self-packaged glutathione S-transferase (GST)-affinity column (2 ml Glutathione Sepharose 4B; Qiagen, Germany), and contaminant proteins were removed with cold phosphate buffer saline (PBS). The fusion protein was then eluted with an elution buffer (Tris-HCl, 25 mM; NaCl, 200 mM; DTT, 2 mM; glutathione, 20 mM, pH 7.5). The eluent was concentrated using an Ultrafree 10,000 molecular weight cut-off filter unit (Millipore, USA). The purified protein was analysed by using SDS-PAGE and Coomassie blue R-250 staining.

The endotoxin contained in the rTgROP17 was removed using a ToxinEraser Endotoxin Removal Kit, and the endotoxin level was measured with a Chromogenic End-point Endotoxin Assay Kit (Chinese Horseshoe Crab Reagent Manufactory, Xiamen, China). Less than 0.1 EU/ml of endotoxin was detected in the final protein preparations. rTgROP17 was dialysed against PBS, quantified with the BCA method, filtered through a 0.2 mm-pore membrane and stored at −70°C.

### Intranasal immunization and sample collection

For the immunization experiments, forty female BALB/c mice were randomly divided into 5 groups (8 mice per group). The mice were immunized nasally with 20 μl PBS containing 15, 25, 35 or 45 μg of rTgROP17, which was instilled slowly into both nostrils (10 μl per nostril) with a micropipette. The control mice were given PBS solution only. All animals were vaccinated three times on days 0, 14, and 21 [24].

The samples, including blood and mucosal washes, were collected two weeks after the last immunization (on day 35). The mice were deprived of food for 8 h to deplete the contents of the intestines and were then anesthetised with sodium pentobarbital (1.5%, 0.1 ml/20 g weight, intraperitoneal injection). The blood, nasal washes, vaginal washes and intestinal washes were collected according to published protocols [25]. All samples were stored at −20°C until assayed for antibody titres and cytokines levels.

### Specific IgG and IgA detection

Serum samples and mucosal washes were tested for the presence of specific IgG and secretory immunoglobulin A (SIgA) using enzyme linked immunosorbent assays (ELISAs). Ninety-six well polystyrene plates (Corning) were coated with 7.5 μg/ml rTgROP17 (100 μl/well) in PBS overnight at 4°C. The plates were washed with PBS containing 0.05% Tween20 (PBST) and blocked for 1 hour at 37°C in PBS containing 5% BSA and then washed with PBS. Serum samples were diluted 1:200 in PBS, mucosal washes without dilutions were incubated at 4°C overnight in different wells of the coated 96 well plates at 100 μl/well. Next day, the wells were washed and incubated with 100 μl of HRP-labelled goat anti-mouse antibody (AbD Serotec; diluted 1:2500 in PBS) for serum specific IgG for 1 h at 37°C. The wells were washed extensively, incubated with orthophenylene diamine (Sigma, USA) and 0.15% $H_2O_2$ for 30 min, and enzyme reactions were terminated by the addition of 50 μl of 1 M $H_2SO_4$. Optical density (OD) was measured in an ELISA plate reader (Epoch Multi-Volume Spectrophotometer System, Biotek, USA) at 492 nm. All samples were analysed in triplicate for at least three independent experiments.

### Cytokines assays

Spleen cells were obtained as described previously [24,25] and cultured in flat-bottom 24-well microtitre plates at $1.5 \times 10^6$ cells/well. Following 24 or 96 h of stimulation with rTgROP17 at 10 μg/ml, tissue culture media were collected and assayed for interleukin-2 (IL-2), IL-4 and IL-5 at 24 h [26] and for interferon-gamma (IFN-γ) at 96 h. The concentrations of IL-2, IL-4, IL-5 and IFN-γ were determined with a commercial ELISA kit (PeproTech, USA) according to the manufacturer's instructions. All assays were performed in triplicates. The sensitivity limits of detection for IL-2, IL-4, IL-5 and IFN-γ were 16, 16, 32 and 23 pg/ml, respectively.

## Lymphocyte proliferation assay

Splenocyte proliferations upon the stimulation of rTgROP17 and controls were determined with the colorimetric Cell Counting Kit-8 (CCK-8, Dojindo Laboratories; Kumamoto, Japan) according to the manufacture's instruction. Isolated spleen cells were plated in flat-bottom 96-well microtitre plate at a density of $5 \times 10^5$ cells/well, and cultured in the presence of rTgROP17 (10 µg/ml), Concanavalin A (Con A; 5 µg/ml; Sigma; positive control) or PRIM 1640 medium alone (negative control) at 37°C in a 5% $CO_2$ incubator. After 72 h, 10 µl of CCK-8 reagent was added to each well, the cells were cultured for further 4 h, and absorbance at 450 nm was determined to evaluate cell proliferation. The stimulation index (SI) was calculated as the ratio of the average $OD_{450}$ value of the wells containing antigen-stimulated cells to the average $OD_{450}$ value of the wells containing cells with medium only. All assays were performed in triplicates for at least three independent experiments.

## Challenge procedure

Challenge procedure was performed in two sets of independent experiments. In each set of experiments, forty female BALB/c mice at 6 weeks of age were randomly divided into two groups (20 mice per group) and vaccinated intranasally with 35 µg of rTgROP17 or GST control (Sigma, USA) in 20 µl volumes on days 0, 14, and 21 as described above. On day 14 after the final immunization, 8 mice from each group were orally challenged via a feeding needle with $1 \times 10^4$ tachyzoites of the RH strain for chronic assay, and 12 mice from each group were similarly challenged with $4 \times 10^4$ tachyzoites for acute infection. One month later, the chronically infected mice were anaesthetized with sodium pentobarbital, and the numbers of tachyzoites in the murine livers and brains were measured using real-time PCR assays. For the acute challenge infection, the times until death of the challenged mice were monitored and recorded for 30 days following the parasite challenge to compute the survival rates. During this period, the infected mice were closely monitored and recorded at 8 am, 2 pm and 8 pm daily for their physical appearance such as rough coat, decrease in appetite, weakness/inability to obtain feed or water, depression. When any of these syndromes happened, the mice would be moved to an isolated cage for further husbandry. If obvious sufferings such as struggling or whining were observed, the mice would be anesthetized by inhaling ether and euthanized.

## DNA extraction and real-time PCR assay

To evaluate the tachyzoite loads, genomic DNA from the liver and brain samples (100 mg each) of the chronically infected mice and purified parasites were extracted using a UniversalGen DNA Kit (CWBIO, China) according to the manufacturer's instructions. The forward and reverse primer sequences for the surface antigen 1 (SAG1) gene of the tachyzoite were 5'-CTGATGTCGTTCTTGCGATGTGGC-3' and 5'-GTGAAGTGGTTCTCCGTCGGT GT-3', respectively. PCR was performed using the Applied Biosystems Real-Time PCR Instrument and SYBR green fluorescence detection. Each reaction mixture contained 10 µl of UltraSYBR Mixture (CWBIO, China), 0.4 µl of each primer (20 µM), 1 µl of DNA template and 8.2 µl of sterile distilled water. Sterile water was used as a negative control, and the DNA extracted from 500 tachyzoites of the *T. gondii* RH strain was used as a positive control. All reactions were performed in triplicates and incubated at 95°C for 1 min, followed by 40 cycles of incubation at 95°C for 5 s, 60°C for 15 s and 72°C for 10 s [27].

The qPCR threshold cycle (Ct) value obtained with DNA samples from a range of serial 10-fold dilutions ($5 \times 10^0$–$5 \times 10^7$/ml) of RH strain tachyzoites was used to calculate the numbers of parasites in the brain and liver samples. The tachyzoite loads were presented as the estimated mean values of the quantities of tachyzoites per gram of tissue.

## Statistical analysis

Data were expressed as mean ± standard error for three or more independent experiments. Statistical comparisons between the test and control groups were performed with one-way analyses of variance (ANOVA) using the SPSS13.0 Data Editor software (SPSS Inc., Chicago, IL). $P$ values $<0.05$ were taken as statistically significant.

## Results

### Purification of rTgROP17 with GST-affinity chromatography

Expression of rTgROP17 was successfully achieved in *E. coli* [22]. To obtain the purified recombinant protein, we performed a GST-affinity chromatography. The isolated protein with a GST Tag at its N-terminal was approximately 96 kDa, water-soluble, and 95% pure as determined by SDS-PAGE gel analysis (Fig. 1).

### rTgROP17 vaccination induces systemic immune responses

To evaluate the systemic immune responses in rTgROP17-immunized mice, the levels of antigen-specific IgG, IgG1 and IgG2a antibodies in the sera and in the spleen cell supernatants were detected using ELISA. Splenic lymphocyte proliferation was determined with the CCK-8 method on day 14 after the final vaccination.

As shown in Fig. 2A, the total IgG antibody productions of the mice immunized with 35 and 45 µg rTgROP17 were significantly higher than those of the control group ($P<0.01$) but not significantly different from each other ($P>0.05$). The results also revealed that 25 µg rTgROP17 but not 15 µg rTgROP17 elicited elevated IgG antibody levels compared to the control groups ($P<0.05$). Moreover, both IgG1 and IgG2a were detected in the sera of all the mice immunized with rTgROP17, particularly in those immunized with 35 and 45 µg rTgROP17 (Fig. 2B). While 25 µg TgROP17 induced significantly higher levels of IgG2a but not IgG1 than those of the control groups ($P<0.05$). No difference of IgG1 or IgG2a responses was found between 15 µg TgROP17 and the control groups ($P>0.05$). In general, greater levels of TgROP17-specific IgG2a were detected than those of its IgG1 counterpart. These results indicate that rTgROP17 immunization trigers a Th1-type systemic immune responses in the mice.

We next analysed the cell-mediated immunity produced in the immunized mice by indirectly evaluating the amounts of cytokines (IFN-γ, IL-2, IL-4 and IL-5) released in the culture media of isolated spleen cells from the rTgROP17-immunized mice. Compared to the PBS controls, the spleen cells from the mice vaccinated with 35 and 45 µg rTgROP17 released significantly higher levels of IFN-γ and IL-2 ($P<0.01$), but there was no significant difference between the 35 and 45 µg groups ($P>0.05$) (Table 1). While 25 µg rTgROP17 also stimulated the production of IFN-γ and IL-2 over the PBS control group ($P<0.05$), 15 µg rTgROP17 did not stimulate IFN-γ or IL-2 production ($P>0.05$). Similarly, the levels of IL-4 secreted by the splenocytes of the 35 and 45 µg rTgROP17-immunized mice were also significantly higher than those of the PBS control group ($P<0.05$) but not the 15 and 25 µg groups (Table 1). However, no significant difference

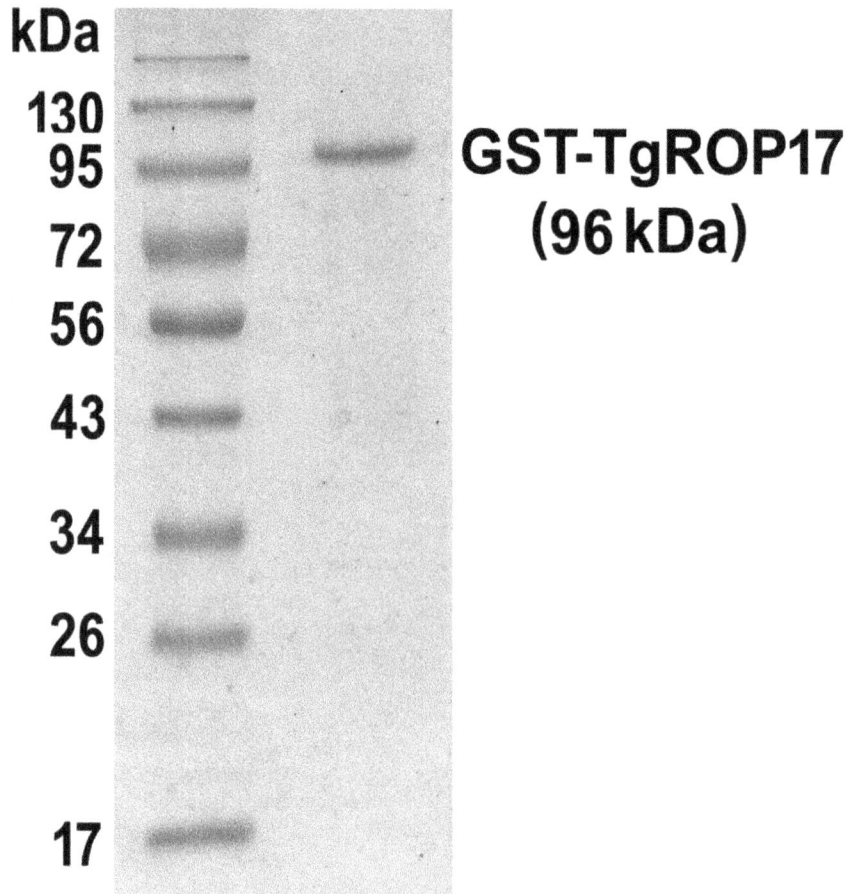

**Figure 1. SDS-PAGE analysis of the rTgROP17 protein.** Purified rTgROP17 protein was analysed by 12% SDS-PAGE and Coomassie blue staining. The purity of the rTgROP17 was greater than 95%.

was observed for the levels of IL-5 between the rTgROP17-vaccinated and PBS control mice ($P>0.05$). Therefore, these data further suggest that rTgROP17 induces a Th1-type immune responses that is correlated with expressions of IFN-γ and IL-2 but not IL-4 and IL-5.

To assess the proliferative immune responses to rTgROP17, splenocytes from the immunized mice were also prepared two weeks after the third immunization. The splenocyte stimulation indices (SIs) of the mice immunized with 35 and 45 μg rTgROP17 were significantly greater than those of the mice immunized with 15 μg rTgROP17 or PBS control ($P<0.01$) (Table 1). While the

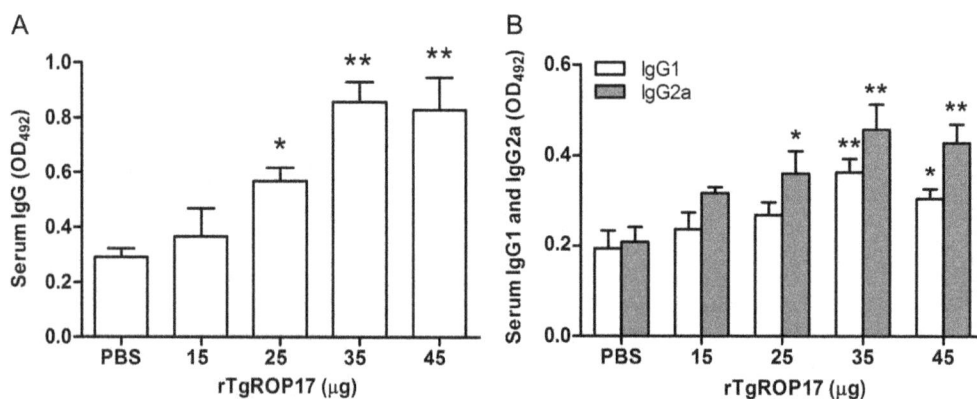

**Figure 2. Determination of specific anti-rTgROP17 antibody responses in serum samples from vaccinated and control mice by ELISA.** The levels of both specific total IgG (A) and IgG isotype (B) antibodies in the sera of the BALB/c mice were determined by ELISA with rTgROP17 as the bound target two weeks after the final immunization. The results are expressed as the mean value ± SD of OD$_{492}$ (n = 8) from three independent experiments. *$P<0.05$, **$P<0.01$ relative to PBS control groups.

**Table 1.** Lymphocyte proliferation and cytokine production by splenocytes stimulated with rTgROP17.

| Groups[a] | Lymphocyte SI | Cytokine production (pg/ml)[b] | | | |
|---|---|---|---|---|---|
| | | IFN-γ | IL-2 | IL-4 | IL-5 |
| PBS | 1.07±0.15 | 49.63±7.42 | 80.81±4.09 | 78.91±8.72 | 81.48±9.75 |
| 15 µg rTgROP17 | 1.28±0.10 | 64.84±6.59 | 96.05±23.88 | 86.22±13.64 | 86.37±7.97 |
| 25 µg rTgROP17 | 1.63±0.13* | 79.46±12.38* | 119.76±12.25* | 93.15±10.25 | 85.29±16.41 |
| 35 µg rTgROP17 | 2.38±0.16** | 125.45±14.12** | 136.69±16.89** | 122.17±13.08* | 89.47±10.52 |
| 45 µg rTgROP17 | 2.29±0.19** | 117.43±11.29** | 133.66±12.26** | 118.20±11.78* | 83.59±11.26 |

[a]n = 8 per group.
[b]Splenocytes were harvested from the mice 2 weeks after the final immunization. The results are presented as the arithmetic means ± the standard errors of three independent experiments. The values of IFN-γ are for 96 h, and the values of IL-2, IL-4 and IL-5 are for 24 h.
*$P<0.05$ and **$P<0.01$ compared to the PBS group.

SI of the 25 µg group was also greater than that of the PBS control ($P<0.05$), no significant difference was found for the SIs of the 15 µg group and PBS controls ($P> 0.05$). Of note, comparable levels of proliferation were observed in response to ConA for all the splenocytes from all experimental and control mice mice (data not shown). These results demonstrate that nasal administration of rTgROP17 is able to trigger systemic cell-mediated immunity in mice.

## rTgROP17 vaccination induces mucosal immune responses

To determine whether the mice intranasally immunized with rTgROP17 developed mucosal immune responses, specific SIgA levels in their mucosal washes were tested with ELISA two weeks after the final immunization. As shown in Fig. 3, higher levels of SIgA were detected in the nasal, vaginal and intestinal washes of rTgROP17-immunized mice compared to those of their PBS controls. In particular, 35 and 45 µg rTgROP17 elicited much higher SIgA antibody titres than those of the controls in all three different mucosal washes ($P<0.01$), and the SIgA antibody titres of the 25 µg group was also higher than those of the PBS controls ($P<0.05$). These data indicate that nasal rTgROP17 immunization in mice can evoke strong mucosal immune responses at their nasal, intestinal and vaginal mucosal sites.

## Protection of mice against oral challenge with T. gondii RH strain

To evaluate whether rTgROP17 immunization could potentially provide protection against T. gondii chronic infection, the numbers of liver and brain tachyzoites in the mice were tested one month after peroral challenge ($1\times10^4$ tachyzoites of RH strain) using a real-time PCR assay. The tachyzoite loads in the brains and livers of the mice immunized with 35 µg rTgROP17 were significantly reduced to 59.17% ($P<0.01$) and 49.08% ($P<0.05$) of the loads found in the GST-treated control mice, respectively (Fig. 4A).

To further determine whether rTgROP17 vaccination protected the mice against lethal T. gondii RH strain infection, the mice were orally challenged with $4\times10^4$ tachyzoites 2 weeks after the final immunization, and their survival rates were recorded daily for 30 days following the challenge. As shown in Fig. 4B, the

survival rates of the mice immunized with 35 µg rTgROP17 were significantly increased (75%) on the 30th day after the challenge when compared to those of the GST-treated control mice (25%) ($P<0.01$). Survival of rTgROP17-immunized mice was three times over the GST-treated mice. The death time of the control mice ranged from 5 to 9 days, while the mice immunized with rTgROP17 died between days 10 and 14 after the challenge. These results suggest that rTgROP17 vaccination can protect the mice against T. gondii infection through reductions in their tissue loads of the tachyzoites and increases of their survival rate and time.

## Discussion

Toxoplasma is an obligate intracellular parasite that requires active invasion of a host cell for replication [28]. The invasion process is punctuated by waves of parasite secretion [2,3]. During the invasive stages of Toxoplasma infection, there are three sets of secretory organelles, i.e., micronemes, rhoptries and dense granules, which release proteins that are essential to intracellular survival [29]. Rhoptry proteins (ROPs) are introduced into a host cell during the invasion and are involved in the determination of the virulence of parasites [30]. Thus far, most ROPs have been found to show clear homology with protein kinases [18,31,32]. Some ROPs, such as ROP16 and ROP18, are active kinases, while others are pseudokinases. ROP16 is a tyrosine kinase that phosphorylates the key tyrosine for the activation of STATs to regulate many immune response genes [11,33–35]. ROP18 is a serine/threonine kinase that can phosphorylate the immunity-related GTPases (IRGs) and maintain the integrity of the PVM to protect the inside parasites from being destroyed [20,36]. Due to the key biological role of rhoptries, some ROPs, including ROP16 and ROP18, have recently become candidate vaccines for the prevention of toxoplasmosis [37]. Moreover, ROP16- and ROP18-immunised mice can induce cellular and humoral immune responses that protect the mice against T. gondii infection [14–17].

ROP17 is predicted to be a serine/threonine kinase because it possesses the catalytic aspartate and the glycine loop responsible for the stabilisation of the αβ-phosphate of adenosine triphosphate (ATP) [11,18,19]. Our previous study showed that ROP17 can phosphorylate c-Jun in HEK 293T cells and thus demonstrated its

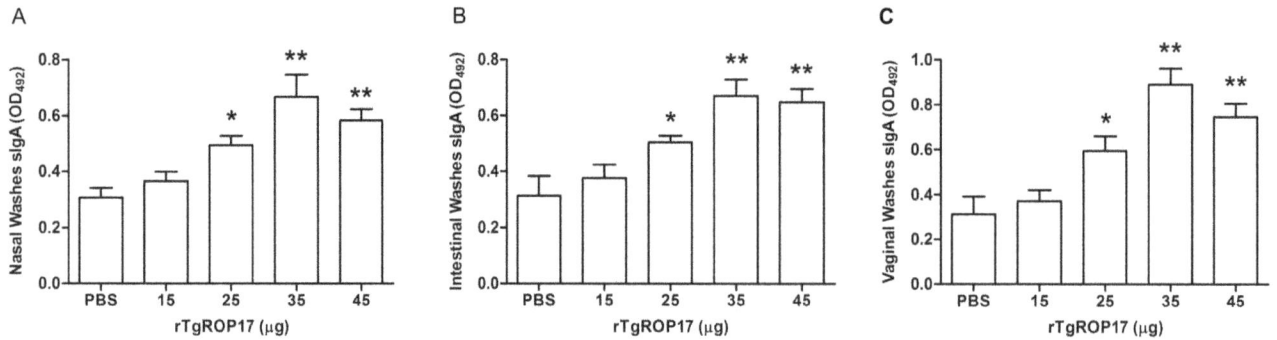

**Figure 3. Nasal rTgROP17 immunization of mice induces specific SIgA antibody responses in their mucosal washes.** The SIgA antibody titres in the mucosal washes of the rTgROP17-immunised mice were determined by ELISA two week after the final immunization. High levels of SIgA in (A) the nasal washes, (B) intestinal washes and (C) vaginal washes were detected in rTgROP17-immunized mice compared to those of the PBS control mice. The results are expressed as the mean value $\pm$ SD of $OD_{492}$ (n = 8) from three independent experiments. *P<0.05, **P<0.01 relative to PBS control groups.

kinase activity [21] In the present study, we demonstrated that nasal delivery of the purified recombinant *Toxoplasma gondii* ROP17 (rTgROP17) protein to BABL/c mice effectively reduced tachyzoite loads in their brain and liver tissues, prolonged their survival time and increased their survival rate following oral challenges with the virulent RH strain of *T. gondii* in association with enhanced systemic and mucosal immune responses and also cellular-mediated immune response of the rTgROP17-immunised mice.

*Toxoplasma* infects humans and animals primarily through the mucosal surfaces of the digestive tracts. Additionally, *Toxoplasma* can invade all nucleic cells and disseminate throughout the hosts. Therefore, the development of a potential vaccine against *Toxoplasma* infection requires the induction of protective responses at mucosal surfaces and also the induction of systemic protective immune responses in the systemic compartment. Recently, vaccination at local mucosal surfaces, particularly the nasal surface, has been proven to be a promising and rational approach for vaccine development [25,38]. A number of lines of evidence also indicate that specific SIgA plays a protective role against many pathogens which colonise mucosal tissues or invade the host

organism by crossing mucous membranes [39–41]. Indeed, SIgA serves as the first line of defence in protecting the intestinal epithelium from enteric toxins and pathogenic microorganisms like *Toxoplasma* [42,43]. Our results showed that the antigen-specific SIgA antibody responses in the nasal, vaginal and intestinal washes were signifantly increased in the mice which were immunized nasally with 35 μg rTgROP17. These findings demonstrated that nasal immunization of mice with rTgROP17 evokes strong SIgA immune responses in mucosal sites for the initiation or generation of protective immunity against *Toxoplasma*.

One of the goals of a vaccination procedure is to be able to appropriately direct the T helper response. Naturally, the Th1 immune response is predominant when *T. gondii* infects a host [44]. Immunisation with rTgROP17 induced a mixed Th1/Th2 response with a slightly greater increase in the anti-*T. gondii* antibodies of the Ig2a isotype (Th1) than the IgG1 (Th2) isotype (Fig. 2B). Spleen cells from rTgROP17-immunized mice exhibited enhanced proliferation potential and secreted more IFN-γ, IL-2 and IL-4 but not IL-5 than those of the PBS control mice, suggesting that rTgROP17 enhanced Th1 and Th2 cell-mediated

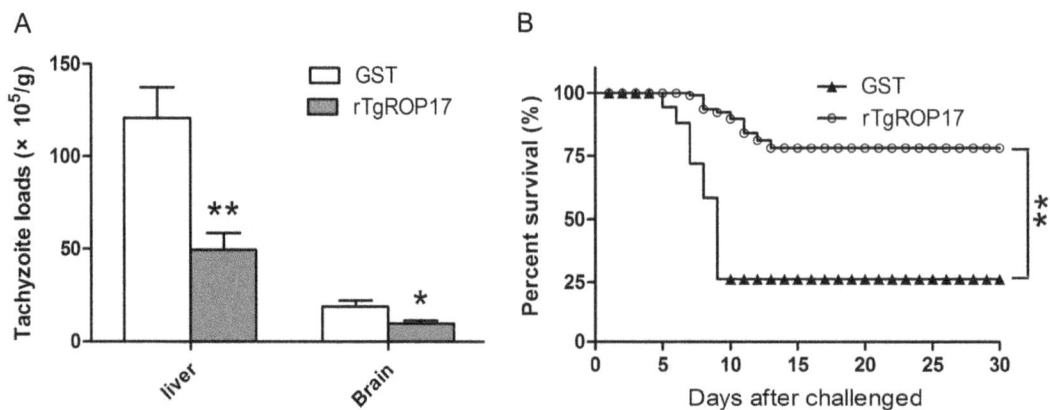

**Figure 4. The tachyzoite burdens and survival rates of rTgROP17-immunised mice after oral challenge with *T. gondii*.** The mice were nasally immunized with rTgROP17 or GST control. Two weeks after the final immunization, the mice were orally challenged with $1 \times 10^4$ tachyzoites (A) or $4 \times 10^4$ tachyzoites (B) of the *T. gondii* RH strain. (A) The liver and brain tachyzoite burdens of two groups of mice (n = 8 per group) were evaluated one month after the challenge, and (B) the survival rates of other two groups of mice (n = 12 per group) were monitored daily for 30 days following the challenge. The values are presented as mean $\pm$ SD from two independent experiments. *P<0.05, **P<0.01 relative to GST control groups.

immunity with a Th1 type predominance (Table 1). These findings demonstrate that intranasal immunization with rTgROP17 can potentiate humoral and cellular Th1 immune responses in BALB/c mice. As is well known, the modulation of the Th1-type response plays a major deterministic role in the induction of cell-mediated immune responses and the control of acute *T. gondii* infections [45].

There is evidence that recombinant protein vaccine can induce stronger immune responses and better survival protections against acute *toxoplasmosis* than their DNA vaccination counterparts [46]. In recent years, tremendous efforts have been made toward identifying and verifying recombinant protein vaccine candidates against *T. gondii* infection, including recombinant rSAG1 [47,48], rROP2 [49,50], rROP4 [50], rGRA1 [46], rGRA4 [49], and rGRA6 [51] protein vaccines. These protein vaccines produce partial protective efficacy or extend the survival times of mice challenged with a lethal *T. gondii*. Our previous studies have also shown that intranasal immunization with rPDI and rACT increases the survival rates and prolongs life spans of the experimental mice [24,25].

Although encouraging progress has been made, there is still little to offer as protection against toxoplasmosis. One of the problems encountered in the field of research in the development of a vaccine against *T. gondii* is the lack of efficacious protective antigen candidates [52]. In the present study, the protective effect of rTgROP17 protein against *T. gondii* infection was evaluated in mice. The vaccinated mice displayed significant protection against lethal infection of the virulent RH strain with a 50% increase of survival rate over the GST controls. Significant reductions of parasite burdens was also observed in the liver (59.17%) and brain (49.08%) tissues of the rTgROP17-immunised mice (Fig. 4). These data indicate that rTgROP17 offers better protection of mice against *T. gondii* infection compared to the ROP16 and ROP18 used as DNA vaccine candidates which only extended the life span but not the survival rate of *T. gondii*-infected mice [10,14–17].

The antigen delivery pathway is a key parameter for the induction of protective immune responses by vaccines against intracellular pathogens [53]. Most candidate vaccines have been given parenterally to induce systemic immunity, which can control toxoplasmosis efficiently [54,55]. Several studies have shown that intranasal vaccination is an effective regimen for the stimulation of mucosal immune responses in the mucosal effective sites, including the gut, genital tract, and nose cavity [56,57]. In the present study, intranasal immunization with rTgROP17 not only induced strong mucosal immune responses in the nasal, vaginal and intestinal mucosal sites, but also induced cell-mediated immunity that included considerable levels of IFN-γ and IL-2. To further enhance Th1 cell-mediated immunity and IFNs, different vaccine adjuvants, such as cholera toxin (CT) and IL-2 [23], will be considered in further studies.

In addition, it has yet to be determined whether the immune response induced by ROP17 of the type I *T. gondii* could be effective against type II or type  *T. gondii* strains. We compared the amino acid sequences of ROP17s from the RH strain (type I), the ME 49 strain (type II) and the VEG strain (type ). Over 99% sequence identity of the ROP17s was found across the three genotypes of *T. gondii* (Figure S1), suggesting that their genetic variation is very low and that similar ROP17 proteins are expressed by the type I, type II and type  strains; thus, these proteins may stimulate similar immune responses. This issue will be experimentally investigated in further studies.

On the other hand, the recombinant rTgROP17 protein we used in this study was GST- tagged, and therefore GST was used as a control in the immunization and *T. gondii* challenge assays. Although our data demonstrated that the mice immunized with GST alone exhibited higher tachyzoite burdens and lower survival values compared with those immunized with GST-rTgROP17, antibodies directed against the GST portion of the GST-fusion protein have also occasionally be detected [58]. The use of recombinant ROP17 without the GST fusion partner or as a fusion to another small carriers, such as the His-binding protein [24] would - increase the specificity of the vaccine and reduce the number of animals used in similar future studies.

In summary, our study suggests that intranasal immunization of mice with rTgROP17 can induce both systemic and local immune responses to provide protection against lethal *T. gondii* infections through reduction of the tachyzoite burdens in the host tissues and increases of the animal survivals. We conclude that ROP17 is a promising vaccine candidate against infection with *T. gondii*.

## Supporting Information

**Figure S1   Alignment of TgROP17 amino acid sequences in three strains of *T. gondii*.** The amino acid sequences of ROP17s from three strains of *T. gondii* were obtained from the Internet (http://www.ncbi.nlm.nih.gov). The protein accession numbers are as follows: CAJ27112 for the RH strain, type I; EPT29356 for the ME 49 strain, type II; and ESS32210 for the VEG strain, type III. The alignment was performed using DNAMAN software (Lynnon, Quebec, Canada). Homology levels are indicated by the colours of: black, 100%; and blue, 50%. Over 99% of sequence identity is found for the ROP17s across the three genotypes of *T. gondii*.

## Author Contributions

Conceived and designed the experiments: GRY. Performed the experiments: HLW TEZ LG. Analyzed the data: LTY JHZ JZB. Contributed reagents/materials/analysis tools: HLL XLM MP. Wrote the paper: HLW GPZ.

## References

1. Montoya JG, Liesenfeld O (2004) Toxoplasmosis. Lancet 363: 1965–1976.
2. Dubey JP (2009) Toxoplasmosis in pigs–the last 20 years. Vet Parasitol 164: 89–103.
3. Luft BJ, Remington JS (1992) Toxoplasmic encephalitis in AIDS. Clin Infect Dis 15: 211–222.
4. Buxton D, Maley SW, Wright SE, Rodger S, Bartley P, et al. (2007) Toxoplasma gondii and ovine toxoplasmosis: new aspects of an old story. Vet Parasitol 149: 25–28.
5. Nie H, Fang R, Xiong BQ, Wang LX, Hu M, et al. (2011) Immunogenicity and protective efficacy of two recombinant pseudorabies viruses expressing Toxoplasma gondii SAG1 and MIC3 proteins. Vet Parasitol 181: 215–221.
6. Khosroshahi KH, Ghaffarifar F, Sharifi Z, D'Souza S, Dalimi A, et al. (2012) Comparing the effect of IL-12 genetic adjuvant and alum non-genetic adjuvant on the efficiency of the cocktail DNA vaccine containing plasmids encoding SAG-1 and ROP-2 of Toxoplasma gondii. Parasitol Res 111: 403–411.
7. Dziadek B, Gatkowska J, Grzybowski M, Dziadek J, Dzitko K, et al. (2012) Toxoplasma gondii: the vaccine potential of three trivalent antigen-cocktails composed of recombinant ROP2, ROP4, GRA4 and SAG1 proteins against chronic toxoplasmosis in BALB/c mice. Exp Parasitol 131: 133–138.
8. Xiang W, Qiong Z, Li-peng L, Kui T, Jian-wu G, et al. (2009) The location of invasion-related protein MIC3 of Toxoplasma gondii and protective effect of its DNA vaccine in mice. Vet Parasitol 166: 1–7.
9. Bradley PJ, Sibley LD (2007) Rhoptries: an arsenal of secreted virulence factors. Curr Opin Microbiol 10: 582–587.
10. Jensen KD, Hu K, Whitmarsh RJ, Hassan MA, Julien L, et al. (2013) Toxoplasma gondii rhoptry 16 kinase promotes host resistance to oral infection and intestinal inflammation only in the context of the dense granule protein GRA15. Infect Immun 81: 2156–2167.
11. Butcher BA, Fox BA, Rommereim LM, Kim SG, Maurer KJ, et al. (2011) Toxoplasma gondii rhoptry kinase ROP16 activates STAT3 and STAT6

resulting in cytokine inhibition and arginase-1-dependent growth control. PLoS Pathog 7: e1002236.

12. Yamamoto M, Takeda K (2012) Inhibition of ATF6beta-dependent host adaptive immune response by a Toxoplasma virulence factor ROP18. Virulence 3: 77–80.

13. Fentress SJ, Steinfeldt T, Howard JC, Sibley LD (2012) The arginine-rich N-terminal domain of ROP18 is necessary for vacuole targeting and virulence of Toxoplasma gondii. Cell Microbiol 14: 1921–1933.

14. Qu D, Han J, Du A (2013) Enhancement of protective immune response to recombinant Toxoplasma gondii ROP18 antigen by ginsenoside Re. Exp Parasitol 135: 234–239.

15. Yuan ZG, Zhang XX, Lin RQ, Petersen E, He S, et al. (2011) Protective effect against toxoplasmosis in mice induced by DNA immunization with gene encoding Toxoplasma gondii ROP18. Vaccine 29: 6614–6619.

16. Liu Q, Wang F, Wang G, Zhao Q, Min J, et al. (2013) Toxoplasma gondii: Immune response and protective efficacy induced by ROP16/GRA7 multi-component DNA vaccine with a genetic adjuvant B7-2. Hum Vaccin Immunother 10.

17. Yuan ZG, Zhang XX, He XH, Petersen E, Zhou DH, et al. (2011) Protective immunity induced by Toxoplasma gondii rhoptry protein 16 against toxoplasmosis in mice. Clin Vaccine Immunol 18: 119–124.

18. El Hajj H, Demey E, Poncet J, Lebrun M, Wu B, et al. (2006) The ROP2 family of Toxoplasma gondii rhoptry proteins: proteomic and genomic characterization and molecular modeling. Proteomics 6: 5773–5784.

19. Qiu W, Wernimont A, Tang K, Taylor S, Lunin V, et al. (2009) Novel structural and regulatory features of rhoptry secretory kinases in Toxoplasma gondii. EMBO J 28: 969–979.

20. Fentress SJ, Behnke MS, Dunay IR, Mashayekhi M, Rommereim LM, et al. (2010) Phosphorylation of immunity-related GTPases by a Toxoplasma gondii-secreted kinase promotes macrophage survival and virulence. Cell Host Microbe 8: 484–495.

21. Wang HL, Yin LT, Zhang TE, Guan L, Meng XL, et al. (2014) [Construction, expression and kinase function analysis of an eukaryocyte vector of rhoptry protein 17 in Toxoplasma gondii]. Zhongguo Ji Sheng Chong Xue Yu Ji Sheng Chong Bing Za Zhi 32: 29–33.

22. Wang HL, Yin LT, Meng XL, Shen JY, Liu HL, et al. (2012) Gene-cloning, expression and antigenicity analysis of rhoptry protein 17 of Toxoplasma gondii. Zhongguo Ji Sheng Chong Xue Yu Ji Sheng Chong Bing Za Zhi 30: 27–31.

23. Yin GR, Meng XL, Ma GY, Ma XM (2007) [Intranasal immunization with mucosal complex vaccine protects mice against Toxoplasma gondii]. Zhongguo Ji Sheng Chong Xue Yu Ji Sheng Chong Bing Za Zhi 25: 290–294.

24. Wang HL, Li YQ, Yin LT, Meng XL, Guo M, et al. (2013) Toxoplasma gondii protein disulfide isomerase (TgPDI) is a novel vaccine candidate against toxoplasmosis. PLoS One 8: e70884.

25. Yin LT, Hao HX, Wang HL, Zhang JH, Meng XL, et al. (2013) Intranasal Immunisation with Recombinant Toxoplasma gondii Actin Partly Protects Mice against Toxoplasmosis. PLoS One 8: e82765.

26. Taguchi T, McGhee JR, Coffman RL, Beagley KW, Eldridge JH, et al. (1990) Detection of individual mouse splenic T cells producing IFN-gamma and IL-5 using the enzyme-linked immunospot (ELISPOT) assay. J Immunol Methods 128: 65–73.

27. Yu H, Huang B, Zhuo X, Chen X, Du A (2013) Evaluation of a real-time PCR assay based on the single-copy SAG1 gene for the detection of Toxoplasma gondii. Vet Parasitol 197: 670–673.

28. Sibley LD (2010) How apicomplexan parasites move in and out of cells. Curr Opin Biotechnol 21: 592–598.

29. Dubey JP, Lindsay DS, Speer CA (1998) Structures of Toxoplasma gondii tachyzoites, bradyzoites, and sporozoites and biology and development of tissue cysts. Clin Microbiol Rev 11: 267–299.

30. Boothroyd JC, Dubremetz JF (2008) Kiss and spit: the dual roles of Toxoplasma rhoptries. Nat Rev Microbiol 6: 79–88.

31. Bradley PJ, Ward C, Cheng SJ, Alexander DL, Coller S, et al. (2005) Proteomic analysis of rhoptry organelles reveals many novel constituents for host-parasite interactions in Toxoplasma gondii. J Biol Chem 280: 34245–34258.

32. Peixoto L, Chen F, Harb OS, Davis PH, Beiting DP, et al. (2010) Integrative genomic approaches highlight a family of parasite-specific kinases that regulate host responses. Cell Host Microbe 8: 208–218.

33. Ong YC, Boyle JP, Boothroyd JC (2011) Strain-dependent host transcriptional responses to Toxoplasma infection are largely conserved in mammalian and avian hosts. PLoS One 6: e26369.

34. Yamamoto M, Standley DM, Takashima S, Saiga H, Okuyama M, et al. (2009) A single polymorphic amino acid on Toxoplasma gondii kinase ROP16 determines the direct and strain-specific activation of Stat3. J Exp Med 206: 2747–2760.

35. Ong YC, Reese ML, Boothroyd JC (2010) Toxoplasma rhoptry protein 16 (ROP16) subverts host function by direct tyrosine phosphorylation of STAT6. J Biol Chem 285: 28731–28740.

36. Khaminets A, Hunn JP, Konen-Waisman S, Zhao YO, Preukschat D, et al. (2010) Coordinated loading of IRG resistance GTPases on to the Toxoplasma gondii parasitophorous vacuole. Cell Microbiol 12: 939–961.

37. Dlugonska H (2008) Toxoplasma rhoptries: unique secretory organelles and source of promising vaccine proteins for immunoprevention of toxoplasmosis. J Biomed Biotechnol 2008: 632424.

38. Zaman M, Chandrudu S, Toth I (2013) Strategies for intranasal delivery of vaccines. Drug Deliv Transl Res 3: 100–109.

39. Winner L 3rd, Mack J, Weltzin R, Mekalanos JJ, Kraehenbuhl JP, et al. (1991) New model for analysis of mucosal immunity: intestinal secretion of specific monoclonal immunoglobulin A from hybridoma tumors protects against Vibrio cholerae infection. Infect Immun 59: 977–982.

40. Michetti P, Mahan MJ, Slauch JM, Mekalanos JJ, Neutra MR (1992) Monoclonal secretory immunoglobulin A protects mice against oral challenge with the invasive pathogen Salmonella typhimurium. Infect Immun 60: 1786–1792.

41. Hutchings AB, Helander A, Silvey KJ, Chandran K, Lucas WT, et al. (2004) Secretory immunoglobulin A antibodies against the sigma1 outer capsid protein of reovirus type 1 Lang prevent infection of mouse Peyer's patches. J Virol 78: 947–957.

42. Velge-Roussel F, Marcelo P, Lepage AC, Buzoni-Gatel D, Bout DT (2000) Intranasal immunization with Toxoplasma gondii SAG1 induces protective cells into both NALT and GALT compartments. Infect Immun 68: 969–972.

43. Debard N, Buzoni-Gatel D, Bout D (1996) Intranasal immunization with SAG1 protein of Toxoplasma gondii in association with cholera toxin dramatically reduces development of cerebral cysts after oral infection. Infect Immun 64: 2158–2166.

44. Denkers EY, Gazzinelli RT (1998) Regulation and function of T-cell-mediated immunity during Toxoplasma gondii infection. Clin Microbiol Rev 11: 569–588.

45. Yang X, Huang S, Chen J, Song N, Wang L, et al. (2010) Evaluation of the adjuvant properties of Astragalus membranaceus and Scutellaria baicalensis GEORGI in the immune protection induced by UV-attenuated Toxoplasma gondii in mouse models. Vaccine 28: 737–743.

46. Doskaya M, Kalantari-Dehaghi M, Walsh CM, Hiszczynska-Sawicka E, Davies DH, et al. (2007) GRA1 protein vaccine confers better immune response compared to codon-optimized GRA1 DNA vaccine. Vaccine 25: 1824–1837.

47. Letscher-Bru V, Villard O, Risse B, Zauke M, Klein JP, et al. (1998) Protective effect of vaccination with a combination of recombinant surface antigen 1 and interleukin-12 against toxoplasmosis in mice. Infect Immun 66: 4503–4506.

48. Dziadek B, Gatkowska J, Brzostek A, Dziadek J, Dzitko K, et al. (2011) Evaluation of three recombinant multi-antigenic vaccines composed of surface and secretory antigens of Toxoplasma gondii in murine models of experimental toxoplasmosis. Vaccine 29: 821–830.

49. Martin V, Supanitsky A, Echeverria PC, Litwin S, Tanos T, et al. (2004) Recombinant GRA4 or ROP2 protein combined with alum or the gra4 gene provides partial protection in chronic murine models of toxoplasmosis. Clin Diagn Lab Immunol 11: 704–710.

50. Dziadek B, Gatkowska J, Brzostek A, Dziadek J, Dzitko K, et al. (2009) Toxoplasma gondii: the immunogenic and protective efficacy of recombinant ROP2 and ROP4 rhoptry proteins in murine experimental toxoplasmosis. Exp Parasitol 123: 81–89.

51. Golkar M, Shokrgozar MA, Rafati S, Musset K, Assmar M, et al. (2007) Evaluation of protective effect of recombinant dense granule antigens GRA2 and GRA6 formulated in monophosphoryl lipid A (MPL) adjuvant against Toxoplasma chronic infection in mice. Vaccine 25: 4301–4311.

52. Liu Q, Singla LD, Zhou H (2012) Vaccines against Toxoplasma gondii: status, challenges and future directions. Hum Vaccin Immunother 8: 1305–1308.

53. Liu CS, Sun Y, Hu YH, Sun L (2010) Identification and analysis of a CpG motif that protects turbot (Scophthalmus maximus) against bacterial challenge and enhances vaccine-induced specific immunity. Vaccine 28: 4153–4161.

54. Brinkmann V, Remington JS, Sharma SD (1993) Vaccination of mice with the protective F3G3 antigen of Toxoplasma gondii activates CD4+ but not CD8+ T cells and induces Toxoplasma specific IgG antibody. Mol Immunol 30: 353–358.

55. Lunden A, Lovgren K, Uggla A, Araujo FG (1993) Immune responses and resistance to Toxoplasma gondii in mice immunized with antigens of the parasite incorporated into immunostimulating complexes. Infect Immun 61: 2639–2643.

56. Wang H, He S, Yao Y, Cong H, Zhao H, et al. (2009) Toxoplasma gondii: protective effect of an intranasal SAG1 and MIC4 DNA vaccine in mice. Exp Parasitol 122: 226–232.

57. Igarashi M, Zulpo DL, Cunha IA, Barros LD, Pereira VF, et al. (2010) Toxoplasma gondii: humoral and cellular immune response of BALB/c mice immunized via intranasal route with rTgROP2. Rev Bras Parasitol Vet 19: 210–216.

58. Parmley S, Slifer T, Araujo F (2002) Protective effects of immunization with a recombinant cyst antigen in mouse models of infection with Toxoplasma gondii tissue cysts. J Infect Dis 185 Suppl 1: S90–95.

# 2

# *Toxoplasma gondii* Is Dependent on Glutamine and Alters Migratory Profile of Infected Host Bone Marrow Derived Immune Cells through SNAT2 and CXCR4 Pathways

**I-Ping Lee**[1]*, **Andrew K. Evans**[1,ヲ], **Cissy Yang**[1,ヲ], **Melissa G. Works**[1], **Vineet Kumar**[2], **Zurine De Miguel**[1], **Nathan C. Manley**[1,3,4], **Robert M. Sapolsky**[1,3,4,5]

1 Department of Biology, Stanford University, Stanford, California, United States of America, 2 School of Biological Sciences, Nanyang Technological University, Singapore, Republic of Singapore, 3 Department of Neurosurgery, Stanford University School of Medicine, Stanford, California, United States of America, 4 Stanford Stroke Center and Stanford Institute for Neuro-Innovation and Translational Neurosciences, Stanford University School of Medicine, Stanford, California, United States of America, 5 Department of Neurology and Neurological Sciences, Stanford University School of Medicine, Stanford, California, United States of America

## Abstract

The obligate intracellular parasite, *Toxoplasma gondii*, disseminates through its host inside infected immune cells. We hypothesize that parasite nutrient requirements lead to manipulation of migratory properties of the immune cell. We demonstrate that 1) *T. gondii* relies on glutamine for optimal infection, replication and viability, and 2) *T. gondii*-infected bone marrow-derived dendritic cells (DCs) display both "hypermotility" and "enhanced migration" to an elevated glutamine gradient *in vitro*. We show that glutamine uptake by the sodium-dependent neutral amino acid transporter 2 (SNAT2) is required for this enhanced migration. SNAT2 transport of glutamine is also a significant factor in the induction of migration by the small cytokine stromal cell-derived factor-1 (SDF-1) in uninfected DCs. Blocking both SNAT2 and C-X-C chemokine receptor 4 (CXCR4; the unique receptor for SDF-1) blocks hypermotility and the enhanced migration in *T. gondii*-infected DCs. Changes in host cell protein expression following *T. gondii* infection may explain the altered migratory phenotype; we observed an increase of CD80 and unchanged protein level of CXCR4 in both *T. gondii*-infected and lipopolysaccharide (LPS)-stimulated DCs. However, unlike activated DCs, SNAT2 expression in the cytosol of infected cells was also unchanged. Thus, our results suggest an important role of glutamine transport via SNAT2 in immune cell migration and a possible interaction between SNAT2 and CXCR4, by which *T. gondii* manipulates host cell motility.

**Editor:** Matthew Bogyo, Stanford University, United States of America

**Funding:** This work was supported by The Stanley Medical Research Institute (grant number: 06R-1463). The funders had no role in study design, data collection and analysis, decision to publish, or preparation of the manuscript.

**Competing Interests:** The authors have declared that no competing interests exist.

* Email: ipinglee@stanford.edu

ヲ These authors contributed equally to this work.

## Introduction

*Toxoplasma gondii*, an obligate intracellular protozoan that is capable of infecting nearly all warm-blooded animals, depends on the host to meet its glucose and energy requirements. Recent studies have shown that apicomplexan parasites catabolize both glucose and the non-essential amino acid glutamine via the tricarboxylic acid (TCA) cycle to generate energy; however, disrupting the entry of glucose-derived intermediates into the TCA cycle has no significant effect on the growth of asexual stages of *Plasmodium falciparum*, indicating that glutamine serves as an alternative carbon source for the cycle [1]. *T. gondii* has also been shown to rely on glutamine for energy during glucose starvation, and glutaminolysis is indispensable to its survival [2–4]. Rapidly dividing tachyzoites of *T. gondii* may therefore impose a heavy glutamine burden on the host, and we here propose that such glutamine dependency may generate a specific means by which the parasite can manipulate the function of host cells to induce

them to leave the circulation and enter tissues with high glutamine level for the benefit of the parasite. *T. gondii* has been shown to migrate to and encyst in brain, retina and muscle during chronic infection, all places with abundant glutamine supply [5–9].

*T. gondii* has been demonstrated to manipulate the migratory properties of infected immune cells, inducing a hypermotility state in infected dendritic cells (DCs), which has been hypothesized to promote dissemination of parasites throughout the body [10,11]. This phenomenon is dependent on G*i*-protein signaling but the specific G*i* protein-coupled receptor has not been identified. C-X-C chemokine receptor 4 (CXCR4), the unique receptor for small cytokine stromal cell-derived factor-1 (SDF-1) on the surface of immune cells, may be a strong candidate. The SDF-1-CXCR4 complex activates the downstream phosphoinositide 3-kinase (PI3K) pathway involved in cell motility [12,13]. Importantly, the PI3K pathway can also be induced by the glutamine transporter (sodium-dependent neutral amino acid transporter 2, SNAT2)/amino acid substrate complex in muscle cells [14,15].

SNAT2, the ubiquitously expressed subtype of the system A transporter in mammalian cells, is the primary mediator of uptake of aliphatic amino acids such as glutamine, glycine, and asparagine [16]. Recent studies on cultured rat myoblasts have demonstrated that SNAT2 exhibits a hybrid transporter-receptor (transceptor) function that can sense extracellular amino acid concentration and activate downstream PI3K pathway during nutrient stress [14,15]. Such amino acid transceptors are well documented in *Drosophila* and yeast like *Saccharomyces cerevisiae*, but the role of SNAT2 in mammalian immune cells has not been extensively studied. On the other hand, it is well known that lymphocytes, monocytes and macrophages have high utilization rates of glutamine, which is involved in cell proliferation, expression of surface activation markers and the production of cytokines [17,18].

In this study we hypothesized that *T. gondii*, due to its high demand for glutamine, manipulates the dual transporter/receptor function of SNAT2 in the host, and via interaction with the CXCR4 signaling pathway causes enhanced migration of infected immune cells to glutamine-enriched environments. In order to determine the specific role of glutamine and the transporter SNAT2 as variables influencing this migration, we used an *in vitro* transwell system to investigate cell migration of normal and *T. gondii*-infected murine bone marrow-derived DCs to a range of L-glutamine concentrations. We examined the role of SNAT2 and CXCR4-PI3K-Rho kinase pathways in this migration. Our data show that CXCR4 signaling-dependent migration is SNAT2-dependent in normal DCs, and antagonizing both CXCR4 and SNAT2 blocks the induced migration in *T. gondii*-infected cells.

## Results

### *T. gondii* is dependent on glutamine for optimal infection, replication and viability *in vitro*

Type II strains of *T. gondii* (e.g., Prugniaud) are the most prevalent in Europe and North America [19,20], are less virulent in the mouse model but highly associated with human diseases [21,22], and they alter host physiology and behavior [23,24]. Therefore, we first investigated if type II tachyzoites maintained in human foreskin fibroblast (HFFs) monolayers depend on glutamine for infection, replication, and viability *in vitro*. INFECTION: By using live staining of SAG1 (red), the major surface antigen of *T. gondii*, we were able to distinguish extra- (red and green) and intracellular (green) tachyzoites and quantify the efficiency in infecting HFFs (Fig. 1A). In the absence of glutamine, no obvious change of the morphology of parasites was observed. However, without glutamine, there were more extracellular parasites in the examined fields of the coverslips over the course of 6 hours and the average of infection efficiency of *T. gondii* was reduced approximately two-thirds (Fig. 1B). REPLICATION: HFF monolayers were firstly infected with *T. gondii* tachyzoites in the presence of glutamine for 6 hours. Extracellular parasites were then washed away with phosphate-buffered saline (PBS) and infected HFFs were cultured in fresh medium containing 0 mM or 2 mM of L-glutamine for another 16 hours. In the presence of glutamine, *T. gondii* was able to duplicate up to 4 times in 22 hours (Fig. 1C) while the replication ability was impaired with ~50% of the intracellular tachyzoites only having divided twice after glutamine deprivation (Fig. 1D). The mean number of replication cycles was quantified and figure 1E shows that the replication ability of *T. gondii* was reduced about 25% in the absence of glutamine. VIABILITY: Parasite viability was evaluated by a tetrazolium dye reduction (MTT) assay, and after 4 hours of glutamine starvation, the viability of *T. gondii* decreased 29% and could not be rescued by the stereoisomer D-

glutamine, which suggests that glutamine serves as an energy source for *T. gondii* metabolic activity rather than a modulator of extracellular redox environments (Fig. 1F). MTT assays showed no significant effect of L-glutamine depletion on the HFF host cells (Fig. S1), indicating that the virulence and growth of *T. gondii* are not compromised by the host viability during glutamine starvation.

### Infected DCs exhibit hypermotility and an enhanced migration specific to high glutamine *in vitro*

We next used rat bone marrow-derived DC culture to study the effect of *T. gondii* infection on cell migration. Bone marrow-derived cells were isolated and cultured in complete medium supplemented with cytokines for seven days (see methods), and the phenotype of DCs was confirmed by flow cytometry (Fig. S2). DCs were infected by tachyzoites (multiplicity of infection, MOI, of 1) for 4 hours and then migration was examined by *in vitro* transwell system. Flow cytometric results show that the infection frequency of DCs is about 75% (Fig. 2A and 2B). *T. gondii* infection resulted in a roughly 2.5-fold increase in the number of DCs migrating to control medium, a result that we will refer to as a "hypermotility state" of infected DCs and one that has been reported previously [10] (Fig. 2C). Moreover, the infected cells exhibited an additional "enhanced migration" towards medium containing glutamine concentration above 0.5 mM (Fig. 2C). This enhanced migration was beyond that of the hypermotility induced by infection alone and was specific to glutamine relative to other amino acids, including the amino acids that utilize the same transporter (Fig. 2D). Importantly, uninfected DCs did not show induced migration to any of the tested glutamine concentrations or amino acids. This hypermotility and enhanced migration could not be explained simply by the maturation of the DCs because challenge with lipopolysaccharide (LPS) had no effect on migration to glutamine or other amino acids (Table S1).

### Enhanced migration of infected DCs towards glutamine is dependent on the glutamine transport via SNAT2 and availability of intracellular glutamine

A glutamine analog (2-(methylamino)-isobutyrate, MeAIB), which is a specific substrate of SNAT2 and decreases the cellular uptake of glutamine [25,26], was able to block the enhanced migration to 0.5 mM glutamine but had no effect on the hypermotility of infected DCs (Fig. 3A). A glutamine synthetase inhibitor (methionine sulfoximine, MSO) also blocked the enhanced migration to glutamine but not the hypermotility. Extracellular glutamine was always present in the media during the experiment (see methods), and thus, these results suggest that glutamine uptake by SNAT2 and glutamine synthetase are both required for the enhanced migration. In contrast, complete glutamine starvation of the infected DCs for 2 hours prior to the assay abolished both enhanced migration to glutamine and hypermotility, indicating that glutamine resource is essential for both hypermotility and enhanced migration of infected DCs.

### Glutamine transport via SNAT2 is important for CXCR4-dependent migration of uninfected DCs to SDF-1α

In order to test if SNAT2 serves as a general factor influencing immune cell migration, we examined another scenario in which uninfected DCs migrate, in this case to SDF-1α, a migration that has been shown to be dependent on CXCR4 signaling. Examining SDF-1α-induced transmigration in uninfected DCs with the same treatments described above, we demonstrated that glutamine starvation, MeAIB or MSO significantly reduced the uninfected DC migration to SDF-1α (Fig. 3B). Both a CXCR4 antagonist

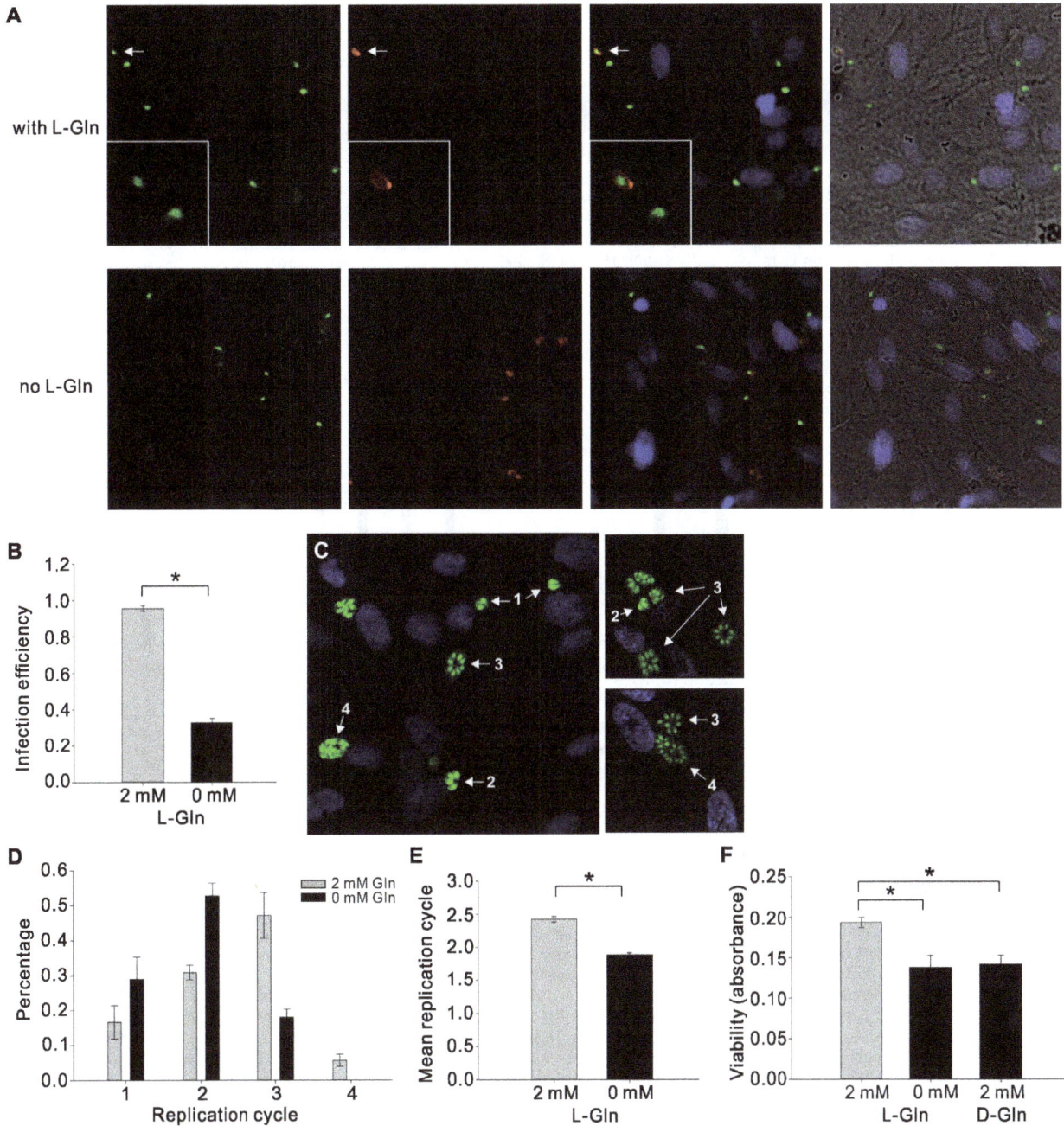

**Figure 1. _T. gondii_ is dependent on glutamine for optimal infection, replication and viability _in vitro_.** Experiments were conducted in human foreskin fibroblast (HFF) monolayers. (A) Representative confocal images on the top row show the intracellular tachyzoites expressing GFP (green) and one extracellular parasite staining with SAG1 (red, also indicated by the white arrow) in the presence of L-glutamine. A 63× oil objective was used to examine the SAG1-positive parasite at higher magnification (bottom left corner). Pictures on the second row represent the SAG1 staining of extracellular _T. gondii_ in the absence of L-glutamine. The nuclei of HFFs were stained with DAPI (blue), and the merged images of the fluorescence and the bright field shows that the HFF monolayer was confluent when the experiment was carried out. (B) Infection efficiency is defined as the ratio of intracellular parasites to total parasites in the examined field on the coverslip after 6 hours of infection. Infection efficiency was reduced ~65% by omitting L-glutamine (Student's t-test, $P < 0.001$). (C) Examples of the confocal images showing _T. gondii_ replication in HFFs after 22 hours in the presence of glutamine. The replication cycle numbers were indicated in white. (D) In the presence of L-glutamine, _T. gondii_ tachyzoites were able to duplicate up to 4 times in 22 hours, but without L-glutamine, over 50% of parasites only divided twice. (E) In the absence of L-glutamine, mean replication cycle number declined approximately 25% in 22 hours (Student's t-test, $P < 0.05$). (F) MTT assay was used to evaluate parasite viability. Glutamine starvation significantly decreased the viability of _T. gondii_ by 29%, and D-glutamine was unable to substitute for L-glutamine (one-way ANOVA, $F_{(2,89)} = 45.036$, $P < 0.001$). Asterisks indicate significant difference from medium containing 2 mM L-glutamine (post hoc Dunnett's test, $P < 0.001$). Data show mean value ± SEM from three independent experiments.

(AMD3100) and a PI3K inhibitor (LY294002) have previously been shown to block the SDF-1α-induced migration in bone marrow-derived mesenchymal stem cells, T lymphocytes and platelets [27–30]. We demonstrated that either of these two drugs,

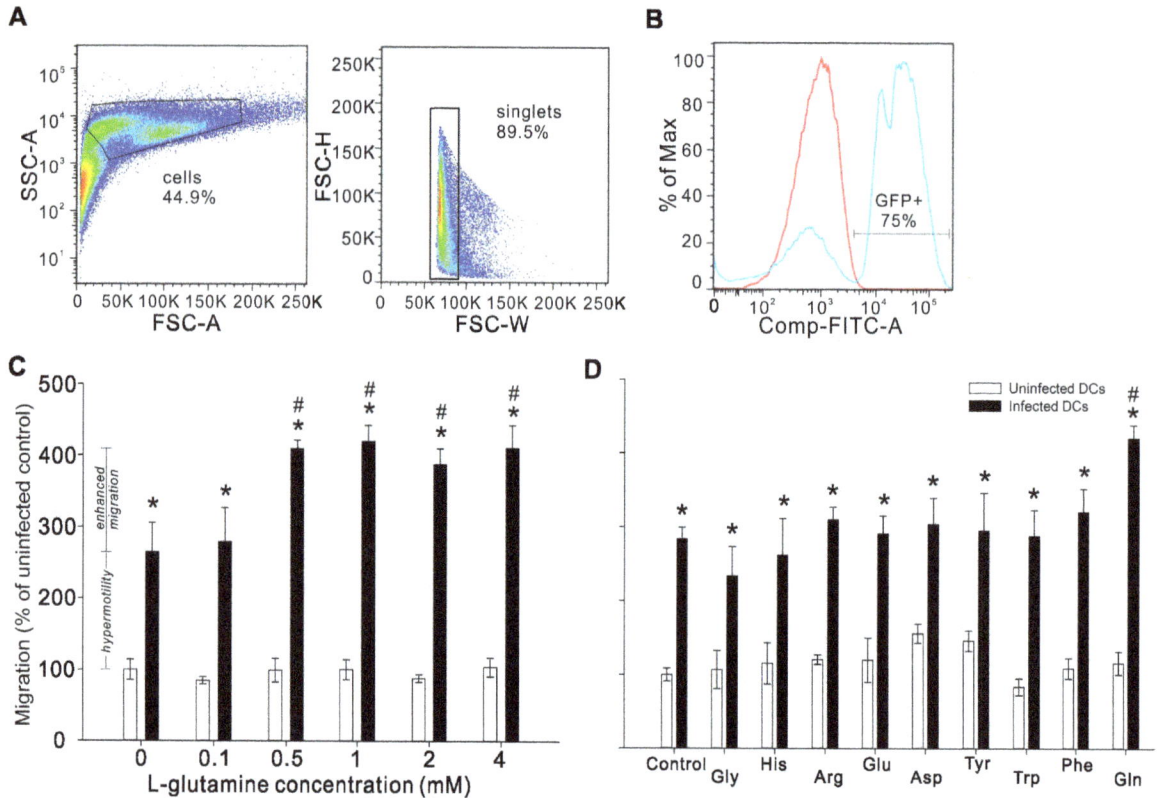

**Figure 2. Transmigration assay of *T. gondii*-infected DCs *in vitro*.** Infection frequency of *T. gondii*-infected DC culture was analyzed by flow cytometery. (A) In the gating strategy, SSC (side scatter) measures intracellular granularity and FSC (forward scatter) measures cell sizes. (B) The representative histogram shows the mean fluorescent intensity (x-axis) versus percentage of gated population (y-axis) for GFP. The pink histogram represents uninfected and the blue indicates infected DC culture. The infection frequency of DCs is about 75%. (C and D) DCs were infected with *T. gondii* for 4 hours and then cell migration was tested by Costar Transwell System. White bars represent uninfected DCs and black bars represent *T. gondii*-infected cells. Bar graphs depict mean values of migration ± SEM from three independent experiments performed in triplicate. Migration value is defined as the number of migrated cells in each condition normalized by the number of migrated cells in spontaneous migration, the migration of uninfected cells in the absence of factors (% of uninfected control). (C) "Hypermotility" state is defined as the increased migration after infection comparing to the spontaneous migration, the migration of uninfected control cells in the absence of factors. "Enhanced migration" to glutamine is the increased migration responding to increasing glutamine concentrations (between 0.5 and 4.0 mM) and is beyond that of the hypermotility induced by infection alone. Two-way ANOVA revealed an interaction effect between *T. gondii* infection and glutamine concentration ($F_{(5,37)} = 4.426$; $P = 0.003$). Further one-way ANOVA examining effects of glutamine within infected groups revealed an effect of glutamine in *T. gondii*-infected ($P < 0.01$; indicated by asterisks) but not uninfected DCs. Post-hoc pairwise comparison of means for glutamine concentrations within the *T. gondii*-infected DCs revealed an increase in migration relative to control at 0.5, 1.0, 2.0, and 4.0 mM concentrations (Fisher's protected LSD, $P < 0.02$; indicated by #), but not at 0.1 mM glutamine. (D) The specificity of enhanced migration to glutamine was tested by examining the cell migration in the presence of glutamine verse other amino acids such as glycine (Gly), histidine (His), arginine (Arg), glutamic acid (Glu), aspartic acid (Asp), tyrosine (Tyr), tryptophan (Trp) or phenylalanine (Phe). Two-way ANOVA revealed an interaction effect between *T. gondii* infection and amino acids ($F_{(9,70)} = 2.765$; $P = 0.008$). Further one way ANOVA examining effects of amino acids within infected groups revealed an effect of amino acids in *T. gondii*-infected ($P < 0.005$; indicated by asterisks) but not uninfected DCs. Post-hoc pairwise comparison of means for amino acid condition within the infected DCs revealed an increase in migration relative to the control within only the increased glutamine condition (Fisher's protected LSD, $P < 0.001$; indicated by #). The y-axis scale for both (C) and (D) is the same and shown at the far left.

as well as the specific Rho kinase inhibitor (Y27632), also dramatically decreased the migratory ability of uninfected DCs to SDF-1α (Fig. 3B). Thus, SDF-1α-induced migration of uninfected DCs is dependent on SNAT2 transport of glutamine, functional glutamine synthetase as well as the CXCR4-PI3K-Rho kinase pathways.

## Both hypermotility and enhanced migration to glutamine of infected DCs are dependent on the CXCR4-PI3K-Rho kinase pathways

We then tested if these CXCR4-PI3K-Rho kinase-related inhibitors could affect hypermotility and enhanced migration of *T. gondii*-infected DCs in the presence of glutamine. After 2 hours

of *T. gondii* infection, DCs were treated for an additional 2 hours with the same concentration of these inhibitors of the CXCR4-PI3K-Rho kinase pathways described previously. Inhibiting PI3K or Rho kinase blocked both the hypermotility and the enhanced migration to glutamine in infected cells (Fig. 3C). The CXCR4 inhibitor, AMD3100, inhibited the enhanced migration to glutamine and partially decreased the hypermotility, which suggested that CXCR4 is not the only activator of the downstream PI3K pathway that contributes to the migratory state of infected cells. In order to determine if glutamine transport through SNAT2 could be coactivating this pathway and thereby contributing to migration, we used the combination of MeAIB and AMD3100 to completely block enhanced migration and reduced hypermotility to close to the control level. The viability of intracellular *T. gondii*

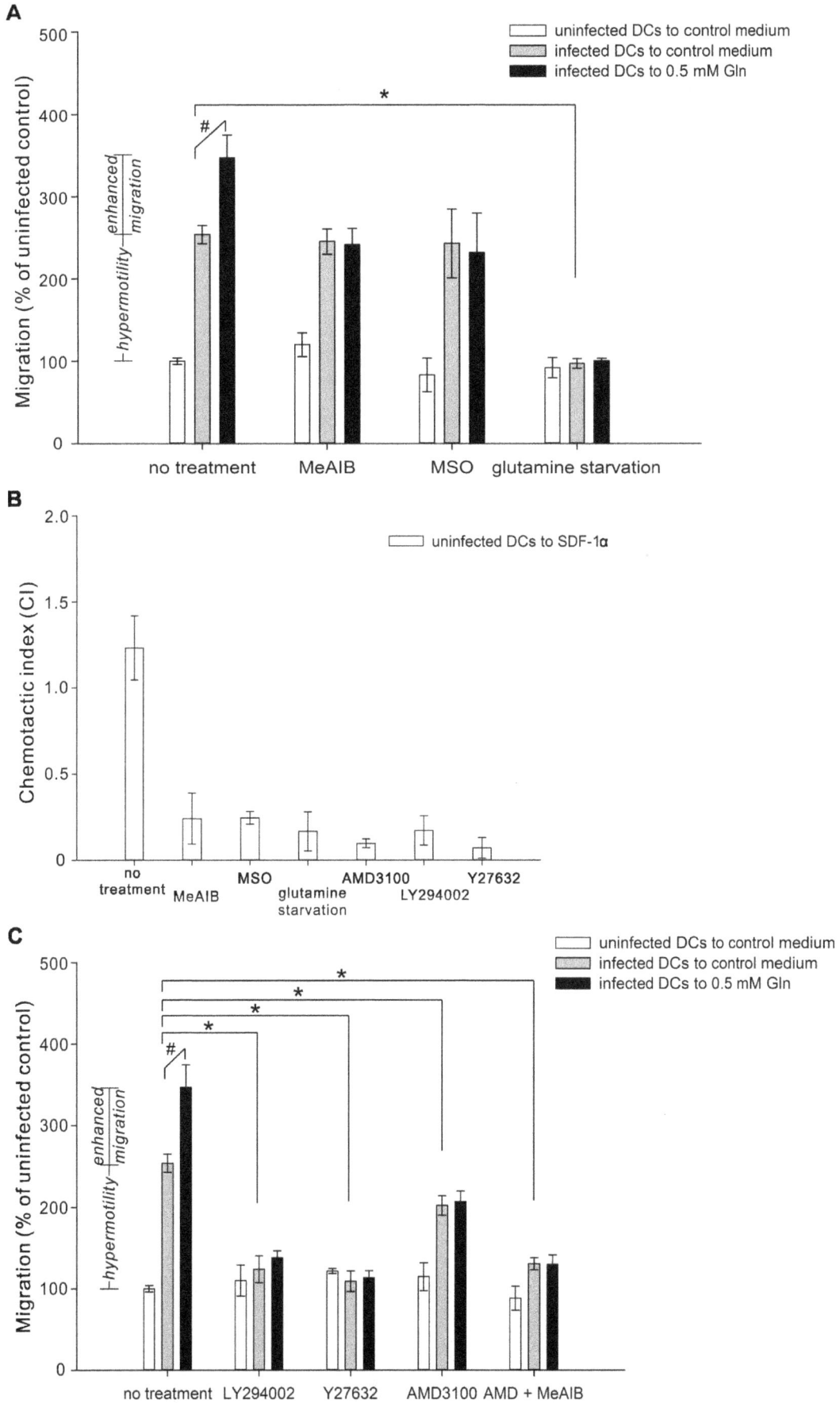

**A**

**B**

**C**

**Figure 3. Pharmacological manipulation study of the induced migration of the *T. gondii*-infected and uninfected DCs.** (A) Uninfected and infected DCs were left untreated (no treatment) or were treated with MeAIB (the competitive inhibitor of glutamine transport via SNAT2) or MSO (the inhibitor of glutamine synthetase), and cell migration to control medium (without glutamine) or to medium containing 0.5 mM glutamine was studied. A pretreatment starving the DCs of glutamine was also used to examine the importance of glutamine source in hypermotility and enhanced migration. (B) Uninfected control DCs were treated with MeAIB, MSO, inhibitors of CXCR4 (AMD3100), PI3K (LY294002) or Rho kinase (Y27632), or Gln starvation for 2 hours before assessing migration to 100 ng/ml SDF-1α. Chemotactic index (CI) is defined as the fold increase in the number of migrating DCs to SDF-1α over the spontaneous migration. One-way ANOVA reveals an effect of pharmacological treatments on the SDF-1α-induced migration ($F_{(6,44)} = 6.700$, $P < 0.001$). Asterisks indicate $P < 0.05$ (Dunnett's post hoc). (C) LY294002, Y27632, AMD3100 and the combination of AMD3100 and MeAIB were used to study the potential mechanisms contributed to the hypermotility and the enhanced migration of *T. gondii*-infected DCs. White bars represent uninfected DCs, and gray and black bars represent infected cells in response to 0 mM and 0.5 mM glutamine, respectively. Bar graphs depict mean values of migration ± SEM from three independent experiments performed in triplicate. "No treatment" bars in (A) and (C) represent the same data. One way ANOVA examining effects of treatment across the uninfected control DCs revealed no effect of treatment on cell migration ($P > 0.5$). One-way ANOVA examining effects of pharmacological treatments on hypermotility within infected groups (gray bars in A and C) revealed an effect of pharmacological treatments in *T. gondii*-infected DCs ($F_{(7,49)} = 16.541$, $P < 0.001$). Asterisks indicate $P < 0.003$ (Fisher's protected LSD). Within each pharmacological treatment, # indicates significant effects of 0.5 mM glutamine (enhanced migration to glutamine) relative to control conditions (hypermotility) (Student's t-test, $P < 0.01$).

following these pharmacological treatments was measured by plaque assays, and none of the inhibitors affected virulence of the parasites (Fig. S3), indicating that the decrease of induced migration of infected DCs was not due to the loss of parasite viability or stalled replication. Based on these results, we developed a hypothesis that hypermotility and enhanced migration of *T. gondii*-infected DCs may be dependent on an interaction between CXCR4 and glutamine transport via SNAT2 in activating the PI3K-Rho kinase pathways.

## *T. gondii* infection activates DCs and the SNAT2 and CXCR4 expression in the host cell remains unchanged

To determine whether *T. gondii* affected CXCR4 and SNAT2 protein expression in host cells, we used Western blot to compare the protein expression on surface and in cytosol in DCs either exposed to live *T. gondii* or an alternative immune activator (LPS + tumor necrosis factor α, TNFα). We also examined CD80 protein expression as a marker of immune cell activation/maturity. Figure 4A shows that both LPS + TNFα treatment and *T. gondii* infection increased CD80 protein expression on the surface of DCs, which suggests that both stimuli result in cell activation and maturation. The quantification of the density of the bands in the blots shows that although cell maturation with either LPS + TNFα or *T. gondii* infection did not affect either surface or cytosolic CXCR4 expression or surface SNAT2 expression, LPS + TNFα treatment decreased SNAT2 in cytosol while *T. gondii*-infected DCs possessed similar levels as uninfected DCs (Fig.4B).

## Discussion

*T. gondii* is an obligate intracellular pathogen that acquires glutamine from the host to carry out functions necessary for its metabolic activity, invasion, and replication. Here, we report that infection, replication and viability in *T. gondii* are greatly facilitated by glutamine availability. We demonstrate that in infected immune cells (DCs) *in vitro*, *T. gondii* induces a) a hypermotility state and, b) a specific enhanced migration to environments with high glutamine levels ($\geq 0.5$ mM). This hypermotility and enhanced migration are dependent on SNAT2 transport of glutamine, the presence of functional glutamine synthetase and the CXCR4-PI3K-Rho kinase pathways. Whereas activation with LPS + TNFα results in downregulation of SNAT2 protein expression in DCs, this expression is unchanged with *T. gondii* infection and may contribute to the induced migration of *T. gondii*-infected cells. These findings support the hypothesis that *T. gondii* manipulates the migratory properties of the host cell to

increase the probability of migrating to places with sufficient glutamine resource to meet its metabolic requirements.

The glutamine dependency of *T. gondii* is well documented. The Toxoplasma TCA cycle, which utilizes glucose and glutamine, generates energy for intracellular growth and replication; inhibition of the cycle leads to the death of parasites [2,4]. However, mutation of the *T. gondii* glucose transporter only has a minor effect on the growth of the parasite, and glutamine, but not glucose, is indispensable to its survival [3]. Therefore, glutaminolysis is essential for the parasitic requirements of *T. gondii*. Recently a GABA shunt was observed for regulating the catabolism of glutamine/glutamate in the Toxoplasma TCA cycle [2], and DCs infected with live *T. gondii* have been demonstrated to secrete GABA and exhibit hypermotility; this induced migration may increase the chance of wide dissemination of the parasites during acute infection [10,31,32]. Glutamine serves as the precursor of GABA; thus it is plausible that the rapidly dividing *T. gondii* imposes a high glutamine burden on the host, which may manipulate the migratory phenotype of the infected immune cells in order to sustain intracellular glutamine concentrations. Our results support this idea as infected DCs exhibit the hypermotility phenotype and enhanced migration to glutamine (Fig. 2). On the other hand, although glutamine-induced migration of *T. gondii*-infected DCs is a new finding, it is not the first amino acid that was reported to be a chemotaxis-inducing factor for immune cells, as glutamate has been shown to attract neutrophils to migrate to the inflamed/wound tissue via activating the downstream PI3K pathway [33,34].

The hypermotility of *T. gondii*-infected DCs is related to the parasite strain, and type II strain, as used in this study, exhibits the strongest induction of hypermotility [11]. This hypermotility has been suggested to be independent of CCR5, CCR7 or Toll/interleukin-1 pathways but to involve a Gi protein signaling transduction [10,32]. Chemokine-receptor complexes are known to initiate signal transduction events leading to cellular responses such as leukocyte chemotaxis and adhesion. CXCR4, one of 19 known chemokine receptors in mammals, couples primarily through Gi proteins, and PI3K and Rho kinase pathways involved in cell motility and orientation are downstream to activation of CXCR4 [12,13,35,36]. While we also demonstrate the crucial roles of PI3K and Rho kinase in this migration, as inhibition of either pathway blocked enhanced migration to glutamine and drastically reduced the hypermotility in *T. gondii*-infected DCs, AMD3100, an antagonist that binds CXCR4 without engaging receptor signaling [37,38], blocked enhanced migration to glutamine but only partially decreased hypermotility. On the other hand, the combination of AMD3100 and MeAIB, a specific

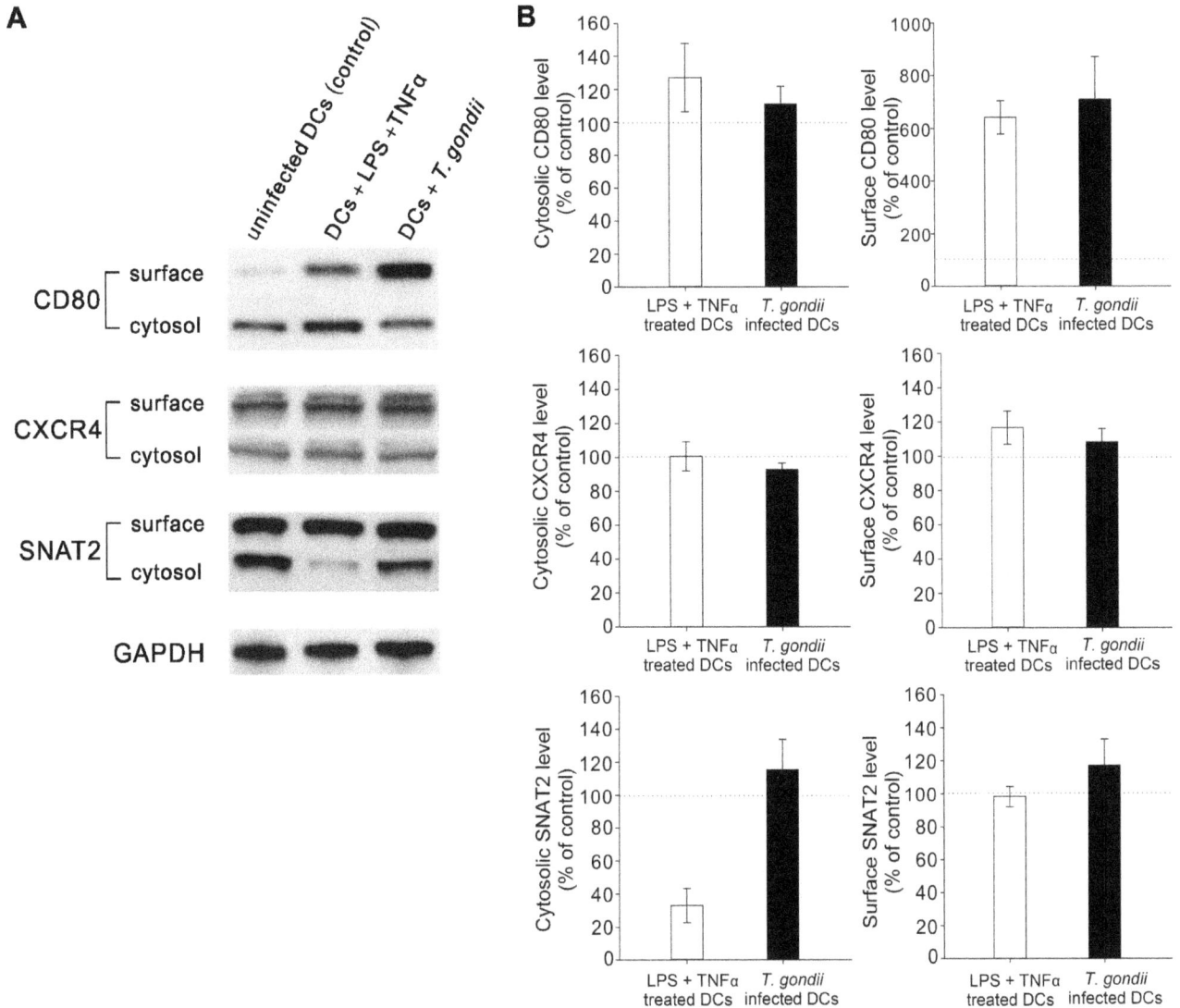

**Figure 4. Western blot analysis of CXCR4 and SNAT2 expression of *T. gondii*-infected DCs *in vitro*.** (A) DCs were treated with 100 ng/ml LPS+50 ng/ml TNFα or were infected with *T. gondii* for 2 hours in the presence of 2 mM glutamine and then the protein expression of CD80, CXCR4 and SNAT2 in cytosol and on the surface were studied. GAPDH is the loading control. (B) The band intensity in the blots was quantified to show the relative protein level of CD80, CXCR4 and SNAT2 in tested groups compared to control (dotted line). Experiments were repeated three times and quantified by Image J.

SNAT2 inhibitor, completely blocked enhanced migration and significantly reduced hypermotility, indicating that CXCR4 and SNAT2 may cumulatively regulate the altered migratory phenotype (Fig. 3). While the exact role of CXCR4 and SNAT2 in the induced migration of infected DCs is unclear, CXCR4 promotes the activation of PI3K pathway, and SNAT2, which has been suggested to possess a dual transporter/receptor (transceptor) function in cultured muscle cells and fibroblasts, activates PI3K pathway and regulates/senses changes of the intracellular amino acid pool and the availability of extracellular amino acids [14,15,39,40]. The transmigration results in Figure 3 support an interaction between SNAT2 and CXCR4 pathways through which *T. gondii* may orient increased migration of infected DCs to environments with high glutamine level.

In this study, western blot analysis showed that whereas DCs challenged with LPS + TNFα downregulate cytosolic SNAT2 protein, *T. gondii*-infected DCs possessed a similar level of the cytosolic pool of SNAT2 protein as uninfected control cells. Downregulation of SNAT2 following LPS + TNFα challenge is, to our knowledge, a novel finding and further study will be required to understand the role of SNAT2 in activated immune cells. It has been shown that during the acute phase of amino acid deprivation, the adaptive regulation of SNAT2 activity relies on recruitment of preformed proteins from a cytosolic pool to the cell membrane [41–43], which may explain why *T. gondii*-infected cells with unchanged SNAT2 expression in cytosol and on cell surface possessed the altered migratory phenotype, whereas LPS-treated immune cells did not (Table S1). CXCR4 protein remains unchanged with both LPS + TNFα treatment and *T. gondii* infection, indicating that expression of CXCR4, along with SNAT2, may be permissive for the migratory phenotype even if expression does not increase with enhanced migration.

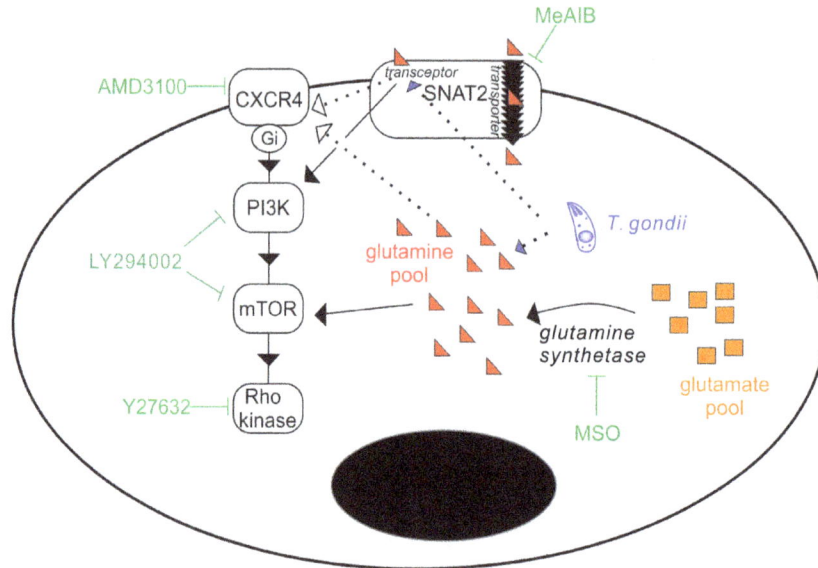

**Figure 5. Schematic diagram of the hypothetical model.** *T. gondii* increases the likelihood that infected immune cells will migrate to glutamine-rich environment by "turning on" the transceptor function of SNAT2 and modulating the CXCR4-PI3K-Rho kinase signaling pathways which are known to regulate migration. Dotted lines denote the new possible interactions between *T. gondii*, SNAT2 and CXCR4.

Based on these studies, it is unclear if the enhanced migration to glutamine is caused by *T. gondii* directly or by the host cell trying to rebalance the glutamine homeostasis. New evidence also shows that LY294002 targets mammalian target of rapamycin (mTOR) at the same concentrations that inhibit PI3K, and mTOR is involved in amino acid-sensing pathways [44–46]. However, based on our results, a new hypothetical mechanism of how *T. gondii* induces migration in infected DCs was proposed (Fig. 5). Firstly, a new on-and-off theory about the transceptor role of SNAT2 was proposed in 2009, and SNAT2 is thought to be "closed off" when sufficient concentrations of its substrates exist in the intracellular compartment, and to be activated ("on" state) when intracellular amino acid concentrations drop [47]. The off state of the transceptor might explain why immature DCs expressing SNAT2 do not show hypermotility or enhanced migration to glutamine in this study. We hypothesized that *T. gondii* may "turn on" the transceptor function of SNAT2. A drop in intracellular glutamine concentration secondary to parasite infection may be sufficient to turn on SNAT2; however, we cannot rule out the potential influence of *T. gondii*-related proteins. One of the ways in which *T. gondii* has been shown to manipulate host cell function is by secretion of rhoptry proteins or dense granule proteins into host cell cytoplasm and nucleus during invasion [48,49]; dense granule protein GRA5 has been shown to trigger the migration of human dendritic cells toward CCL19 [50], and rhoptry proteins and dense granule proteins of *T. gondii* interfere with host signaling pathways including immunity-related GTPases (IRGs) and nuclear factor-κB (NF-κB) pathways [49,51,52]. Therefore, *T. gondii* might directly or indirectly turn on the transceptor role of SNAT2 by modulating the intracellular glutamine metabolism and/or secreting modulatory protein(s) in the infected immune cells.

In summary, we hypothesize that *T. gondii* metabolic requirements may lead to parasite-host interactions that benefit the parasite by increasing the probability that infected-immune cells will carry the parasite out of circulation into tissues with sufficient glutamine (e.g. brain, eye, muscle, lung…). Technical challenges moving forward include measuring intracellular concentration of

radioactive labeled glutamine in *T. gondii*-infected DCs, delivering RNAi targeting CXCR4 and/or SNAT2 to *T. gondii*-infected immune cells without altering *T. gondii* or immune cell viability, manipulating the extracellular glutamine concentration *in vivo* to redirect the migration of infected cells, and the lack of knockout animal models of CXCR4 or SNAT2 to study the dissemination of the parasites. We here provide evidence for the CXCR4 and SNAT2 pathways as novel mechanisms for immune cell migration and potential targets for prevention of parasite dissemination and chronic infection in rodents and humans.

## Materials and Methods

### Parasites

Prugniaud stain of *T. gondii* expressing firefly luciferase and green fluorescent protein (GFP) was a gift from Dr. John Boothroyd. Tachyzoites were maintained by serial 2-day passage in human foreskin fibroblast (HFF) monolayers cultured in complete medium (DMEM plus 10% fetal bovine serum, 1% penicillin-streptomycin, and 2 mM L-glutamine) (Life Technology).

### Animals

Pregnant female Sprague-Dawley rats (single caged; Charles River Laboratories) were used for generation of pup bone marrow derived immune cells. All procedures were reviewed and approved by the Stanford University Administrative Panel on Laboratory Care and the Association for Assessment of Laboratory Animal Care.

### *T. gondii* Invasion Assay

HFFs were cultured on 12 mm coverslips in complete medium 24 hours before the experiment. Tachyzoites were added to HFFs (MOI = 2) and invasion in the absence (0 mM) or presence (2 mM) of glutamine was allowed for 6 hours at 37°C in the incubator with 5% $CO_2$ and then washed by PBS three times and stained with anti-SAG1 (1:10,000) for 15 minutes, which labels extracellular parasites before 4% paraformaldehyde fixation. Coverslips were then blocked for 30 minutes and incubated in Alexa Fluor 555 (1:1000) for 2 hours. A Zeiss confocal laser scanning microscope (LSM) with a 40× objective was used to count the number of parasites and total of 15 fields per coverslip were examined. The ratio of intracellular tachyzoites to total tachyzoites in the examined field on the coverslip is defined as infection efficiency.

## T. gondii Replication Ability Assay

HFF culture conditions were identical to the invasion assay. HFFs were challenged with tachyzoites (MOI = 1) in complete medium (which contains 2 mM L-glutamine) for 6 hours and then washed with PBS to remove glutamine-rich medium and extracellular tachyzoites before incubating in media with or without 2 mM L-glutamine for another 16 hours. SAG1 antibody was used to label extracellular parasites and stained with Alexa Fluor 555 for 2 hours. The number of replication cycles undergone by intracellular T. gondii was quantified for 15 fields under a Zeiss confocal LSM with a 63× objective, and the mean replication cycle was presented as the replication ability. No parasites underwent more than 4 rounds of replication cycle in the time period that we chose.

## T. gondii Viability Assay

Free tachyzoites were incubated in the medium containing 0 or 2 mM of L-glutamine or 2 mM of D-glutamine for 4 hours at 37°C in the incubator with 5% $CO_2$, and then yellow MTT (which is reduced to purple formazan by mitochondrial enzymes) was added in living cells and the absorbance of colored solution was quantified by a spectrophotometer.

## Bone Marrow-Derived Dendritic Cell Culture

The bone barrow-derived DC culture was generated as previously described [53], with slight modifications. Briefly, bone marrow from 16 day-old Sprague-Dawley rat pups was isolated and stored in liquid nitrogen until use. Upon thawing, cells were washed with RPMI 1640 (Catalog number 21870, Life Technology) and placed in T-75 flasks at a density of approximately 2.5-$3.5 \times 10^6$ cells/ml in complete medium: RPMI 1640, 10% fetal bovine serum, 2 mM L-glutamine, 1% nonessential amino acids, 1 mM sodium pyruvate, 50 units/ml penicillin +50 ug/ml streptomycin (all from Life Technology) and the cytokines, interleukin-4, granulocyte-macrophage colony stimulating factor (GM-CSF), Flt-3 ligand (5 ng/ml per cytokine, R&D Systems). On the third day, each flask was given 5 ml of fresh complete medium containing GM-CSF only, and then three quarters of medium containing non-adherent cells were replaced with fresh complete medium with GM-CSF on day 5. Cells were used on day 7.

## Flow Cytometry

DCs were infected with freshly egressed T. gondii (MOI = 1) for 4 hours and then collected and washed in FACS buffer (1 mM EDTA+1% FBS in PBS) on Day 7. An LSR II flow cytometer (BD Biosciences) was used to investigate the infection frequency and the results were analyzed with FlowJo Software (Tree Star, Inc.).

## In Vitro Transmigration Assay

Cell migration was tested using Transwell assays (Costar 24-well plate with inserts, 5 μm pore). Briefly, 1 million cells challenged with freshly egressed T. gondii (MOI = 1) for 4 hours were resuspended in culture medium consisting of RPMI 1640 with 1%

bovine serum albumin (BSA) and 2 mM L-glutamine in a 100-μl volume. Another group of cells were infected with T. gondii the same way for 2 hours following another 2 hours of inhibitor treatments before being resuspended in culture medium as described above. Infected and inhibitor-treated cells were then counted by trypan blue staining to confirm the cell viability before being transferred to the transwell system. Tested amino acids or SDF-1α (final working concentration: 100 ng/ml) diluted in the control medium (RPMI 1640+1% BSA) were added to the bottom well in a 600-μl volume when the same amount of live cells in each condition were added to the upper wells in a 100-μl volume. The migration was assayed for 1.5 hours in a 37°C incubator with 5% $CO_2$ and then migrated cells were collected and counted by trypan blue staining with a hemocytometer.

The tested amino acids were diluted in control medium at the concentrations as following: 0.133 mM glycine, 0.0968 mM histidine, 1.15 mM arginine, 0.136 mM glutamic acid, 0.15 mM aspartic acid, 0.129 mM tyrosine, 0.0245 mM tryptophan, 0.0909 mM phenylalanine (Sigma). Because the formula of RPMI 1640 contains a mixture of amino acids with the concentrations described above, adding another equal amount of amino acids in the migration medium created 2× of tested amino acids, which were comparable to 4 mM of glutamine (2× Gln). Inhibitors were used at following final concentrations: 40 mM MeAIB (Sigma), 5 mM MSO (Sigma), 50 μM LY294002 (Cell Signaling), 10 μM Y27632 (Sigma), 50 μM AMD3100 (Abcam).

## Cell Surface Protein Biotinylation and Western Blot

Surface proteins of DCs were isolated by using the Pierce Cell Surface Protein Isolation kit (Thermo Scientific) according to the manufacturer's protocol. In brief, cells grown in Falcon T-75 flasks were washed with PBS, incubated with EZ-LINK Sulfo-NHS-SSbiotin for 30 minutes at 4°C and then lysed with lysis buffer containing protease inhibitor. The biotinylated surface proteins were trapped within NeutrAvidin agarose gel when cytosol proteins were eluted, and then surface proteins were eluted by Bio-Rad sample buffer containing DTT.

Protein samples (25 μg) were separated on 10% SDS-PAGE gel and then transferred to Bio-Rad PVDF membranes and immunoblotted with antibodies against SNAT2 (1:500, Santa Cruz biotechnology), CXCR4 (1:500, Abcam) and CD80 (1:10000, Abcam). Glyceraldehyde 3-phosphate dehydrogenase (GAPDH) was used as a loading control (1:10000, Sigma). The density of bands on 3 independent blots was quantified by Image J.

## Statistical Analysis

Statistical analyses were performed using IBM SPSS statistics software (version 20).

## Supporting Information

**Figure S1  Viability assay of _T. gondii_-infected HFFs _in vitro_.**

**Figure S2  Characterization of rat bone marrow-derived DC cultures on Day 7.**

**Figure S3  Viability assay of intracellular _T. gondii_ in HFFs following inhibitor treatments.**

**Table S1   Transmigration assay *in vitro*: cell migration of activated DCs to glutamine or other tested amino acids.**

## Author Contributions

Conceived and designed the experiments: IL. Performed the experiments: IL AKE CY MGW VK ZDM NCM. Analyzed the data: IL AKE CY MGW. Wrote the paper: IL AKE RMS.

## Acknowledgments

We thank Anita Koshy and John Boothroyd for the strain of *T. gondii* and the SAG1 antibody. Advice and feedback was provided in all phases of this project by John Boothroyd and Michelle Cheng.

## References

1. MacRae JI, Dixon MW, Dearnley MK, Chua HH, Chambers JM, et al. (2013) Mitochondrial metabolism of sexual and asexual blood stages of the malaria parasite Plasmodium falciparum. BMC Biol 11: 67.
2. MacRae JI, Sheiner L, Nahid A, Tonkin C, Striepen B, et al. (2012) Mitochondrial metabolism of glucose and glutamine is required for intracellular growth of Toxoplasma gondii. Cell Host Microbe 12: 682–692.
3. Blume M, Rodriguez-Contreras D, Landfear S, Fleige T, Soldati-Favre D, et al. (2009) Host-derived glucose and its transporter in the obligate intracellular pathogen Toxoplasma gondii are dispensable by glutaminolysis. Proc Natl Acad Sci U S A 106: 12998–13003.
4. Sheiner L, Vaidya AB, McFadden GI (2013) The metabolic roles of the endosymbiotic organelles of Toxoplasma and Plasmodium spp. Curr Opin Microbiol 16: 452–458.
5. Broer S, Brookes N (2001) Transfer of glutamine between astrocytes and neurons. J Neurochem 77: 705–719.
6. Kalloniatis M, Tomisich G (1999) Amino acid neurochemistry of the vertebrate retina. Prog Retin Eye Res 18: 811–866.
7. Tjader I, Berg A, Wernerman J (2007) Exogenous glutamine—compensating a shortage? Crit Care Med 35: S553–S556.
8. Tanhoffer RA, Yamazaki RK, Nunes EA, Pchevozniki AI, Pchevozniki AM, et al. (2007) Glutamine concentration and immune response of spinal cord-injured rats. J Spinal Cord Med 30: 140–146.
9. Albrecht J, Sidoryk-Wegrzynowicz M, Zielinska M, Aschner M (2010) Roles of glutamine in neurotransmission. Neuron Glia Biol 6: 263–276.
10. Lambert H, Hitziger N, Dellacasa I, Svensson M, Barragan A (2006) Induction of dendritic cell migration upon Toxoplasma gondii infection potentiates parasite dissemination. Cell Microbiol 8: 1611–1623.
11. Lambert H, Vutova PP, Adams WC, Lore K, Barragan A (2009) The Toxoplasma gondii-shuttling function of dendritic cells is linked to the parasite genotype. Infect Immun 77: 1679–1688.
12. Procko E, McColl SR (2005) Leukocytes on the move with phosphoinositide 3-kinase and its downstream effectors. Bioessays 27: 153–163.
13. Cain RJ, Ridley AJ (2009) Phosphoinositide 3-kinases in cell migration. Biol Cell 101: 13–29.
14. Hyde R, Cwiklinski EL, MacAulay K, Taylor PM, Hundal HS (2007) Distinct sensor pathways in the hierarchical control of SNAT2, a putative amino acid transceptor, by amino acid availability. J Biol Chem 282: 19788–19798.
15. Evans K, Nasim Z, Brown J, Clapp E, Amin A, et al. (2008) Inhibition of SNAT2 by metabolic acidosis enhances proteolysis in skeletal muscle. J Am Soc Nephrol 19: 2119–2129.
16. Mackenzie B, Erickson JD (2004) Sodium-coupled neutral amino acid (System N/A) transporters of the SLC38 gene family. Pflugers Arch 447: 784–795.
17. Li P, Yin YL, Li D, Kim SW, Wu G (2007) Amino acids and immune function. Br J Nutr 98: 237–252.
18. Newsholme P (2001) Why is L-glutamine metabolism important to cells of the immune system in health, postinjury, surgery or infection? J Nutr 131: 2515S–2522S.
19. Saeij JP, Boyle JP, Boothroyd JC (2005) Differences among the three major strains of Toxoplasma gondii and their specific interactions with the infected host. Trends Parasitol 21: 476–481.
20. Darde ML (2004) Genetic analysis of the diversity in Toxoplasma gondii. Ann Ist Super Sanita 40: 57–63.
21. Saeij JP, Boyle JP, Grigg ME, Arrizabalaga G, Boothroyd JC (2005) Bioluminescence imaging of Toxoplasma gondii infection in living mice reveals dramatic differences between strains. Infect Immun 73: 695–702.
22. Boothroyd JC, Grigg ME (2002) Population biology of Toxoplasma gondii and its relevance to human infection: do different strains cause different disease? Curr Opin Microbiol 5: 438–442.
23. Vyas A, Kim SK, Giacomini N, Boothroyd JC, Sapolsky RM (2007) Behavioral changes induced by Toxoplasma infection of rodents are highly specific to aversion of cat odors. Proc Natl Acad Sci U S A 104: 6442–6447.
24. Vyas A (2013) Parasite-augmented mate choice and reduction in innate fear in rats infected by Toxoplasma gondii. J Exp Biol 216: 120–126.
25. Maroni BJ, Karapanos G, Mitch WE (1986) System A amino acid transport in incubated muscle: effects of insulin and acute uremia. Am J Physiol 251: F74–F80.
26. Rae C, Hare N, Bubb WA, McEwan SR, Broer A, et al. (2003) Inhibition of glutamine transport depletes glutamate and GABA neurotransmitter pools: further evidence for metabolic compartmentation. J Neurochem 85: 503–514.
27. Song C, Li G (2011) CXCR4 and matrix metalloproteinase-2 are involved in mesenchymal stromal cell homing and engraftment to tumors. Cytotherapy 13: 549–561.
28. Lee JY, Buzney CD, Poznansky MC, Sackstein R (2009) Dynamic alterations in chemokine gradients induce transendothelial shuttling of human T cells under physiologic shear conditions. J Leukoc Biol 86: 1285–1294.
29. Badillo AT, Zhang L, Liechty KW (2008) Stromal progenitor cells promote leukocyte migration through production of stromal-derived growth factor 1alpha: a potential mechanism for stromal progenitor cell-mediated enhancement of cellular recruitment to wounds. J Pediatr Surg 43: 1128–1133.
30. Kraemer BF, Borst O, Gehring EM, Schoenberger T, Urban B, et al. (2010) PI3 kinase-dependent stimulation of platelet migration by stromal cell-derived factor 1 (SDF-1). J Mol Med (Berl) 88: 1277–1288.
31. Fuks JM, Arrighi RB, Weidner JM, Kumar MS, Jin Z, et al. (2012) GABAergic signaling is linked to a hypermigratory phenotype in dendritic cells infected by Toxoplasma gondii. PLoS Pathog 8: e1003051.
32. Lambert H, Barragan A (2010) Modelling parasite dissemination: host cell subversion and immune evasion by Toxoplasma gondii. Cell Microbiol 12: 292–300.
33. Gupta R, Chattopadhyay D (2009) Glutamate is the chemotaxis-inducing factor in placental extracts. Amino Acids 37: 359–366.
34. Gupta R, Palchaudhuri S, Chattopadhyay D (2013) Glutamate induces neutrophil cell migration by activating class I metabotropic glutamate receptors. Amino Acids 44: 757–767.
35. Rot A, von Andrian UH (2004) Chemokines in innate and adaptive host defense: basic chemokinese grammar for immune cells. Annu Rev Immunol 22: 891–928.
36. Tan W, Martin D, Gutkind JS (2006) The Galpha13-Rho signaling axis is required for SDF-1-induced migration through CXCR4. J Biol Chem 281: 39542–39549.
37. Fricker SP, Anastassov V, Cox J, Darkes MC, Grujic O, et al. (2006) Characterization of the molecular pharmacology of AMD3100: a specific antagonist of the G-protein coupled chemokine receptor, CXCR4. Biochem Pharmacol 72: 588–596.
38. Hatse S, Princen K, Bridger G, De CE, Schols D (2002) Chemokine receptor inhibition by AMD3100 is strictly confined to CXCR4. FEBS Lett 527: 255–262.
39. Evans K, Nasim Z, Brown J, Butler H, Kauser S, et al. (2007) Acidosis-sensing glutamine pump SNAT2 determines amino acid levels and mammalian target of rapamycin signalling to protein synthesis in L6 muscle cells. J Am Soc Nephrol 18: 1426–1436.
40. Gazzola RF, Sala R, Bussolati O, Visigalli R, Dall'Asta V, et al. (2001) The adaptive regulation of amino acid transport system A is associated to changes in ATA2 expression. FEBS Lett 490: 11–14.
41. Hyde R, Christie GR, Litherland GJ, Hajduch E, Taylor PM, et al. (2001) Subcellular localization and adaptive up-regulation of the System A (SAT2) amino acid transporter in skeletal-muscle cells and adipocytes. Biochem J 355: 563–568.
42. Jones HN, Ashworth CJ, Page KR, McArdle HJ (2006) Expression and adaptive regulation of amino acid transport system A in a placental cell line under amino acid restriction. Reproduction 131: 951–960.
43. Ling R, Bridges CC, Sugawara M, Fujita T, Leibach FH, et al. (2001) Involvement of transporter recruitment as well as gene expression in the substrate-induced adaptive regulation of amino acid transport system A. Biochim Biophys Acta 1512: 15–21.
44. Pinilla J, Aledo JC, Cwiklinski E, Hyde R, Taylor PM, et al. (2011) SNAT2 transceptor signalling via mTOR: a role in cell growth and proliferation? Front Biosci (Elite Ed) 3: 1289–1299.
45. Hay N, Sonenberg N (2004) Upstream and downstream of mTOR. Genes Dev 18: 1926–1945.
46. Tokunaga C, Yoshino K, Yonezawa K (2004) mTOR integrates amino acid- and energy-sensing pathways. Biochem Biophys Res Commun 313: 443–446.
47. Hundal HS, Taylor PM (2009) Amino acid transceptors: gate keepers of nutrient exchange and regulators of nutrient signaling. Am J Physiol Endocrinol Metab 296: E603–E613.

48. Boothroyd JC, Dubremetz JF (2008) Kiss and spit: the dual roles of Toxoplasma rhoptries. Nat Rev Microbiol 6: 79–88.

49. Bougdour A, Tardieux I, Hakimi MA (2013) Toxoplasma exports dense granule proteins beyond the vacuole to the host cell nucleus and rewires the host genome expression. Cell Microbiol.

50. Persat F, Mercier C, Ficheux D, Colomb E, Trouillet S, et al. (2012) A synthetic peptide derived from the parasite Toxoplasma gondii triggers human dendritic cells' migration. J Leukoc Biol 92: 1241–1250.

51. Fleckenstein MC, Reese ML, Konen-Waisman S, Boothroyd JC, Howard JC, et al. (2012) A Toxoplasma gondii pseudokinase inhibits host IRG resistance proteins. PLoS Biol 10: e1001358.

52. Hunter CA, Sibley LD (2012) Modulation of innate immunity by Toxoplasma gondii virulence effectors. Nat Rev Microbiol 10: 766–778.

53. Manley NC, Caso JR, Works MG, Cutler AB, Zemlyak I, et al. (2013) Derivation of injury-responsive dendritic cells for acute brain targeting and therapeutic protein delivery in the stroke-injured rat. PLoS One 8: e61789.

# 3

# Understanding the Mechanism of Atovaquone Drug Resistance in *Plasmodium falciparum* Cytochrome b Mutation Y268S Using Computational Methods

**Bashir A. Akhoon**[1,9]**, Krishna P. Singh**[1,9]**, Megha Varshney**[2]**, Shishir K. Gupta**[3]**, Yogeshwar Shukla**[4,5]**, Shailendra K. Gupta**[1,5*¤]

1 Department of Bioinformatics, Systems Toxicology Group, CSIR-Indian Institute of Toxicology Research, Lucknow, India, 2 Interdisciplinary Biotechnology Unit, Aligarh Muslim University, Aligarh, India, 3 Department of Bioinformatics, Biocenter, Am Hubland, University of Würzburg, Würzburg, Germany, 4 Department of Proteomics, CSIR-Indian Institute of Toxicology Research, Lucknow, India, 5 Academy of Scientific and Innovative Research (AcSIR), New Delhi, India

## Abstract

The rapid appearance of resistant malarial parasites after introduction of atovaquone (ATQ) drug has prompted the search for new drugs as even single point mutations in the active site of Cytochrome b protein can rapidly render ATQ ineffective. The presence of Y268 mutations in the Cytochrome b (Cyt b) protein is previously suggested to be responsible for the ATQ resistance in *Plasmodium falciparum* (*P. falciparum*). In this study, we examined the resistance mechanism against ATQ in *P. falciparum* through computational methods. Here, we reported a reliable protein model of Cyt bc1 complex containing Cyt b and the Iron-Sulphur Protein (ISP) of *P. falciparum* using composite modeling method by combining threading, *ab initio* modeling and atomic-level structure refinement approaches. The molecular dynamics simulations suggest that Y268S mutation causes ATQ resistance by reducing hydrophobic interactions between Cyt bc1 protein complex and ATQ. Moreover, the important histidine contact of ATQ with the ISP chain is also lost due to Y268S mutation. We noticed the induced mutation alters the arrangement of active site residues in a fashion that enforces ATQ to find its new stable binding site far away from the wild-type binding pocket. The MM-PBSA calculations also shows that the binding affinity of ATQ with Cyt bc1 complex is enough to hold it at this new site that ultimately leads to the ATQ resistance.

**Editor:** Adrian J.F. Luty, Institut de Recherche pour le Développement, France

**Funding:** This work was supported by the Council of Scientific & Industrial Research (CSIR) - network project GENESIS (BSC0121) and INDEPTH (BSC0111). The funder had no role in study design, data collection and analysis, decision to publish, or preparation of the manuscript.

**Competing Interests:** The authors have declared that no competing interests exist.

* Email: skgupta@iitr.res.in

¤ Current address: Department of Systems Biology and Bioinformatics, University of Rostock, Rostock, Germany

9 These authors contributed equally to this work.

## Introduction

Studies revealed that human malaria is caused by protozoan parasites of the genus *Plasmodium*. The four most common *Plasmodium* species that infect human are *P. vivax*, *P. ovale*, *P. malariae*, and *P. falciparum*. Additionally, a fifth one *P. knowlesi* has also been identified as responsible for infection in human [1] often in many countries of Southeast Asia [2]. According to the latest World malaria report (2012) by World health organization, there were about 219 million cases of malaria in 2010 and an estimated 660 000 deaths. *P. falciparum* predominates in Africa and is the most deadly form leading to death due to malaria. 90% of malaria occurs in Africa and among which 85% deaths happen in children under the age of 5 [3].

*Plasmodium* species can acquire drug resistance through several mechanisms, like change in drug permeability, increased expression of the drug target, or changes in the enzyme target [4]. ATQ drug acts against malarial parasites by inhibiting mitochondrial electron transport [5] and collapsing mitochondrial membrane potential [6]. Based on its structural similarity to ubiquinol, it has been postulated that ATQ binds to parasite Cyt b protein [7]. It is

supported by experimental findings that mutations at 268th position in *P. falciparum* Cyt b are unambiguously associated with acquired ATQ resistance [8]. However, the mechanism of ATQ resistance is still not well understood. Thus, there is an urgent need to develop novel disease management strategies against various *Plasmodium sp* induced malaria.

In several studies [9–11], researchers have modeled some *P. falciparum* mutations including Y268S using *in silico* methods, however none of them have completely modeled the *P. falciparum* Cyt bc1 complex rather they rely on the ATQ-bound yeast Cyt bc1 complex. Moreover, none of the study has examined the dynamics of the Cyt bc1 ATQ-bound complex. Therefore in the present study, we exploited the *in silico* approaches to identify molecular basis of ATQ drug resistance in the Y286S mutation model of Cyt b protein of *P. falciparum*. To best of our knowledge, this is the first study to report the modeling and molecular dynamics simulation of ATQ-bound *P. falciparum* Cyt bc1 complex in both wild and mutant-type models for nanoseconds time scale.

## Materials and Methods

### Computational model building and quality assessment

The ubiquinol oxidation (Qo) site of the Cytochrome bc1 complex serves as a pocket for ATQ binding [11] and two subunits of the complex (Cyt b and ISP) are involved in ATQ binding [11,12]. To model the whole complex, amino acid sequences of *P. falciparum* Cyt b (Genbank accession no: NP_059668.1) and ubiquinol-Cyt C reductase ISP subunit (Genbank accession no: XP_001348547.1) were retrieved from the Entrez protein database available at NCBI (http://www.ncbi.nlm.nih.gov). In the process of protein modeling, we observed that no single template was able to satisfy ~100% query coverage. Hence, the composite modeling which combines various techniques such as threading, *ab initio* modeling and atomic-level structure refinement approaches [13–16] implemented in the iterative threading assembly refinement (I-TASSER) server was preferred to build the full-length protein structure of both the protein chains. I-TASSER generates 3D atomic models from multiple threading alignments and iterative structural assembly simulations. The full methodology of the server has been described elsewhere [17]. The template modeling score (TM-score) calculation [18] was used to assess the structural similarity of model and template protein structures [Eq. i].

$$\text{TM - score} = \text{Max} \left[ \frac{1}{L} \sum_{i=1}^{L_{ali}} \frac{1}{1 + d_i^2 / d_0^2} \right] \quad (1)$$

where $L$ is the length of the target protein, $L_{ali}$ is the number of the equivalent residues in two proteins, $di$ is the distance of the $i^{th}$ pair of the equivalent residues between the two structures, which depends on the superposition matrix; the 'max' means the procedure to identify the optimal superposition matrix that superposition matrix that maximizes the sum in Eq. i. The scale $d_0 = \sqrt[3]{(L - 15)} - 1.8d$ is defined to normalize the TM-score in a way that the magnitude of the average TM-score for random protein pairs is independent on the size of the proteins.

Confidence score (C-score) was taken into consideration to determine the accuracy of the predicted structure. The score is defined based on the quality of the threading alignments and the convergence of the I-TASSER's structural assembly refinement simulations [Eq. ii].

$$C \text{ - } score = \ln \left[ \frac{M}{M_{tot}} \times \frac{1}{\langle RMSD \rangle} \times \frac{1}{7} \sum_{i=1}^{7} \frac{Z(i)}{Z_0(i)} \right] \quad (2)$$

Where $M$ is the number of structure decoys in the cluster and $M_{tot}$ is the total number of decoys generated during the I-TASSER simulations. $\langle RMSD \rangle$ is the average RMSD of the decoys to the cluster centroid. $Z(i)$ is the Z-score of the best template generated by $i^{th}$ threading in the seven LOMETS programs and $Z_0(i)$ is a program-specified Z-score cutoff for distinguishing between good and bad templates.

The geometry of the theoretical model was improved by side-chain geometry optimization using the ChiRotor algorithm [19]. The modeled structures were further subjected to energy minimization followed by model quality estimation. In order to further design the Cyt b-ISP complex from the individually modeled structures of Cyt b and ISP subunits from *P. falciparum*, we superimposed them to the *S. cerevisiae* bc1 complex already available in Protein Data Bank (pdb entry: 3CX5). From this

superimposed structure, we got the coordinates of modelled Cyt b and ISP subunits of *P. falciparum* in the orientation similar to the one found in Cyt bc1 complex of *S. cerevisiae*. Since the yeast bc1 complex also strongly interacts with water molecules in the vicinity of Glu 272 [20], the water molecule was also added to the modeled *P. falciparum* Cyt b-ISP complex before performing the docking experiments.

### Mutation mapping of Cyt b protein at 268$^{th}$ position

All the known point mutations observed at position 268 of *P. falciparum* Cyt b were individually incorporated in the modeled 3D protein structure to scan their impact on ATQ binding. For this, all the mutant models of the Cyt b protein of *P. falciparum* were generated using the mutational modeling protocol of DS3.1. The Build Mutants protocol mutates selected residues to specified types and optimizes the conformation of the mutated residues and their neighbors using MODELER program.

### Retrieval of ATQ structure and modeling of Cyt bc1 complex

PubChem Compound, one of the linked databases within the NCBI's Entrez information retrieval system was accessed for the retrieval of ATQ structure. Both the protein and ligand molecules were prepared before being subjected to docking analysis using Prepare Protein and Prepare Ligand protocols of DS3.1 respectively. Prepare Protein protocol rectify the protein for various problems, such as missing atoms in incomplete residues; missing loop regions; alternate conformations (disorder); nonstandard atom names; incorrect protonation state of titratable residues etc. The generation of 3D conformation of ATQ was attained by the Prepare Ligands tool.

Molecular docking experiments of the ATQ into the Qo site of both mutated and non-mutated variants of the Cyt b protein was performed by CDOCKER, a molecular dynamics (MD) simulated-annealing-based algorithm [22]. The ATQ was assumed to bind in the same binding pocket to that of ligand stigmatellin, a known inhibitor of Qo site of the Cyt bc1 complex. Water molecules were removed, except HOH7187 which has been reported to play important role in the observed hydrogen bonding network [9]. General-purpose all-atom force field (CHARMm) with a wide coverage for proteins, nucleic acids and general organic molecules was included in the random structure generation. 10 orientations were generated for the ATQ, improved by performing simulated annealing method and finally refined by applying low, but most accurate full potential as a refined pose minimization method.

### Molecular dynamics simulations

Molecular dynamics simulations were performed with Gromacs ver. 4.5.3. The ligand topology and parameterization was attained with SwissParam (http://www.swissparam.ch/), an automatic tool that generates topology and parameters based on the Merck molecular force field. The wild and mutant-type Cytbc1 complexes (Cyt b-ISP/ATQ) were subjected to molecular dynamics simulation with explicit TIP3P water solvation model in the Isothermal–isobaric (NPT) ensemble using the AM-BER99SB force field. Each system was minimized with the steepest descent method to relax unfavorable contacts between molecules and equilibrated for 150 ps before production runs to achieve stability during production dynamics. Simulations were performed at a constant temperature of 310 K and pressure of 1 atm, using Particle Mesh Ewald method [23] for long-range electrostatic [24] and van der Waals (vdW) [25,26] interactions with a cut-off of 1.4 nm while constraints were applied on all bonds using the LINCS [27] algorithm. All systems were simulated in the NPT (fixed number of atoms N, pressure P, and temperature T) ensemble using the v-rescale coupling algorithm [28] and the Parrinello–Rahman coupling algorithm [29] for 90 ns and time step of 2 fs without any position restraints.

## MM/PBSA calculations

The MM/PBSA approach [30] was applied to perform the binding free energy calculations. The binding free energy of a protein to a ligand ($\Delta G_{bind}$) is defined from the complex, protein and ligand free energies ($G_{complex}$, $G_{protein}$ and $G_{ligand}$, respectively) as [Eq. iii]

$$\Delta G_{bind} = G_{complex} - (G_{protein} + G_{ligand}) \qquad (3)$$

Each free energy term is obtained from a MD-derived ensemble of structures as the sum of six terms as mentioned in [Eq. iv].

$$\begin{aligned} <G> \;=\; &<G_{int}> + <G_{vdW}> + <G_{coul}> \\ &+ <G_{ps}> + <G_{nps}> - T<SMM> \end{aligned} \qquad (4)$$

Where $G_{int}$, $G_{vdW}$ and $G_{coul}$ indicate the internal (including bond, angle, and torsional angle energies), van der Waals and coulombic energy terms, respectively, collectively defined "gas phase terms". $<G_{ps}>$ and $<G_{nps}>$ are the polar and nonpolar solvation energy terms, respectively. $<SMM>$ is the entropic term. Angle brackets denote the average along the structures.

The single trajectory method (STM) has been used for both the wild-type and the mutant systems. The STM requires the trajectory of the complex to be run only. The structures of the free forms of the protein and ligand species were obtained by stripping the partner molecule from the structure of the complex. Thus, zeroing out the $<G_{int}>$ term in the STM analysis.

For the MM/PBSA calculations, the GMXAPBS tool was used [31]. In particular: (1) the van der Waals term was calculated with Gromacs, (2) the coulombic term was calculated using the APBS accessory program coulomb, (3) the polar solvation term was calculated via APBS [32], using the non-linearized Poisson Boltzmann equation. Internal and external dielectric constants were set to 1 and 80, respectively; temperature was set to 310 K; the salt concentration was defined as 0.15 M; grid spacing was set to an upper limit of 0.5 Å, (4) The nonpolar solvation term was considered proportional to the solvent accessible surface area (SASA) as shown in [Eq. v]

$$<G_{nps}> \;=\; <SASA> \gamma + \beta \qquad (5)$$

where $\gamma = 0.0227$ kJ mol$^{-1}$ Å$^{-2}$ and $\beta = 0$ kJ mol$^{-1}$ [33]. The dielectric boundary was defined using a probe of radius 1.4 Å.

The equilibrium phase (70–90 ns) of the two molecular dynamics simulations (i.e., 151 equally time-distant frames for each system) was considered for MM/PBSA calculations. The standard errors (SE) were calculated as [Eq. vi]

$$SE = SD/sqrt\,(151) \qquad (6)$$

where SD is the standard deviation.

## Principal component analysis

The Principal Component Analysis (PCA) was used to characterize and compare the overall motions of the two complexes. We calculated the principal components on the converged simulation and focused on the movement of the 731 Cα atoms of the protein that resulted in 2193 dimensional displacement vectors. The 2193×2193 covariance matrix was then diagonalized to obtain its eigenvalues and eigenvectors. The PCA method decomposes the overall protein motion into a set of modes (eigenvectors) that are ordered from largest to smallest contributions to the protein fluctuations. The contribution of atom j to the $i^{th}$ mode's fluctuation was obtained using the following equation [Eq. vii]:

$$|m_i^j| = \sqrt{\left(m_i^{jx}\right)^2 + \left(m_i^{jy}\right)^2 + \left(m_i^{jz}\right)^2} = component_i^j \qquad (7)$$

The $m_i^j = \left(m_i^{jx}, m_i^{jy}, m_i^{jz}\right)$ term represents the component vectors of the $j^{th}$ atom for the $i^{th}$ mode.

Each of the eigenvectors depicts a collective motion of particles and their respective amount of participation is represented by eigenvalues. Usually, the first ten eigenvectors are sufficient to describe almost all of the conformational subspace accessible to the protein.

## Results and Discussion

The Y268 residue of Cyt b in *Plasmodium* is known to play a key role in ATQ drug resistance [8] and thus, can be used as a potential resistance marker [34]. Studies of *P. falciparum* resistance to ATQ revealed 3 point mutations at 268th position (Y268N, Y268S, Y268C) [8,21,35]. The substitution of one or several amino acid residues in a protein often lead to substantial changes in properties such as thermodynamic stability, catalytic activity, or binding affinity [36–38]. As point mutations at Y268 have already been identified for ATQ resistance, it is obvious that this substitution should affect the fitting and binding of the drug. ATQ drug, an analogue of coenzyme Q (ubiquinone), interrupts electron transport and leads to loss of the mitochondrial membrane potential [6].

## Molecular modeling of Cyt b-ISP complex of *P. falciparum*

The 3D structure details of proteins are of major importance in providing insights into their molecular functions. Since computational methods not only help in directing the selection of key experiments, but also in the formulation of new testable hypotheses [39]. Therefore, in the absence of X-ray structures, the 3D theoretical models were built using the I-TASSER. I-TASSER [12,13], a hierarchical protein structure modeling approach based on 2 protein structure prediction methods i.e., threading and *ab initio* prediction, was used to build 3D models of the Cyt b and ISP sequences from *P. falciparum*. I-TASSER uses restraints from templates identified by multiple threading programs to build full length model using replica-exchange Monte-carlo simulations. Cyt b chain was modeled by I-TASSER using restraints from PDB templates 2IBZ, 3CX5, 1EZV, 1BCC and 3H1J while the ISP subunit used restrains from PDBs 1KB9, 3CX5, 1EZV, 1BCC, 3CWB, 1BGY and 3L72. No significant similarity was observed with any of the PDB templates against N-terminal sequence of ISP, therefore, this part was modeled by I-TASSER using ab-initio approach. Although, including the long

N-terminal sequence, modeled by ab-initio method can compromise the overall model accuracy, in the present study, we used full length model because of multiple reasons, (i) N-terminal 158 residues (modeled) of ISP subunit may also contribute in the reliable folding of tertiary structure as we observed several secondary structure elements in that region (Figure S1). Several studies have shown that the removal of N-terminal building blocks from the structure may contribute in error during protein folding [40–42] as the protein may acquire a non-native stable conformation due to mis-association of the adjunct building blocks. (ii) After modeling of the Cyt b-ISP complex, we observed the N-terminal sequence of chain moves back and forth over the active-site cleft (Figure S2) therefore this part was also taken into consideration. (iii) Since, even a single amino acid residue may significantly affect the conformation of binding site if present around 4.5 Å radius of the ligand, and may also alter the binding efficacy [43], thus we preferred to include this region in our model, so that the possible impact of N-terminal sequence in ATQ binding may be evaluated during simulations.

The computation of a structural alignment of 2 protein structures is critical in modeling, as in contrast to sequence alignment methods, structure alignment methods aim directly on optimizing the structural similarity of the input proteins [44]. The implemented TM-score in I-TASSER is a sensitive scale to the global topology for measuring the structural similarity between 2 proteins. Statistically, a TM-score <0.17 means a randomly selected protein pair with the gapless alignment taken from PDB. The better TM-score of our Cyt b model (0.99±0.04) and ISP subunit model (0.35±0.12) indicates much better structural match of the target sequence with the templates. I-TASSER provides C-score to estimate the quality of the predicted models, and is calculated based on the significance (i.e. Z-score) of the threading alignments in LOMETS and the convergence parameters (i.e. cluster density) of the I-TASSER structure assembly simulations. The C-score scheme has been extensively tested in large-scale benchmarking tests [13,18] and is typically in the range (−5, 2), where a higher score reflects a model of better quality. The C-score of the best predicted model of Cyt b and ISP subunit model was +2.00 and −3.25.

The Side-Chain Refinement protocol of DS was used to optimize the protein side-chain conformations. This protocol uses ChiRotor algorithm and CHARMm force field to systematically search for optimal side-chain conformation of all residues and generates a model structure with the best side-chain conformation. Model quality estimation is critically important in computational protein modeling, since the accuracy of a model determines its suitability for specific biological and biochemical experimental design [45]. The fitness of a protein sequence in its current 3D environment before and after side chain refinement was evaluated by Verify Protein (Profiles-3D). The Verify score of the protein is the sum of the scores of all residues in the protein and has been used by several researchers for structural assessment of theoretical models [46–48]. If the overall quality is lower than the expected low score, the structure is certainly misfolded. The verify score of the Cyt b model before and after energy minimization was 89.2 and 93.5, with the expected low score of 77.1, showing that the structure after side chain refinement was much better than the non-refined one. Similarly, we observed the verify score of the ISP model as 89.4, with the expected low score of 72.7906. The exact percentage of amino acids located in the core region was calculated by Procheck program. Generally, the atomic resolution structures have over 90% of their residues in the most favorable regions and for lower resolution structures resolved at 3.0–4.0 Å, the core percentage is around 70% [49]. The Ramachandran plot

showed that 90.3% residues were located in the core region of Cyt b chain and 83.33% residues were in the core region of ISP chain, indicating the reliability of models for further studies (Figure S3).

Moreover, we also looked into the RMSD of the modelled Qo site and the templates chosen for modeling of wild-type protein of *P. falciparum*. The calculated residuals for 1EZV, 2IBZ, 3CX5, 3H1J and 1BCC templates from our modeled protein were 0.48 Å, 0.44 Å, 0.46 Å, 0.56 Å and 1.02 Å respectively. The insignificant RMSD of Qo site from the respective templates further supports the model accuracy.

## Mutagenesis of Cyt b protein

Most of the reported mutations in Cyt b either destabilize the important hydrophobic interactions between ATQ and the amino acid residues in the binding site of the protein, or are responsible for the change of pocket volume [9,50]. The conserved bulky Tyrosine (T) residue at 268th position forms hydrophobic contact with the ATQ drug in the Qo region of the ubiquinol oxidation site. Substitution of the hydrophilic and less bulky asparagine (N) at position 268 not only reduces the volume of the binding pocket but it also decreases the affinity and binding of ATQ [51]. Besides, substitution of serine (S), a hydrophilic amino acid, limits hydrophobic contact with ATQ resulting in marked decrement of ATQ susceptibility in mutated malaria parasites [21,35]. Moreover, a role for cysteine (C) in impairment of ATQ binding has also been observed [8]. Hence, *P. falciparum* Cyt b protein mutations (Y268S, Y268N and Y268C) were implemented in the 3D structure of the parent model using mutational modeling protocol of DS.

## Active site selection and docking calculations of ATQ

ATQ is very likely to bind in a manner similar to stigmatellin, a known inhibitor of Qo site of the Cyt bc1 complex [10] and hence the potential binding site for stigmatellin as proposed by Solmaz and Hunte (PDB acquisition code 3CX5) [20] was chosen as the biologically favorable site for ATQ docking. After checking the conservancy of the active site residues of Cyt bc1 complex of *P. falciparum* (CYTB: MET116, ILE119, VAL120, PHE123, VAL124, MET133, TRP136, GLY137, VAL140, ILE141, THR142, LEU144, LEU145, ILE155, PHE169, LEU172, ILE258, VAL259, PRO260, GLU261, PHE264, PHE267, TYR268, LEU271, VAL284, LEU285; ISP: HIS104, LEU302, CYS319) with the Cyt b protein of *S. cerevisiae*, we observed that with the exception of 8 amino acid residues, all other residues were conserved between these 2 species (Figure 1). Moreover, we observed the non-identical residues of *P. falciparum* were also showing strongly similar properties (scoring>0.5 in the Gonnet PAM 250 matrix) with the amino acid residues present in *S. cerevisiae*. *Plasmodium* Cyt b is unusual in the sense that cd2 helix (a critical structural component of the catalytic Qo site) contains a 4-residue deletion that is not found in non-Apicomplexan sequences. It would seem very likely that this would alter the fold of the Qo site when compared to the Cyt b structural data available in the PDB. Therefore, we aligned the 3D structures of Cyt b from *P. falciparum* and *S. cerevisiae*. The structural overlay of the homology model of the *P. falciparum* Cyt b with the yeast Cyt b has been presented for comparison in Figure 2. Our model suggests that the 4 residue deletion in the cd2 helix results in a 0.83 Å displacement of this structural element compared with the yeast Cyt b (Figure 2). Likewise catalytically essential 'PEWY' motif of the ef helix was observed to be displaced by 0.35 Å from the yeast enzyme. To perform docking analysis, the ATQ structure was modeled properly using the Prepare/Filter Ligand tool of DS. Docking is a potentially powerful and inexpensive method for the

```
Cyt b_P. falciparum    MNFYS----INLVKAHLINYPCPLNINFLWNYGFLLGIIFFIQIITGVFLASRYTPDVSY  56
Cyt b_S. cerevisiae    MAFRKSNVYLSLVNSYIIDSPQPSSINYWWNMGSLLGLCLVIQIVTGIFMAMHYSSNIEL  60

Cyt b_P. falciparum    AYYSIQHILRELWSGWCFRYMHATGASLVFLLTYLHILRGLNYSYMYLP--LSWISGLIL  114
Cyt b_S. cerevisiae    AFSSVEHIMRDVHNGYILRYLHANGASFFFMVMFMHMAKGLYYGSYRSPRVTLWNVGVII  120

Cyt b_P. falciparum    FMIFIVTAFVGYVLPWGQMSYWGATVITNLLSSIPVA----VIWICGGYTVSDPTIKRFF  170
Cyt b_S. cerevisiae    FILTIATAFLGYCCVYGQMSHWGATVITNLFSAIPFVGNDIVSWLWGGFSVSNPTIQRFF  180

Cyt b_P. falciparum    VLHFILPFIGLCIVFIHIFFLHLHGSTNPLGYDTALK-IPFYPNLLSLDVKGFNNVIILF  229
Cyt b_S. cerevisiae    ALHYLVPFIIAAMVIMHLMALHIHGSSNPLGITGNLDRIPMHSYFIFKDLVTVFLFMLIL  240

Cyt b_P. falciparum    LIQSLFGIIPLSHPDNAIVVNTYVTPSQIVPEWYFLPFYAMLKTVPSKPAGLVIVLLSLQ  289
Cyt b_S. cerevisiae    ALFVFYSPNTLGHPDNYIPGNPLVTPASIVPEWYLLPFYAILRSIPDKLLGVITMFAAIL  300

Cyt b_P. falciparum    LLFLLAE-QRSLTTIIQFKMIFGARDYSVPIIWFMCAFYALLWIGCQLPQDIFILYGRLF  348
Cyt b_S. cerevisiae    VLLVLPFTDRSVVRGNTFKVLS-----KFFFFIFVFNFVLLGQIGACHVEVPYVLMGQIA  355

Cyt b_P. falciparum    IVLFFCSGLFVL--VHYRRTHYDYSSQANI  376
Cyt b_S. cerevisiae    TFIYFAYFLIIVPVISTIENVLFYIGRVNK  385
```

```
ISP_P. falciparum    MNNIKYVELFYKCKIFRKNGLNRIIRRNGGTFNHNIKENERIPPASEDPS  50
ISP_S. cerevisiae    MLGIR------------------------------------SSVKTC  11

ISP_P. falciparum    YKNLFDHAEDIKLWEIEEKQNVSHKKVEDLSELVEPSNHPHQYEGIFART  100
ISP_S. cerevisiae    FK---------------PMSLTSKRLISQSLLASKS------------  32

ISP_P. falciparum    RYAHYNQTAEPVFPRKPDLEKGELASGANVTRTDVWHNPKEPAIVSIGKF  150
ISP_S. cerevisiae    --TYRTPNFDDVLKENNDADKG--------------------------  52

ISP_P. falciparum    EPRNFRPAGYAENCPNPESINSDHHPDFREYRLRSGNEDRRSFMYFISAS  200
ISP_S. cerevisiae    ----------------------------------------RSYAYFMVGA  62

ISP_P. falciparum    YFFIMSSIMRSAICKSVHFFWISKDLVAGGTTELDMRTVNPGEHVVIKWR  250
ISP_S. cerevisiae    MGLLSSAGAKSTVETFISSMTATADVLAMAKVEVNLAAIPLGKNVVVKWQ  112

ISP_P. falciparum    GKPVFVKHRTPEDIQRAKEDDKLIQTMRDPQLDSDRTIKPEWLVNIGICT  300
ISP_S. cerevisiae    GKPVFIRHRTPHEIQEANSVD--MSALKDPQTDADRVKDPQWLIMLGICT  160

ISP_P. falciparum    HLGCVPA-QGGNYSGYFCPCHGSHYDNSGRIRQGPAPSNLEVPPYEFVDE  349
ISP_S. cerevisiae    HLGCVPIGEAGDFGGWFCPCHGSHYDISGRIRKGPAPLNLEIPAYEFDGD  210

ISP_P. falciparum    NTIKIG  355
ISP_S. cerevisiae    -KVIVG  215
```

**Figure 1. Sequence alignment of Cyt b protein and ISP chain of *P. falciparum* and *S. cerevisiae*.** Amino acid residues involved in the formation of Qo site are highlighted in yellow tinted color.

discovery of binary interactions. The Qo site residues were chosen to define the binding site in our modeled Cyt bc1 complex of *P. falciparum* based on known Qo site inhibitor interactions for *S. cerevisiae* available in protein data bank (pdb id: 3CX5, 2IBZ). A total of 10 random ligand conformations were generated from the ATQ structure through high temperature molecular dynamics, followed by random rotations. These conformations were refined by grid-based (GRID 1) simulated annealing and a final full force field minimization method. We observed that ATQ was showing less binding affinity towards all the mutant variants when compared to the wild-type.

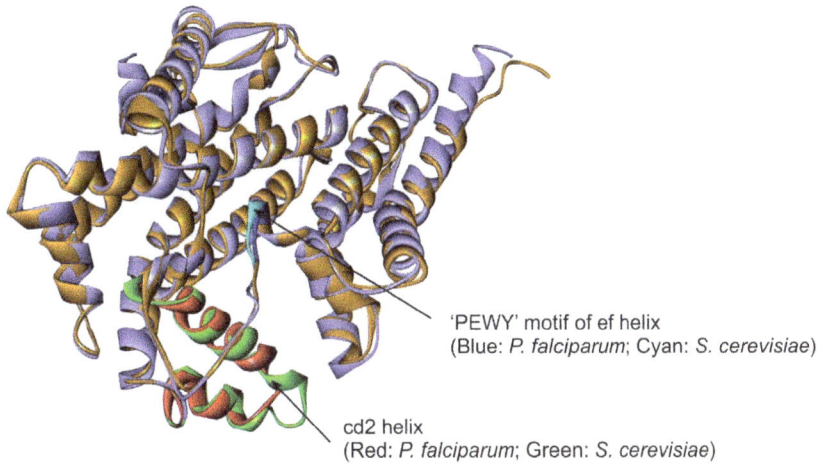

**Figure 2. Structural overlay of the homology model of Cyt b protein of *P. falciparum* (blue) with the Cyt b unit of *S. cerevisiae* (golden) (PDB ID: 3CX5).** A total of 4 amino acid residues deletion in cd2 helix (red) of *P. falciparum* resulted in structural displacement when compared with the same domain of *S. cerevisiae* (green). Also the structural changes in 'PEWY' motif of ef helix are shown.

While examining all the contact amino acid residues within 5 Å of the ATQ, as shown in the Figure 3, we observed the presence of Y268 within the ATQ binding site in the wild-type protein. Surprisingly when Y268 was mutated to any of the 3 possible amino acid mutations, i.e.Y268N, Y268S, Y268C, this position shifted far away from ATQ binding site. We feel the shift of amino acid residue at 268th position after point mutation might be the main reason of ATQ resistance in the mutant models. In order to understand the mechanistic insight of ATQ resistance in the mutant models, we further performed detailed molecular dynamics simulation studies by considering only the most prevalent

mutant variant (Y268S) identified in various experimental settings [9].

## Dynamic insights into the Cyt bc1 modeled complexes

Several simulation studies have already shown nice correlation between computational and experimental measurements of macromolecular dynamics [52–55]. As molecular dynamics based techniques can provide more precise protein–ligand models in the state close to natural conditions therefore to get detailed insights into the molecular basis of ATQ resistance in malaria, we individually simulated ATQ-bound wild-type and the most prevalent mutant variant (Y268S) of *P. falciparum* Cyt b protein

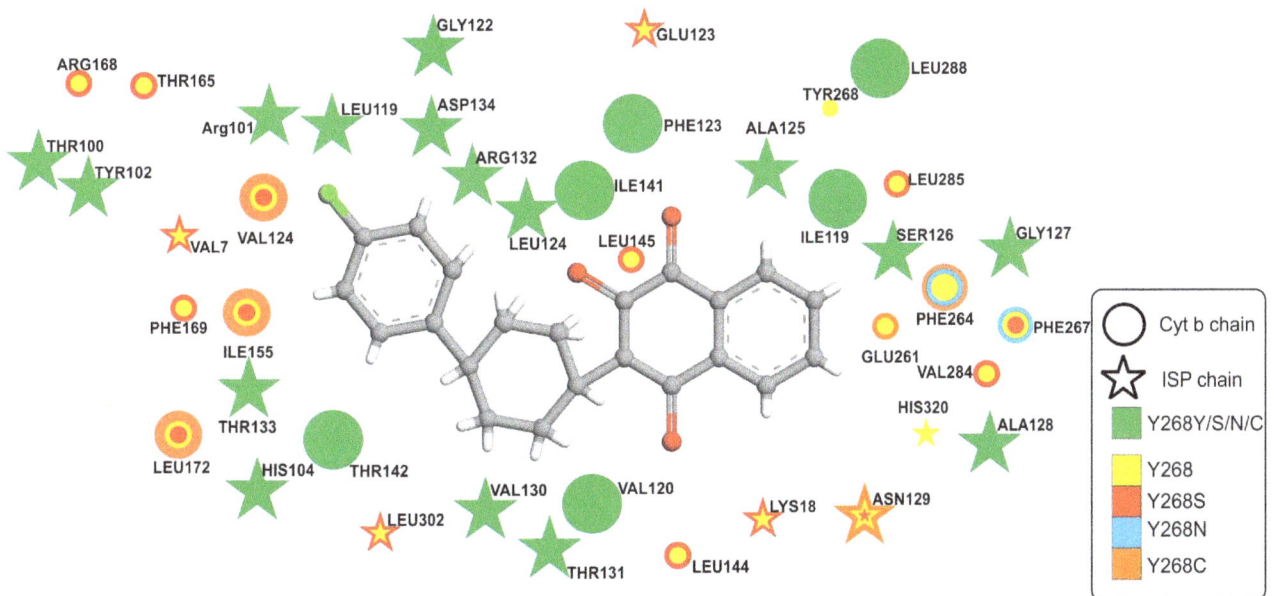

**Figure 3. Two dimensional contact plots of amino acid residues from the wild and all screened mutant models of Cyt bc1 complex from P. falciparum in the vicinity of 5 Å radius around ATQ.** It may be noted that in the mutant models the position 268 shifted away from the 5 Å radius of ATQ Binding site. Whereas green color indicates that the particular amino acid residue is present in the wild as well as in all mutant models in the observed area; yellow, red, blue, cyan color shows amino acid residues present only in wild type, Y268S, Y268N and Y268C mutant models respectively.

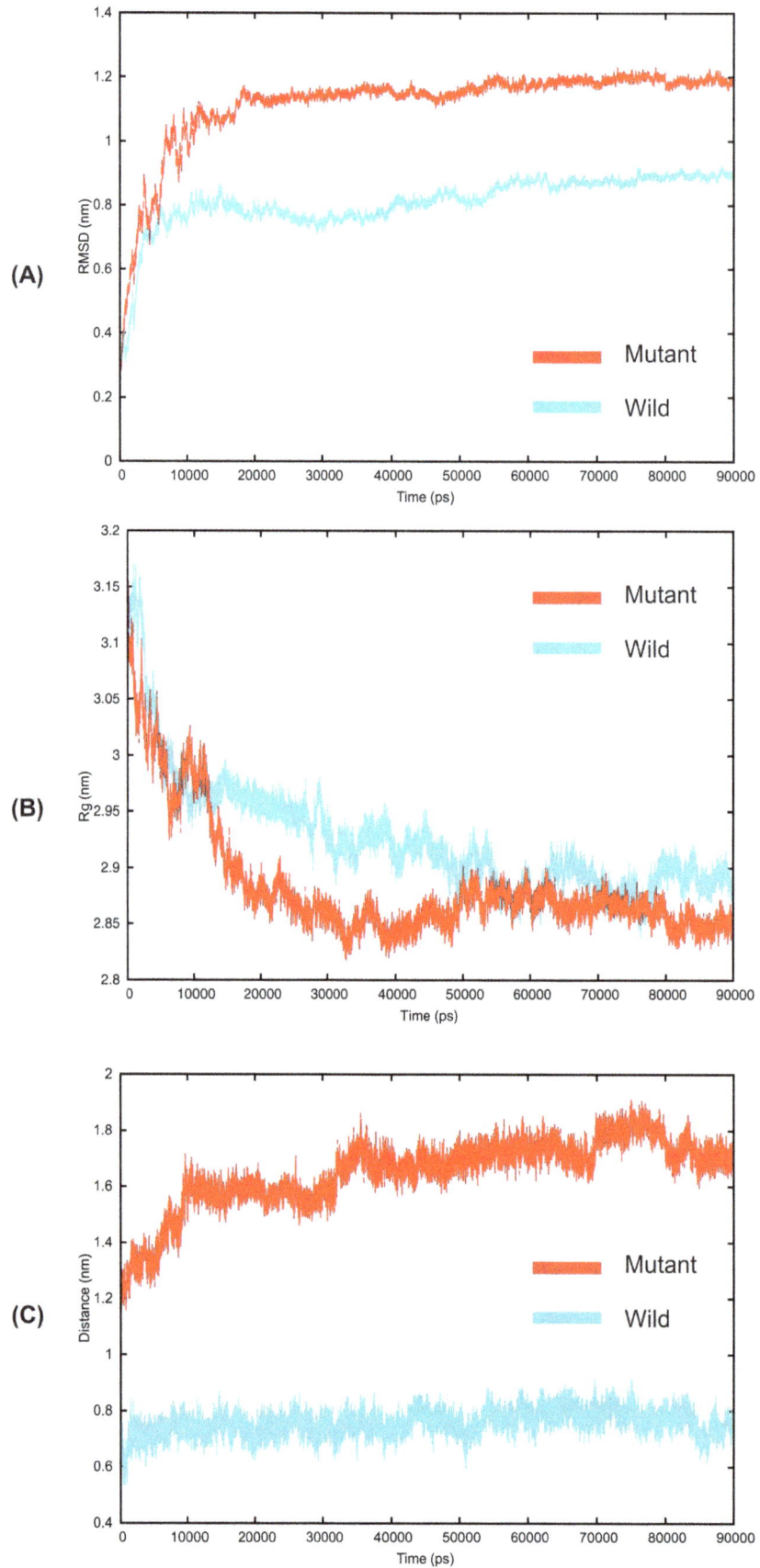

**(A)**

**(B)**

**(C)**

**Figure 4. Molecular dynamics simulation graph of wild (Y268) and mutant (Y268S) Cyt bc1 complexes of *P. falciparum* in solution.** (A) Root mean square deviation (RMSD) of backbone atoms with respect to their initial complexes over a period of 90 ns simulation time. (B) Radius of gyration (Rg) graph and (C) Distance of ATQ from the Qo site over the whole simulation in wild as well as mutant type.

[9] till we attained the convergence of MD simulation around 70–90 ns of production run. To explore the dynamic stability of the systems, root mean square deviation (RMSD) of the Cyt b-ISP backbone atoms (both wild-type and mutant complexes) was computed with reference to their respective initial structures as a function of simulation run time. MD simulation shows that the wild-type complex undergoes less structural changes when compared with the mutant complex over a period of 90 ns simulation time (Figure 4A, in Cyan) and RMSD of both the complexes remain almost stagnant after ~65 ns time period. The plotted graph in Figure 4A (red) shows that although the RMSD of mutant complex was initially in agreement with the wild-type complex but the complex underwent significant deviation after 7ns and reaches at ~1.1 nm RMSD (consistent throughout the whole dynamics run), showing the comparatively unstable behavior of the mutant complex. We further extended our study to analyze the compactness (radius of gyration) of both the wild-type and mutant

complexes. Though we did not find much significant difference in Rg values, however we noticed that at the equilibrium state mutant structure was more compact than the wild-type (Figure 4B). We were also interested to check the ATQ distance from the Qo site during the entire dynamics run. After analyzing the results, we noticed that ATQ retained its position throughout the whole dynamics run in the wild-type however it showed significant fluctuations during the entire 90ns simulation in mutant case (Figure 4C). Even as clear from the Figure 4C, the distance between the Qo site and ATQ was increasing with respect to time. We may attribute this behavior to the change in binding pocket configuration and also because of some steric clashes of ATQ near the binding site.

In order to understand how the resistant mutations affect the interaction of ATQ in mutant protein, we calculated root mean square inner product (RMSIP) from both the complexes to ascertain the convergence of conformational sampling instep-wise

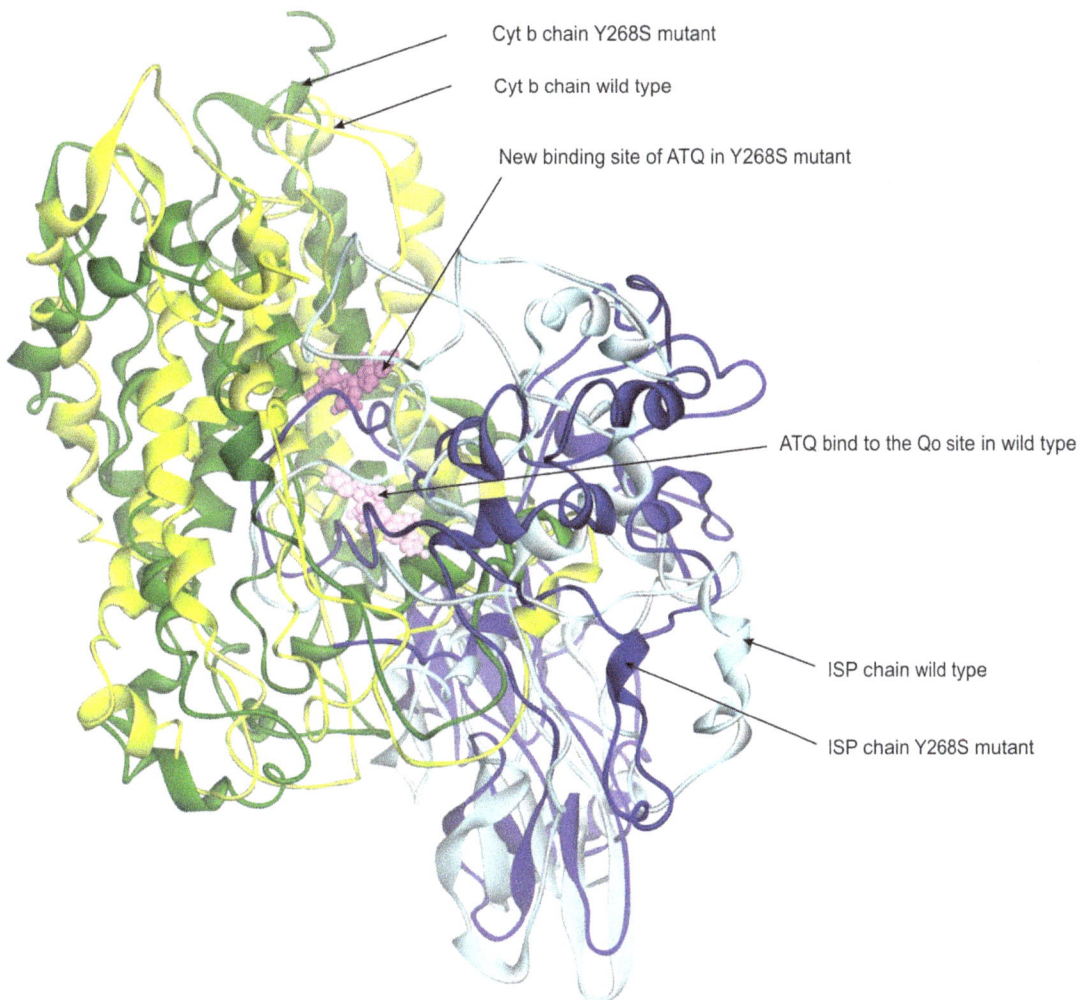

**Figure 5. ATQ binding site in the wild and mutant (Y268S) Cyt bc1 complex of *P. falciparum*.** The figure indicates that ATQ binds to a new site in the mutant model which is around 12 Å distant from the Qo site. The structure was captured from the average structure of the Cyt bc1 complex of *P. falciparum* over 70–90 ns (converged part of the trajectory).

**Table 1.** MM/PBSA binding free energies (kJ/mol) of wild-type and mutant Cyt bc1/ATQ complexes.

| Binding free energies (kJ/mol) | Y268S | Y268 |
|---|---|---|
| $\Delta G_{coul}$[1] | $-71.438 \pm 1.993$ | $-49.334 \pm 1.527$ |
| $\Delta G_{vdW}$[2] | $-188.381 \pm 1.118$ | $-209.546 \pm 1.023$ |
| $\Delta G_{ps}$[3] | $190.650 \pm 2.357$ | $246.728 \pm 2.430$ |
| $\Delta G_{nps}$[4] | $-17.924 \pm 0.070$ | $-19.962 \pm 0.037$ |
| $\Delta G_{polar}$[5] | $119.212$ | $197.394$ |
| $\Delta G_{nonpolar}$[6] | $-206.305$ | $-229.508$ |
| $\Delta G_{bind}$[7] | $-87.094 \pm 2.178$ | $-32.113 \pm 2.653$ |

[1]: coulombic term;
[2]: van der Waals term;
[3]: polar solvation term;
[4]: nonpolar solvation term;
[5]: polar term (sum of coulombic and polar solvation terms);
[6]: nonpolar term (sum of van der Waals and nonpolar solvation terms);
[7]: computational binding free energy.

manner till we find the convergence. In this process, we increased the production run from initial 20 to 90 ns in the block of 10 ns. In general, RMSIP values between 0.5–0.7 represent adequate convergence [56] and here in our case, we observed an acceptable convergence measure of 0.55 from 70–90 ns trajectory. Therefore, this part of simulation was used for the computation of average structure of the complex. To remove the crudeness of the average structure, the structure was subjected to 1000 steps of energy minimization using the Smart Minimizer (SM) available in DS. SM begins with the Steepest Descent method, followed by the Conjugate Gradient method for faster convergence towards a local minimum. We observed the ATQ interactions were

predominantly hydrophobic, although certain hydrophilic interactions exist temporarily during the MD simulation. Histidine residue 181 in the ISP (Yeast) is reported to form a strong hydrogen-bond with certain classes of Qo-bound inhibitors such as stigmatellin ('b-distal' inhibitors). We noticed that in *P. falciparum*, HIS104 of ISP chain forms such stable interaction with the ATQ. However, we did not find the stability of the ATQ hydrogen bond that was supposed to be formed via a water molecule with Glu (Glu-272 in yeast) of Cyt b. In mutant case, both these interactions were altogether absent. In the mutant model irrespective of ATQ binding at Qo site, it was found to get stabilized at a new site (site II) which is around 12 Å apart from

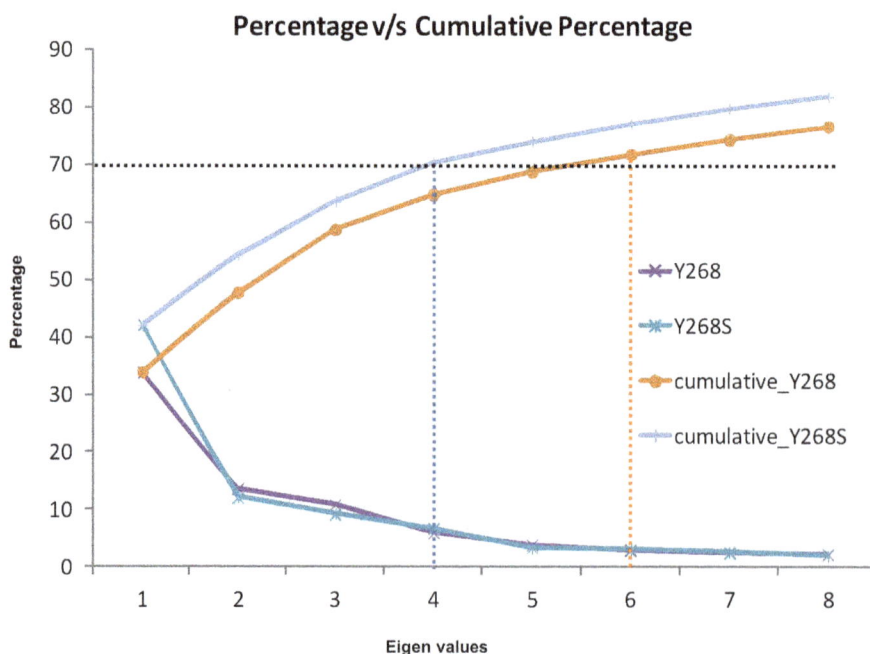

**Figure 6. Proportion of variance and cumulative proportion of total variance of first ten eigenvalues of the wild (Y268) and mutant variant (Y268S) of *P. falciparum* proteins.**

the Qo site. The two different binding modes of ATQ in wild and mutant-type are shown in Figure 5. Moreover, we also noticed that the induced mutation makes the active site to undergo significant conformational changes that reduce both the active site volume (7174.6 Å$^3$) and its surface area (4783.4 Å$^2$) than the wild-type (11388 Å$^3$; 7269.3 Å$^2$). The shrinkage in both volume and surface area (also reflected by Rg values, Figure 4C) might be also one of the probable reasons for different binding site selectivity of ATQ in mutant-type. Overall, all the above consequences might be the causing agents for ATQ to have different selectivity in the mutant protein than the wild-type.

## MD-based binding free energy calculations of ATQ-bound Cyt bc1 complex

To acquire an estimate of the binding free energies in the two systems (i.e., Y268 and Y268S) and to inspect the differences in terms of polar and non-polar interactions, we performed binding free energy calculations on the converged MD trajectories (70–90 ns) using the MM/PBSA approach. For this purpose, we took advantage of the GMXAPBS tool [30]. Here we would like to mention that in this investigation we were interested in highlighting the differences of binding free energies of similar complexes. As already reported for several other investigations [30,57,58], we will thus ignore the entropic term and focus only on the binding enthalpy, defined as the sum of 4 terms, namely coulombic, van der Waals, polar solvation and non-polar solvation. The results of our calculations are shown in Table 1.

Both systems present a similar pattern: the van der Waals, coulombic and non-polar solvation terms are negative, meaning that they favor the formation of the complex. On the contrary, the polar solvation term is positive, which indicates that it antagonizes the binding process. This observation might be due to the cost of desolvating the polar moieties in the protein residues and in the ligand. It is noteworthy that in both cases the polar contribution, defined as the sum of the coulombic and the polar solvation terms, is positive. Overall, the MM/PBSA analysis reveals that the complex stabilization is promoted by the non-polar contribution only. Such a behavior has been described in the past both for protein-ligand [59] and for protein-protein [31] interactions. The Y268S mutation significantly affects all terms.

Our study demonstrates that the binding of ATQ with the site II (newly identified binding site) of P. falciparum Cyt bc1 complex was energetically favorable compare to the site I (Qo site). We assume the sufficient binding energy of ATQ at site II might be preventing ATQ to bind its native site (Qo) thereby resulting in the loss of the anti-malarial efficiency of ATQ in the Y268S mutant-type.

## Essential dynamics analyses of the Cyt bc1 complexes

To support the MD results, we performed essential dynamics study of both the simulated complexes and our results shows that the cumulative variance captured by the first few eigenvectors or principal components of the wild-type complex was comparatively lower than the mutant-type (Figure 6).

The analysis indicates that the Y268S/ATQ complex pertain more motions than the wild-type complex. This is in consistent with the earlier dynamic results where we observed some fluctuations and far away displacement of ATQ from the Qo site in the Y268S/ATQ complex.

## Conclusion

The emerging acquired drug resistance because of mutations has presented a challenge to follow the traditional drug discovery pipelines. Even a single point mutation has potential to produce the drug resistance. In this study, we evaluated the drug-resistance mechanism of ATQ in the mutated (Y268S) Cyt b protein of P. falciparum using the potential of in silico methods. We observed that the interaction between ATQ and Cyt b is mainly stabilized by the hydrophobic contacts and after mutation of Y268S ATQ contacts with the Qo site are greatly reduced. Such findings have also been reported by other authors [9]. We presume that this reduction in Qo contacts and also change in the volume and surface area of the binding pocket (similar observations have been made by Kessl et al. 2005 [9] in I269M mutation of ATQ-bound yeast bc1 complex) enforce ATQ to find its desirable contacts at distant location from its wild active site. Moreover, the MM-PBSA calculations firmly proved the tighter binding of ATQ with the mutant-type at additional binding site (first time observation), present at ~12 Å faraway from the active site, thereby raising no choice for ATQ to bind its native site. This might be the probable reason for ATQ anti-malarial efficacy loss in Y268S mutants. We hope the structural details presented in this study would aid the experimental plan to design new suitable selective ligands that could have correct size to fit properly in the active site even after mutation. Such ligands might be able to resist the mutation effect and can be used as future effective drugs against malaria.

## Supporting Information

**Figure S1** ISP subunit of Cyt bc1 complex of P. falciparum with predicted secondary structure elements. It is important to note that the initial 158 N-terminal residues, which were not present in the S. cerevisiae Cyt bc1 complex subunit in PDB file 3CX5, are also involved in critical secondary structure confirmations. This is the reason why we consider full length ISP subunit in our analysis.

**Figure S2** The Qo site (ATQ binding site) of Cyt bc1 complex of P. falciparum is shown. The N-terminal residues of ISP chain are involved in the formation of active site (Qo) cleft (red color and shown as surface model). Cyt b subunit is shown in green color and ISP subunit as blue.

**Figure S3** The Ramachandran plots of modeled Cyt b protein and ISP subunit of P. falciparum in Cyt bc1 complex are shown. The plots indicate the quality of the modeled structure was satisfactory.

## Acknowledgments

We would like to thank Giovanna Musco, Cristina Paissoni and Dimitrios Spiliotopoulos from S. Raffaele Scientific Institute, Biomolecular NMR Laboratory, Milano, Italy for their support in MM/PBSA calculations and fruitful discussion to improve this manuscript. KPS and SKG acknowledge Council of scientific and industrial Research (CSIR) India network project GENESIS (BSC 0121) and INDEPTH (BSC 0111). BAA is recipient of Senior Research Fellowship from CSIR India. CSIR-IITR Manuscript Communication number 3246.

## Author Contributions

Conceived and designed the experiments: Shailendra Gupta BAA. Performed the experiments: BAA KPS MV Shishir Gupta. Analyzed the data: Shailendra Gupta BAA Shishir Gupta KPS YS. Contributed reagents/materials/analysis tools: Shailendra Gupta. Wrote the paper: Shailendra Gupta BAA KPS YS.

# References

1. Singh B, Kim Sung L, Matusop A, Radhakrishnan A, Shamsul SS, et al. (2004) A large focus of naturally acquired Plasmodium knowlesi infections in human beings. Lancet 363: 1017–1024.

2. Daneshvar C, Davis TM, Cox-Singh J, Rafa'ee MZ, Zakaria SK, et al. (2009) Clinical and laboratory features of human Plasmodium knowlesi infection. Clin Infect Dis 49: 852–860.

3. Crowther GJ, Napuli AJ, Gilligan JH, Gagaring K, Borboa R, et al. (2011) Identification of inhibitors for putative malaria drug targets among novel antimalarial compounds. Mol Biochem Parasitol 175: 21–29.

4. Garcia LS (2010) Malaria. Clin Lab Med 30: 93–129.

5. Fry M, Pudney M (1992) Site of action of the antimalarial hydroxynaphtho-quinone, 2-[trans-4-(4'-chlorophenyl) cyclohexyl]-3-hydroxy-1,4-naphthoqui-none (566C80). Biochem Pharmacol 43: 1545–1553.

6. Srivastava IK, Rottenberg H, Vaidya AB (1997) Atovaquone, a broad spectrum antiparasitic drug, collapses mitochondrial membrane potential in a malarial parasite. J BiolChem 272: 3961–3966.

7. Vaidya AB, Lashgari MS, Pologe LG, Morrisey J (1993) Structural features of Plasmodium cytochrome b that may underlie susceptibility to 8-aminoquinolines and hydroxynaphthoquinones. Mol Biochem Parasitol 58: 33–42.

8. Musset L, Bouchaud O, Matheron S, Massias L, Le Bras J (2006) Clinical atovaquone-proguanil resistance of Plasmodium falciparum associated with cytochrome b codon 268 mutations. Microbes Infect 8: 2599–2604.

9. Kessl JJ, Ha KH, Merritt AK, Lange BB, Hill P, et al. (2005) Cytochrome b mutations that modify the ubiquinol-binding pocket of the cytochrome bc1 complex and confer anti-malarial drug resistance in Saccharomyces cerevisiae. J Biol Chem 280: 17142–17148.

10. Kessl JJ, Meshnick SR, Trumpower BL (2007) Modeling the molecular basis of atovaquone resistance in parasites and pathogenic fungi. Trends Parasitol 23: 494–501.

11. Fisher N, Abd Majid R, Antoine T, Al-Helal M, Warman AJ, et al.(2012) Cytochrome b mutation Y268S conferring atovaquone resistance phenotype in malaria parasite results in reduced parasite bc1 catalytic turnover and protein expression. J Biol Chem287: 9731–41.

12. Hill P, Kessl J, Fisher N, Meshnick S, Trumpower BL, et al. (2003) Recapitulation in Saccharomyces cerevisiae of cytochrome b mutations conferring resistance to atovaquone in Pneumocystis jiroveci. Antimicrob Agents Chemother 47: 2725–31.

13. Zhang Y (2008) I-TASSER server for protein 3D structure prediction. BMC Bioinformatics 9: 40.

14. Das R, Qian B, Raman S, Vernon R, Thompson J, et al. (2007) Structure prediction for CASP7 targets using extensive all-atom refinement with Rosetta@home. Proteins 69: 118–28.

15. Zhang Y (2007) Template-based modeling and free modeling by I-TASSER in CASP7. Proteins 69: 108–117.

16. Zhou H, Pandit SB, Lee SY, Borreguero J, Chen H, et al. (2007) Analysis of TASSER-based CASP7 protein structure prediction results. Proteins 69: 90–97.

17. Roy A, Kucukural A, Zhang Y (2010) I-TASSER: a unified platform for automated protein structure and function prediction. Nat Protoc 5: 725–738.

18. Zhang Y, Skolnick J (2004) Scoring function for automated assessment of protein structure template quality. Proteins 57: 702–10.

19. Spassov VZ, Bashford D (1999) Multiple-Site Ligand Binding to Flexible Macromolecules: Separation of Global and Local Conformational Change and an Iterative Mobile Clustering Approach J Comput Chem 2: 1091–1111.

20. Solmaz SR, Hunte C (2008) Structure of complex III with bound cytochrome c in reduced state and definition of a minimal core interface for electron transfer. J Biol Chem 283: 17542–17549.

21. Korsinczky M, Chen N, Kotecka B, Saul A, Rieckmann K, et al. (2000) Mutations in Plasmodium falciparum cytochrome b that are associated with atovaquone resistance are located at a putative drug-binding site. Antimicrob Agents Chemother 44: 2100–2108.

22. Wu G, Robertson DH, Brooks CL 3rd, Vieth M (2003) Detailed analysis of grid-based molecular docking: A case study of CDOCKER-A CHARMm-based MD docking algorithm. J Comput Chem 24: 1549–1562.

23. Essmann U, Perera L, Berkowitz ML, Darden T, Lee H, et al. (1995) A smooth particle mesh Ewald method. J Chem Phys 103: 8577–8593.

24. Merlino A, Mazzarella L, Carannante A, Di Fiore A, Di Donato A, et al. (2005) The importance of dynamic effects on the enzyme activity: X-ray structure and molecular dynamics of onconase mutants. J Biol Chem 280: 17953–1760.

25. Taly JF, Marin A, Gibrat JF (2008) Can molecular dynamics simulations help in discriminating correct from erroneous protein 3D models? BMC Bioinformatics 9: 6.

26. Malek K, Odijk T, Coppens MO (2005) Diffusion of water and sodium counter-ions in nanopores of a β-lactoglobulin crystal: a molecular dynamics study. Nanotechnology 16: S522–530.

27. Hess B, Bekker H, Berendsen HJC, Fraaije JGEM (1997) LINCS: A linear constraint solver for molecular simulations. J Comput Chem 18: 1463–1472.

28. Berendsen HJC, Postma JPM, Van Gunsteren WF, DiNola A, Haak JR (1984) Molecular dynamics with coupling to an external bath. J Chem Phys 81: 3684–3690.

29. Parrinello M, Rahman A (1980) Crystal Structure and Pair Potentials: A Molecular-Dynamics Study. Phys Rev Lett 45: 1196–1199.

30. Massova I, Kollman PA (1999) Computational alanine scanning to probe protein-protein interactions: A novel approach to evaluate binding free energies. J Am Chem Soc 121: 8133–8143.

31. Spiliotopoulos D, Spitaleri A, Musco G (2012) Exploring PHD fingers and H3K4me0 interactions with molecular dynamics simulations and binding free energy calculations: AIRE-PHD1, a comparative study. PLoS One 7: e46902.

32. Baker NA, Sept D, Joseph S, Holst MJ, McCammon JA (2001) Electrostatics of nanosystems: Application to microtubules and the ribosome. Proc Natl Acad Sci U S A 98: 10037–1004.

33. Brown SP, Muchmore SW (2009) Large-scale application of high-throughput molecular mechanics with poisson-boltzmann surface area for routine physics-based scoring of protein-ligand complexes. J Med Chem 52: 3159–3165.

34. Schwöbel B, Alifrangis M, Salanti A, Jelinek T (2003) Different mutation patterns of atovaquone resistance to Plasmodium falciparum in vitro and in vivo: rapid detection of codon 268 polymorphisms in the cytochrome b as potential in vivo resistance marker. Malar J 2: 5.

35. Fivelman QL, Butcher GA, Adagu IS, Warhurst DC, Pasvol G (2002) Malarone treatment failure and in vitro confirmation of resistance of Plasmodium falciparum isolate from Lagos, Nigeria. Malar J 1: 1.

36. Wells JA (1990) Additivity of mutational effects in proteins. Biochemistry 29: 8509–17.

37. Lo TP, Komar-Panicucci S, Sherman F, McLendon G, Brayer GD (1995) Structural and functional effects of multiple mutations at distal sites in cytochrome c. Biochemistry 34: 5259–5268.

38. Schreiber G, Fersht AR (1995) Energetics of protein-protein interactions: analysis of the barnase-barstar interface by single mutations and double mutant cycles. J Mol Biol 248: 478–486.

39. Baloria U, Akhoon BA, Gupta SK, Sharma S, Verma V (2012) In silico proteomic characterization of human epidermal growth factor receptor 2 (HER-2) for the mapping of high affinity antigenic determinants against breast cancer. Amino Acids 42: 1349–60.

40. Ma B, Tsai CJ, Nussinov R (2000) Binding and folding: in search of intramolecular chaperone-like building block fragments. Protein Eng 13: 617–27.

41. Sham YY, Ma B, Tsai CJ, Nussinov R (2001) Molecular dynamics simulation of Escherichia coli dihydrofolate reductase and its protein fragments: relative stabilities in experiment and simulations. Protein Sci 10: 135–148.

42. Kumar S, Sham YY, Tsai C-J, Nussinov R (2001) Protein folding and function: the N-terminal fragment in adenylate kinase. Biophys J 80: 2439–2454.

43. Feyfant E, Sali A, Fiser A (2007) Modeling mutations in protein structures. ProtSci 16: 2030–2041.

44. Gupta SK, Srivastava M, Akhoon BA, Gupta SK, Grabe N (2012) Insilico accelerated identification of structurally conserved CD8+ and CD4+ T-cell epitopes in high-risk HPV types. Infect Genet Evol 12: 1513–8.

45. Srivastava M, Gupta SK, Abhilash PC, Singh N (2012) Structure prediction and binding sites analysis of curcin protein of Jatrophacurcas using computational approaches. J Mol Model. 18: 2971–9.

46. Akhoon BA, Gupta SK, Verma V, Dhaliwal G, Srivastava M, et al. (2010) In silico designing and optimization of anti-breast cancer antibody mimetic oligopeptide targeting HER-2 in women. J Mol Graph Model 28: 664–9.

47. Cherkis KA, Temple BR, Chung EH, Sondek J, Dangl JL (2012) AvrRpm1 missense mutations weakly activate RPS2-mediated immune response in Arabidopsis thaliana. PLoS One 7:e42633.

48. Frauer C, Rottach A, Meilinger D, Bultmann S, Fellinger K, et al. (2012) Different binding properties and function of CXXC zinc finger domains in Dnmt1 and Tet1. PLoS One 6:e16627.

49. Laskowski RA (2003) Structural quality assurance. In Structural Bioinformatics. Edited by Bourne P, Weissig H. Wiley-Liss, Inc. 292 p.

50. Fisher N, Meunier B (2008) Molecular basis of resistance to cytochrome bc1 inhibitors. FEMS Yeast Res 8: 183–192.

51. Khositnithikul R, Tan-Ariya P, Mungthin M (2008) In vitro atovaquone/proguanil susceptibility and characterization of the cytochrome b gene of Plasmodium falciparum from different endemic regions of Thailand. Malar J 28: 7: 23.

52. Friedman SH, DeCamp DL, Sijbesma RP, Srdanov G, Wudl F, et al. (1993) Inhibition of the HIV-1 protease by fullerene derivatives: model building studies and experimental verification. J Am Chem Soc 115: 6506–6509.

53. Friedman SH, Ganapathi PS, Rubin Y, Kenyon GL (1998) Optimizing the binding of fullerene inhibitors of the HIV-1 protease through predicted increases in hydrophobic desolvation. J Med Chem 41: 2424–9.

54. Cheng Y, Li D, Ji B, Shi X, Gao H (2010) Structure-based design of carbon nanotubes as HIV-1 protease inhibitors: atomistic and coarse-grained simulations. J Mol Graph Model 29: 171–7.

55. Lee VS, Nimmanpipug P, Aruksakunwong O, Promsri S, Sompornpisut P, et al. (2007) Structural analysis of lead fullerene-based inhibitor bound to human immunodeficiency virus type 1 protease in solution from molecular dynamics simulations. J Mol Graph Model 26: 558–70.

56. Laberge M, Yonetani T (2008) Molecular dynamics simulations of hemoglobin A in different states and bound to DPG: Effector-linked perturbation of tertiary conformations and HbA concerted dynamics. Biophys J 94: 2737–2751.

57. Huo S, Massova I, Kollman PA (2002) Computational alanine scanning of the 1:1 human growth hormone-receptor complex. J Comput Chem 23: 15–27.

58. Bradshaw RT, Patel BH, Tate EW, Leatherbarrow RJ, Gould IR (2011) Comparing experimental and computational alanine scanning techniques for probing a prototypical protein-protein interaction. Protein Eng Des Sel 24: 197–207.

59. Chiappori F, Merelli I, Milanesi L, Marabotti A (2013) Static and dynamic interactions between GALK enzyme and known inhibitors: Guidelines to design new drugs for galactosemic patients. Eur J Med Chem 63C: 423–434.

**4**

# Safety and Comparability of Controlled Human *Plasmodium falciparum* Infection by Mosquito Bite in Malaria-Naïve Subjects at a New Facility for Sporozoite Challenge

Angela K. Talley[1], Sara A. Healy[2], Olivia C. Finney[1], Sean C. Murphy[3], James Kublin[4], Carola J. Salas[5], Susan Lundebjerg[1], Peter Gilbert[4], Wesley C. Van Voorhis[6], John Whisler[1], Ruobing Wang[1], Chris F. Ockenhouse[7], D. Gray Heppner[7], Stefan H. Kappe[1], Patrick E. Duffy[2]*

1 Malaria Clinical Trials Center, Seattle Biomedical Research Institute, Seattle, Washington, United States of America, 2 Laboratory for Malaria Immunology and Vaccinology, National Institute of Allergy and Infectious Diseases, National Institutes of Health, Bethesda, Maryland, United States of America, 3 Department of Laboratory Medicine, University of Washington Medical Center, Seattle, Washington, United States of America, 4 Fred Hutchinson Cancer Research Center, Seattle, Washington, United States of America, 5 United States Naval Medical Research Unit Number 6, Lima, Peru, 6 Department of Medicine, University of Washington Medical Center, Seattle, Washington, United States of America, 7 United States Military Malaria Vaccine Program, Walter Reed Army Institute of Research, Silver Spring, Maryland, United States of America

## Abstract

*Background:* Controlled human malaria infection (CHMI) studies which recapitulate mosquito-borne infection are a critical tool to identify protective vaccine and drug candidates for advancement to field trials. In partnership with the Walter Reed Army Institute of Research, the CHMI model was established at the Seattle Biomedical Research Institute's Malaria Clinical Trials Center (MCTC). Activities and reagents at both centers were aligned to ensure comparability and continued safety of the model. To demonstrate successful implementation, CHMI was performed in six healthy malaria-naïve volunteers.

*Methods:* All volunteers received NF54 strain *Plasmodium falciparum* by the bite of five infected *Anopheles stephensi* mosquitoes under controlled conditions and were monitored for signs and symptoms of malaria and for parasitemia by peripheral blood smear. Subjects were treated upon diagnosis with chloroquine by directly observed therapy. Immunological (T cell and antibody) and molecular diagnostic (real-time quantitative reverse transcriptase polymerase chain reaction [qRT-PCR]) assessments were also performed.

*Results:* All six volunteers developed patent parasitemia and clinical malaria. No serious adverse events occurred during the study period or for six months post-infection. The mean prepatent period was 11.2 days (range 9–14 days), and geometric mean parasitemia upon diagnosis was 10.8 parasites/µL (range 2–69) by microscopy. qRT-PCR detected parasites an average of 3.7 days (range 2–4 days) earlier than blood smears. All volunteers developed antibodies to the blood-stage antigen merozoite surface protein 1 (MSP-1), which persisted up to six months. Humoral and cellular responses to pre-erythrocytic antigens circumsporozoite protein (CSP) and liver-stage antigen 1 (LSA-1) were limited.

*Conclusion:* The CHMI model was safe, well tolerated and characterized by consistent prepatent periods, pre-symptomatic diagnosis in 3/6 subjects and adverse event profiles as reported at established centers. The MCTC can now evaluate candidates in the increasingly diverse vaccine and drug pipeline using the CHMI model.

*Trial Registration:* ClinicalTrials.gov NCT01058226

**Editor:** Lorenz von Seidlein, Mahidol-Oxford Tropical Medicine Research Unit, Thailand

**Funding:** Funding for the initial MCTC infrastructure and for this clinical trial funding was provided by the PATH Malaria Vaccine Initiative (MVI) and the United States Department of Defense. MVI staff (Florence Kaltovich and Cynthia Lee) provided technical support on the chemistry, manufacturing, and control section of the IND application and the clinical operations aspects associated with this study, the latter through MVI's clinical quality assurance consultants at The Total Approach. C. F. O. is now employed by MVI/PATH and contributed to the review of the manuscript. This work was also supported by the Intramural Research Program of National Institute of Allergy and Infectious Diseases, National Institutes of Health (S. A. H., P. E. D.), and the Global Emerging Infections Surveillance and Response System of the United States Department of Defense (AFHSC/GEIS) under work unit number 847705.82000.256B.B0016 (C. J. S.). COPYRIGHT STATEMENT: Several authors are employees of the United States Government. This work was prepared as part of their official duties. Title 17 U.S.C. §105 provides that 'Copyright protection under this title is not available for any work of the United States Government.' Title 17 U.S.C. §101 defines a United States Government work as a work prepared by a military service member or employee of the United States Government as part of that person's official duties. The views expressed in this article are those of the authors and do not necessarily reflect the official policy or position of the Department of the Navy, Department of Defense, nor the United States Government.

**Competing Interests:** The authors have declared that no competing interests exist.

* Email: patrick.duffy@nih.gov

## Introduction

In the absence of defined immune correlates of protection and consistently predictive animal models, controlled human malaria infections (CHMI) have become the most effective means of assessing early-stage efficacy of candidate pre-erythrocytic and erythrocytic vaccines and anti-malarial drugs. Under this model, malaria parasite-infected mosquitoes are allowed to bite human volunteers to inoculate them with *Plasmodium* sporozoites under controlled conditions. To date, CHMI has mostly been performed using *Plasmodium falciparum*-infected mosquitoes [1–8], although a few *Plasmodium vivax* studies have recently taken place as well [9,10]. In all trials, the subjects are monitored closely for development of patent blood-stage parasitemia and treated with standard doses of anti-malarial medications defined by the known sensitivity profile of the parasite. The parasite densities reported from CHMI studies can be used for modeling parasite growth kinetics [11,12].

CHMI studies have been conducted in the United States of America (USA) and elsewhere for decades. Reviews of the published literature indicate the model is safe, reproducible and well-tolerated by subjects participating in clinical challenge trials with *P. falciparum* [6,7,13]. Over the last 20 years, over 900 subjects have been experimentally infected or "challenged" with malaria at the Walter Reed Army Institute of Research (WRAIR), with an outstanding safety record and no occurrence of complicated malaria or Serious Adverse Events (SAEs). Importantly, the CHMI model plays a critical role in early phase testing of candidate interventions in order to prioritize only the most promising candidates for further product development efforts.

Despite an expanding pipeline of candidate malaria interventions in development, the number of clinical research centers capable of conducting CHMI is limited [14]. To expand the global capacity for CHMI studies, Seattle Biomedical Research Institute (Seattle BioMed) established the Malaria Clinical Trials Center (MCTC) as an innovative center for the integration of basic science and clinical research with an immediate mandate to support the development of an effective malaria vaccine. This study (ClinicalTrials.gov NCT01058226) was performed to demonstrate the successful implementation of the CHMI model at Seattle BioMed, prior to conducting efficacy trials of candidate malaria vaccines.

## Methods

### Objectives

The primary objective of the study was to demonstrate the safety, tolerability and infectivity of CHMI in a newly-established facility. The secondary objectives were to assess the malaria-specific immune response following CHMI in healthy malaria-naïve adults. Additionally, we sought to evaluate the detection and quantification of prepatent parasitemia by real-time quantitative reverse transcriptase polymerase chain reaction (qRT-PCR).

### Study Site

The study was conducted at the MCTC at Seattle BioMed in Seattle, WA. The institute also houses the Center for Mosquito Production and Malaria Infection Research (CeMPMIR), an entomology facility that was upgraded for the production of malaria sporozoites and malaria-infected mosquitoes under phase-appropriate current Good Manufacturing Practices (cGMPs) to support MCTC trials. The Clinical and Translational Research Laboratory (CTRL) was simultaneously established to support

specimen processing and to perform standardized assays according to Good Clinical Laboratory Practices (GCLPs) for clinical trials.

Seattle BioMed and WRAIR established a Cooperative Research and Development Agreement to institute the CHMI model at the MCTC. WRAIR research and clinical staff from the US Military Malaria Vaccine Program served in an advisory capacity throughout preparation and conduct of the trial to ensure alignment with WRAIR practices and participant safety.

### Study design

This was a prospective, open-label, single intervention study of six healthy malaria-naïve adult volunteers undergoing CHMI. All subjects were enrolled as a single cohort and followed the same study schedule (**Figure 1**). On the day of enrollment (Day 0), six eligible subjects underwent CHMI ('challenge') according to standard procedures [5]. Briefly, subjects were infected with *P. falciparum* sporozoites by the bites of five infected *Anopheles stephensi* mosquitoes under controlled conditions. Volunteers were closely monitored in the post-challenge period and treated with standard oral doses of chloroquine phosphate upon diagnosis of malaria parasitemia by positive thick blood films. Periodic clinical assessments, physical examinations and laboratory monitoring were performed for evaluation of protocol-defined safety, infectivity and immunology endpoints.

### Ethical conduct

This study was conducted in accordance with the International Conference of Harmonization (ICH) Good Clinical Practices (GCP) and applicable Food and Drug Administration (FDA) regulations. Based on evolving regulatory standards for human challenge studies, the P. falciparum sporozoite was considered an investigational product and for the first time, the CHMI procedures and study protocol were conducted under an investigational new drug application (IND) filed with the FDA. Prior to study initiation, the protocol and study-related documents were approved by the Western Institutional Review Board. The study was registered on ClinicalTrials.gov (Identifier NCT01058226; see http://clinicaltrials.gov/show/NCT01058226). The protocol for this trial and supporting CONSORT checklist are available as supporting information; see **Checklist S1** and **Protocol S1**.

### Study population

Subjects were recruited using IRB-approved messaging from the greater Seattle area. Subjects provided written acknowledgment of informed consent and were required to pass an assessment of understanding quiz covering key study concepts and risks related to participation. Adult males and non-pregnant females aged 18–50 years were eligible to participate in the study if they were in good general health as demonstrated by detailed medical history review, physical examination and laboratory assessment performed within 56 days of enrollment. Subjects were required to have a low risk of coronary heart disease based on the NHANES 1 screening criteria [15] and a normal or essentially normal electrocardiogram as read by an independent cardiologist. Laboratory assessment of eligibility included measurements of hemoglobin (Hb), white blood cells (WBC), platelets, serum creatinine, alanine aminotransferase (ALT), aspartate aminotransferase (AST), total bilirubin and alkaline phosphatase, along with urine dipstick or full urinalysis, and screening for active infection with HIV-1/-2, hepatitis B and hepatitis C viruses. Subjects with a prior diagnosis of malaria or recent travel to a malaria-endemic area were excluded, as were those who had previously received an

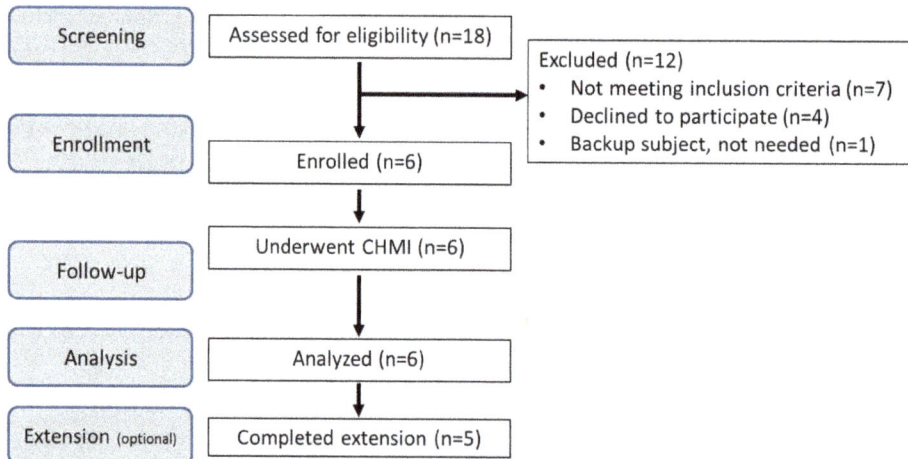

**Figure 1. Study flow diagram.** Eighteen subjects were screened for eligibility to participate in the trial and 7 healthy volunteers were considered eligible and willing to participate. On the day of enrollment, 6 subjects were enrolled and one backup subject was discharged from the study. The six subjects underwent CHMI and completed the 56 day study. Five subjects returned for optional long term safety and immunology follow up assessments at 3 and 6 months post-challenge.

investigational malaria vaccine, recent malaria chemoprophylaxis or antibiotics with anti-malarial properties. Inclusion and exclusion criteria are available in the IRB-approved protocol (**Protocol S1**).

## Investigational product (*P. falciparum* sporozoites)

Wild-type NF54 strain *P. falciparum*-infected *A. stephensi* mosquitoes for use in CHMI were reared in compliance with phase-appropriate cGMPs in the secure CeMPMIR entomology facility according to standard procedures adapted from WRAIR. The *A. stephensi* mosquitoes were a laboratory-reared strain originally received from WRAIR. Mosquitoes were infected with *P. falciparum* by standard membrane feeding methods on parasite cultures derived from the WRAIR master cell bank (MCB) containing a sufficient proportion of mature gametocytes. Prior to the CHMI, sensitivity testing of the parasite MCB to the anti-malarial drugs chloroquine, doxycycline, atovaquone and quinine was performed. The results confirmed the sensitivity of the NF-54 parasite MCB to these drugs at published values and were virtually identical to the sensitive type strain, *P. falciparum* 3D7, a clone of NF-54. On the day of challenge, five infected mosquitoes were placed in a screen-covered carton and allowed to feed on the subject's forearm for five minutes. Following the feed, mosquitoes were evaluated to confirm both an adequate blood meal (as evidenced by the presence of a blood meal in the abdomen) and the presence of adequate numbers of sporozoites in their salivary glands (by individual mosquito dissections). The sporozoite load was rated microscopically according to a semi-quantitative scale: 0 (no sporozoites observed), +1 (1–10), +2 (11–100), +3 (101–1000) and +4 (>1000) [5]. Only mosquitoes with a salivary gland rating of +2 or greater were considered infective. Feeding iterations were repeated using additional mosquitoes as needed to ensure each subject received a total of five infective bites.

## Post-challenge monitoring and follow up

Subjects were closely monitored for 30–45 minutes post-challenge for acute reactogenicity and were issued a symptom diary and thermometer to record symptoms and oral temperatures daily for five days. Subjects were evaluated in clinic on Day 1 post-challenge, contacted by phone daily on Days 2-4 and evaluated

again in clinic daily on Days 5–8 post-challenge. From Day 9 post-challenge, subjects were housed in a local hotel with study staff for close observation during the time they were expected to develop patent parasitemia and symptoms of clinical malaria infection. Clinical assessments conducted at each visit included symptom review, vital signs and physical examination. Peripheral blood smears were examined for detection of malaria parasites at least daily beginning on Day 5 post-infection and continuing through diagnosis and treatment. Blood samples for safety endpoint assays were collected prior to challenge, on the day of the first positive blood smear and on Days 35 and 56 post-challenge. Blood samples for immunology endpoint assays were collected at baseline, on the day of the first positive blood smear, on Days 1, 5, 35 and 56 post-challenge.

Malaria treatment decisions were based on diagnosis of patent (microscopic) parasitemia by a positive peripheral blood smear. Subjects were treated upon detection of patent parasitemia with a standard oral regimen of oral chloroquine phosphate (600 mg chloroquine base initially, followed by 300 mg chloroquine base at 6, 24 and 48 hours later) under direct observation. Subjects continued their overnight hotel stays until completion of treatment and confirmation of three consecutive daily negative blood smears, after which they were followed weekly for eight weeks. Subjects were invited to return for optional long-term immunology assessments at three and six months post-infection.

Primary safety endpoints were assessed by clinical and laboratory evaluations following the challenge and clinically significant changes from baseline status were reported as adverse events (AEs). Solicited AEs were captured on a diary card and/or by direct questioning during the 28 days post-challenge. A solicited AE was defined as a predetermined event that may reflect safety concerns related to the investigational product or events that could be reasonably expected to occur as part of the intervention. These included local and systemic signs and symptoms related to the challenge and/or clinical malaria. (**Table 1**). While not previously established to be related to mild malaria infection, for the purpose of ensuring safety in this initial challenge study, cardiac AEs were solicited by symptom review including chest pain. Local and systemic reactogenicity related to the challenge was collected from the day of challenge (Day 0) through Day 5 post-challenge.

**Table 1.** Incidence of adverse events during the 28 days following CHMI.

| Adverse Event | N (%) | Highest Grade | | | Mean duration[2] days (range) |
| | | Mild | Moderate | Severe[1] | |
|---|---|---|---|---|---|
| **Solicited Local[3]** | | | | | |
| Pruritus | 3 (50) | 2 | 1 | | 3.3 (2–5) |
| Pain | 1 (16.7) | 1 | | | 1 (1–1) |
| **Solicited Systemic[4]** | | | | | |
| Abdominal Pain | 3 (50) | 3 | | | 1.25 (1–2) |
| Arthralgia | 4 (66.7) | 4 | | | 2 (1–3) |
| Chills | 3 (50) | 1 | 1 | 1 | 1.67 (1–2) |
| Diarrhea | 1 (16.7) | 1 | | | 1 (1–1) |
| Fever | 4 (66.7) | 3 | 1 | | 1 (1–1) |
| Headache | 5 (83.3) | 4 | | 1 | 2 (1–4) |
| Low Back Pain | 2 (33.3) | | 2 | | 6.5 (4–9) |
| Malaise | 4 (66.7) | 3 | | 1 | 2.25 (1–6) |
| Myalgia | 5 (83.3) | 2 | 2 | 1 | 2.8 (1–4) |
| Nausea | 4 (66.7) | 1 | 3 | | 1.5 (1–2) |
| Vomiting | 1 (16.7) | 0 | 1 | | 1 (1–1) |
| Chest Pain | 0 (0) | | | | 0 (0) |
| **Unsolicited[5]** | | | | | |
| Decreased Appetite | 1 (16.7) | 1 | | | 4 (4–4) |
| Dizziness | 1 (16.7) | 1 | | | 2 (2–2) |
| Insomnia | 1 (16.7) | 1 | | | 1 (1–1) |
| Cough | 2 (33.3) | 2 | | | 3 (3–3) |
| Exertional Dyspnea | 1 (16.7) | 1 | | | 1(1–1) |

[1]A single subject accounted for all severe AEs which appeared on the day of positive blood smear, and all of which decreased in severity within 24 hours of treatment.
[2]Per episode (number of episodes/subjects reporting episodes).
[3]Only symptoms reported from day 0 through day 5 are included.
[4]Only symptoms reported from day 6 through day 28 and determined by the Investigator to be malaria-related are included.
[5]Collected throughout the 56 day study.

Systemic symptoms related to clinical malaria were collected from Day 5 through Day 28 post-challenge. All other AEs reported at any time during the study (Day 0 to Day 56 post-challenge) were recorded as unsolicited AEs, including any worsening or exacerbation of pre-existing conditions. All subjects were contacted by phone six months post-malaria challenge to assess for the occurrence of SAEs, chronic illnesses, or other medically significant conditions. AEs were graded by the Investigator for severity and for relationship to the investigational product. Severity was graded according to a protocol-defined toxicity grading scale, adapted from toxicity grading scales used in similar trials (**Protocol S1**). AEs were coded using the Medical Dictionary for Regulatory Activities (MedDRA) and reported using System Organ Class and Preferred Term.

## Malaria diagnosis

Development of patent parasitemia was monitored by microscopic evaluation of peripheral blood smears. All subjects undergoing CHMI were expected to develop patent parasitemia. Blood smears were assessed by microscopy daily from Day 5 post-infection through diagnosis and after treatment until three consecutive daily blood smears were negative. Additional blood smears were prepared at other times as needed for diagnosis in symptomatic subjects. The prepatent period was defined as the time from CHMI to microscopic diagnosis, whereas the incuba-

tion period was the time from CHMI to the onset of malaria-related symptoms.

Malaria microscopy was conducted according to standard challenge microscopy procedures adapted from WRAIR. Giemsa-stained thick blood smears were prepared in duplicate from 10 μL of blood sample spread over a 1×2 cm rectangle and examined under a high power oil immersion objective (100X) by microscopists certified to read thick blood smears for CHMI studies. Smears were considered positive if at least two unambiguous parasites per slide were identified by a study microscopist and confirmed by the lead microscopist. For asymptomatic or treated subjects, up to five passes were read for an area equivalent to ~290–320 total high-power fields (hpf), allowing for detection of a parasite density of approximately three parasites/μL before declaring a blood smear negative. For symptomatic volunteers, up to 15 passes (~870–960 hpf) were scanned before declaring a smear to be negative. Quantification of microscopic parasite densities on slides from the day of diagnosis was performed by two independent microscopists who examined a minimum total of 1 μL of blood for each subject. Average parasitemia density was calculated from the number of parasites observed divided by the volume of blood examined microscopically (using conversion of volume per hpf) (**Table S1**).

At the conclusion of the trial, blood samples were tested using a first-generation *P. falciparum* 18S rRNA qRT-PCR assay by the

University of Washington Department of Laboratory Medicine as described [16]. Samples tested by qRT-PCR included whole blood obtained twice daily from Day 5 until the first positive blood smear as well as blood collected daily after the first positive blood smear until hotel discharge and a final sample at Day 28. Briefly, 50 µL aliquots of whole blood were preserved in 2 mL of NucliSens lysis buffer (bioMérieux) and stored at −80°C until testing. Samples were extracted on a semi-automated instrument (EasyMag, bioMérieux), and total nucleic acids were subjected to qRT-PCR to detect the *P. falciparum* A-type 18S rRNA and a competitive internal control [16]. The limit of quantification for the qRT-PCR assay was 20 parasites/mL of whole blood; some samples below this threshold could be classified as 'low positive' by melting curve analysis as described [16]. Time to detection of parasites by molecular means was calculated from day of challenge to the day that parasites were detected quantitatively by qRT-PCR at ≥20 parasites/mL.

## Immune responses

Blood sampling for immunology endpoints occurred prior to challenge (baseline) and at regular intervals after challenge including Day 1 and Day 5 (corresponding to the liver stage of parasite development), the day of the first positive blood smear (corresponding to the blood stage) and post-treatment at Days 35 and 56 following challenge. In addition, subjects had the option to provide separate written consent to participate in a long-term immunology follow-up assessment with collection of serum samples on at three and six months post-challenge.

Levels of antibodies against liver- and blood-stage antigens including circumsporozoite protein (CSP), apical membrane antigen 1 (AMA-1) and the 42 kDa fragment of merozoite surface protein 1 (MSP-1$_{42}$) were evaluated by enzyme-linked immunosorbent assay (ELISA) as described [17]. The antigens were produced by the National Institutes of Health Malaria ELISA Laboratory. ELISA plates (Immulon 4 HBX microtiter plates, Thermo Scientific) were coated with the above antigens (10 µg per complete 96-well plate), blocked with a non-reactive blocking buffer (5% non-fat dry milk) and incubated with 100 µL of volunteer samples or malaria-positive sera (pool of 20 sera from malaria-exposed Tanzanian adults) or malaria-naïve sera (pool of 16 sera from malaria-naïve US donors) for two hours at room temperature; all sera were pre-diluted to 1:200 in 5% milk. Negative controls consisted of two wells per plate that did not receive any sera, but were otherwise subjected to all other ELISA steps. Overall ELISA results were accepted if the negative control wells had an average optical density (OD) ≤0.1 with the %CV ≤ 30% and positivity cutoffs were determined as described [17]. Individual samples were retested with dilution if ODs exceeded 3.5 or if the %CV exceeded 30% for a given sample.

Cell-mediated immune responses to CSP and liver-stage antigen 1 (LSA-1) were evaluated by interferon-gamma (IFNγ) ELISpot assay using peripheral blood mononuclear cells (PBMCs) as previously described [18]. Briefly, MultiScreen IP filter plates (Millipore) were coated overnight with mouse anti-human IFNγ capture antibody (clone 1-D1K, 10 µg/mL, Mabtech), washed and blocked. PBMCs (200,000/well) were incubated with media only (negative control), CSP- or LSA-1-derived peptide pools (1 µg/mL), cytomegalovirus (CMV) peptide controls (1 µg/mL) or phytohemagglutinin (PHA, 1 µg/mL, positive control) in 125 µL RPMI supplemented with 10% fetal bovine serum, 2 mM L-glutamine, 100 U/mL penicillin and 100 µg/mL streptomycin for 18–22 hours at 37°C. Media only wells (no cells) were used as a plate control. After incubation, plates were washed and incubated for 2–4 hours with biotinylated mouse anti-human IFNγ antibody

(clone 7-b6-1, 1 µg/mL, Mabtech) at room temperature. Plates were again washed and incubated with alkaline-phosphatase-conjugated anti-biotin antibody (1:750, Vector Labs) for 2–3 hours at room temperature. Plates were developed for 5–10 min using NBT/BCIP (nitro blue tetrazolium/5-bromo-4-chloro-3-indolyl-phosphate) substrate according to manufacturer instructions (Pierce). Plates were dried overnight and counted on an Immuno-Spot reader (C.T.L., Shaker Heights, OH). Sample results were accepted if the mean negative control wells spot count for a given sample was ≤20 spots/well, if the PHA control was ≥400 spots/well and if the mean spot count in media only wells (no cells) per plate was ≤6 spots/well. Positivity criteria was ≥55 spots per $10^6$ PBMCs and ≥4-fold above the mean of the negative control wells as described [18,19].

## Statistics

The study was an open-label single intervention trial to demonstrate the reproducibility of CHMI at a new center. Primary and secondary analyses involved summaries and descriptive statistics for safety, efficacy (infectivity) and immunology endpoints. Comparative analyses of the primary (microscopy) and exploratory (qRT-PCR) infectivity endpoints was undertaken for the safety population and is presented as the difference in the cumulative distribution of the number of days between challenge and the first positive test (blood smear minus qRT-PCR) using a Kaplan-Meier estimator. Analyses of immunology endpoints (by ELISA and ELISpot) were performed for all subjects for whom data was available at a specific time point. Qualitative assay data analysis was performed by tabulating the frequency of positive responses for each assay at each timepoint an assessment was performed.

## Results

### Subject disposition

A total of 18 subjects were screened from January – March 2010 for eligibility to enroll in the study (**Figure 1**). Seven subjects were screen failures based on eligibility criteria, four subjects withdrew consent and six of the remaining seven eligible subjects were enrolled in the trial, including four male and two female subjects aged 19–28 years (mean 23.5 years). All six subjects were included in the safety and efficacy analyses. No subject discontinued the trial. All six subjects completed the trial through the six month follow up phone call. Additionally, five of the six subjects consented to participate in the optional blood draws for immunology samples at 3 and 6 months post-challenge.

### Malaria challenge

Six subjects successfully completed the experimental infection, receiving a minimum of five invective bites. One subject inadvertently received a total of six infective bites, rather than the five defined in the protocol. This deviation was reviewed by the Medical Monitor and Safety Monitoring Committee. After reviewing safety data, both agreed that the additional infective bite did not put the subject at increased risk. As noted in **Table 2**, all subjects required more than one feeding iteration in order to receive a total of five infective mosquito bites. The average number of feeding iterations was 4.6 (range 2–7). In this initial study, the overall prevalence of sporozoite containing mosquitoes was 50% (**Figure S1**).

### Safety

No SAEs were reported for any subject through Day 56 of the trial or during six months following the CHMI. The incidence of

**Table 2.** Subject infectivity summary.

| Subject ID | Feeding Iterations[1] | Time to qRT-PCR positive[2] (days) | Prepatent period[3] (days) | Incubation period[4] (days) | Peak parasite density by method[5] (parasites/mL) |
|---|---|---|---|---|---|
| A | 4 | 7.0 | 11.0 | 7 | 9460 RTPCR |
| | | | | | 12450 BS |
| B | 2 | 7.0 | 11.0 | 12 | 50170 RTPCR |
| | | | | | 53700 BS |
| C | 4 | 7.0 | 11.0 | 8 | 120680 RTPCR |
| | | | | | 39000 BS |
| D | 7 | 7.0 | 11.0 | 11 | 5350 RTPCR |
| | | | | | 7270 BS |
| E | 4 | 7.0 | 9.0 | 6 | 3730 RTPCR |
| | | | | | 5150 BS |
| F | 7 | 10.0 | 14.0 | 14 | 12880 RTPCR |
| | | | | | 2330 BS |
| Mean | 4.7 | 7.5 | 11.2 | 9.7 | 33712 RTPCR |
| | | | | | 19983 BS |
| Median | 4 | 7.0 | 11 | 9.5 | 11170 RTPCR |
| | | | | | 9860 BS |
| SD | 2.0 | 1.2 | 1.6 | 3.1 | 45956 RTPCR |
| | | | | | 21206 BS |

[1]Number of rounds of mosquito feeding exposures required to achieve a total of 5 infective bites, as demonstrated by evidence of a blood meal in the mosquito abdomen and a post-feed sporozoite salivary gland score of +2 or higher. Subject D received a total of 6 infective bites while all other subjects received 5 bites.
[2]Number of days from CHMI to qRT-PCR-positive.
[3]Number of days from CHMI to peripheral blood smear-positive.
[4]Number of days from CHMI to symptomatic.
[5]BS, blood smear.

solicited and unsolicited AEs reported as related to the malaria challenge or infection is summarized in **Table 1**. The mean incubation period (time to first malaria symptom) was 9.7 days (range 6–14 days; 95% CI 6.4–13.0 days; **Table 2**). Three of six subjects (50%) experienced at least one post-challenge local and/or systemic reactogenicity symptom between the day of challenge (Day 0) and Day 5 post-CHMI. Similarly, 50% of subjects reported at least one symptom of clinical malaria prior to the diagnosis of patent parasitemia. Two subjects (33.3%) had onset of clinical symptoms on the same day of blood smear diagnosis and one subject (16.7%) had no systemic symptoms until after diagnosis of patent parasitemia (**Figure 2**).

During the acute reactogenicity period (Day 0 to Day 5 post-challenge), the most common AE was pruritus localized to the site of inoculation, occurring in 3 (50%) subjects. No other AE occurred in >1 subject during this period. All subjects experienced at least one systemic AE during the 28 days post-challenge. The most frequently reported systemic AEs related to malaria were headache and myalgia each occurring in 5 (83.3%) subjects, followed by fever, malaise, nausea and arthralgia each occurring in 4 (66.7%) subjects (**Table 1**). Most AEs were classified as mild or moderate in severity. One subject developed severe chills, headache, malaise and myalgia on the day of first positive blood smear. The symptoms peaked in severity on the day of diagnosis and decreased in severity within 24 hours of treatment initiation. The constellation of symptoms in this subject was consistent with uncomplicated clinical malaria, and all symptoms resolved within 48–72 hours of completion of treatment.

All subjects reported at least one unsolicited AE. The majority of unsolicited AEs were mild, and most were considered unrelated

to challenge procedures or malaria infection. Related events included cough, decreased appetite, dizziness, insomnia and exertional dyspnea (**Table 1**). Of these, only cough (2/6) was reported by more than one subject. All AEs resolved by Day 56. Subjects were followed for six months post-CHMI, and study follow-up ended in September 2010.

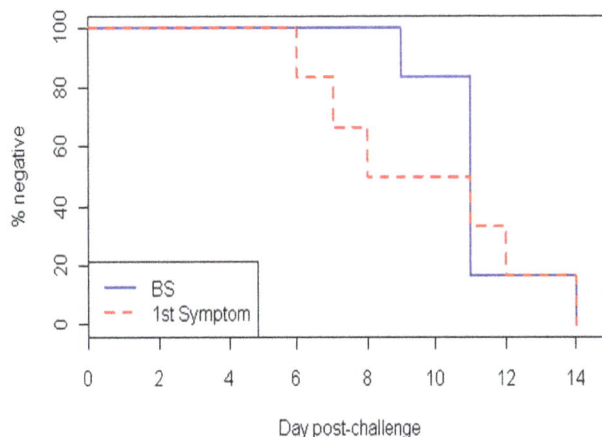

**Figure 2. Comparison of prepatent and incubation periods.** Kaplan-Meier survival curve showing the percentage of subjects without patent parasitemia by blood smear (blue line) or without symptoms (dashed red line) following challenge.

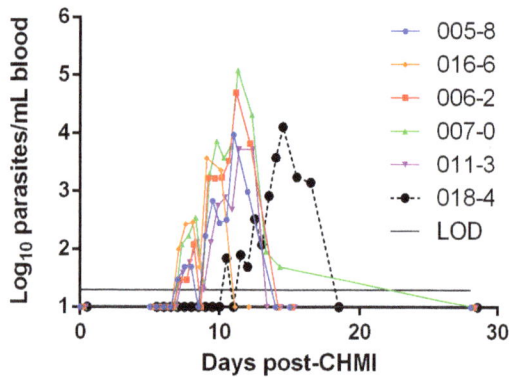

**Figure 3. qRT-PCR-based course of parasitemia.** Parasite density based on qRT-PCR measurements are presented individually for each participant.

Four of six subjects had no laboratory abnormalities throughout the trial. One subject had an asymptomatic mild-moderate increase in liver transaminases (ALT 160 U/L [2.5 xULN] and AST 105 U/L [2.6 x ULN]) occurring on the same day as diagnosis of patent parasitemia (Day 11 post-challenge). The other subject experienced a mild, asymptomatic decrease in Hb from baseline (1.8 g/dL) occurring on Day 10 post-challenge that remained within the normal range. Both abnormalities resolved by Day 35.

## Infectivity

All subjects undergoing CHMI developed microscopic parasitemia with a mean prepatent period of 11.2 days (range 9–14 days; SD 1.6 days) (**Table 2**). Four of six subjects (67%) had a prepatent period of 11 days. The earliest detection of patent parasitemia was on Day 9 post-challenge and the latest was at 14 days. As noted above, the mean incubation period was 9.7 days (range 7–14 days; SD 3.1 days). Similar to reports from other centers, there was no apparent correlation between prepatent and incubation periods (**Figure 2**).

## Parasitemia

The geometric mean parasite density (by microscopy) at the time of microscopic diagnosis was 10.8 parasites/μL across all slides examined (range 2–69 parasites/μL; SD 21.2 parasites/μL) (**Table S1**). Similar to reports from other centers, there was no apparent correlation between incubation or prepatent periods and density of parasitemia at diagnosis.

We previously reported the performance characteristics of our qRT-PCR assay, and its comparability to microscopic diagnosis in this trial [16]. All subjects became qRT-PCR positive with a mean time to positivity of 7.5 days (range 7.0–10.0 days; SD 1.2 days) (**Table 2**). Detection of parasites by qRT-PCR occurred an average of 3.7 days (range 2.0–4.0 days; SD 0.8 days) earlier than by peripheral blood smears [16]. Five of six subjects (83%) were positive by qRT-PCR four days prior to blood smear diagnosis including four subjects who become qRT-PCR positive on Day 7 and one subject who became qRT-PCR positive on Day 10 (**Figure 3**). The sixth subject became positive by qRT-PCR on Day 7 post-CHMI, two days prior to blood smear diagnosis. For most individuals, the qRT-PCR findings were consistent with the expected timing for release of merozoites from the liver into the bloodstream. With the exception of one subject, qRT-PCR and microscopic parasite density measurements aligned within $0.5 \log_{10}$ parasites/mL (**Table S2**). The parasite density by qRT-PCR on the day of corresponding microscopic diagnosis ranged from 3,730 to 120,700 parasites/mL, which was in agreement with reports from other CHMI centers [13].

Subjects were treated with chloroquine upon microscopic diagnosis and were released from the hotel after three consecutive negative daily blood smears. All subjects were blood smear-negative within two days of initiating treatment (**Table 3**, mean 1.67 days). As previously reported [16], except for one subject, qRT-PCR remained positive even after blood smears became negative after treatment. Two subjects were qRT-PCR-negative within two days of treatment, but the remaining subjects were still positive by qRT-PCR at the time of discharge from the hotel (**Table 3**). By five days post-treatment, the qRT-PCR signal was undetectable in three subjects and reduced by more than two orders of magnitude in the remaining three subjects. All subjects were negative by qRT-PCR at the Day 28 follow-up clinic visit.

## Immunology

Humoral and cellular responses to *P. falciparum* antigens were analyzed for all six subjects through Day 56 and, for five of the six subjects who consented to the extension portion of the protocol, at three and six months post-challenge. Humoral responses to *P. falciparum* were measured by ELISA for pre-erythrocytic (CSP) and blood-stage (AMA-1 and MSP-$1_{42}$) antigens (**Figure 4**). Two subjects initially seronegative for CSP became seropositive, and of these two, one remained seropositive at 6 months. One subject was seropositive to CSP at baseline and remained so throughout the 6 month study period; this could be due to cross-reactive response to epitopes shared between CSP and other common antigens or remote exposure to malaria, although this was not reported by the subject. All five subjects were seronegative to MSP-$1_{42}$ at baseline,

**Table 3.** Time from anti-malarial treatment to clearance of peripheral parasitemia.

| Subject | Time to first negative blood smear (days) | Time to first negative qRT-PCR (days) |
|---|---|---|
| A | 2 | * |
| B | 2 | * |
| C | 1 | * |
| D | 2 | 2 |
| E | 1 | 2 |
| F | 2 | 4 |

*Data not available. Subjects were qRT-PCR positive and peripheral blood smear negative at the time of discharge from the hotel. No further daily sampling was performed after discharge until Day 28, at which point all subjects in the study were blood smear and qRT-PCR negative.

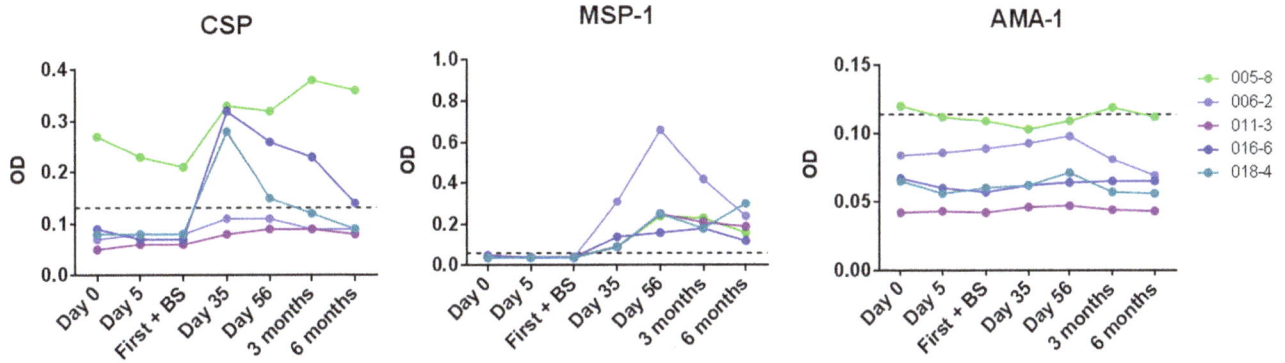

**Figure 4. Humoral immune responses to *P. falciparum* antigens.** ELISAs were performed on the indicated days post-CHMI to test for responses against the indicated *P. falciparum* antigens. The positivity cut-off (dotted line) was calculated per ELISA plate as three standard deviations above the mean of the two negative control wells. All samples with an OD higher than the calculated cut-off were deemed positive. Data are presented for the five subjects completing the follow up at 3 and 6 months. Data to Day 56 for the sixth subject did not differ considerably from the five subjects in the graph.

but became seropositive on Days 35 and remained so throughout the six month study period (**Figure 4**). No subject was seropositive to AMA-1 at any timepoint, except for one subject who exhibited low responses on Day 0 and at three months post-CHMI, reporting an average OD of 0.12 for both timepoints (positivity cut-off was 0.11).

Only one subject showed cellular antigen-specific cell-mediated immune responses to CSP, detected on the day of the first positive blood smear (**Figure 5**). The CSP-specific response was negative at Day 35 and Day 56. Responses to later timepoints were therefore anticipated to be negative. As a result, PBMC samples from extension visits were not tested by ELISpot in all subjects. For LSA-1, no positive responses were detected in any subject at any timepoint (**Figure 5**). The mean LSA-1 response was 7.78 spot forming units (SFU)/million PBMCs (95% CI 6.08–9.47 SFU/million), similar to the negative control wells (mean 8.5 SFU/million). Based on the lack of response seen by Day 56, the LSA-1 cellular responses at three months and six months were not evaluated.

## Discussion

Therapeutic and experimental malaria infection of humans has a long history, dating back nearly a century [20–22]. Chulay et al. first described the production of infected mosquitoes from *in vitro* cultured NF54 strain *P. falciparum* gametocytes for the experimental challenge of six healthy volunteers under controlled

conditions [5]. Since that time, the CHMI mosquito bite model has continued to be an important and powerful tool in the clinical development path towards effective malaria vaccines and drugs. This study demonstrates that CHMI at the Seattle-based MCTC is safe, reproducible and well tolerated. Recent efforts at standardization of practices across centers [23] and compliance with phase-appropriate cGMP for the production of the infected mosquitoes ensures the continued safety and integrity of the CHMI model and allows for comparability in data sets across multiple centers. To this end, this is the first CHMI study conducted under an IND where the *P. falciparum* sporozoite was considered an investigational product. Similarly, efforts to align and standardize supportive diagnostic tools such as qRT-PCR [24] will further ensure comparability of data between centers.

Similar to reports from other centers, the local, systemic and laboratory AEs observed following CHMI in this study were consistent with mosquito exposure and subsequent uncomplicated clinical malaria episodes. Solicited adverse events peaked within 48 hours of blood smear diagnosis and most resolved within 72 hours. Severe symptoms (chills, headache, malaise, myalgia) observed in one subject were likewise consistent with uncomplicated malaria and resolved within 24 hours of initiating antimalarial treatment.

Consistent with the literature [13], all subjects developed signs and symptoms of malaria infection during the trial. Both the mean incubation and prepatent periods (9.7 days and 11.2 days,

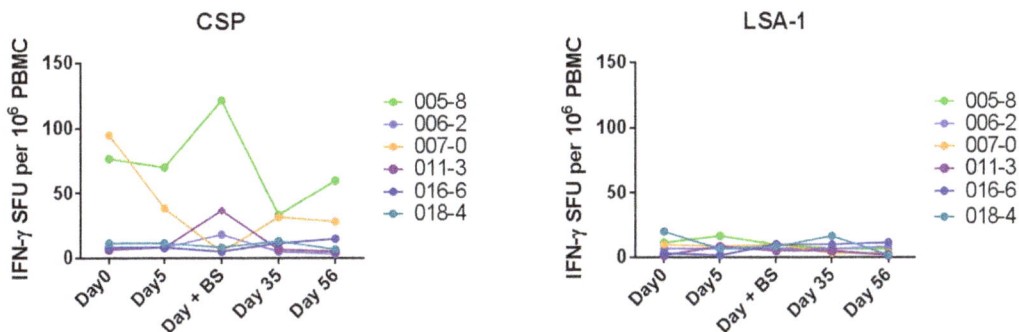

**Figure 5. Cellular immune response to CSP and LSA-1.** IFNγ ELISpot assays were performed using the indicated CSP and LSA-1 peptide pools on the indicated days post-CHMI. Spot forming units (SFU) per million PBMC are shown per individual subject for all six subjects.

respectively) were consistent with those reported in other CHMI studies (8.8 and 11.5 days) [1,3,4,6], and there appeared to be no significant relationship between incubation and prepatent periods and parasitemia upon diagnosis. Half of the subjects reported clinical symptoms of malaria prior to blood smear diagnosis, one remained asymptomatic until the first day of treatment and the remaining two had symptoms concurrent with blood smear diagnosis, which was generally consistent with the published literature [1,6]. Similarly, there was no correlation between peak parasitemia or duration of parasitemia and severity of clinical illness.

In this study, prepatent parasitemia was detected by qRT-PCR an average of 3.7 days before blood smear diagnosis, and modeling of the qRT-PCR data suggested that 28–60 (range 8–111) hepatocytes were infected per subject as previously reported [16]. In addition, there was good quantitative agreement between qRT-PCR- and microscopy-derived parasite density results on the day of patency. This study was the first to demonstrate accurate parasite detection and quantification by this qRT-PCR method. Because the 18S rRNA is biologically enriched in malaria-infected erythrocytes relative to the parent 18S rDNA by a factor of 3500–5000 [16,25], the qRT-PCR format offers robust sensitivity relative to qPCR for samples of the same volume.

In this study, mosquito infectivity was approximately 50% as assessed by the post-challenge dissection and sporozoite rating system [5], and multiple mosquito bite iterations were required to achieve a total of five infectious bites. Ultimately, 100% of subjects were infected and developed patent parasitemia. Mosquito infectivity has since been optimized, and in Seattle BioMed trials since this study, 81–91% of mosquitoes had sporozoite ratings of 2+ or higher (Jen Hume, personal communication). High sporozoite loads have also been recently noted for aseptic cGMP-grade mosquitoes used to induce infections with less than the standard five bites [3].

As expected from the limited antigenic exposure with only five to six infected mosquito bites, limited humoral and cellular immune responses to pre-erythrocytic and erythrocytic antigens were observed. Although two subjects had transient measurable antibody levels to CSP, neither maintained a detectable response at six months post-challenge. A third subject had pre-existing humoral and cellular responses to CSP that persisted at all timepoints, did not markedly increase after infection and remains unexplained; anomalous cross-reactive responses to CSP have also been reported by others [26].

Likewise, a single challenge with *P. falciparum* in our study was not sufficient to induce significant antibody response to AMA-1 as detected by our standardized assay. The peak cellular response was expected at Day 35+/−3 days, as previously reported in experimental malaria infection models [27]. In this study, cellular responses to pre-erythrocytic antigens were limited, except for two subjects with pre-existing responses to CSP that persisted at many timepoints. For CSP, there was no difference between timepoints when all subjects were analyzed together. Only one subject showed new antigen-specific responses to CSP, detected on the day of the first positive blood smear but not on Day 35 or 56. For LSA-1, no positive responses were detected in any subject at any timepoint.

In summary, this study demonstrates the successful implementation of the CHMI mosquito bite model at Seattle BioMed, thereby expanding the global capacity for conducting these important studies. Subjects were successfully infected with *P. falciparum* sporozoites, and all subjects developed patent parasitemia and were successfully treated and cured with standard doses of chloroquine. The CHMI was safe and well tolerated. The AE profile, prepatent and incubation periods were consistent with the expected clinical course and response to CHMI reported in previous malaria studies conducted at other centers worldwide. All volunteers developed antibody responses to the blood-stage antigen MSP-1, which was detectable up to six months post-challenge. Both humoral and cellular responses to pre-erythrocytic antigens CSP and LSA-1 were limited. Results from a diagnostic qRT-PCR assay correlated with blood smear density results and identified prepatent infection nearly four days before peripheral blood smears. Seattle BioMed plans to support additional CHMI studies to accelerate development of effective malaria drugs and vaccines.

## Supporting Information

**Figure S1  Cumulative mosquito sporozoite ratings for all mosquitoes used to challenge each subject.** The sporozoite load was rated microscopically according to a semi-quantitative scale: 0 (no sporozoites observed), +1 (1–10), +2 (11–100), +3 (101–1000) and +4 (>1000) [5]. Only mosquitoes with a salivary gland rating of +2 or greater were considered infective.

**Table S1  Microscopy-based parasite density assessment.** [1]Volume per 1 cm pass based on the field number for each microscope, corresponding to 290–320 hpf. Microscope/equipment numbers: Nikon Eclipse 50i, (FN 22), Nikon microscope E200, (FN20)

**Table S2  Comparison of microscopy and qRT-PCR-based parasite density assessments for first smear-positive sample.** [1]qRT-PCR measurements as reported in [1] (converted from parasites/mL to parasites/µL). [2]Quantitative agreement denoted if microscopic and qRT-PCR parasite density measurements were within 0.5 $\log_{10}$ parasites/mL of each other.

## Acknowledgments

The authors thank Sharon Wong-Madden for project management of this study. Jack Williams, Megan Dowler and Peter Weina provided scientific and technical advice for execution of the mosquito feeds and clinical study.

DISCLAIMER: The views expressed in this article are those of the authors and do not necessarily reflect the official policy or position of the Department of the Navy, Department of Defense, nor the U.S. Government.

## Author Contributions

Conceived and designed the experiments: AKT SHK PED. Performed the experiments: AKT SAH OCF SCM CJS RW. Analyzed the data: AKT OCF SCM JK PG WCVV RW PED. Contributed reagents/materials/analysis tools: SL JW CFO DGH. Wrote the paper: AKT SCM DGH PED.

# References

1. Church LW, Le TP, Bryan JP, Gordon DM, Edelman R, et al. (1997) Clinical manifestations of Plasmodium falciparum malaria experimentally induced by mosquito challenge. J Infect Dis 175: 915–920.
2. Roestenberg M, McCall M, Hopman J, Wiersma J, Luty AJ, et al. (2009) Protection against a malaria challenge by sporozoite inoculation. N Engl J Med 361: 468–477.
3. Lyke KE, Laurens M, Adams M, Billingsley PF, Richman A, et al. (2010) Plasmodium falciparum malaria challenge by the bite of aseptic Anopheles stephensi mosquitoes: results of a randomized infectivity trial. PLoS One 5: e13490.
4. Verhage DF, Telgt DS, Bousema JT, Hermsen CC, van Gemert GJ, et al. (2005) Clinical outcome of experimental human malaria induced by Plasmodium falciparum-infected mosquitoes. Neth J Med 63: 52–58.
5. Chulay JD, Schneider I, Cosgriff TM, Hoffman SL, Ballou WR, et al. (1986) Malaria transmitted to humans by mosquitoes infected from cultured Plasmodium falciparum. Am J Trop Med Hyg 35: 66–68.
6. Epstein JE, Rao S, Williams F, Freilich D, Luke T, et al. (2007) Safety and clinical outcome of experimental challenge of human volunteers with Plasmodium falciparum-infected mosquitoes: an update. J Infect Dis 196: 145–154.
7. Hoffman SL (1997) Experimental challenge of volunteers with malaria. Ann Intern Med 127: 233–235.
8. Spring M, Polhemus M, Ockenhouse C (2014) Controlled human malaria infection. J Infect Dis 209 Suppl 2: S40–45.
9. Herrera S, Solarte Y, Jordan-Villegas A, Echavarria JF, Rocha L, et al. (2011) Consistent safety and infectivity in sporozoite challenge model of Plasmodium vivax in malaria-naive human volunteers. Am J Trop Med Hyg 84: 4–11.
10. Herrera S, Fernandez O, Manzano MR, Murrain B, Vergara J, et al. (2009) Successful sporozoite challenge model in human volunteers with Plasmodium vivax strain derived from human donors. Am J Trop Med Hyg 81: 740–746.
11. Douglas AD, Edwards NJ, Duncan CJ, Thompson FM, Sheehy SH, et al. (2013) Comparison of modeling methods to determine liver-to-blood inocula and parasite multiplication rates during controlled human malaria infection. J Infect Dis 208: 340–345.
12. Bejon P, Andrews L, Andersen RF, Dunachie S, Webster D, et al. (2005) Calculation of liver-to-blood inocula, parasite growth rates, and preerythrocytic vaccine efficacy, from serial quantitative polymerase chain reaction studies of volunteers challenged with malaria sporozoites. J Infect Dis 191: 619–626.
13. Roestenberg M, O'Hara GA, Duncan CJ, Epstein JE, Edwards NJ, et al. (2012) Comparison of clinical and parasitological data from controlled human malaria infection trials. PLoS One 7: e38434.
14. Ballou WR, Arevalo-Herrera M, Carucci D, Richie TL, Corradin G, et al. (2004) Update on the clinical development of candidate malaria vaccines. Am J Trop Med Hyg 71: 239–247.
15. Gaziano TA, Young CR, Fitzmaurice G, Atwood S, Gaziano JM (2008) Laboratory-based versus non-laboratory-based method for assessment of cardiovascular disease risk: the NHANES I Follow-up Study cohort. Lancet 371: 923–931.
16. Murphy SC, Prentice JL, Williamson K, Wallis CK, Fang FC, et al. (2012) Real-time quantitative reverse transcription PCR for monitoring of blood-stage Plasmodium falciparum infections in malaria human challenge trials. Am J Trop Med Hyg 86: 383–394.
17. Miura K, Orcutt AC, Muratova OV, Miller LH, Saul A, et al. (2008) Development and characterization of a standardized ELISA including a reference serum on each plate to detect antibodies induced by experimental malaria vaccines. Vaccine 26: 193–200.
18. McElrath MJ, De Rosa SC, Moodie Z, Dubey S, Kierstead L, et al. (2008) HIV-1 vaccine-induced immunity in the test-of-concept Step Study: a case-cohort analysis. Lancet 372: 1894–1905.
19. Dubey S, Clair J, Fu TM, Guan L, Long R, et al. (2007) Detection of HIV vaccine-induced cell-mediated immunity in HIV-seronegative clinical trial participants using an optimized and validated enzyme-linked immunospot assay. J Acquir Immune Defic Syndr 45: 20–27.
20. James SP, Shute PG (1926) Malaria Commission: report on the first results of laboratory work on malaria in England. Geneva: League of Nations Health Organization.
21. Nicol WD (1927) The care and management of induced malaria. J Ment Sci 6: 1–9.
22. James SP (1931) Some general results of a study of induced malaria in England. Trans R Soc Trop Med Hyg 24: 477–538.
23. Laurens MB, Duncan CJ, Epstein JE, Hill AV, Komisar JL, et al. (2012) A consultation on the optimization of controlled human malaria infection by mosquito bite for evaluation of candidate malaria vaccines. Vaccine 30: 5302–5304.
24. Murphy SC, Hermsen CC, Douglas AD, Edwards N, Petersen I, et al. (2014) External quality assurance of malaria nucleic acid testing for clinical trials and eradication surveillance. PLoS One 9: e97398.
25. Murphy SC, Daza G, Chang M, Coombs R (2012) Laser cutting eliminates nucleic acid cross-contamination in dried-blood-spot processing. J Clin Microbiol 50: 4128–4130.
26. Zevering Y, Amante F, Smillie A, Currier J, Smith G, et al. (1992) High frequency of malaria-specific T cells in non-exposed humans. Eur J Immunol 22: 689–696.
27. Todryk SM, Walther M, Bejon P, Hutchings C, Thompson FM, et al. (2009) Multiple functions of human T cells generated by experimental malaria challenge. Eur J Immunol 39: 3042–3051.

# 5

# Characterization of the Commercially-Available Fluorescent Chloroquine-BODIPY Conjugate, LynxTag-CQ$_{GREEN}$, as a Marker for Chloroquine Resistance and Uptake in a 96-Well Plate Assay

**Cheryl C. Y. Loh**[1♥]**, Rossarin Suwanarusk**[2♥]**, Yan Quan Lee**[1,3♥]**, Kitti W. K. Chan**[1]**, Kit-Ying Choy**[4]**, Laurent Rénia**[2]**, Bruce Russell**[1]**, Martin J. Lear**[5]**, François H. Nosten**[6,7,8]**, Kevin S. W. Tan**[1,3]*****, Larry M. C. Chow**[4]*****

**1** Department of Microbiology, Yong Loo Lin School of Medicine, National University of Singapore, Singapore, Singapore, **2** Singapore Immunology Network, Agency for Science Technology and Research, Biopolis, Singapore, Singapore, **3** NUS Graduate School for Integrative Sciences and Engineering, National University of Singapore, Singapore, Singapore, **4** Department of Applied Biology and Chemical Technology, The State Key Laboratory of Chirosciences, The Hong Kong Polytechnic University, Hung Hom, Kowloon, Hong Kong SAR, PR China, **5** Department of Chemistry, Graduate School of Science, Tohoku University, Aza Aramaki, Aoba-ku, Sendai, Japan, **6** Centre for Tropical Medicine, Nuffield Department of Medicine, University of Oxford, Oxford, United Kingdom, **7** Shoklo Malaria Research Unit, Mahidol-Oxford Tropical Medicine Research Unit, Faculty of Tropical Medicine, Mahidol University, Mae Sot, Thailand, **8** Centre for Tropical Medicine, University of Oxford, Churchill Hospital, Oxford, United Kingdom

## Abstract

Chloroquine was a cheap, extremely effective drug against *Plasmodium falciparum* until resistance arose. One approach to reversing resistance is the inhibition of chloroquine efflux from its site of action, the parasite digestive vacuole. Chloroquine accumulation studies have traditionally relied on radiolabelled chloroquine, which poses several challenges. There is a need for development of a safe and biologically relevant substitute. We report here a commercially-available green fluorescent chloroquine-BODIPY conjugate, LynxTag-CQ$_{GREEN}$, as a proxy for chloroquine accumulation. This compound localized to the digestive vacuole of the parasite as observed under confocal microscopy, and inhibited growth of chloroquine-sensitive strain 3D7 more extensively than in the resistant strains 7G8 and K1. Microplate reader measurements indicated suppression of LynxTag-CQ$_{GREEN}$ efflux after pretreatment of parasites with known reversal agents. Microsomes carrying either sensitive- or resistant-type PfCRT were assayed for uptake; resistant-type PfCRT exhibited increased accumulation of LynxTag-CQ$_{GREEN}$, which was suppressed by pretreatment with known chemosensitizers. Eight laboratory strains and twelve clinical isolates were sequenced for PfCRT and Pgh1 haplotypes previously reported to contribute to drug resistance, and *pfmdr1* copy number and chloroquine IC$_{50}$s were determined. These data were compared with LynxTag-CQ$_{GREEN}$ uptake/fluorescence by multiple linear regression to identify genetic correlates of uptake. Uptake of the compound correlated with the logIC$_{50}$ of chloroquine and, more weakly, a mutation in Pgh1, F1226Y.

**Editor:** Georges Snounou, Université Pierre et Marie Curie, France

**Funding:** As part of the Oxford Tropical Medicine Research Program of Wellcome Trust–Mahidol University, Shoklo Malaria Research Unit (SMRU) is funded by the Wellcome Trust of Great Britain. The authors further thank the National Research Foundation (NRF2009NRF-POC002–102), the National Medical Research Council (NMRC/1310/2011; NMRC/EDG/1038/2011), and the Agency for Science, Technology and Research (A*STAR, Singapore) for their generous support. The funders had no role in study design, data collection and analysis, decision to publish, or preparation of the manuscript.

**Competing Interests:** KSWT and MJL are founding directors of BioLynx Technologies (Singapore), a private company that specializes in fluorophore-conjugated drug surrogates including LynxTag-CQGREEN. Other authors declare no competing interests. KSWT and MJL own minority shares in BioLynx Technologies (Singapore).

* Email: kevin_tan@nuhs.edu.sg (KSWT); larry.chow@polyu.edu.hk (LMCC)

♥ These authors contributed equally to this work.

## Introduction

Despite years of intense global effort to eradicate it, malaria is still one of the deadliest infectious diseases, killing more than 600 000 people in 2010 alone [1,2]. The severest form of malaria is caused by the protozoan parasite *Plasmodium falciparum*. Chloroquine (CQ), once a spectacularly successful antimalarial drug, was first discovered by the German chemist Johann Andersag but was mistakenly thought to be too toxic for therapeutic purposes, an incident which became known as "the resochin error" (resochin being the name given to the compound by Andersag) [3,4]. CQ was so effective that it inspired optimism for the eradication of malaria. However, resistance soon arose, first appearing along the Thai-Cambodian border in the 1950s. By the 1970s, CQ resistance had spread throughout the world [5,6]. This resistance is generally attributed to mutations in the *pfcrt* (*P.*

*falciparum* chloroquine resistance transporter) gene, which codes for a transporter situated on the membrane of the parasite digestive vacuole (DV).

During parasite development in the intraerythrocytic cycle, haemoglobin is digested in the DV and the toxic heme moiety is released, which the parasite crystalizes into non-toxic hemozoin [7]. CQ is generally thought to kill the parasite by inhibiting the formation of hemozoin and thus preventing the detoxification of free heme [8–10]. In wild-type parasites CQ diffuses through the DV membrane and is diprotonated in the acidic environment of the DV, acquiring a net positive charge which prevents it from escaping the DV; however, mutant PfCRT found in CQ-resistant parasites effluxes this charged CQ out of the DV, removing it from its site of action [11]. Although the current first line artemisinin-combination therapies are effective in clearing parasitaemia, resistance against artemisinins has emerged [12–17]. There is therefore an urgent need to develop novel antimalarial strategies. Several research groups, including our own, have tried different approaches to tackle the problem of CQ resistance by either reversing CQ resistance with a PfCRT inhibitor or synthesizing "reversed" CQ analogues that cannot be effluxed by PfCRT [18–24]. The ultimate goal is to reintroduce CQ as a viable treatment for malaria. Both development of PfCRT inhibitors and synthesis of "reversed" CQ analogues require a sensitive assay for CQ uptake which is typically performed by the use of radiolabelled CQ [22,25–28]. Such methods are difficult to adopt in a high-throughput screen and may raise concerns of safety. To overcome this technical difficulty, fluorescent derivatives of chloroquine have recently been developed and used for this purpose; fluorophores used include 6-(N-(7-nitrobenz-2-oxa-1, 3-diazol-4-yl)amino)hexanoic acid (NBD) [29], coumarin [21,30], and 4, 4-difluoro-4-bora-3a, 4a-diaza-s-indacene (BODIPY) [31].

BODIPY derivatives typically exhibit strong fluorescence and are relatively inert in biological conditions [32]. Furthermore, their maximum emission wavelengths are in the green-red region [32], allowing them to be used with many DNA dyes that fluoresce blue, such as the DAPI and Hoechst stains. These properties make BODIPY a promising candidate as a marker for CQ uptake in *P. falciparum*. We therefore present here the characterization of a commercially-available BODIPY-CQ conjugate, LynxTag-CQ$_{GREEN}$, in several laboratory strains and clinical isolates.

## Methods

### Parasite culture and synchronization

*P. falciparum* laboratory strains 3D7 (MRA-102), K1 (MRA-159), 7G8 (MRA-154), HB3 (MRA-155), CS2 (MRA-96), T9-94 (MRA-153), and Dd2 (MRA-156) were obtained from MR4, ATCC Manassas Virginia. Strain T9/96 was obtained from The European Malaria Reagent Repository. A further twelve clinical isolates were collected from the Mae Sot district, Tak Province, in northwest Thailand at the Shoklo Malaria Research Unit; these isolates are prefixed 'SMRU'. Parasites were continuously cultured in complete malaria culture media (MCM) consisting of RPMI 1640 (Life Technologies) supplemented with 0.5% (w/v) Albumax I (Invitrogen), 0.005% (w/v) hypoxanthine, 0.03% (w/v) L-glutamate, 0.25% (w/v) gentamycin, with human erythrocytes at 2.5% haematocrit. Cultures were gassed with a mixture of 3% $CO_2$, 4% $O_2$ and 93% $N_2$ and incubated at 37°C. Synchronization of parasite cultures was performed by resuspending erythrocytes in 5% (w/v) D-sorbitol and incubating at 37°C for 10 min, after which the erythrocytes were washed twice, resuspended in MCM and returned to culture conditions. Thin Giemsa smears

were made before each experiment to determine parasitemia and parasite stage.

### Compound preparation

For work involving parasites, chlorpheniramine maleate salt, chlorpromazine hydrochloride, desipramine hydrochloride, promethazine hydrochloride, verapamil hydrochloride and CQ diphosphate (all from Sigma-Aldrich) were dissolved in PBS to a working concentration of 1 mM. LynxTag-CQ$_{GREEN}$ (BioLynx Technologies, Singapore; hereafter abbreviated to 'CQ$_{GREEN}$') was dissolved in DMSO to the same concentration. All compounds were stored at $-20$°C and protected from light. For microsome uptake assays, methiothepin mesylate salt, metergoline, loperamide hydrochloride, octoclothepin maleate salt, mibefradil dihydrochloride hydrate, L703,606 oxalate salt hydrate, and chlorprothixene hydrochloride (all from Sigma-Aldrich) were dissolved in DMSO to 10 mM and stored at 4°C. Verapamil hydrochloride, adenosine triphosphate (ATP), and CQ diphosphate (all from Sigma-Aldrich) were dissolved in water to 7.5 mM, 50 mM and 0.1 M respectively and stored at $-20$°C. Triitated CQ ($^3$H-CQ; from Moravek Biochemicals and Radiochemicals) was diluted in water to 5.32 µM and stored at $-20$°C; specific activity was 4.7 Ci/mmol.

### Reinvasion half-maximal inhibitory concentration (IC$_{50}$)

Synchronized ring-stage cultures at 1–2% parasitemia, 1.25% haematocrit were incubated with either CQ or CQ$_{GREEN}$ at a range of concentrations for 48 h in 96-well flat-bottomed plates at culture conditions. Following this, cells were stained with 1 µg/ml of Hoechst 33342 (Invitrogen) for 30 min at 37°C, washed twice and resuspended in PBS. Parasitemia was then assessed with the CyAn ADP flow cytometer (Beckman Coulter). IC$_{50}$s were determined by plotting the measurements in Graphpad Prism 5 using a variable slope logistic curve.

### Confocal imaging

200 µl cultures of 3D7 at 3% parasitemia, 1.25% haematocrit were incubated with CQ$_{GREEN}$ for 2 h at 2 µM in a 96-well plate format. Erythrocytes were then washed twice and stained with Hoechst 33342 as previously. Wet mounts of stained parasites were visualized under ×100 magnification with the Fluoview FV1000 confocal microscope (Olympus). Hoechst and CQ$_{GREEN}$ were excited at 405 nm and 488 nm with emissions captured at 430–470 nm and 505–525 nm respectively.

### Parasite CQ$_{GREEN}$ uptake assay

Synchronized trophozoite-stage cultures at 3–5% parasitemia were resuspended in 200 µl of MCM with 2 µM of CQ$_{GREEN}$ to 2.5% haematocrit in a 96-well plate format. The parasites were then incubated for 2 h at culture conditions, after which they were washed twice and resuspended in PBS. Cells were allowed to settle in a Nunc F96 MicroWell black non-treated polystyrene plate (Thermo Scientific) for 1 h. Fluorescence was then measured with the Infinite M200 microplate reader (Tecan) with excitation and emission wavelengths of 488 nm and 520 nm respectively. K1 chemoreversal assays were performed by pretreatment with 10 µM of the reversal agents for 30 min prior to the addition of CQ$_{GREEN}$.

### Preparation of microsomes carrying PfCRT

PfCRT originating from *P. falciparum* strains Dd2 or 3D7 were expressed in *Pichia pastoris* KM71 and microsomes harvested as described previously [33]. Microsomal levels of PfCRT were determined by western blot with standard curves generated from blots of purified PfCRT.

## Uptake kinetics in microsomes

In order to assess the Michaelis-Menten kinetics of $CQ_{GREEN}$ uptake by the microsomes, total or non-specific uptake was measured. Non-specific uptake was determined by pretreating microsomes with unlabelled CQ at 1000 times of the concentration of $CQ_{GREEN}$ used, for 15 min at 37°C, before adding $CQ_{GREEN}$; total uptake was determined without the pretreatment. Reactions were carried out in accumulation buffer (0.25 M sucrose, 10 mM Tris–HCl, 5 mM $MgCl_2$, and 3 mM ATP, pH 7.5). Following this, microsomes were washed twice in accumulation buffer, then lysed in lysis buffer (0.75 M HCl, 1% Triton X-100, 77.5% isopropanol) on ice for at least one hour. Unlabelled CQ was excluded from the washing step as the addition of CQ was observed to cause displacement of $CQ_{GREEN}$ from the microsomes, possibly due to the higher affinity of CQ for PfCRT. Measurement of the fluorescence intensity was performed using the FLUOstar Galaxy microplate reader (BMG Labtech) with excitation and emission wavelengths of 480 nm and 520 nm respectively. For each experiment, measurements were made in triplicate and the mean calculated. Specific uptake was then calculated by subtracting non-specific uptake from total uptake. Non-linear regression analysis with the Michaelis-Menten model (GraphPad Prism 5) was then applied to obtain the $V_{max}$ and $K_m$ of specific uptake.

## $CQ_{GREEN}$ uptake in PfCRT microsomes

Unless stated otherwise, microsomes were incubated with 15 μM $CQ_{GREEN}$ at 37°C for 15 min. For uptake inhibition assays, microsomes were pre-incubated with chemoreversal compounds [21] for 15 min at 37°C prior to the addition of $CQ_{GREEN}$. All data presented are specific uptake based on the calculations stated above.

## $^3$H-CQ uptake in PfCRT microsomes

$^3$H-CQ uptake was measured as described previously [33], with some modifications. Non-specific uptake was determined by pre-incubation with 200-fold unlabelled CQ. Incubation was performed with $^3$H-CQ at 308 nM for 5 min, after which 200-fold unlabelled CQ was added to stop the reaction. Microsomes were then precipitated by the addition of polyethylene glycol (PEG) 8000 and washed twice with accumulation buffer containing 200-fold unlabelled CQ to remove excess $^3$H-CQ. Microsomes were then resuspended in scintillation buffer and agitated overnight. Radioactivity was measured using the LS 5600 Scintillation Counter (Beckman). All data presented are specific uptake.

## Genotyping of strains and isolates

To assess *pfmdr1* polymorphisms, parasite DNA from *in vitro* cultures was extracted with the QIAamp DNA Mini kit (Qiagen) as per the manufacturer's instructions. For *pfcrt* polymorphisms, total RNA was extracted with the RNeasy Mini Kit (Qiagen) and reverse transcription performed with SuperScript III (Invitrogen) as per manufacturers' instructions. Polymerase chain reaction (PCR) mixtures were made with 200 μM of each dNTP, 0.5 μM forward primer, 0.5 μM reverse primer, 0.02 U/μl Phusion DNA polymerase (Thermo Scientific), 6 μl of 5× Phusion HF buffer, and 1 μl of genomic DNA or cDNA to a total reaction volume of 30 μl. Thermocycler parameters were as follows: 98°C for 30 s, followed by 35 cycles of 98°C for 10 s, 60°C for 30 s, and 72°C for 1 min. Primers used for *pfmdr1* sequencing were 5'- ATGGG-

TAAAGAGCAGAAAGA and 5'- TCCACAATAACTTGCAA-CAGT, or 5'- GTCAAGCGGAGTTTTTGC and 5'- TAT-TCTCTGTTTTTGTCCAC. *Pfcrt*-specific primers were 5'-GACGAGCGTTATAGAGAAT and 5'- CTTCGGAATCTT-CATTTTCT. PCR products were purified with the QIAquick PCR purification kit as per manufacturer's instructions. Purified PCR products were sequenced by a commercial vendor (AIT Biotech, Singapore). Copy number of *pfmdr1* was assessed by real-time PCR as previously reported [34]. Briefly, reaction mixtures were prepared with TaqMan universal PCR master mix (Applied Biosystems), 5.5 mM $MgCl_2$, 300 nM dNTPs, 300 nM each of forward and reverse primers, and 100 nM of the probe. Thermocycler parameters were 95°C for 10 min, then 40 cycles of 95°C for 15 s and 60°C for 1 min. Forward and reverse primers used were 5'- TGCATCTATAAAACGATCAGACAAA and 5'-TCGTGTGTTCCATGTGACTGT respectively, and TaqMan probe was 5'- 6FAM-TTTAATAACCCTGATCGAAATG-GAACCTTTG-TAMRA. A reference gene, β-tubulin, was also included; primers were 5'- TGATGTGCGCAAGTGATCC and 5'- TCCTTTGTGGACATTCTTCCTC, while the probe was 5'- VIC-TAGCACATGCCGTTAAATATCTTCCATGTCT-TAMRA. The threshold cycle ($C_t$) was analysed by the comparative $C_t$ method, based on DNA amplification efficiencies of the *pfmdr1* and β-tubulin genes. *Pfmdr1* copy number was calculated according to the following formula: $\Delta C_t = C_t R - C_t G$, where $C_t R$ is the reference β-tubulin $C_t$, and $C_t G$ is that of *pfmdr1*. Each TaqMan run included three reference DNA samples from clones 3D7, K1, and Dd2 having *pfmdr1* copy numbers of 1, 1, and 3 respectively.

## Statistical analyses

All statistical analyses were performed with SPSS 21. Chemoreversal assays were assessed with Student's t test, 2-tailed. Multiple linear regression was performed with the stepwise method, using the log of $IC_{50}$s and with dummy coded values for the respective amino acid residues.

## Ethics statement

The blood collection protocol for *in vitro* malaria culture was approved by the Institutional Review Board (NUS-IRB Reference Code: 11–383, Approval Number: NUS-1475) of the National University of Singapore (NUS). All participants provided written informed consent. The clinical isolates used were obtained under ethical guidelines in the approved protocol: OXTREC Reference Number 29–09 (Center for Clinical Vaccinology and Tropical

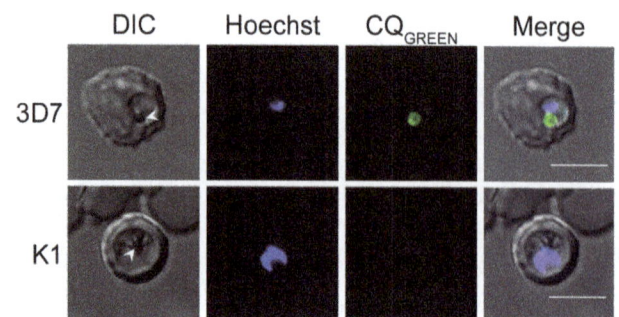

**Figure 1. $CQ_{GREEN}$ localization in *P. falciparum* 3D7.** Parasites were stained with $CQ_{GREEN}$ and Hoechst and visualized via confocal microscopy under a 100× objective. $CQ_{GREEN}$ accumulates in the DV but also slightly stains parasite cytosol; erythrocyte cytosol is not stained. Arrowheads denote the DV. Scale bars represent 5 μm.

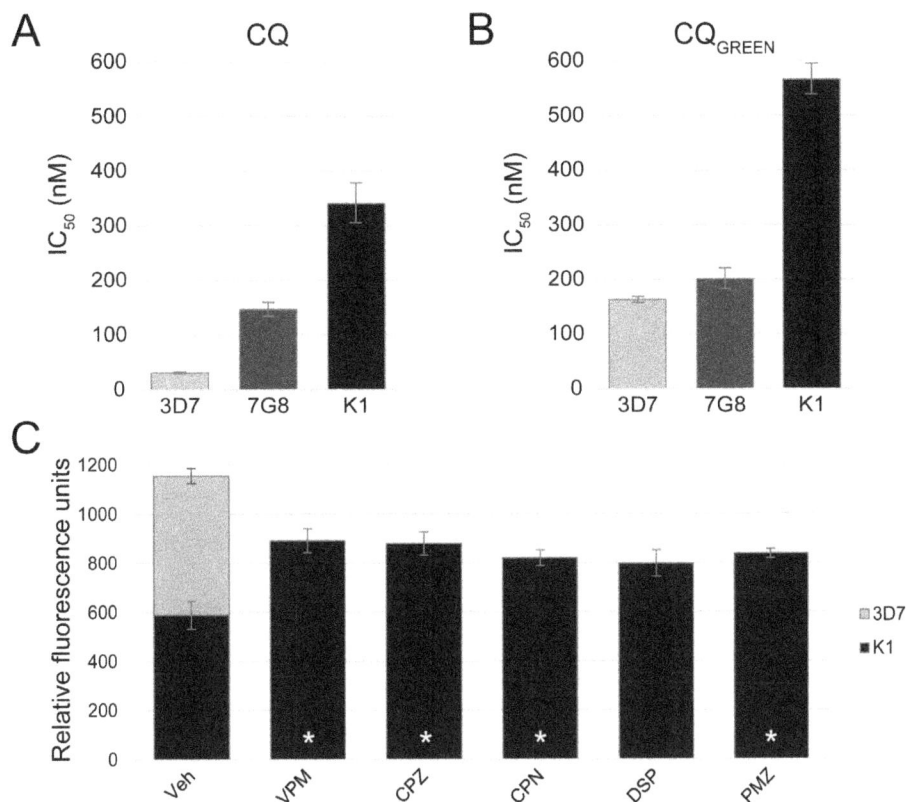

**Figure 2. Validation of CQ$_{GREEN}$ activity and uptake in laboratory strains.** (A, B) CQ$_{GREEN}$ IC$_{50}$s recapitulates CQ IC$_{50}$s in three *P. falciparum* laboratory strains, 3D7, 7G8 and K1. 3D7 is a CQ-susceptible strain, 7G8 is moderately resistant, while K1 is highly CQ-resistant. Data shown are geometric means from at least 3 experiments. As the error from IC$_{50}$s are not symmetrical, error bars indicate 95% confidence intervals instead of the standard error of the mean (S.E.M.). (C) CQ$_{GREEN}$ uptake as measured by fluorescence is increased in CQ-resistant K1 when pretreated with chemoreversal agents. 3D7 exhibits highest uptake of CQ$_{GREEN}$. Data are means from at least 3 experiments; error bars are S.E.M. Veh: vehicle control; VPM: verapamil; CPZ: chlorpromazine; CPN: chlorpheniramine; DSP: desipramine; PMZ: promethazine. *: $p<0.05$.

Medicine, University of Oxford, Oxford, United Kingdom). Use of field isolates in NUS was in accordance with NUS IRB (Reference Code: 12–369E).

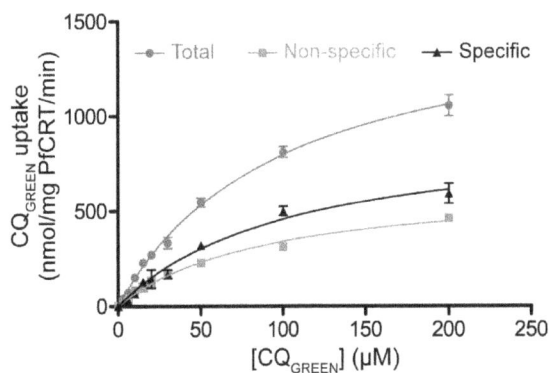

**Figure 3. CQ$_{GREEN}$ uptake by Dd2 PfCRT microsomes.** Total uptake of CQ$_{GREEN}$ was measured at various CQ$_{GREEN}$ concentrations in microsomes carrying Dd2 PfCRT. Non-specific uptake of CQ$_{GREEN}$ was measured with pre-treatment of excess unlabelled CQ. Specific uptake was estimated as the difference between total and non-specific uptake. V$_{max}$ and K$_m$ of the specific uptake was 938.5 nmol/mg PfCRT/min and 105.1 µM respectively. Data are means ± S.E.M.; n≥3.

## Results and Discussion

### Validation of CQ$_{GREEN}$ localization and antimalarial activity

CQ is generally believed to accumulate in the DV as a result of ion-trapping [35,36]. Confocal microscopy was therefore performed to ascertain the localization of CQ$_{GREEN}$ in the parasite. Cultures of *P. falciparum* 3D7 were co-stained with Hoechst dye and CQ$_{GREEN}$, revealing a preferential accumulation of CQ$_{GREEN}$ in the parasite DV (Fig. 1). Interestingly, CQ$_{GREEN}$ fluorescence was also observed in the parasite cytosol but not in the erythrocyte cytosol. As CQ$_{GREEN}$ is a CQ analog, the antimalarial potency of CQ$_{GREEN}$ should be similar to that of CQ. To assess this, reinvasion IC$_{50}$s of CQ and CQ$_{GREEN}$ on the laboratory strains 3D7, 7G8 and K1 were determined. CQ$_{GREEN}$ showed the same general trend of antimalarial activity as CQ, in that it is most potent against 3D7, followed by 7G8, then K1 (Fig. 2A, 2B).

### CQ$_{GREEN}$ fluorescence as a proxy for CQ uptake in parasites

Next, we determined if CQ$_{GREEN}$ uptake by the highly CQ-resistant strain K1 can be increased by pre-treatment with chemosensitizers. Verapamil, chlorpromazine, chlorpheniramine, desipramine, and promethazine have previously been reported to reverse CQ resistance and increase CQ uptake in CQ-resistant

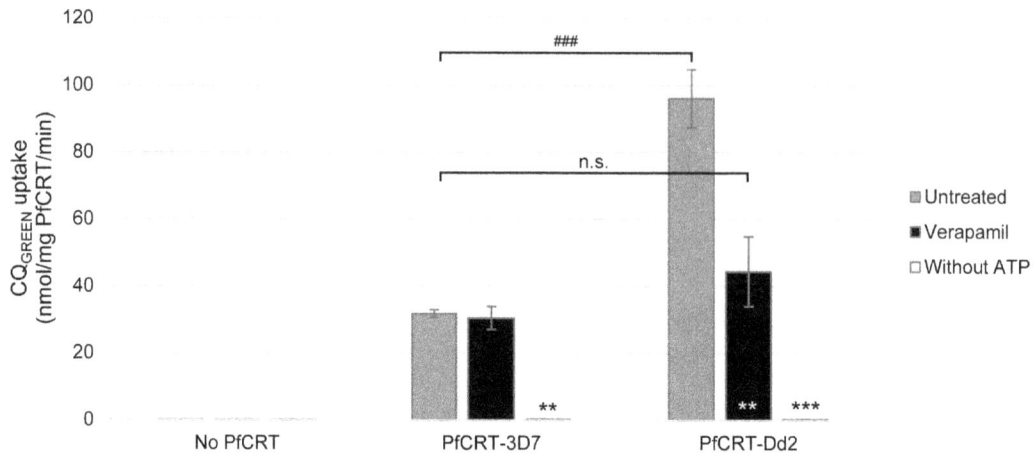

**Figure 4. ATP-dependent, verapamil-sensitive uptake of CQ$_{GREEN}$ in microsomes.** Yeast microsomes expressing CQ-sensitive or -resistant PfCRT ("PfCRT-3D7" and "PfCRT-Dd2" respectively), or microsomes from plasmid vector control ("No PfCRT"), were incubated with CQ$_{GREEN}$ under different conditions. Preincubation with 150 μM verapamil abrogated CQ$_{GREEN}$ uptake from PfCRT-Dd2 but did not affect uptake in PfCRT-3D7 microsomes. Removal of ATP from buffer abolished CQ$_{GREEN}$ uptake entirely. **, ***: $p < 0.005$ and $p < 0.001$ respectively, against untreated control. ###: $p < 0.001$. N.s.: not significant. Data presented are means ± S.E.M.; $n \geq 3$.

strains [37–41]. CQ-sensitive 3D7 was included as a reference for complete reversal. All reversal agents except desipramine induced a significant increase in CQ$_{GREEN}$ fluorescence (Fig. 2C). Desipramine is in fact a less potent reversal agent compared to verapamil when tested in the resistant strain Dd2, and two CQ-resistant field isolates [42]; this may explain why desipramine's effect on CQ$_{GREEN}$ uptake did not achieve statistical significance. Taken together with the CQ$_{GREEN}$ IC$_{50}$ data, we believe that the reversibility of CQ$_{GREEN}$ uptake by known chemoreversal agents suggests that CQ$_{GREEN}$ shares similar structural properties with CQ.

## Uptake of CQ$_{GREEN}$ in microsomes bearing PfCRT

To test if CQ$_{GREEN}$ can be transported by PfCRT, we have expressed PfCRT in *Pichia pastoris* and used the microsomes derived to study CQ$_{GREEN}$ uptake. Figure 3 shows that CQ$_{GREEN}$

uptake in Dd2 PfCRT-expressing microsomes is specific. At the highest concentration of CQ$_{GREEN}$ used (200 μM), uptake was close to saturated. Michaelis-Menten approximation of CQ$_{GREEN}$ uptake kinetics in Dd2 microsomes (Figure 3) yields a V$_{max}$ and K$_m$ of 938.5 nmol/mg PfCRT/min and 105.1 μM respectively, which are approximately 2000 and 500 times higher compared to when $^3$H-CQ was used [33]. Conjugation of CQ with the BODIPY fluorophore may have altered the affinity of PfCRT for the molecule. However, both the high (micromolar) K$_m$ and non-saturation of CQ$_{GREEN}$ transport are consistent with a previous report in a *Xenopus* oocyte system using $^3$H-CQ [43]. We have also compared CQ$_{GREEN}$ uptake in microsomes with PfCRT originating from either CQ-sensitive 3D7 or CQ-resistant Dd2. CQ$_{GREEN}$ uptake in Dd2 PfCRT-expressing microsome was 96.07 nmol/mg PfCRT/min, which was about three times that of 3D7 PfCRT (31.64 nmol/mg PfCRT/min) (Figure 4). PfCRT-

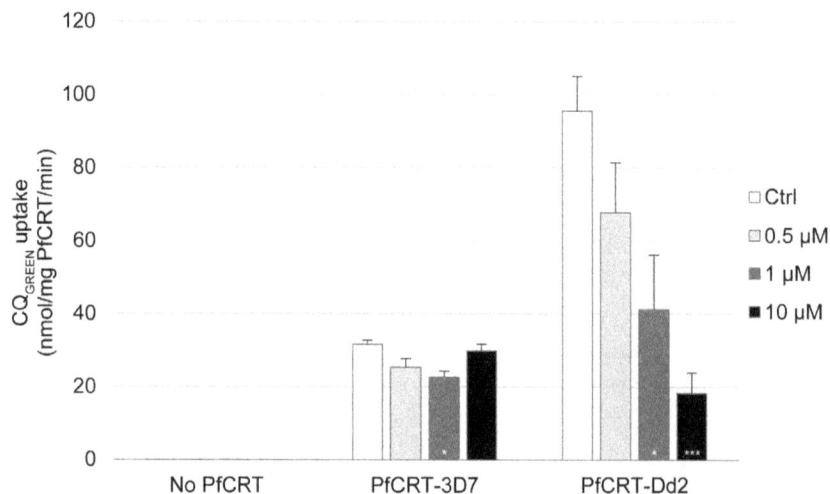

**Figure 5. CQ$_{GREEN}$ uptake by resistant-type PfCRT is inhibited by mibefradil in a dose-dependent manner.** Microsomes were preincubated with varying concentrations of the PfCRT inhibitor mibefradil prior to addition of CQ$_{GREEN}$. At the highest concentration of 10 μM, mibefradil drastically suppressed CQ$_{GREEN}$ uptake in PfCRT-Dd2 microsomes but had no significant effect on uptake in PfCRT-3D7 microsomes. *, ***: $p < 0.05$ and $p < 0.001$ respectively, against no mibefradil control (Ctrl). Data presented are means ± S.E.M.; $n \geq 3$.

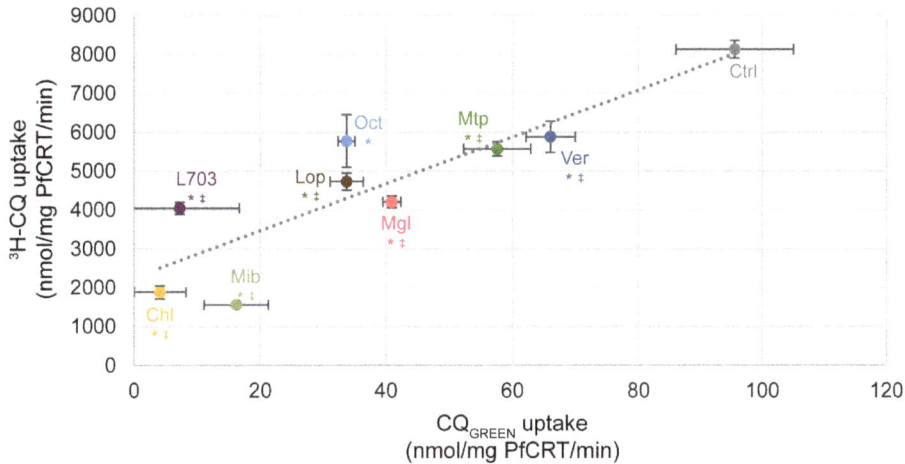

**Figure 6. Accumulation of CQ$_{GREEN}$ and $^3$H-CQ in PfCRT-Dd2 microsomes.** Microsomes were incubated with 10 µM of chemosensitizers before addition of CQ$_{GREEN}$ or $^3$H-CQ. Ctrl: negative control; Ver: verapamil; Mtp: methiothepin; Mgl: metergoline; Lop: loperamide; Oct: octoclothepin; Mib: mibefradil; L703: L703,606; Chl: chlorprothixene. *: $p<0.05$, comparing CQ$_{GREEN}$ uptake against control. $^‡$: $p<0.05$, comparing $^3$H-CQ uptake against control. Data presented are means ± S.E.M.; $n \geq 3$.

mediated transport of CQ is thought to be ATP-dependent [25,44]. Here we demonstrated that removal of ATP completely abolished CQ$_{GREEN}$ uptake in both Dd2 and 3D7 PfCRT microsomes (Figure 4). Verapamil, a known CQ resistance chemosensitizer which has no effect on CQ-sensitive strains, can reverse CQ$_{GREEN}$ uptake in Dd2 PfCRT to almost that of 3D7 level but has almost no effect on 3D7 PfCRT (Figure 4). These results suggest that CQ$_{GREEN}$ is similar to CQ in that (1) it is differentially recognized by CQ-resistant versus CQ-sensitive PfCRT, (2) its uptake by PfCRT is ATP-dependent, and (3) its uptake by CQ-resistant PfCRT is verapamil-reversible. Others have shown that extensive CQ side-chain modifications can render the CQ analogues not transportable by PfCRT and abolish their verapamil sensitivity [45,46]. Our findings show that the additional BODIPY moiety in CQ$_{GREEN}$ still allows CQ$_{GREEN}$ to be differentially recognized by resistant versus sensitive PfCRT and these transport activities are still sensitive to verapamil. To demonstrate the usefulness of CQ$_{GREEN}$ in screening for PfCRT inhibitors, we found that CQ$_{GREEN}$ uptake by Dd2 PfCRT microsomes can be inhibited by mibefradil, a novel potent chemosensitizer [21], in a dose-dependent manner (Figure 5). Inhibition of CQ$_{GREEN}$ uptake by a panel of known chemoreversal agents was compared to that using $^3$H-CQ (Figure 6). Mean uptake of CQ$_{GREEN}$ was positively correlated with that of $^3$H-CQ (β of 60.24, $p = 0.002$), showing moderately good agreement between CQ$_{GREEN}$ and $^3$H-CQ uptake ($R^2 = 0.766$). However, $^3$H-CQ uptake was roughly 100 times greater than that of CQ$_{GREEN}$.

## Polymorphisms and copy number variation in *pfcrt* and *pfmdr1*

Eight laboratory strains and twelve clinical isolates were sequenced for polymorphisms in PfCRT and Pgh1, the proteins encoded by the genes *pfcrt* and *pfmdr1* respectively. Residues examined were 72, 74, 75, 76, 220, 271, 326, 356, and 371 for PfCRT, and 86, 184, 1034, 1042, 1226, and 1246 for Pgh1. These residues were chosen for analysis as they were previously implicated in modulating multidrug resistance as well as resistance against CQ [47,48]. Copy number of *pfmdr1* was also determined for each strain and isolate. All clinical isolates showed Dd2-type

PfCRT mutations, whereas Pgh1 mutations and *pfmdr1* copy numbers were more varied (Table 1).

## Genetic correlates of CQ$_{GREEN}$ uptake

All strains and isolates were assayed for CQ$_{GREEN}$ uptake in the trophozoite stage, and their CQ IC$_{50}$s determined with the standard reinvasion assay. For the entire data set, CQ$_{GREEN}$ uptake was inversely correlated to the log of CQ IC$_{50}$, with an $R^2$ of 0.53 (Fig. 7A). However, multiple linear regression with sequencing and copy number data revealed that CQ$_{GREEN}$ uptake was significantly correlated with not only CQ logIC$_{50}$ but also a F1226Y substitution in Pgh1 (β of -587.32 and 178.70, $p<0.001$ and $p = 0.024$ respectively; adjusted $R^2$ of 0.615). None of the other mutations was significantly correlated with CQ$_{GREEN}$ uptake. Separating the data set to two subpopulations improved the $R^2$, to 0.72 in the Pgh1 1226F group and 0.676 in the Pgh1 1226Y group (Fig. 7B). It is tempting to conclude that the Pgh1 F1226Y substitution plays a significant causal role in modulating CQ$_{GREEN}$ uptake, given that it is also correlated with resistance to artemisinin, mefloquine and lumefantrine [48] and Pgh1's putative sequestration of cytosol-active drugs in the DV [49]. However, it must be kept in mind that the F1226Y mutation was only detected in the clinical isolates, and given the localized collection of these isolates within a small geographical region, F1226Y is likely to be strongly correlated with other undiscovered mutations. In fact, the PfCRT and Pgh1 haplotypes of the F1226Y mutants examined were identical, apart from SMRU0501 which had an additional Pgh1 Y184F substitution (Table 1). It is notable that for the F1226Y mutants, CQ$_{GREEN}$ uptake could range as high as that of the CQ-susceptible strains (Fig. 7B) while still maintaining CQ resistance. One possible gene modulating CQ$_{GREEN}$ uptake could be *pfmrp*, which has been proposed to efflux drugs across the parasite plasma membrane into the parasitophorous vacuolar lumen [49], which would still contribute to CQ$_{GREEN}$ fluorescence but sequester the drug from its site of action. Knock-out mutants of this gene exhibit increased susceptibility to CQ, quinine, artemisinin, piperaquine, and primaquine [50]. Alternative mechanisms besides efflux of CQ could perhaps also contribute to CQ resistance, such as decreased susceptibility to CQ-induced apoptosis-like cell death [51].

**Table 1.** CQ IC$_{50}$, PfCRT and Pgh1 polymorphisms, and *pfmdr1* copy number.

| Laboratory strains | CQ IC$_{50}$ (nM) | PfCRT residue no. | | | | | | | | | Pgh1 residue no. | | | | | | *pfmdr1* copy number |
|---|---|---|---|---|---|---|---|---|---|---|---|---|---|---|---|---|---|
| | | 72 | 74 | 75 | 76 | 220 | 271 | 326 | 356 | 371 | 86 | 184 | 1034 | 1042 | 1226 | 1246 | |
| T9/96 | 24 | C | M | N | K | A | Q | N | I | R | N | Y | S | N | F | D | 1 |
| 3D7 | 31 | C | M | N | K | A | Q | N | I | R | N | Y | S | N | F | D | 1 |
| HB3 | 42 | C | M | N | K | A | Q | N | I | R | N | **F** | S | **D** | F | D | 1 |
| CS2 | 115 | C | **I** | **E** | **T** | **S** | **E** | **S** | I | **I** | **Y** | Y | S | N | F | D | 3 |
| T9-94 | 146 | C | **I** | **E** | **T** | **S** | **E** | **S** | I | **I** | **Y** | Y | S | N | F | D | 3 |
| 7G8 | 146 | **S** | M | N | **T** | **S** | Q | **D** | **L** | R | N | **F** | **C** | **D** | F | **Y** | 1 |
| Dd2 | 276 | C | **I** | **E** | **T** | **S** | **E** | **S** | **T** | **I** | **Y** | Y | S | N | F | D | 3 |
| K1 | 340 | C | **I** | **E** | **T** | **S** | **E** | **S** | I | **I** | **Y** | Y | S | N | F | D | 1 |

| Clinical isolates | CQ IC$_{50}$ (nM) | 72 | 74 | 75 | 76 | 220 | 271 | 326 | 356 | 371 | 86 | 184 | 1034 | 1042 | 1226 | 1246 | *pfmdr1* copy number |
|---|---|---|---|---|---|---|---|---|---|---|---|---|---|---|---|---|---|
| SMRU0233 | 142 | C | **I** | **E** | **T** | **S** | **E** | **S** | **T** | **I** | N | Y | S | N | **Y** | D | 2 |
| SMRU0116 | 147 | C | **I** | **E** | **T** | **S** | **E** | **S** | **T** | **I** | N | Y | S | N | **Y** | D | 2 |
| SMRU0101 | 156 | C | **I** | **E** | **T** | **S** | **E** | **S** | **T** | **I** | N | Y | S | N | F | D | 2 |
| SMRU1116 | 172 | C | **I** | **E** | **T** | **S** | **E** | **S** | **T** | **I** | N | Y | S | N | F | D | 1 |
| SMRU0270 | 178 | C | **I** | **E** | **T** | **S** | **E** | **S** | **T** | **I** | N | Y | S | N | F | D | 2 |
| SMRU0402 | 196 | C | **I** | **E** | **T** | **S** | **E** | **S** | **T** | **I** | N | Y | S | N | **Y** | D | 2 |
| SMRU1093 | 231 | C | **I** | **E** | **T** | **S** | **E** | **S** | **T** | **I** | N | Y | S | N | **Y** | D | 2 |
| SMRU0201 | 249 | C | **I** | **E** | **T** | **S** | **E** | **S** | **T** | **I** | N | Y | S | N | F | D | 1 |
| SMRU0501 | 287 | C | **I** | **E** | **T** | **S** | **E** | **S** | **T** | **I** | N | **F** | S | N | **Y** | D | 1 |
| SMRU0272 | 332 | C | **I** | **E** | **T** | **S** | **E** | **S** | **T** | **I** | N | Y | S | N | **Y** | D | 3 |
| SMRU0002 | 377 | C | **I** | **E** | **T** | **S** | **E** | **S** | **T** | **I** | N | Y | S | N | **Y** | D | 1 |
| SMRU0279 | 473 | C | **I** | **E** | **T** | **S** | **E** | **S** | **T** | **I** | N | Y | S | N | **Y** | D | 1 |

Bolded residues indicate deviation from 3D7 haplotype. CQ IC$_{50}$s are geometric means of at least 3 measurements.

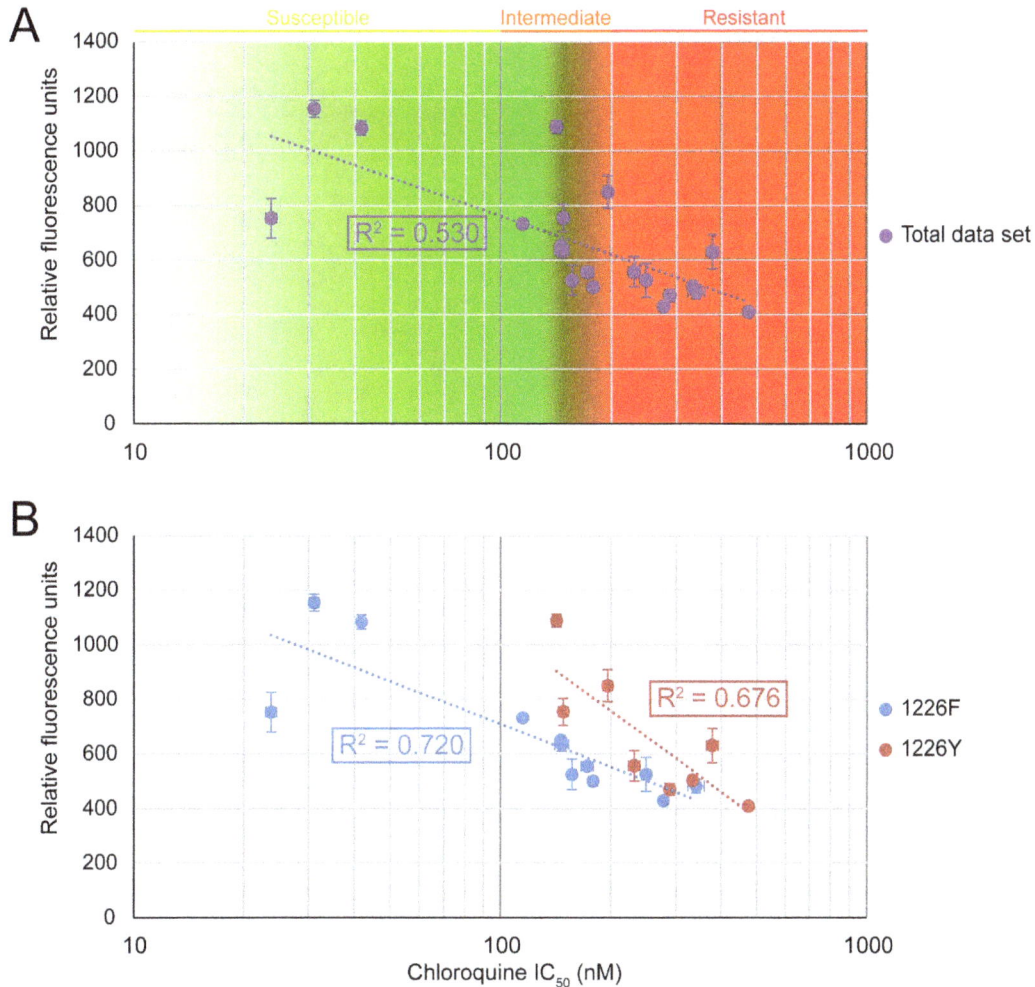

**Figure 7. CQ$_{GREEN}$ uptake correlates with CQ resistance.** (A) CQ$_{GREEN}$ fluorescence is inversely correlated with CQ logIC$_{50}$. Total data set shows moderate $R^2$ of 0.53. (B) When split into subpopulations on the basis of Pgh1 residue 1226, $R^2$ is improved. Data shown are means of at least 3 experiments. Error bars represent S.E.M.

## Conclusions

CQ$_{GREEN}$ is a commercially available fluorescent CQ analog that interacts with the parasite in a similar fashion to CQ. CQ$_{GREEN}$ presents some advantages over traditional radiolabelled CQ in uptake studies: it is safer to handle as it is not radioactive, and its fluorescence properties allows it to be monitored by common fluorescence equipment. Using a defined microsomal platform, we showed that CQ$_{GREEN}$ interacts with PfCRT in a manner similar to CQ. Its use as a predictor of CQ susceptibility is enhanced if residue 1226 of Pgh1 is known. Unlike a typical reinvasion assay which may require 48 h or more, the use of CQ$_{GREEN}$ allows for measurement of CQ susceptibility within several hours. We believe that CQ$_{GREEN}$ could be a valuable tool in future drug discovery projects or used in the identification of factors involved in drug resistance.

## Acknowledgments

We thank the patients and staff of Shoklo Malaria Research Unit (SMRU) for their contributions to this project. We would also like to extend our gratitude to the following for *P. falciparum* strains obtained through MR4's BEI Resources Repository, NIAID, NIH: DJ Carucci (3D7, MRA-102), TE Wellems (HB3, MRA-155; Dd2, MRA-156), SJ Rogerson (CS2, MRA-96), D Walliker (T9-94, MRA-153), and DE Kyle (7G8, MRA-154; K1, MRA-159). *P. falciparum* T9/96 was obtained from The European Malaria Reagent Repository (http://www.malariaresearch.eu).

## Author Contributions

Conceived and designed the experiments: KSWT BR FHN LR MJL LMCC. Performed the experiments: YQL CCYL KWKC KYC RS. Analyzed the data: YQL CCYL RS. Contributed reagents/materials/analysis tools: KSWT LR FHN. Wrote the paper: YQL KSWT BR FHN LR MJL LMCC CCYL KWKC KYC RS.

## References

1. Murray CJ, Rosenfeld LC, Lim SS, Andrews KG, Foreman KJ, et al. (2012) Global malaria mortality between 1980 and 2010: a systematic analysis. The Lancet 379: 413–431. doi:10.1016/S0140-6736(12)60034-8.

2. World Health Organization (2012) World Malaria Report 2012. Available: http://www.who.int/malaria/publications/world_malaria_report_2012/en/index.html. Accessed 2013 Mar 11.

3.  Krafts K, Hempelmann E, Skórska-Stania A (2012) From methylene blue to chloroquine: a brief review of the development of an antimalarial therapy. Parasitol Res 111: 1–6. doi:10.1007/s00436-012-2886-x.

4.  Coatney GR (1963) Pitfalls in a Discovery: The Chronicle of Chloroquine. Am J Trop Med Hyg 12: 121–128.

5.  Butler A, Khan S, Ferguson E (2010) A brief history of malaria chemotherapy. J R Coll Physicians Edinb 40: 172–177. doi:10.4997/JRCPE.2010.216.

6.  Wellems TE, Plowe CV (2001) Chloroquine-Resistant Malaria. J Infect Dis 184: 770–776. doi:10.1086/322858.

7.  Sullivan DJ Jr, Gluzman IY, Goldberg DE (1996) Plasmodium hemozoin formation mediated by histidine-rich proteins. Science 271: 219–222.

8.  Hempelmann E (2007) Hemozoin Biocrystallization in *Plasmodium falciparum* and the antimalarial activity of crystallization inhibitors. Parasitol Res 100: 671–676. doi:10.1007/s00436-006-0313-x.

9.  Gildenhuys J, Roex T le, Egan TJ, de Villiers KA (2013) The Single Crystal X-ray Structure of β-Hematin DMSO Solvate Grown in the Presence of Chloroquine, a β-Hematin Growth-Rate Inhibitor. J Am Chem Soc 135: 1037–1047. doi:10.1021/ja308741e.

10. Rathore D, Jani D, Nagarkatti R, Kumar S (2006) Heme detoxification and antimalarial drugs – Known mechanisms and future prospects. Drug Discov Today Ther Strateg 3: 153–158. doi:10.1016/j.ddstr.2006.06.003.

11. Summers RL, Martin RE (2010) Functional characteristics of the malaria parasite's "chloroquine resistance transporter": implications for chemotherapy. Virulence 1: 304–308. doi:10.4161/viru.1.4.12012.

12. Pradines B, Bertaux L, Pomares C, Delaunay P, Marty P (2011) Reduced *in vitro* susceptibility to artemisinin derivatives associated with multi-resistance in a traveller returning from South-East Asia. Malar J 10: 268. doi:10.1186/1475-2875-10-268.

13. Miotto O, Almagro-Garcia J, Manske M, MacInnis B, Campino S, et al. (2013) Multiple populations of artemisinin-resistant *Plasmodium falciparum* in Cambodia. Nat Genet 45: 648–655. doi:10.1038/ng.2624.

14. Takala-Harrison S, Clark TG, Jacob CG, Cummings MP, Miotto O, et al. (2013) Genetic loci associated with delayed clearance of *Plasmodium falciparum* following artemisinin treatment in Southeast Asia. Proc Natl Acad Sci 110: 240–245. doi:10.1073/pnas.1211205110.

15. Ariey F, Witkowski B, Amaratunga C, Beghain J, Langlois A-C, et al. (2014) A molecular marker of artemisinin-resistant *Plasmodium falciparum* malaria. Nature 505: 50–55. doi:10.1038/nature12876.

16. Mok S, Imwong M, Mackinnon MJ, Sim J, Ramadoss R, et al. (2011) Artemisinin resistance in *Plasmodium falciparum* is associated with an altered temporal pattern of transcription. BMC Genomics 12: 391. doi:10.1186/1471-2164-12-391.

17. Witkowski B, Amaratunga C, Khim N, Sreng S, Chim P, et al. (2013) Novel phenotypic assays for the detection of artemisinin-resistant *Plasmodium falciparum* malaria in Cambodia: *in-vitro* and *ex-vivo* drug-response studies. Lancet Infect Dis. doi:10.1016/S1473-3099(13)70252-4.

18. Peyton DH (2012) Reversed chloroquine molecules as a strategy to overcome resistance in malaria. Curr Top Med Chem 12: 400–407.

19. Martin SK, Oduola AMJ, Milhous WK (1987) Reversal of Chloroquine Resistance in *Plasmodium falciparum* by Verapamil. Science 235: 899–901.

20. Burgess SJ, Selzer A, Kelly JX, Smilkstein MJ, Riscoe MK, et al. (2006) A Chloroquine-like Molecule Designed to Reverse Resistance in *Plasmodium falciparum*. J Med Chem 49: 5623–5625. doi:10.1021/jm060399n.

21. Ch'ng J-H, Mok S, Bozdech Z, Lear MJ, Boudhar A, et al. (2013) A Whole Cell Pathway Screen Reveals Seven Novel Chemosensitizers to Combat Chloroquine Resistant Malaria. Sci Rep 3. doi:10.1038/srep01734.

22. Martin RE, Butterworth AS, Gardiner DL, Kirk K, McCarthy JS, et al. (2012) Saquinavir Inhibits the Malaria Parasite's Chloroquine Resistance Transporter. Antimicrob Agents Chemother 56: 2283–2289. doi:10.1128/AAC.00166-12.

23. Adovelande J, Delèze J, Schrével J (1998) Synergy between two calcium channel blockers, verapamil and fantofarone (SR33557), in reversing chloroquine resistance in *Plasmodium falciparum*. Biochem Pharmacol 55: 433–440. doi:10.1016/S0006-2952(97)00482-6.

24. Kelly JX, Smilkstein MJ, Cooper RA, Lane KD, Johnson RA, et al. (2007) Design, Synthesis, and Evaluation of 10-N-Substituted Acridones as Novel Chemosensitizers in *Plasmodium falciparum*. Antimicrob Agents Chemother 51: 4133–4140. doi:10.1128/AAC.00669-07.

25. Sanchez CP, Stein W, Lanzer M (2003) Trans Stimulation Provides Evidence for a Drug Efflux Carrier as the Mechanism of Chloroquine Resistance in *Plasmodium falciparum*. Biochemistry (Mosc) 42: 9383–9394. doi:10.1021/bi034269h.

26. Sanchez CP, McLean JE, Stein W, Lanzer M (2004) Evidence for a Substrate Specific and Inhibitable Drug Efflux System in Chloroquine Resistant *Plasmodium falciparum* Strains. Biochemistry (Mosc) 43: 16365–16373. doi:10.1021/bi048241x.

27. Martin RE, Marchetti RV, Cowan AI, Howitt SM, Broer S, et al. (2009) Chloroquine Transport via the Malaria Parasite's Chloroquine Resistance Transporter. Science 325: 1680–1682. doi:10.1126/science.1175667.

28. Zishiri VK, Hunter R, Smith PJ, Taylor D, Summers R, et al. (2011) A series of structurally simple chloroquine chemosensitizing dibemethin derivatives that inhibit chloroquine transport by PfCRT. Eur J Med Chem 46: 1729–1742. doi:10.1016/j.ejmech.2011.02.026.

29. Cabrera M, Natarajan J, Paguio MF, Wolf C, Urbach JS, et al. (2009) Chloroquine Transport in *Plasmodium falciparum*. 1. Influx and Efflux Kinetics for Live Trophozoite Parasites Using a Novel Fluorescent Chloroquine Probe. Biochemistry (Mosc) 48: 9471–9481. doi:10.1021/bi901034r.

30. Ch'ng J-H, Kotturi SR, Chong AG-L, Lear MJ, Tan KS-W (2010) A programmed cell death pathway in the malaria parasite *Plasmodium falciparum* has general features of mammalian apoptosis but is mediated by clan CA cysteine proteases. Cell Death Dis 1: e26. doi:10.1038/cddis.2010.2.

31. Alcantara LM, Kim J, Moraes CB, Franco CH, Franzoi KD, et al. (2013) Chemosensitization potential of P-glycoprotein inhibitors in malaria parasites. Exp Parasitol 134: 235–243. doi:10.1016/j.exppara.2013.03.022.

32. Tram K, Yan H, Jenkins HA, Vassiliev S, Bruce D (2009) The synthesis and crystal structure of unsubstituted 4, 4-difluoro-4-bora-3a, 4a-diaza-s-indacene (BODIPY). Dyes Pigments 82: 392–395. doi:10.1016/j.dyepig.2009.03.001.

33. Tan W, Gou DM, Tai E, Zhao YZ, Chow LMC (2006) Functional reconstitution of purified chloroquine resistance membrane transporter expressed in yeast. Arch Biochem Biophys 452: 119–128. doi:10.1016/j.abb.2006.06.017.

34. Price RN, Uhlemann A-C, Brockman A, McGready R, Ashley E, et al. (2004) Mefloquine resistance in *Plasmodium falciparum* and increased *pfmdr1* gene copy number. The Lancet 364: 438–447. doi:10.1016/S0140-6736(04)16767-6.

35. Homewood CA, Warhurst DC, Peters W, Baggaley VC (1972) Lysosomes, pH and the anti-malarial action of chloroquine. Nature 235: 50–52.

36. Yayon A, Cabantchik ZI, Ginsburg H (1984) Identification of the acidic compartment of *Plasmodium falciparum*-infected human erythrocytes as the target of the antimalarial drug chloroquine. EMBO J 3: 2695–2700.

37. Van Schalkwyk DA, Walden JC, Smith PJ (2001) Reversal of Chloroquine Resistance in *Plasmodium falciparum* Using Combinations of Chemosensitizers. Antimicrob Agents Chemother 45: 3171–3174. doi:10.1128/AAC.45.11.3171-3174.2001.

38. Kyle DE, Oduola AMJ, Martin SK, Milhous WK (1990) *Plasmodium falciparum*: modulation by calcium antagonists of resistance to chloroquine, desethylchloroquine, quinine, and quinidine *in vitro*. Trans R Soc Trop Med Hyg 84: 474–478. doi:10.1016/0035-9203(90)90004-X.

39. Martiney JA, Cerami A, Slater AFG (1995) Verapamil Reversal of Chloroquine Resistance in the Malaria Parasite *Plasmodium falciparum* Is Specific for Resistant Parasites and Independent of the Weak Base Effect. J Biol Chem 270: 22393–22398. doi:10.1074/jbc.270.38.22393.

40. Oduola AM, Sowunmi A, Milhous WK, Brewer TG, Kyle DE, et al. (1998) *In vitro* and *in vivo* reversal of chloroquine resistance in *Plasmodium falciparum* with promethazine. Am J Trop Med Hyg 58: 625–629.

41. Egan T, Kaschula C (2007) Strategies to reverse drug resistance in malaria. Curr Opin Infect Dis 20: 598–604. doi:10.1097/QCO.0b013e3282f1673a.

42. Bayoumi RAL, Babiker HA, Arnot DE (1994) Uptake and efflux of chloroquine by chloroquine-resistant *Plasmodium falciparum* clones recently isolated in Africa. Acta Trop 58: 141–149. doi:10.1016/0001-706X(94)90053-1.

43. Summers RL, Dave A, Dolstra TJ, Bellanca S, Marchetti RV, et al. (2014) Diverse mutational pathways converge on saturable chloroquine transport via the malaria parasite's chloroquine resistance transporter. Proc Natl Acad Sci 111: E1759–E1767. doi:10.1073/pnas.1322965111.

44. Krogstad DJ, Gluzman IY, Herwaldt BL, Schlesinger PH, Wellems TE (1992) Energy dependence of chloroquine accumulation and chloroquine efflux in Plasmodium falciparum. Biochem Pharmacol 43: 57–62. doi:10.1016/0006-2952(92)90661-2.

45. De D, Krogstad FM, Cogswell FB, Krogstad DJ (1996) Aminoquinolines That Circumvent Resistance in *Plasmodium falciparum in Vitro*. Am J Trop Med Hyg 55: 579–583.

46. Lakshmanan V, Bray PG, Verdier-Pinard D, Johnson DJ, Horrocks P, et al. (2005) A critical role for PfCRT K76T in Plasmodium falciparum verapamil-reversible chloroquine resistance. EMBO J 24: 2294–2305. doi:10.1038/sj.emboj.7600681.

47. Valderramos SG, Valderramos J-C, Musset L, Purcell LA, Mercereau-Puijalon O, et al. (2010) Identification of a Mutant PfCRT-Mediated Chloroquine Tolerance Phenotype in *Plasmodium falciparum*. PLoS Pathog 6: e1000887. doi:10.1371/journal.ppat.1000887.

48. Veiga MI, Ferreira PE, Jörnhagen L, Malmberg M, Kone A, et al. (2011) Novel Polymorphisms in *Plasmodium falciparum* ABC Transporter Genes Are Associated with Major ACT Antimalarial Drug Resistance. PLoS ONE 6: e20212. doi:10.1371/journal.pone.0020212.

49. Sanchez CP, Dave A, Stein WD, Lanzer M (2010) Transporters as mediators of drug resistance in *Plasmodium falciparum*. Int J Parasitol 40: 1109–1118. doi:10.1016/j.ijpara.2010.04.001.

50. Raj DK, Mu J, Jiang H, Kabat J, Singh S, et al. (2009) Disruption of a *Plasmodium falciparum* Multidrug Resistance-associated Protein (PfMRP) Alters Its Fitness and Transport of Antimalarial Drugs and Glutathione. J Biol Chem 284: 7687–7696. doi:10.1074/jbc.M806944200.

51. Ch'ng J-H, Liew K, Goh AS-P, Sidhartha E, Tan KS-W (2011) Drug-induced permeabilization of parasite's digestive vacuole is a key trigger of programmed cell death in *Plasmodium falciparum*. Cell Death Dis 2: e216. doi:10.1038/cddis.2011.97.

# 6

# Exacerbation of Autoimmune Neuro-Inflammation in Mice Cured from Blood-Stage *Plasmodium berghei* Infection

Rodolfo Thomé[1], André Luis Bombeiro[1], Luidy Kazuo Issayama[1], Catarina Rapôso[2], Stefanie Costa Pinto Lopes[3], Thiago Alves da Costa[1], Rosária Di Gangi[1], Isadora Tassinari Ferreira[1], Ana Leda Figueiredo Longhini[4], Alexandre Leite Rodrigues Oliveira[1], Maria Alice da Cruz Höfling[2], Fábio Trindade Maranhão Costa[3], Liana Verinaud[1]*

1 Department of Structural and Functional Biology, Institute of Biology, University of Campinas, Campinas, Brazil, 2 Department of Histology and Embryology, Institute of Biology, University of Campinas, Campinas, Brazil, 3 Department of Genetics, Evolution and Bioagents, Institute of Biology, University of Campinas, Campinas, Brazil, 4 Department of Hematology, Faculdade de Ciências Médicas, University of Campinas, Campinas, Brazil

## Abstract

The thymus plays an important role shaping the T cell repertoire in the periphery, partly, through the elimination of inflammatory auto-reactive cells. It has been shown that, during *Plasmodium berghei* infection, the thymus is rendered atrophic by the premature egress of $CD4^+CD8^+$ double-positive (DP) T cells to the periphery. To investigate whether autoimmune diseases are affected after *Plasmodium berghei* NK65 infection, we immunized C57BL/6 mice, which was previously infected with *P.berghei* NK65 and treated with chloroquine (CQ), with $MOG_{35-55}$ peptide and the clinical course of Experimental Autoimmune Encephalomyelitis (EAE) was evaluated. Our results showed that NK65+CQ+EAE mice developed a more severe disease than control EAE mice. The same pattern of disease severity was observed in $MOG_{35-55}$-immunized mice after adoptive transfer of *P.berghei*-elicited splenic DP-T cells. The higher frequency of $IL-17^+$- and $IFN-\gamma^+$-producing DP lymphocytes in the Central Nervous System of these mice suggests that immature lymphocytes contribute to disease worsening. To our knowledge, this is the first study to integrate the possible relationship between malaria and multiple sclerosis through the contribution of the thymus. Notwithstanding, further studies must be conducted to assert the relevance of malaria-induced thymic atrophy in the susceptibility and clinical course of other inflammatory autoimmune diseases.

**Editor:** Claudio Romero Farias Marinho, Instituto de Ciências Biomédicas/Universidade de São Paulo – USP, Brazil

**Funding:** This work was supported by São Paulo Research Foundation (FAPESP, grant number #2011/17965-3). RT received a FAPESP scholarship (#2011/13191-3). MACH, FTMC and LV are fellows of the National Council of Technological and Scientific Development (CNPq). The funders had no role in study design, data collection and analysis, decision to publish, or preparation of the manuscript.

**Competing Interests:** The authors have declared that no competing interests exist.

* Email: verinaud@unicamp.br

## Introduction

The thymus plays a crucial role in the development and maturation of T lymphocytes. Inside the organ, thymocytes (T cells in the thymus) are provided with a wealthy microenvironment consisting of cytokines, chemokines and cell-cell interactions that ultimately leads to proliferation, T-cell receptor gene rearrangements and thymocyte differentiation into mature T cells [1]. However, the thymus undergoes weight and cellular loss caused by ageing and environmental disturbances, such as infections and malnutrition [2]. We have previously shown that infection with *Plasmodium berghei* NK65 infection, the non-cerebral malaria pathological agent, renders the thymus atrophic through the enhanced thymocyte death by apoptosis and premature egress of $CD4^+CD8^+$ (Double-positive, DP) T cells to the periphery [3–5].

It is already known that some viral and bacterial infections can promote the development of autoimmunity by inducing the breakdown of T cell tolerance and development of effector T cells reactive with the self-antigens or by the phenomenon called molecular mimicry, where a foreign antigen shares sequence or structural similarities with self-antigens [6,7]. For instance, acute rheumatic fever, where antibodies attack the heart, can occur after the body makes immune responses against Group A β-hemolytic streptococci [8,9]. In addition, it has been proposed that the prematurely egressed DP-T cells observed during *Trypanosoma cruzi* infection play an important role in the autoimmune cardio-inflammation [10].

Experimental Autoimmune Encephalomyelitis is a T cell-driven inflammation of the Central Nervous System (CNS) that presents similar characteristics to human Multiple Sclerosis [11]. In this model, following an inflammatory stimulus containing neuro-peptides T cells migrate from the peripheral immune system towards the CNS where they promote inflammation through the release of inflammatory mediators such as cytokines and chemokines [12,13]. Cells from the Th1 and Th17 subsets are important for disease establishment, as evidenced by previous reports [14–16]. Both in the human and animal diseases T cells play a major role. Therefore, changes in the subpopulations of T cells influence the outcome and susceptibility to autoimmune development.

In this context, we aimed to evaluate whether the previous infection with *Plasmodium berghei* NK65 would interfere with the clinical course of Experimental Autoimmune Encephalomyelitis, a mouse model for human Multiple Sclerosis (MS). We observed that EAE-susceptible mice cured from malaria developed an aggravated form of EAE, with increased infiltration of DP-T cells in the Central Nervous System (CNS). Further analyses showed that thymic-prematurely egressed DP-T cells were important for the enhanced clinical manifestation of the disease. To our knowledge, this is the first study to demonstrate the possible integration between malaria and EAE through the contribution of the thymus.

## Materials and Methods

### Animals

Six- to eight-week-old female C57BL/6 mice from the Multidisciplinary Center for Biological Research, University of Campinas, were used in this study. Mice were kept in specific-pathogen free conditions, in a controlled temperature and photoperiod environment, with free access to autoclaved food and water throughout the experiment. All protocols involving laboratory animals were approved and performed in accordance with the guidelines of the Institutional Committee on the Use and Care of Animals (CEUA, #2687–1).

### Infection and treatment

For these experiments, we used the NK65 strain of *Plasmodium berghei*, because this parasite do not accumulate in the CNS of mice to cause cerebral malaria. Mice (n = 6 mice/group) were intra-peritoneally injected with $10^6$ *Plasmodium berghei*-infected red blood cells (iRBCs) obtained from a source mouse. The frequency of iRBCs was assessed daily by examination with Giemsa-stained thin blood smear. Ten days after infection, animals of each group started the treatment with chloroquine (5 mg.kg$^{-1}$, via i.p. for five consecutive days). Three days after the last dose of the drug, mice were immunized for the induction of EAE.

### EAE induction and evaluation

EAE was induced and evaluated in mice according to a previous published paper [17]. Briefly, each mouse was subcutaneously injected with 100 μg MOG$_{35–55}$ (MEVGWYRSPFSRVVH-LYRNGK, Genemed Syn, USA) emulsified with Complete Freunds Adjuvant (CFA, Sigma-Aldrich, USA). 200 ηg Pertussis toxin (Ptx, Sigma-Aldrich, USA) was administrated via i.p. at 0 and 48 h after MOG$_{35–55}$ inoculation. Clinical signs were followed and graded daily according to a score method, where 0: no sign, 1: flaccid tail, 2: hind limbs weakness, 3: hind limbs paralysis, 4: hind paralysis and fore limbs weakness, 5: full paralysis/dead. Disease severity was evaluated for thirty days. At the end of the experiment, results were analyzed by linear regression, thinner lines indicate 95% confidence interval.

### Determination of the thymic index

The calculations to determine the thymic index were previously described [3]. Briefly, at the indicated time-points, the gross weight of each mouse was recorded and then the thymuses were collected and weighed. The thymic index was calculated using the formula: organ weight (g)/body weight (g) ×100.

### Cell sorting and flow cytometry

Spleen cells were obtained from *P.berghei*-infected mice at ten days of infection. Total splenic T lymphocytes were isolated using Dynabeads following the manufacturers instructions (Mouse Pan T cell isolation kit, Life Technologies, USA). For DP-T cell isolation, spleen-derived single cell suspension was incubated with FITC-conjugated anti-CD4 (clone H129.19) and PE-conjugated anti-CD8a (clone 53–6.7) for 20 minutes. CD4$^+$, CD8$^+$ and CD4$^+$CD8$^+$ (double-positive) T lymphocytes from malaria-bearing mice at ten days of infection were sorted in FACSAria II cell sorter (BD Biosciences, USA). The purity of isolations was assessed and accounted for 98% pure subpopulations. For adoptive transfer experiments, $1,5×10^5$ cells were adoptively transferred (via i.v.) to each group of mice at the onset of EAE (around the tenth day after immunization). For activation experiments, sorted cells were seeded in U-bottom 96 well/plates ($2×10^4$ cells/well) and incubated for 72h with lipopolysaccharide (1 ng/mL, from *E. coli*, Sigma-Aldrich, USA), *Plasmodium berghei* extracts (PbX, 50 μg/mL) or MOG$_{35–55}$ peptide (20 μg/mL, Genscript, USA). At the end of culture period, the supernatants were collected and assayed for detection of mouse IFN-γ and IL-17 by Cytometric Bead Assay (CBA, BD Biosciences, USA). In some set of experiments, total T cells isolated from naïve and malaria-bearing mice were transferred ($1×10^6$ cells/mouse) at the onset of EAE.

### Preparation of Plasmodium berghei extracts

The production of extracts from *P. berghei*-infected RBCs (iRBC) followed a previously published recommendation [18]. Briefly, iRBC-enriched suspension was submitted to 20 cycles of freeze/thawing in liquid nitrogen and warm bath (37°C). The protein concentration was determined using the Bradford Protein Assay following the manufacturers instructions (Sigma-Aldrich, St Louis, MO, USA).

### Analysis of cellular infiltration in the CNS

Fourteen days after EAE induction, mice were anesthetized, perfused with ice cold PBS and half of the spinal cords were removed and stored at −80°C until use for RT-PCR assays; the remaining tissue was prepared for the enrichment of infiltrating leukocytes according to a previously described methodology and analyzed by flow cytometry [34].

### RT-PCR assays

Mice were killed at the indicated time-points and frozen tissues were used for RNA extraction (Trizol reagent, Life Technologies, USA) and cDNA synthesis according to the manufacturers recommendations (High Capacity RNA-to-cDNA converter kit, Life Techcnologies, USA). RNA extraction from DP-T cells was carried our immediately after sorting, using an extraction kit according to the manufacturers instruction (RNeasy Micro kit, Qiagen USA). Expression of AIRE (Mm00477461_m1), IL-7 (Mm01295803_m1), IL-6 (Mm00446190_m1), IL-17 (Mm00439618_m1), IFNg (Mm01168134_m1), FOXP3 (Mm00475162_m1) and RAR-related orphan receptor C (RORc) (Mm01261022_m1) were analyzed in comparison to GAPDH (Mm99999915_g1, house-keeping gene) levels. RT-PCR reactions were performed using Taqman reagents according to manufacturers recommendations (Applied Biosystems, USA). Expression levels of genes were represented as a relative copy numbers by using the method of delta threshold ($2^{-\Delta\Delta Ct}$).

## Histopathology and immunofluorescence

At the indicated time-points mice were killed, spinal cords were removed and snap frozen; 12 μm thin slices were made in cryostat and stained with haematoxylin and eosin (H&E) for histopathological analysis. For the characterization of the infiltrating cells, the reactive sites of the slices were blocked with Phosphate-Buffer (0,1M, pH7,2)-BSA 3%. The cells were stained with FITC-conjugated rat anti-mouse CD4 (clone H129.19, at a 1:100 dilution, BD Biosciences, USA) and PE-conjugated rat anti-mouse CD8a (clone 53–6.7, at a 1:100 dilution, BD Biosciences, USA). In some set of experiments, the lumbar spinal cords were incubated with purified anti-mouse GFAP, COX-2, IL-6, NF-κB, phosphorylated iKB and iNOS. Later, secondary Cy3- or TRITC-conjugated antibodies were added. The reaction was analyzed under epifluorescence microscope (Leica, GER).

## In vitro re-stimulation and cytokine dosage

Splenic cells were aseptically collected from mice after 10 days of MOG$_{35-55}$ immunization. Single cell suspensions were stained with Carboxyfluorescein succinimidyl ester (CFSE, 2,5 μM, Sigma-Aldrich, USA) following the manufacturers instructions. Cells ($5 \times 10^5$/well) were diluted in RPMI 1640 media supplemented with Fetal Calf Serum (FCS; 10% vol/vol), guaramicine (50 μg/mL), 2-Mercaptoethanol (2 mM) and myelin oligodendrocyte glycoprotein peptide (MOG$_{35-55}$; 20 μg/mL), plated in U-bottom plates and incubated for 96 h. After the incubation period, cells were stained with PercPCy5-conjugated anti-CD3e, PE-conjugated anti-CD8a and PECy7-conjugated anti-CD4 antibodies and fixed in 1% paraformaldehyde prior to flow cytometer analysis. CFSE$^{low}$ cells inside each population were considered proliferating T cells. Culture supernatants were collected and assayed for cytokines (IL-4, IL-6, IL-10, IL-17, IFN-γ and TNF-α) secretion using the Cytometric Bead Array (CBA, BD Biosciences, USA) according to manufacturers instructions.

## T cell co-cultures

For the co-cultures, total spleen T cells were isolated from malaria-bearing and EAE-inflicted mice after ten days of immunization. T cells derived from EAE mice were stained with CFSE according to the manufacturers instructions (2,5 μM, Sigma-Aldrich, USA) and seeded in U-bottom 96-well plates ($5 \times 10^5$ cells/well). T cells derived from naïve or P.berghei-infected mice were added afterwards at a 1:1 proportion. MOG$_{35-55}$ peptide was added to a final concentration of 20 μg/mL. The cultures were incubated for 96 h at 37°C. At the end of culture time, the cells were collected and stained with PercP-conjugated anti-CD3e antibodies and fixed in 1% paraformaldehyde prior to flow cytometer analysis. CFSE$^{low}$CD3$^+$ cells were considered proliferating responder T cells.

## Statistical analysis

Clinical score comparisons between control and experimental groups were done by Two-Way ANOVA and post-tested with Bonferroni. Other analyses among two and three (or more) groups

were carried out with Students t test and One-Way ANOVA, respectively. Results are expressed as mean ± standard error mean (SEM) and p<0,05 value were defined as significant.

## Results

### Exacerbation of Experimental Autoimmune Encephalomyelitis in mice cured from P.berghei infection

The present study aimed to evaluate whether the previous infection with P.berghei NK65 would interfere with the clinical outcome of Experimental Autoimmune Encephalomyelitis. For that purpose, C57BL/6 mice were infected with parasitized erythrocytes and at the tenth day of infection, parasites were eliminated following chloroquine (CQ) treatment. Three days after the last dose of CQ, mice were immunized with MOG$_{35-55}$ peptide for the induction of EAE (NK+CQ+EAE group). As a control, besides EAE-bearing mice (EAE group), we used naïve mice injected with non-parasitized erythrocytes and treated with chloroquine before EAE induction (CQ+EAE group). The results showed that mice from the NK+CQ+EAE group developed an aggravated disease course with higher clinical score when compared to EAE group (Figure 1A). Also noteworthy is the mild clinical scores developed by mice from CQ+EAE group. As recently published by our group, CQ reduces the severity of EAE probably by induction of regulatory T cells [17]. Thus, our results also show that P.berghei infection was able to overcome the suppressive effect of chloroquine.

### The aggravation of EAE correlates with DP-T lymphocytes reactivity to neuro-peptides

Since EAE model is highly dependent on the cellular response against neuro-antigens, the cellular immune response against MOG peptide was evaluated. Spleen cells from EAE-bearing mice were harvested ten days after neuro-peptide immunization and cultivated in the presence of MOG$_{35-55}$. The data obtained showed that NK+CQ+EAE-derived cells proliferated significantly more when compared to cells from EAE control group (Figure 1B). Among the proliferating cells, the supopulation of DP-T cells had the most active proliferation, while CQ+EAE mice presented no DP-T cells in the periphery. Cells from CQ+EAE group did not present significant proliferation or the presence of DP-T cells in the spleens (Figure 1B). The cytokine secretion was altered as well. Cell culture supernatants from NK+CQ+EAE mice presented significant higher levels of pro-inflammatory cytokines (TNF-α, IL-17 and IFN-γ) compared with the EAE group, whereas the levels of IL-10, IL-6 and IL-4 remained similar between the two groups (Figure 1C). Interestingly, cell cultures from CQ+EAE mice showed significantly higher levels of IL-10 and IL-4, and lower levels of IL-17, IL-6 and TNF-α than the other two groups (Figure 1C).

### Plasmodium berghei NK65 infection promotes thymic alterations and the premature egress of CD4$^+$CD8$^+$ double-positive lymphocytes to the periphery

The results presented so far show that the exacerbated clinical score from NK+CQ+EAE mice correlates with an enhanced cellular immune response towards neuro-antigens orchestrated, partly, by DP-T cells. The presence of DP-T cells in the spleens of NK+CQ+EAE mice is interesting. It was previously shown that during P.berghei infection, the thymus of BALB/c mice undergoes structural and phenotypical changes that together with the uncontrolled parasitemia culminates in death of the host fourteen days after infection [3]. To investigate whether the

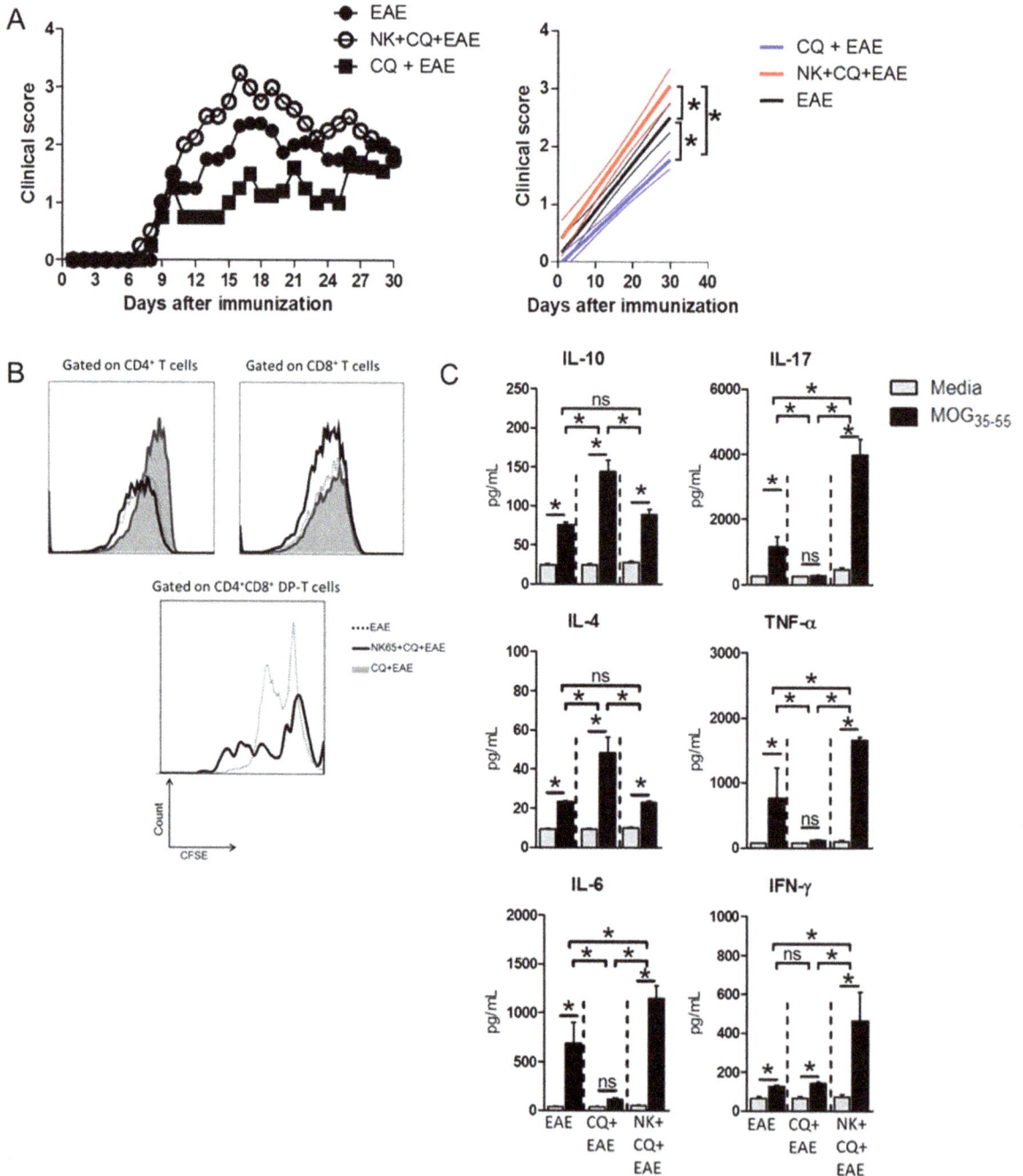

**Figure 1. Aggravation of EAE in mice cured from malaria correlates with increased cellular immune response towards myelin.**
C57BL/6 mice (n = 6 mice/group) were intraperitoneally (i.p.) infected with $1 \times 10^6$ *P.berghei*-infected Red Blood Cells and treated with chloroquine (CQ, 5 mg/Kg) for five consecutive days starting at the $10^{th}$ day after infection. Three days after the last dose of CQ, mice were immunized with 100 µg of $MOG_{35-55}$ peptide and Pertussis toxin was administrated (via i.p.) at 0 and 48h after peptide immunization for EAE induction. A) The clinical course of EAE was then monitored. Linear regression analyses are exposed in the side panels, thinner lines indicate 95% confidence interval. B) At the $10^{th}$ day after MOG-immunization, the spleens of mice were collected and dissociated. Total leukocytes ($5 \times 10^5$/well) were CFSE-stained (2,5 µM) and cultured in the presence of $MOG_{35-55}$ (10µg/mL) peptide for 96h. At the end of culture period, the cells were surface stained with anti-CD3/CD4/CD8 antibody cocktail and events were acquired in a flow cytometer. The proliferation was analyzed inside each T cell population. C) The culture supernatants were assayed for the secreted cytokines IL-10, IL-4, IL-6, IL-17, TNF-$\alpha$ and IFN-$\gamma$. Data was analyzed by One-Way Anova and post-tested with Bonferroni. In all analyses, *: $p < 0,05$; ns: not significant. Representative data of three independent experiments.

infection with *P.berghei* would promote similar alterations in the spleens and thymuses of a different mouse strain, C57BL/6 (B6) mice were injected with *Plasmodium berghei*-infected Red Blood Cells. In this strain, the infection is controlled when chloroquine

(5 mg.kg$^{-1}$) is administrated for five consecutive days starting at day 10 after infection (Figure 2A). Similarly as in the BALB/c model, the thymus of B6 mice is reduced with loss of the thymic index, starting at the third day of infection (Figure 2B).

Interestingly, the expression of the Autoimmune Regulator (AIRE) gene was found up-regulated compared with control mice as well as the expression of IL-7 and IL-6 (Figure 2C). The gene expression within the thymus of B6 mice infected with other *Plasmodium* species, such as *P.yoelli* and *P.chabaudi* was also evaluated and the results showed that the gene expression of cytokines and AIRE differs between groups (Figure S1).

It is already clear nowadays that disarranged thymic structure associated with an inflamed microenvironment result in altered maturation of thymocytes [19]. Accordingly, the thymocyte subpopulations were altered in the thymus of *P.berghei*-infected

mice, with reduced numbers of CD4$^+$CD8$^+$ (Double-positive, DP) and increased frequency of CD4$^+$ and CD8$^+$ thymocytes (Figure 2D). In the spleen, results showed an increased frequency of DP-T lymphocytes as well as CD4$^+$ and CD8$^+$ T cells. Naïve mice treated with chloroquine showed no alterations regarding the frequency of DP-T cells in the spleen and in thymic T cell subpopulations in comparison to mice without treatment (*data not shown*). Thus, it is possible to suppose that the responding DP-T cells observed in the NK+ CQ+EAE mice was elicited during *P.berghei* NK65 infection

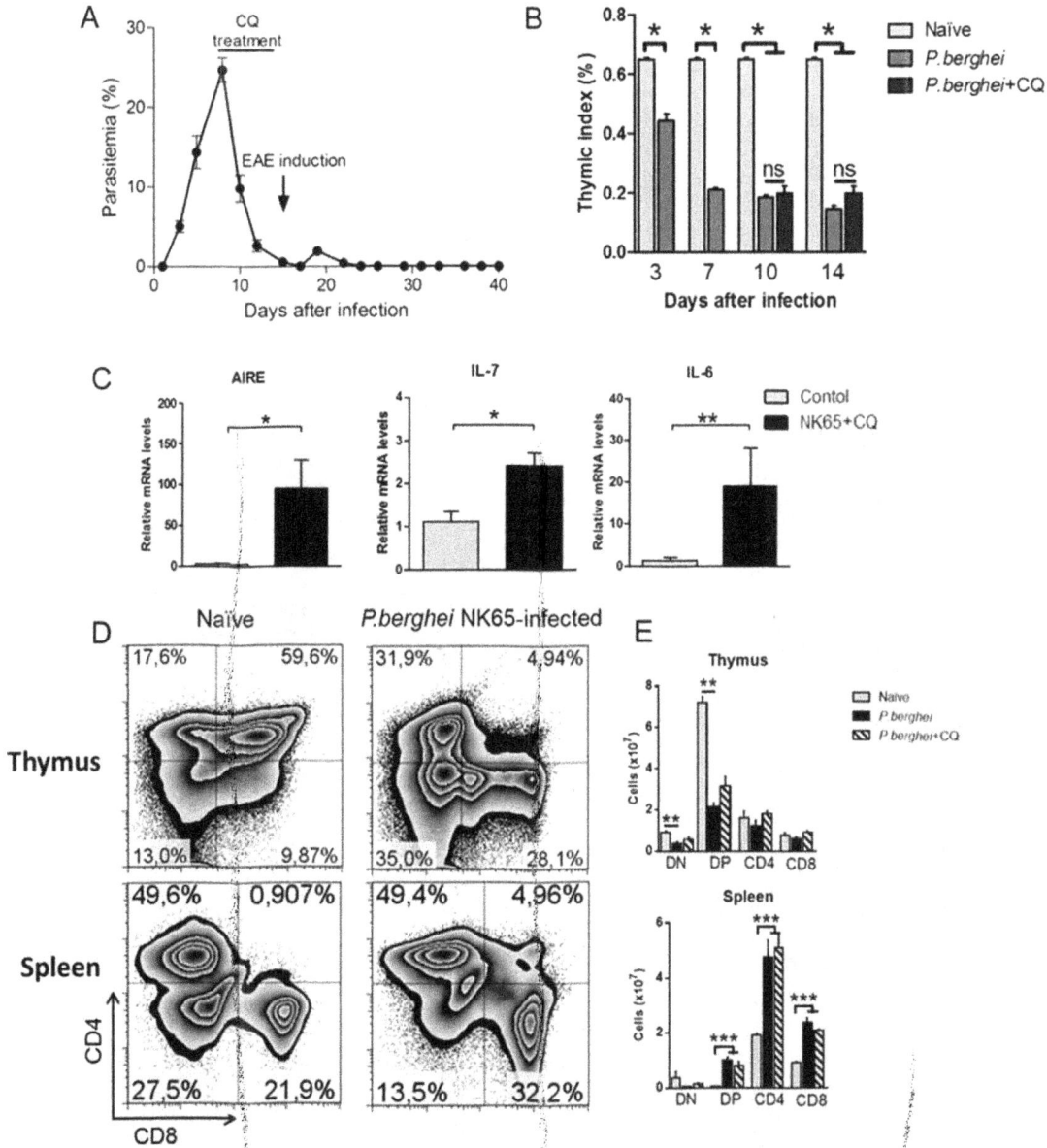

**Figure 2. *P. berghei* infection provokes thymic alterations and the premature egress of double-positive T lymphocytes in the spleens.** C57BL/6 mice (n = 6 mice/group) were intraperitoneally (i.p.) infected with 1x10$^6$ *P. berghei*-infected Red Blood Cells and treated with chloroquine (CQ, 5mg/Kg) for five consecutive days starting at the 10$^{th}$ day after infection. A) Mice treated with CQ showed decrease in the parasite burden. B) The thymuses were collected at different time-points and the relative body weight was determined. C) The gene expression of AIRE, IL-7 and IL-6 was assessed in the thymuses of *P. berghei* NK65-infected mice at the fifteenth day of infection. Flow cytometry analysis of the frequency (in D) and absolute numbers (in E) of T lymphocytes subpopulations (CD4$^+$ and CD8$^+$) in the thymuses and spleens. Data was analyzed by Student's t test. In all analyses, *: p<0,05 and **: p<0,01. Ns: not significant. #: p<0,05 in comparison with untreated mice. Representative data of four independent experiments.

and that these cells are related to the exacerbated EAE outcome.

## Infiltration of inflammatory cells in the central nervous system is characterized by the presence of DP-T cells and the local production of inflammatory cytokines

To investigate the possible association of DP-T cells with the exacerbation of EAE, the frequency of these cells in the Central Nervous System was evaluated. Fourteen days after EAE induction, mice were killed and spinal cords were prepared for histological analyses. The phenotype of the infiltrating T cells was determined by direct immunofluorescence labeling technique using FITC-conjugated anti-CD4 and PE-conjugated anti-CD8 antibodies. Results showed that in EAE group, among the infiltrating cells, a high frequency of CD4$^+$ T cells was observed, while no infiltration was observed in the CNS of CQ+EAE mice (Figure 3). As expected, an elevated frequency of CD4$^+$CD8$^+$ (DP) T cells among the infiltrating cells in the CNS of NK+CQ+EAE mice was identified (Figure 3). These data indicate that the double-positive T cells that prematurely migrated from the thymus during *P.berghei* infection are antigen-responsive (Figure 1B) and capable to migrate to the CNS during neuro-inflammation (Figure 3). In addition, the CNS of NK+CQ+EAE mice showed an increased production of IL-6, NF-κB, phosphorylated iκB (iκBα-p), iNOS, COX-2 and GFAP compared with EAE group (Figure S2).

In the CNS, an increased gene expression of IL-17 in tissue from NK+CQ+EAE mice was observed in comparison with mice from EAE group (Figure 4A). In contrast, the expression of Foxp3 and IL-10 was not significantly altered in comparison with the naïve group (Figure 4A). IFN-γ expression was not significantly altered. To investigate whether DP-T cells found in the CNS tissue were able to produce inflammatory mediators, mice were killed at the fourteenth day of EAE and the infiltrating cells were enriched from the CNS. The intracellular staining results showed that NK+CQ+EAE mice had higher frequency of IFN-γ- and IL-17-producing DP-T cells than EAE mice. (Figure 4B).

## Transfer of lymphocytes from *P.berghei*-infected mice increases the severity of EAE

In order to better characterize the relationship among malaria-elicited DP-T lymphocytes and the exacerbation of EAE, a series of experiments to demonstrate that the mechanism of disease aggravation is T-cell dependent was conducted. First, splenic T cells isolated from *P.berghei*-infected or naïve mice were co-cultured with splenic T cells from EAE-bearing mice in the presence of MOG$_{35-55}$ peptide for four days. Results showed that the cultures conducted in the presence of T cells from malaria-infected mice proliferated significantly more than cells from co-cultures with naïve T cells (Figure 5A).

Then, it was investigated whether the adoptive transfer of splenic T cells from malaria mice would alter the course of EAE. For that purpose, B6 mice were infected with *Plasmodium berghei*

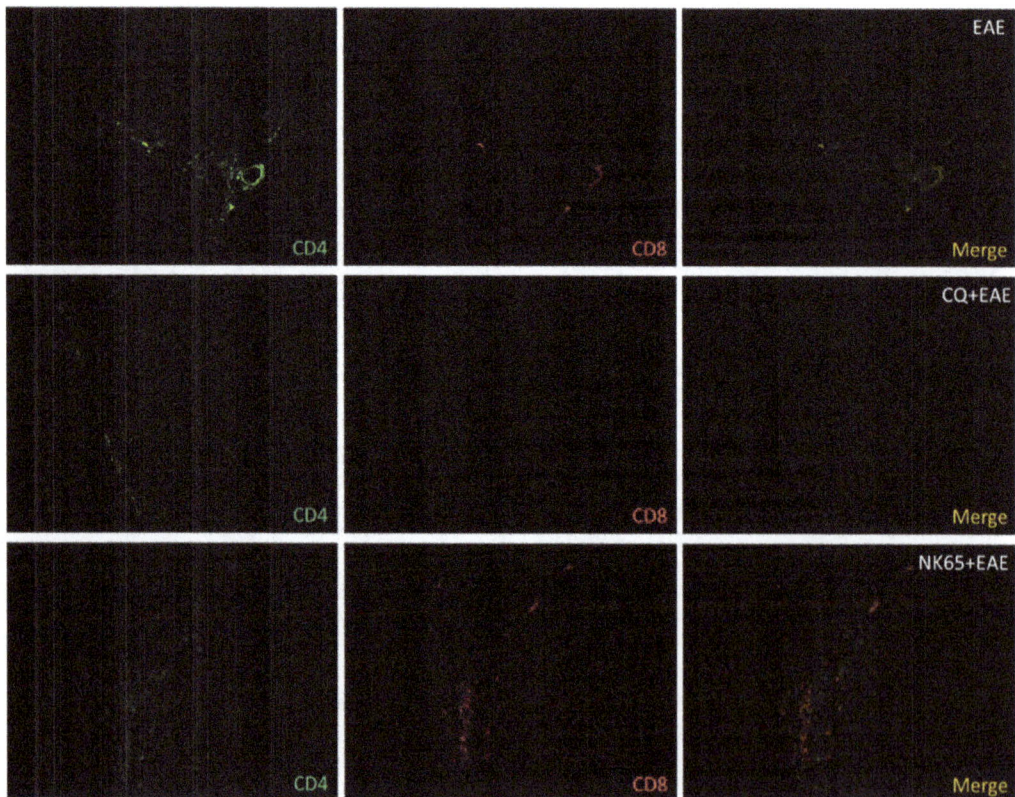

**Figure 3. Central Nervous System of malaria-cured EAE mice show increased cellular infiltration of DP-T cells.** C57BL/6 mice (n = 6 mice/group) were intraperitoneally (i.p.) infected with 1×10$^6$ *P.berghei*-infected Red Blood Cells and treated with chloroquine (CQ, 5 mg/Kg) for five consecutive days starting at the 10$^{th}$ day after infection. Three days after the last dose of CQ, EAE was induced. As controls, naïve mice were treated with CQ or vehicle before EAE induction. The spinal cords of EAE-inflicted mice were collected fourteen days after MOG-immunization. Frozen thin sections (12 μm) were made and fixed in formalin. Cells were stained with FITC-conjugated anti-CD4 and PE-conjugated anti-CD8 and analyzed in epifluorescence microscope. Figures are representative of three independent experiments. Magnification: 200X.

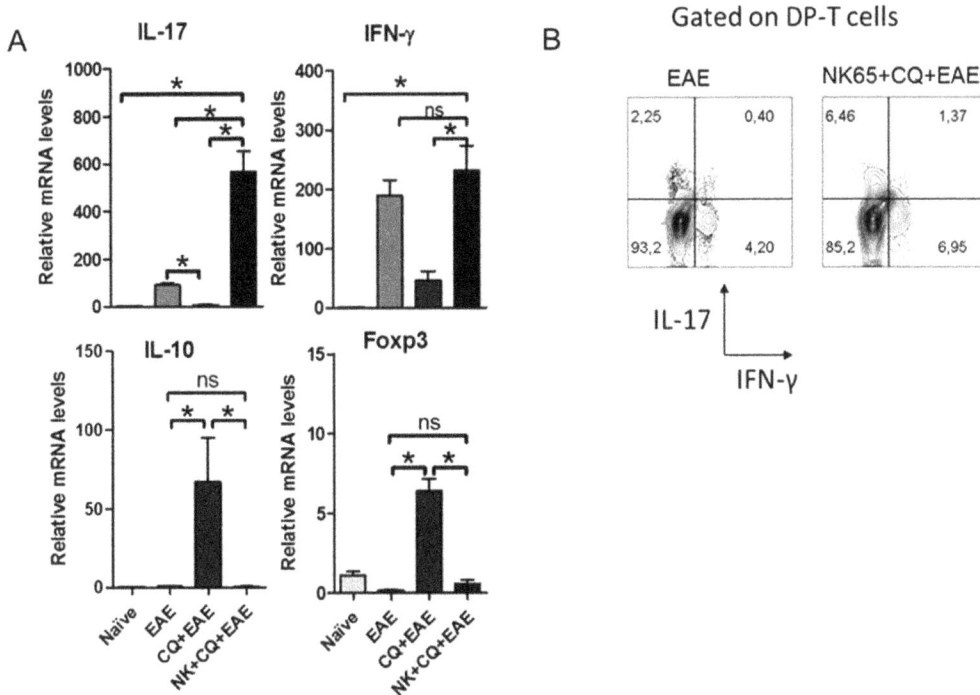

**Figure 4. Inflammation in the CNS of NK+CQ+EAE mice correlates with an increased production of inflammatory cytokines by DP-T cells.** Groups of mice (n = 6 mice/group) subjected to infection and EAE induction. A) At the 10th day after MOG-immunization, mice were killed and spinal cords were removed to analyze the gene expression of IL-17, IFN-γ, Foxp3 and IL-10 in the lumbar spinal cords of mice. Data was analyzed by One-Way Anova and post-tested with Bonferroni. B) The infiltrating cells of the CNS were enriched and stimulated by Phorbol Myristate Acetate and Ionomycin in the presence of Brefeldin A for 4 h. The frequency of IFN-γ- and IL-17-producing cells inside CD4+CD8+ T cell gate was analyzed. In all analyses, *: p<0,05. ns: not significant. Representative data of three independent experiments.

NK65 and the total splenic T cells were isolated ten days after infection. These cells were adoptively transferred to EAE mice at the beginning of the clinical signs (around the 10th day after immunization). As expected, the clinical course of EAE was significantly aggravated in malaria T cells-transferred mice compared to naïve T cell-recipient mice (Figure 5B).

Lastly, since the infiltrating T cells in the CNS of NK+CQ+ EAE mice comprised mostly of DP-T cells, the next goal was to evaluate whether these cells play a role in the aggravation of EAE. Therefore, sorted splenic CD4+, CD8+ and DP-T cells from malaria-infected mice were intravenously transferred to EAE mice as soon as the clinical signs of EAE started to appear (around the 10th day after MOG$_{35-55}$ immunization). The results showed that mice that received DP-T cells developed a more severe disease than mice that received CD4+ and CD8+ T cells (Figure 5C). In addition, we aimed to evaluate whether the cellular response elicited by DP-T cells were antigen-specific or were raised in a by-stander fashion. For that purpose, we sorted DP-T cells from the spleens of naïve and *P. berghei*-infected mice and cultivated the cells in the presence of LPS, *P. berghei* extracts and MOG peptide. At the end of the culture time, the supernatants were assayed for the presence of inflammatory cytokines. As depicted in figure 5D, cultures conducted in the presence of LPS and PbX produced elevated levels of IFN-γ and IL-17 in comparison with DP-T cells from naïve mice. Interestingly, the presence of MOG was able to stimulate the production of IL-17, but not IFN-γ, in DP-T cell cultures. This data shows that the prematurely egressed DP-T cells from *P. berghei*-infected mice are able to produce inflammatory cytokines in the presence of PAMPs (LPS), specific stimulus (PbX) and auto-antigens (MOG).

## Discussion

In this study, we show that, after infection with *Plasmodium berghei* NK 65, MOG-immunized mice develop an aggravated form of Experimental Autoimmune Encephalomyelitis. Further analysis demonstrated that the thymic prematurely egressed CD4+CD8+ (double-positive) T cells play an important role in the enhanced severity of the disease.

The thymus is a key organ for the development of T lymphocytes and plays a major role in the elimination of auto-reactive T cells [20,21]. T cell precursors enter the thymus through the high endothelial venules found in the cortico-medullary junction and lack the expression of antigen receptor and co-receptors CD4 and CD8 [22]. The maturation process consists of both migration of thymocytes through the cortex towards the medulla and expression of T cell receptor and co-receptors [23]. During this process, immature T cells interact with a myriad of cytokines, chemokines and self-peptides presented in association with MHC molecules on thymic epithelial cells (TECs) and cells that recognize antigens with high avidity are deleted [19]. We have previously shown that the lethal *Plasmodium berghei* NK65 infection promotes thymic atrophy and that this phenom-enon is dependent on thymocyte death and its premature egress to the periphery [3–5]. The lymphocytes subpopulations was altered in the thymus as well, with increased frequency of simple positive CD4+ and CD8+ T cells. However, due to the loss of thymic mass, the absolute numbers of cells were reduced. In the spleens, the augmentation in the numbers of CD4 and CD8 T cells may be explained by the clonal expansion observed in the course of malaria [24–26]. In addition, we observed an increased expression

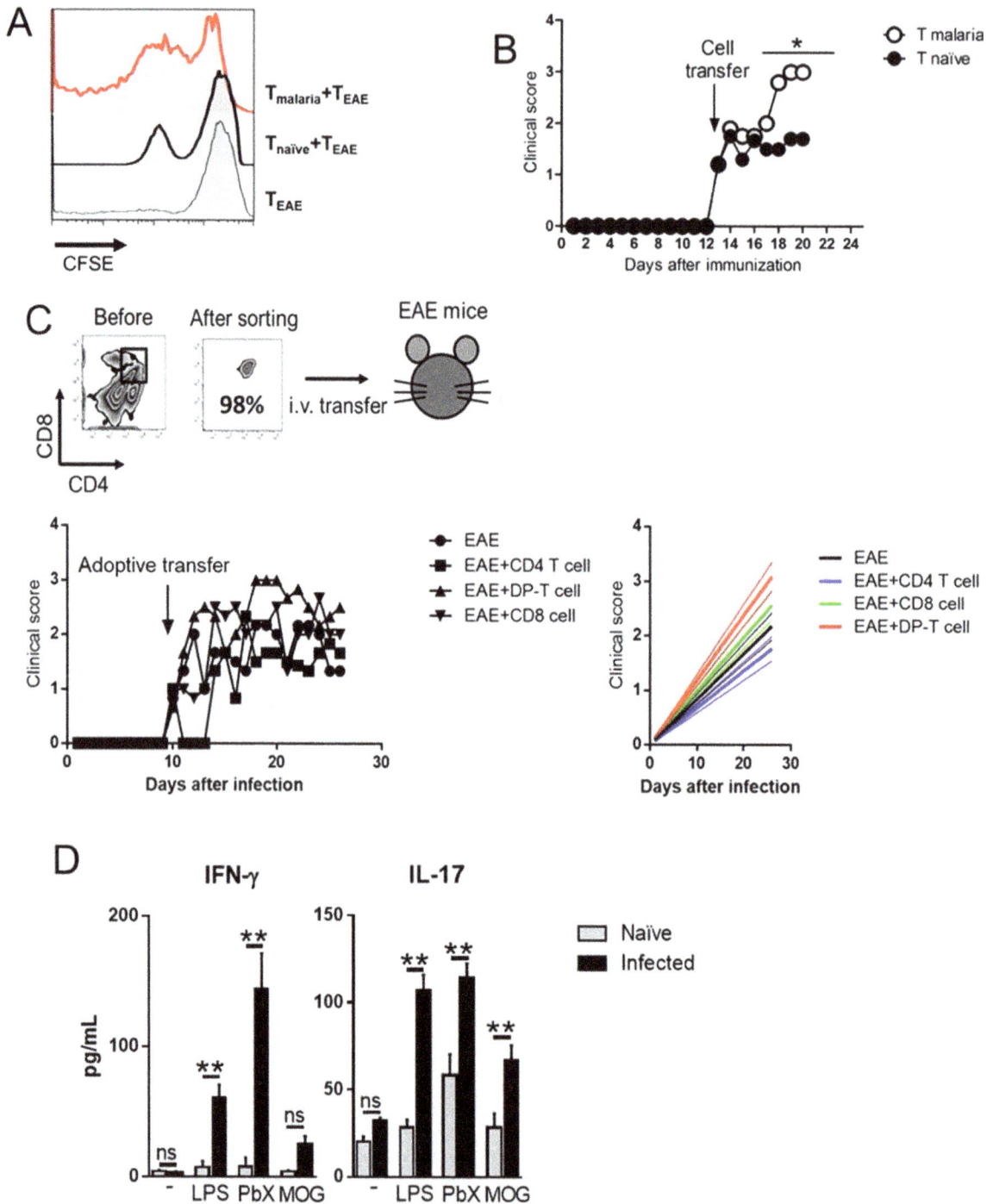

**Figure 5. Exacerbation of EAE in mice transferred with total T from naïve mouse or DP-T cells from malaria-cured mouse.** A) C57BL/6 mice (n = 6) were intraperitoneally (i.p.) infected with $1 \times 10^6$ *P.berghei*-infected Red Blood Cells. At the 10th day of infection, splenocytes were collected and total T cells were isolated using dynabeads (Pan T cell isolation kit). Naïve-derived splenic T cells were used as well. EAE-inflicted mice (n = 6) were killed at the 10th day after immunization and the total splenic T cell were isolated with the same methodology and CFSE-stained (2,5 μM). As controls, T cells from EAE-inflicted mice were cultured without the presence of other cells. B) Total T cells isolated from naïve and malaria-bearing mice were transferred ($1 \times 10^6$ cells/mouse) at the onset of EAE (n = 6 mice/group), and clinical course of the disease was evaluated daily. Data was analyzed by Two-Way ANOVA and post-tested with Bonferroni. C) Sorted splenic CD4+, CD8+ and CD4+CD8+ T cells from malaria-bearing mice were adoptively transferred ($1,5 \times 10^5$ cells/mouse) to EAE-inflicted mice (n = 6 mice/group) at disease onset. The clinical score of the disease was evaluated daily. The side panel contains the linear regression lines with the 95% confidence interval (thinner lines). D) Malaria-bearing mice were killed at 10 days after infection and the DP-T cells were sorted and incubated with LPS (1 ng/mL), *P. berghei* extracts (50 μg/mL) and MOG$_{35-55}$ peptide (20 μg/mL) for 72 h, at the end of culture period the supernatants were removed and assayed for the detection of IFN-γ and IL-17. Analyses were conducted using Student's t test. *: p<0,05. **: p<0,01. Representative data of two independent experiments with similar results.

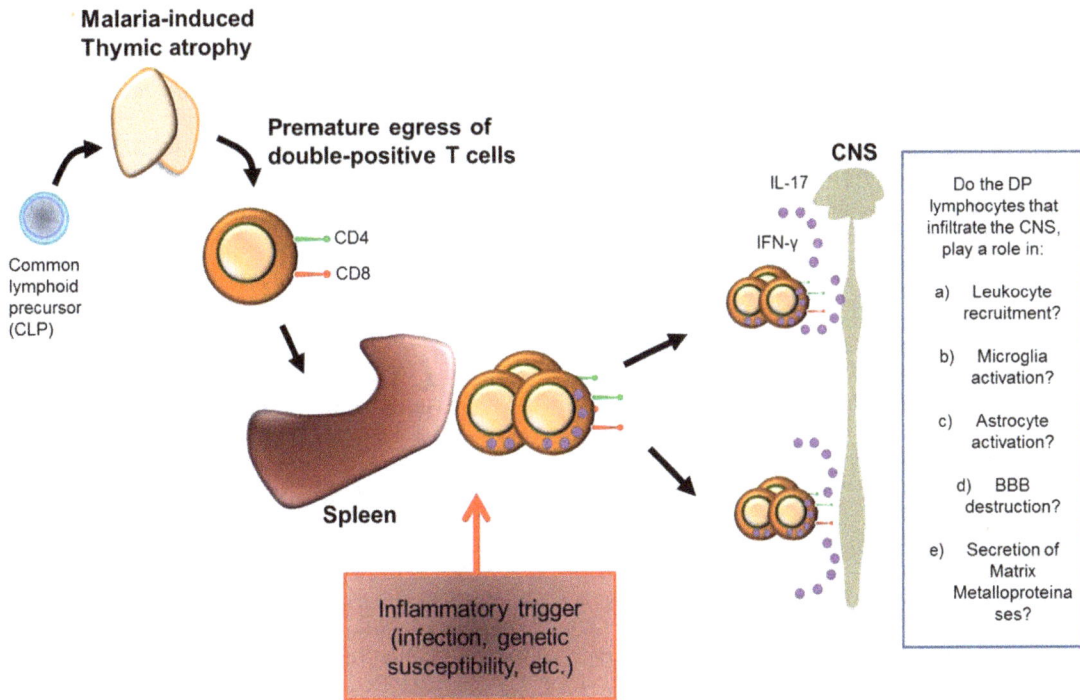

**Figure 6. Hypothesis model for the exacerbation of autoimmune neuro-inflammation.** Based in our observations, we propose a model for EAE exacerbation. Malaria infection promotes thymic atrophy and the premature egress of DP-T cells to the peripheral immune system. After an inflammatory trigger, which can be infection, genetic susceptibility or chronic inflammation, these cells proliferate and migrate to the target organ where they stimulate CNS inflammation by secreting cytokines. However, there is still much to be explored, as for example, whether these DP-T cells are able to induce leukocyte recruitment, microglia and astrocyte activation, and, Blood-Brain Barrier (BBB) destruction.

of AIRE, IL-6 and IL-7 in thymuses from *P.berghei*-infected mice, which indicate that an altered repertoire selection might be occurring. An impaired T cell maturation and selection in the thymus may lead to the release of self-reactive T cells to the periphery and the sub-sequential development of autoimmune disorders [27–29].

Indeed, we observed that mice, which were cured from *P.berghei* infection, developed a more severe form of EAE compared with uninfected animals. When we analyzed the inflammatory infiltration in the CNS, we found that most of the infiltrating cells were double-positive T cells. These DP-T cells were functional as they produced high levels of IL-17, an inflammatory cytokine related to EAE worsening [14,30,31]. As expected, when we adoptively transferred total T cells or sorted DP-T cells derived from malaria-bearing mice towards EAE-inflicted mice, the clinical score arose significantly. Taken together, these set of experiments demonstrated that the DP-T cells elicited in *P.berghei*-infected mice are inflammatory and exacerbate autoimmune neuro-inflammation. Co-cultures showed that malaria-elicited T cells stimulated the proliferation of T cells from MOG-immunized mice. Together with the proliferative response observed in DP-T cells from NK+CQ+EAE, these data suggest that DP-T cells, originated in the course of malaria, exacerbate EAE by, at least, two different mechanisms: (i) DP-T cells self-react and proliferate against neuro-antigens and (ii) DP-T cells stimulate the proliferation of encephalitogenic T cells. Of note, when cultured in the presence of different stimuli, DP-T cells in the spleens produced inflammatory cytokines IFN-γ and IL-17 at high levels. These observations indicate that DP-T cells elicited in *P. berghei* infection respond to Pathogen-Associated Molecular Patterns and specific antigens as well as to unrelated auto-antigens.

Interestingly, infection overcame the suppressive effect of chloroquine. We previously observed that CQ treatment reduces the severity of Experimental Autoimmune Encephalomyelitis through the expansion of regulatory T cells [17]. Later, our results showed that CQ is able to modulate dendritic cells *in vitro* directing them towards a tolerogenic profile that can be used to treat EAE [32]. The mechanisms by which *P. berghei* surpasses the CQ suppressive effect remains to be elucidated. Still, the recently described effect of *Plasmodium* infection in Toll-like receptors (TLR) priming may explain, at least in part, the aggravation of inflammation [33,34].

The data presented here might be concerning as malaria remains the worlds most prevalent infectious disease, with over 250 million people infected [35]. Several reports have correlated autoimmunity prevalence in countries that have eradicated the *Plasmodium* infection [36]. Although the hygiene hypothesis supports the idea of infection-induced autoimmune resistance, this theory may not fully apply to malaria. There is an increasing amount of evidence supporting or rejecting the role of malaria in autoimmune disease both in human and animal models [37–47]. Similarly, it was previously proposed that in *Trypanosoma cruzi* infection, the inflammatory profile of prematurely egressed DP-T cells from the thymus may be involved in the autoimmune process observed in murine and human Chagas' disease [10].

It was demonstrated that *Plasmodium chabaudi* infection increases the frequency of regulatory T cells in spleens, and these cells suppressed the development of EAE in a by-stander fashion [48]. However, *P. berghei* infection was shown to induce the secretion of anti-DNA and anti-nuclear antibodies in a T cell-dependent manner [49]. These observations indicate that different *Plasmodium* species induce distinct types of immune responses.

Indeed, we observed that the gene expression of inflammatory cytokines and AIRE within the thymus vary depending on the type of *Plasmodium* species. Nonetheless, infected Red Blood Cells from *P.falciparum*, *P.yoelli*, *P.berghei* and *P.vivax* induce distinct maturation status of dendritic cells (DCs) [18,50–59]. Indeed, we have shown that DCs modulated with plasmodium extracts are able to reduce the severity of EAE through the suppression of inflammatory responses in the Central Nervous System [18].

Interestingly, it was recently observed that *Plasmodium chabaudi*-infected mice presented exacerbated response to bacterial challenge [60]. In this study, the authors showed that following infection, mice were hyper-responsive to *E. coli*-derived LPS in a mechanism that is dependent on NLRP12/NLRP3 inflammasome activation. The results presented by the authors are in line with others [33,34,61]. In our model, although we did not investigate the role of inflammasomes, we found that the T cell compartment is severely altered in *P. berghei*-infected mice resulting in an exacerbation of autoimmune responses. Still, further studies must be conducted in order to ascertain the possible interplay between innate cells-derived inflammasome and T cells in the hyper-responsive inflammation elicited by *P. berghei* infection.

Based on our observations, we hypothesized that during malaria infection, thymocytes prematurely egress from the thymus towards the peripheral immune system. In the periphery, due to a trigger event (which can be genetic susceptibility, infection, or radiation, for instance) these cells proliferate and migrate to the target organ, where they produce high amounts of inflammatory cytokines leading to exacerbated inflammation (Figure 6).

Further studies must be conducted in order to evaluate how these DP-T cells are activated and whether they are antigen-specific or not. Although an increasing amount of evidences show the presence of extra-thymic DP-T cells in the control of inflammation [62–64], our data clearly shows that in malaria infection, the thymic-derived DP-T cells are inflammatory. To our knowledge, this is the first study that shows the interplay between malaria infection and autoimmune neuro-inflammation through the contribution of the thymus-derived double-positive T cells.

## Supporting Information

**Figure S1 Analysis of the gene expression in the thymus of mice infected with distinct *Plasmodium* species.** C57BL/6 mice were intraperitoneally (i.p.) infected with $1 \times 10^6$ infected Red Blood Cells. Mice were infected with *P.berghei* NK65, *P.chabaudi* AS and *P.yoelli* CL. The gene expression of AIRE, FOXP3, Rorγt, IL-17, IL-7 and IL-6 was evaluated. Results showed that the different *Plasmodium* species triggers distinct gene expression. Data was analyzed by One-Way ANOVA and post-tested with Bonferroni, where *: p<0,05. Representative data of four independent experiments.

**Figure S2 Analysis of the inflammation in the CNS of malaria-cured EAE mice.** C57BL/6 mice were intraperitoneally (i.p.) infected with $1 \times 10^6$ *P.berghei*-infected Red Blood Cells and treated with chloroquine (CQ, 5 mg/Kg) for five consecutive days starting at the 10th day after infection. Three days after the last dose of CQ, mice were immunized with 100 μg of MOG$_{35-55}$ peptide and Pertussis toxin was administrated (via i.p.) at 0 and 48 h after peptide immunization for EAE induction. The spinal cords of EAE mice were collected fourteen days after immunization. Frozen thin sections (12 μm) were made and fixed in formalin. Cells were stained with purified anti-mouse iNOS (in A), IL-6 (in B), COX-2 (in C), NF-κB (in D), phosphorylated iκBα (in E) and GFAP (in F). DAPI was added to stain DNA (blue color). The slices were analyzed in epifluorescence microscope. Figures are representative of three independent experiments. Magnification: 400X.

## Author Contributions

Conceived and designed the experiments: RT FTMC MACH ALRO LV. Performed the experiments: RT ALB LKI CR SCPL TAC RDG ITF ALFL. Analyzed the data: RT MACH FTMC LV. Contributed reagents/materials/analysis tools: LV FTMC MACH ALRO. Wrote the paper: RT LV.

## References

1. Gameiro J, Nagib P, Verinaud L (2010) The thymus microenvironment in regulating thymocyte differentiation. Cell Adh Migr 4: 382–390.
2. Savino W, Dardenne M, Velloso LA, Dayse Silva-Barbosa S (2007) The thymus is a common target in malnutrition and infection. Br J Nutr 98 Suppl 1: S11–16.
3. Andrade CF, Gameiro J, Nagib PR, Carvalho BO, Talaisys RL, et al. (2008) Thymic alterations in Plasmodium berghei-infected mice. Cell Immunol 253: 1–4.
4. Francelin C, Paulino LC, Gameiro J, Verinaud L (2011) Effects of Plasmodium berghei on thymus: high levels of apoptosis and premature egress of CD4(+)CD8(+) thymocytes in experimentally infected mice. Immunobiology 216: 1148–1154.
5. Gameiro J, Nagib PR, Andrade CF, Villa-Verde DM, Silva-Barbosa SD, et al. (2010) Changes in cell migration-related molecules expressed by thymic microenvironment during experimental Plasmodium berghei infection: consequences on thymocyte development. Immunology 129: 248–256.
6. Gilden DH (2005) Infectious causes of multiple sclerosis. Lancet Neurol 4: 195–202.
7. Herrmann I, Kellert M, Schmidt H, Mildner A, Hanisch UK, et al. (2006) Streptococcus pneumoniae Infection aggravates experimental autoimmune encephalomyelitis via Toll-like receptor 2. Infect Immun 74: 4841–4848.
8. Kawakita S, Takeuchi T, Uemura Y, Onishi T, Saito K (1976) Group A streptococcal infections as related to rheumatic fever. Jpn Heart J 17: 592–598.
9. Read SE, Fischetti VA, Utermohlen V, Falk RE, Zabriskie JB (1974) Cellular reactivity studies to streptococcal antigens. Migration inhibition studies in patients with streptococcal infections and rheumatic fever. J Clin Invest 54: 439–450.
10. Mendes-da-Cruz DA, de Meis J, Cotta-de-Almeida V, Savino W (2003) Experimental Trypanosoma cruzi infection alters the shaping of the central and peripheral T-cell repertoire. Microbes Infect 5: 825–832.
11. Mix E, Meyer-Rienecker H, Hartung HP, Zettl UK (2010) Animal models of multiple sclerosis—potentials and limitations. Prog Neurobiol 92: 386–404.
12. Dittel BN, Visintin I, Merchant RM, Janeway CA Jr (1999) Presentation of the self antigen myelin basic protein by dendritic cells leads to experimental autoimmune encephalomyelitis. J Immunol 163: 32–39.
13. Fletcher JM, Lalor SJ, Sweeney CM, Tubridy N, Mills KH (2010) T cells in multiple sclerosis and experimental autoimmune encephalomyelitis. Clin Exp Immunol 162: 1–11.
14. Chu CQ, Wittmer S, Dalton DK (2000) Failure to suppress the expansion of the activated CD4 T cell population in interferon gamma-deficient mice leads to exacerbation of experimental autoimmune encephalomyelitis. J Exp Med 192: 123–128.
15. Domingues HS, Mues M, Lassmann H, Wekerle H, Krishnamoorthy G (2010) Functional and pathogenic differences of Th1 and Th17 cells in experimental autoimmune encephalomyelitis. PLoS One 5: e15531.
16. Peron JP, Yang K, Chen ML, Brandao WN, Basso AS, et al. (2010) Oral tolerance reduces Th17 cells as well as the overall inflammation in the central nervous system of EAE mice. J Neuroimmunol 227: 10–17.
17. Thome R, Moraes AS, Bombeiro AL, Farias AD, Francelin C, et al. (2013) Chloroquine Treatment Enhances Regulatory T Cells and Reduces the Severity of Experimental Autoimmune Encephalomyelitis. PLoS One 8: e65913.
18. Thome R, Issayama LK, Alves da Costa T, Di Gangi R, Ferreira IT, et al. (2014) Dendritic cells treated with crude Plasmodium berghei extracts acquire immune-modulatory properties and suppress the development of autoimmune neuroinflammation. Immunology In press.
19. Savino W, Mendes-da-Cruz DA, Silva JS, Dardenne M, Cotta-de-Almeida V (2002) Intrathymic T-cell migration: a combinatorial interplay of extracellular matrix and chemokines? Trends Immunol 23: 305–313.
20. Asano M, Toda M, Sakaguchi N, Sakaguchi S (1996) Autoimmune disease as a consequence of developmental abnormality of a T cell subpopulation. J Exp Med 184: 387–396.
21. Starr TK, Jameson SC, Hogquist KA (2003) Positive and negative selection of T cells. Annu Rev Immunol 21: 139–176.

22. Lind EF, Prockop SE, Porritt HE, Petrie HT (2001) Mapping precursor movement through the postnatal thymus reveals specific microenvironments supporting defined stages of early lymphoid development. J Exp Med 194: 127–134.

23. Anderson G, Jenkinson EJ (2001) Lymphostromal interactions in thymic development and function. Nat Rev Immunol 1: 31–40.

24. Falanga PB, D'Imperio Lima MR, Coutinho A, Pereira da Silva L (1987) Isotypic pattern of the polyclonal B cell response during primary infection by Plasmodium chabaudi and in immune-protected mice. Eur J Immunol 17: 599–603.

25. Muxel SM, Freitas do Rosario AP, Sardinha LR, Castillo-Mendez SI, Zago CA, et al. (2010) Comparative analysis of activation phenotype, proliferation, and IFN-gamma production by spleen NK1.1(+) and NK1.1(−) T cells during Plasmodium chabaudi AS malaria. J Interferon Cytokine Res 30: 417–426.

26. Sardinha LR, D'Imperio Lima MR, Alvarez JM (2002) Influence of the polyclonal activation induced by Plasmodium chabaudi on ongoing OVA-specific B- and T-cell responses. Scand J Immunol 56: 408–416.

27. Capalbo D, Giardino G, Martino LD, Palamaro L, Romano R, et al. (2012) Genetic basis of altered central tolerance and autoimmune diseases: a lesson from AIRE mutations. Int Rev Immunol 31: 344-362.

28. Hogquist KA, Jameson SC, Heath WR, Howard JL, Bevan MJ, et al. (1994) T cell receptor antagonist peptides induce positive selection. Cell 76: 17–27.

29. Itoh M, Takahashi T, Sakaguchi N, Kuniyasu Y, Shimizu J, et al. (1999) Thymus and autoimmunity: production of CD25+CD4+ naturally anergic and suppressive T cells as a key function of the thymus in maintaining immunologic self-tolerance. J Immunol 162: 5317–5326.

30. O'Connor RA, Prendergast CT, Sabatos CA, Lau CW, Leech MD, et al. (2008) Cutting edge: Th1 cells facilitate the entry of Th17 cells to the central nervous system during experimental autoimmune encephalomyelitis. J Immunol 181: 3750–3754.

31. Willenborg DO, Fordham SA, Staykova MA, Ramshaw IA, Cowden WB (1999) IFN-gamma is critical to the control of murine autoimmune encephalomyelitis and regulates both in the periphery and in the target tissue: a possible role for nitric oxide. J Immunol 163: 5278–5286.

32. Thome R, Issayama LK, Digangi R, Bombeiro AL, da Costa TA, et al. (2014) Dendritic cells treated with chloroquine modulate experimental autoimmune encephalomyelitis. Immunol Cell Biol 92: 124–132.

33. Hartgers FC, Obeng BB, Voskamp A, Larbi IA, Amoah AS, et al. (2008) Enhanced Toll-like receptor responsiveness associated with mitogen-activated protein kinase activation in Plasmodium falciparum-infected children. Infect Immun 76: 5149–5157.

34. McCall MB, Netea MG, Hermsen CC, Jansen T, Jacobs L, et al. (2007) Plasmodium falciparum infection causes proinflammatory priming of human TLR responses. J Immunol 179: 162–171.

35. Bhaumik S (2013) Malaria funds drying up: World Malaria Report 2012. Natl Med J India 26: 62.

36. Sotgiu S, Angius A, Embry A, Rosati G, Musumeci S (2008) Hygiene hypothesis: innate immunity, malaria and multiple sclerosis. Med Hypotheses 70: 819–825.

37. Butcher G (2008) Autoimmunity and malaria. Trends Parasitol 24: 291–292.

38. Butcher GA (1996) Malaria and macrophage function in Africans: a possible link with autoimmune disease? Med Hypotheses 47: 97–100.

39. Daniel-Ribeiro CT (2000) Is there a role for autoimmunity in immune protection against malaria? Mem Inst Oswaldo Cruz 95: 199–207.

40. Daniel-Ribeiro CT, Zanini G (2000) Autoimmunity and malaria: what are they doing together? Acta Trop 76: 205–221.

41. Fawcett PT, Fawcett LB, Doughty RA, Coleman RM (1989) Suppression of malaria-induced autoimmunity by immunization with cryoglobulins. Cell Immunol 118: 192–198.

42. Jarra W (1983) Protective immunity to malaria and anti-erythrocyte autoimmunity. Ciba Found Symp 94: 137–158.

43. Loose LD, Di Luzio NR (1973) Autoimmunity in Plasmodium berghei malaria. Microbios 8: 111–115.

44. Cox HW (1964) Comments on Autoimmunity in Malaria. Am J Trop Med Hyg 13: SUPPL 225–227.

45. Lloyd CM, Collins I, Belcher AJ, Manuelpillai N, Wozencraft AO, et al. (1994) Characterization and pathological significance of monoclonal DNA-binding antibodies from mice with experimental malaria infection. Infect Immun 62: 1982–1988.

46. Phanuphak P, Tirawatnpong S, Hanvanich M, Panmuong W, Moollaor P, et al. (1983) Autoantibodies in falciparum malaria: a sequential study in 183 Thai patients. Clin Exp Immunol 53: 627–633.

47. Adu D, Williams DG, Quakyi IA, Voller A, Anim-Addo Y, et al. (1982) Anti-ssDNA and antinuclear antibodies in human malaria. Clin Exp Immunol 49: 310–316.

48. Farias AS, Talaisys RL, Blanco YC, Lopes SC, Longhini AL, et al. (2011) Regulatory T cell induction during Plasmodium chabaudi infection modifies the clinical course of experimental autoimmune encephalomyelitis. PLoS One 6: e17849.

49. Poels LG, van Niekerk CC, van der Sterren-Reti V, Jerusalem C (1980) Plasmodium berghei: T cell-dependent autoimmunity. Exp Parasitol 49: 97–105.

50. Bettiol E, Carapau D, Galan-Rodriguez C, Ocana-Morgner C, Rodriguez A (2010) Dual effect of Plasmodium-infected erythrocytes on dendritic cell maturation. Malar J 9: 64.

51. Clemente AM, Fadigati G, Caporale R, Marchese DG, Castronovo G, et al. (2013) Modulation of the immune and inflammatory responses by Plasmodium falciparum schizont extracts: role of myeloid dendritic cells in effector and regulatory functions of CD4+ lymphocytes. Infect Immun 81: 1842–1851.

52. Cockburn IA, Chen YC, Overstreet MG, Lees JR, van Rooijen N, et al. (2010) Prolonged antigen presentation is required for optimal CD8+ T cell responses against malaria liver stage parasites. PLoS Pathog 6: e1000877.

53. Giusti P, Urban B, Frascaroli G, Albrecht L, Tinti A, et al. (2011) Plasmodium falciparum-infected erythrocytes and beta-hematin induce partial maturation of human dendritic cells and increase their migratory ability in response to lymphoid chemokines. Infection and immunity 79: 2727–2763.

54. Goncalves RM, Salmazi KC, Santos BA, Bastos MS, Rocha SC, et al. (2010) CD4+ CD25+ Foxp3+ regulatory T cells, dendritic cells, and circulating cytokines in uncomplicated malaria: do different parasite species elicit similar host responses? Infect Immun 78: 4763–4772.

55. Mauduit M, See P, Peng K, Rénia L, Ginhoux F (2012) Dendritic cells and the malaria pre-erythrocytic stage. Immunological research.

56. Skorokhod OA, Alessio M, Mordmuller B, Arese P, Schwarzer E (2004) Hemozoin (malarial pigment) inhibits differentiation and maturation of human monocyte-derived dendritic cells: a peroxisome proliferator-activated receptor-gamma-mediated effect. J Immunol 173: 4066–4074.

57. Urban BC, Ferguson DJ, Pain A, Willcox N, Plebanski M, et al. (1999) Plasmodium falciparum-infected erythrocytes modulate the maturation of dendritic cells. Nature 400: 73–77.

58. Urban BC, Todryk S (2006) Malaria pigment paralyzes dendritic cells. J Biol 5: 4.

59. Wykes MN, Kay JG, Manderson A, Liu XQ, Brown DL, et al. (2011) Rodent blood-stage Plasmodium survive in dendritic cells that infect naive mice. Proc Natl Acad Sci U S A 108: 11205–11210.

60. Ataide MA, Andrade WA, Zamboni DS, Wang D, Souza Mdo C, et al. (2014) Malaria-induced NLRP12/NLRP3-dependent caspase-1 activation mediates inflammation and hypersensitivity to bacterial superinfection. PLoS Pathog 10: e1003885.

61. Franklin BS, Parroche P, Ataide MA, Lauw F, Ropert C, et al. (2009) Malaria primes the innate immune response due to interferon-gamma induced enhancement of toll-like receptor expression and function. Proc Natl Acad Sci U S A 106: 5789–5794.

62. Bang K, Lund M, Wu K, Mogensen SC, Thestrup-Pedersen K (2001) CD4+ CD8+ (thymocyte-like) T lymphocytes present in blood and skin from patients with atopic dermatitis suggest immune dysregulation. Br J Dermatol 144: 1140–1147.

63. Das G, Augustine MM, Das J, Bottomly K, Ray P, et al. (2003) An important regulatory role for CD4+CD8 alpha alpha T cells in the intestinal epithelial layer in the prevention of inflammatory bowel disease. Proc Natl Acad Sci U S A 100: 5324–5329.

64. Tutaj M, Szczepanik M (2007) Epicutaneous (EC) immunization with myelin basic protein (MBP) induces TCRalphabeta+ CD4+ CD8+ double positive suppressor cells that protect from experimental autoimmune encephalomyelitis (EAE). J Autoimmun 28: 208–215.

# 7

# Trypanosoma cruzi Evades the Protective Role of Interferon-Gamma-Signaling in Parasite-Infected Cells

**Philipp Stahl[1], Volker Ruppert[2], Ralph T. Schwarz[1,4], Thomas Meyer[3]***

1 Institut für Virologie, AG Parasitologie, Philipps-Universität Marburg, Marburg, Germany, 2 Klinik für Kardiologie, Philipps-Universität Marburg, Marburg, Germany, 3 Klinik für Psychosomatische Medizin und Psychotherapie, Georg-August-Universität Göttingen, Göttingen, Germany, 4 Unité de Glycobiologie Structurale et Fonctionnelle, UMR CNRS/USTL n° 8576, Université de Lille1 Sciences et Technologies, Villeneuve d'Ascq, France

## Abstract

The protozoan parasite *Trypanosoma cruzi* is responsible for the zoonotic Chagas disease, a chronic and systemic infection in humans and warm-blooded animals typically leading to progressive dilated cardiomyopathy and gastrointestinal manifestations. In the present study, we report that the transcription factor STAT1 (signal transducer and activator of transcription 1) reduces the susceptibility of human cells to infection with *T. cruzi*. Our *in vitro* data demonstrate that interferon -γ (IFNγ) pre-treatment causes *T. cruzi*-infected cells to enter an anti-parasitic state through the activation of the transcription factor STAT1. Whereas stimulation of STAT1-expressing cells with IFNγ significantly impaired intracellular replication of parasites, no protective effect of IFNγ was observed in STAT1-deficient U3A cells. The gene encoding indoleamine 2, 3-dioxygenase (*ido*) was identified as a STAT1-regulated target gene engaged in parasite clearance. Exposure of cells to *T. cruzi* trypomastigotes in the absence of IFNγ resulted in both sustained tyrosine and serine phosphorylation of STAT1 and its increased DNA binding. Furthermore, we found that in response to *T. cruzi* the total amount of intracellular STAT1 increased in an infectious dose-dependent manner, both at the mRNA and protein level. While STAT1 activation is a potent strategy of the host in the fight against the invading pathogen, amastigotes replicating intracellularly antagonize this pathway by specifically promoting the dephosphorylation of STAT1 serine 727, thereby partially circumventing its protective effects. These findings point to the crucial role of the IFNγ/STAT1 signal pathway in the evolutionary combat between *T. cruzi* parasites and their host.

**Editor:** Martin E. Rottenberg, Karolinska Institutet, Sweden

**Funding:** The authors gratefully acknowledge the excellent technical assistance of Anke Gregus from the University of Göttingen and Nancy Mounogou Kouassi from the University of Marburg. The research on this subject was funded by a grant from the Deutsche Forschungsgemeinschaft to TM, a scholarship from the Studienstiftung des Deutschen Volkes to PS and a grant from the Deutsche Gesellschaft für Kardiologie. The funders had no role in study design, data collection and analysis, decision to publish, or preparation of the manuscript.

**Competing Interests:** The authors have declared that no competing interests exist.

* Email: thomas.meyer@med.uni-goettingen.de

## Introduction

*Trypanosoma cruzi* is the causative agent of Chagas disease, an endemic infection in Latin America characterized in the chronic stage by dilated cardiomyopathy and megavisceral syndromes. The disease is transmitted to humans through hematophagous insect vectors called triatomines, which are members of the *Reduviidae* family and the *Triatominae* subfamily. The complex life cycle of *T. cruzi* begins when a triatomine ingests flagellated trypomastigotes during a blood meal from an infected mammalian host [1–3]. The parasite then passes through the triatomine digestive tract and undergoes a number of morphological differentiations that result in the production and multiplication of epimastigote parasites. During the next blood meal, the insect excretes highly infective metacyclic trypomastigotes via faeces, and, subsequently, the parasites can invade their new host directly through the vector's bite or by crossing mucous membranes. After host cell penetration, which coincides with the formation of a nascent parasitophorous vacuole, the parasite is released in the cytoplasm where numerous amastigotes are formed through binary fission [4]. The intracellular amastigotes then transform into trypomastigotes, which burst out of the cell, enter the bloodstream, and disseminate throughout the host. The newly infected host can then serve as a reservoir for further parasite propagation. However, non-vectorial mechanisms of infection have also been identified, such as congenital transmission, blood transfusion, organ transplantation, and recently incidental ingestion of parasite-contaminated food or drink [5].

Both innate and acquired immune responses are crucial for controlling *T. cruzi* dissemination and host survival [6]. The transmembrane Toll-like receptor (TLR) family of pattern recognition receptors (PRRs) plays an important role in the recognition of *T. cruzi* during early infection. Binding of DNA, RNA or glycosylphosphatidyl-inositol (GPI) anchors from trypomastigotes to distinct members of the TLR family initiates a signaling cascade that is dependent on the adaptor molecule myeloid differentiation factor 88 (MyD88) and culminates in the activation of pro-inflammatory genes crucial for the resistance to *T. cruzi* infection, such as interleukin-1β (IL-1β) [7], IL-6 [7,8], IL-12 [9–13], tumor necrosis factor-α (TNFα) [10,11,14], interferon-β (IFNβ) [15–18], and interferon-$\gamma$ (IFN$\gamma$) [7,12,13,16,19,20]. Mice lacking functional MyD88 are highly susceptible to *T. cruzi* infection, possibly because of defects in the production of pro-inflammatory cytokines [21]. In *T. cruzi*-infected macrophages, gene expression of pro-inflammatory cytokines is mediated by the two transcription factors *nuclear factor-κB* (NF-κB) [22–24] and *interferon-regulatory factor 3* (IRF3) [18]. NF-κB additionally activates the *inducible nitric oxide synthase* (iNOS), which catalyzes the production of microbicidal nitric oxide (NO). Mice deficient in iNOS or the IFN$\gamma$ receptor are highly susceptible to *T. cruzi* infections with increased parasite burdens and their macrophages show impaired trypanocidal activities due to a lack of NO production [25].

While the important role of MyD88- and TRIF-dependent signal pathways for the pathogenesis of Chagas disease is well established, much less is known about the contribution of STAT proteins (*signal transducer and activator of transcription*) to the control of *T. cruzi* infection [24,26–28]. The founding member of this family of cytokine-driven transcription factors, STAT1, has been reported to be an important anti-microbial mediator of host resistance to *Toxoplasma gondii*, but data on *T. cruzi* infection are controversial [27,29,30]. Upon binding of IFN$\gamma$ to its receptor, non-covalently receptor-associated *Janus kinases* (JAKs) phosphorylate the cytoplasmic tail of the IFN$\gamma$ receptor, thereby permitting recruitment of non-phosphorylated STAT1 (for a review see, e.g. [31–33]). In the next step, the activated JAKs then phosphorylate STAT1 on a signature tyrosine residue (Y701) in its carboxy-terminus. Tyrosine-phosphorylated STAT1 dimers are then translocated to the nucleus, where they bind to palindromic *gamma-activated sites* (GAS) in the promoter regions of IFN$\gamma$-responsive genes to initiate transcription. Phosphorylation of serine 727 induces nuclear export acceleration [34] which is required for full-fledged transcriptional activation [35–37].

In our study, we sought to investigate the impact of the IFN$\gamma$/STAT1 pathway on *T. cruzi* infection in human cells. Our *in vitro* data demonstrate that *T. cruzi* multiplication resulted in both infectious dose-dependent STAT1 serine 727 and tyrosine 701 phosphorylation which was associated with increased binding to GAS elements. In addition, we revealed that STAT1 signaling functions as a protective factor but is also subjected to inhibition by the parasite.

## Materials and Methods

### Cell lines, transfection, and parasite culture

Human foreskin fibroblasts (HFF, obtained from American Type Culture Collection [ATCC], Manassas, USA), adenocarcinomic alveolar basal epithelial cells (A549, ATCC), Vero E6 cells (ATCC) and STAT1-negative fibrosarcoma cells (U3A) [38] were grown in Dulbecco's modified eagle medium (DMEM; Gibco) supplemented with 10% fetal calf serum (FCS; Biochrom), 1% penicillin, and 1% streptomycin. Cells were cultivated at 37°C in a humidified 5% $CO_2$ atmosphere in cell culture flasks and passaged by trypsinization. Parasites from the CL Brener and Brazilian Y strain, a gift from Dr. M. A. Campos (Research Centre René Rachou, Fiocruz, Belo Horizonte, Brazil) and Dr. T. Jacobs (Bernhard-Nocht-Institut für Tropenmedizin, Hamburg, Germany), were grown in Vero E6 cells in DMEM supplemented with 1% fetal calf serum. Motile and infective trypomastigotes were collected from the culture supernatant, washed twice with phosphate-buffered saline (PBS) and were either used for experiments or propagation of the parasite culture. For stimulation of human cells, IFN$\gamma$ (Biomol) was used at a concentration of 5 ng/ml. The indoleamine 2,3-dioxygenase (IDO) inhibitor 1-methyltryptophan (1-MT, Sigma-Aldrich) was used at a final

concentration of 1.5 mM, and the potent iNOS inhibitor S-methylisothiourea sulfate (SMT, Fluka Analytics) was used at 1 mM. The inhibitors were added to HFF cells together with IFN$\gamma$ and, after 24 h of co-incubation, cells were washed twice with PBS. *T. cruzi* trypomastigotes were then added together with fresh culture medium at a cell-to-parasite ratio of 1:2 and removed after 24 h of incubation. Culture supernatants were removed 72 h p.i., and the fixed cells were stained for light microscopy. In a subset of experiments, the JAK inhibitor AG-490 was added at a concentration of 50 μM. For dephosphorylation assays, U3A cells were transfected with the pSTAT1-GFP vector coding for a carboxy-terminal fusion protein of full-length human STAT1 with green fluorescent protein (GFP), while for gel shift experiments U3A cells were reconstituted with recombinant untagged STAT1 using the plasmid pSTAT1 [39]. For fluorescence microscopical detection of *T. cruzi*, U3A cells were reconstituted with plasmids coding for wild-type STAT1 (pSTAT1-WT-GFP) or its tyrosine-phosphorylation-deficient point mutant Y701F (pSTAT1-Y701F-GFP). Transfection was performed using MegaTran1.0 (Origene) according to the manufacturer's recommendation. Twenty-four hours after transfection, cells were stimulated with 5 ng/ml of recombinant human IFN$\gamma$.

### Cell infection and analysis of infectivity

Confluent HFF or U3A cells were either left untreated or treated with IFN$\gamma$ (5 ng/ml) for 6, 12 or 24 h, as indicated, before the cells were infected with trypomastigotes with a multiplicity of infection (MOI) of 2 or 10. After 24 h of infection, remaining parasites in the culture medium that had not infected cells were removed by washing, and infected cells were replaced in fresh DMEM for an additional 48 h. The cells were then fixed with ice-cold methanol and stained with crystal violet and May-Grünwald (both from Merck). The mean numbers of infected cells and intracellular amastigotes per cell were blindly counted in 10 non-overlapping high power microscopic fields in three independent experiments performed in duplicate.

### Determination of IDO enzymatic activity

To measure enzymatic activity of indoleamine 2,3-dioxygenase (IDO), HFF cells grown in 96-well microtiter plates in DMEM containing 5% FCS supplemented with 0.6 mM L-tryptophan were either treated for 72 h with IFN$\gamma$ or left untreated. Then 160 μl of media were removed from each well and transferred to a 96-well V-bottomed plate. Enzymatic activity of IDO was determined spectrophotometrically by measuring the concentration of kynurenine, which directly correlates to the IDO-produced concentration of N-formyl-kynurenine [40]. After addition of 10 μL 30% trichloroacetic acid to each well, the plates were incubated at 50°C for 30 min to hydrolyze N-formyl-kynurenine to kynurenine and then centrifuged at 500 g for 10 min. One hundred μL of each culture supernatant were transferred to 96-well flat-bottomed plates and mixed with an equal volume of Ehrlich reagent (1.2% (w/v) p-dimethylaminobenzaldehyde in glacial acetic acid). After leaving for 10 min at room temperature, the extinction was determined at 492 nm using a microplate reader (BioTek). Experiments were performed in triplicates, and data are presented relative to those of unstimulated samples, set as 100.

### Measurement of nitric oxide

Nitric oxide (NO) production by untreated and IFN$\gamma$-treated confluent HFF cells was determined by measuring the nitrite concentration in the culture media. Briefly, 100 μl of medium were mixed with an equal volume of Griess reagent (1 part 0.1%

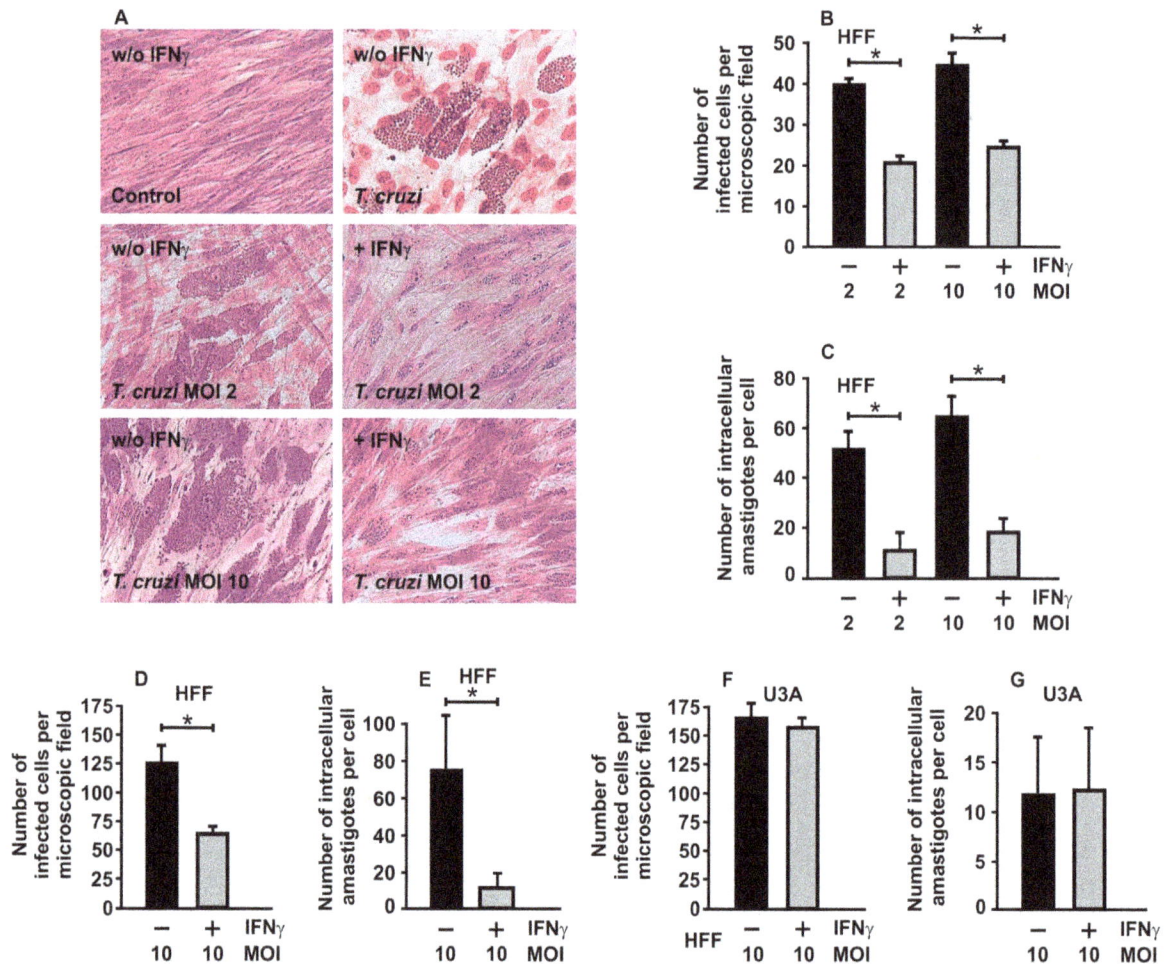

**Figure 1. STAT1 deficiency in U3A cells is associated with high susceptibility to *T. cruzi* infection and impaired cellular responses to IFNγ.** (A) Stimulation of HFF cells with IFNγ (5 ng/ml for 6 h) resulted in reduced numbers of intracellular *T. cruzi* amastigotes as determined by conventional crystal violet and May-Grünwald staining. As indicated, cells were either left untreated (w/o IFNγ) or treated with IFNγ (+IFNγ) before being infected with trypomastigotes of the Brazilian Y strain of *T. cruzi* at MOIs of 0, 2 and 10, respectively (n = 2 in duplicate). (B-E) The histograms depict the average number of infected cells per microscopic field (B,D) and the average number of intracellular amastigotes per cell (C,E) in untreated HFF cells (black columns) and cells pre-treated with IFNγ (grey columns) for 6 h (B,C) and 24 h (D,E), respectively. Fibroblasts were infected with parasites at 2 or 10 MOI before parasite burden was quantified microscopically. (F,G) No inhibitory effects of IFNγ pre-treatment on *T. cruzi*-infected U3A cells lacking STAT1 expression. Similar experiment with respect to stimulation of cells with IFNγ, incubation with parasites and measurement of parasite invasion as described in (D,E), except that U3A cells were used.

naphthylethylenediamine dihydrochloride to 1 part 1% sulfanil-amide in 5% phosphoric acid) and absorbance was measured at 540 nm after 10 minutes incubation in the dark.

## Immunocytochemistry

Adherent HFF and U3A cells grown on 8-well chamber slides were either left uninfected or incubated with *T. cruzi* parasites for 18 h. Cells were fixed with methanol at −20°C for 20 min and, after two washes in PBS, permeabilized with 1.0% Triton X-100 in PBS for 20 min. Non-specific binding was blocked by incubation with 25% FCS/PBS for 45 min at RT before the samples were incubated for 45 min with anti-STAT1 antibody C-24 (Santa Cruz) diluted 1:1000 in 25% FCS/PBS. After three washes in PBS, the specimens were incubated with Cy3-conjugated secondary antibody (Dianova), diluted 1:500 in PBS, for an additional 45 min. Nuclei of human cells and intracellular parasites were detected by staining with Hoechst 33258 dye at a final concentration of 5 µg/ml. Samples were mounted in

fluorescence mounting medium (Southern Biotech) and visualized using an Axiovert 200 M microscope (Carl Zeiss) equipped with appropriate fluorescence filters. Images were obtained with a CCD camera and further processed with the Image-Pro MDA5.1 (Media Cybernetics) software. For microscopic localization of GFP-tagged STAT1, cells were fixed with 4% paraformaldehyde in PBS and stained with Hoechst dye.

## Cell lysis

Mock-infected and *T. cruzi*-infected cells grown on 6-well dishes were lysed for 5 min on ice in 50 µl cytoplasmic extraction buffer (20 mM Hepes, pH 7.4, 10 mM KCl, 10% (v/v) glycerol, 1 mM EDTA, 0.1 mM $Na_3VO_4$, 0.1% IGEPAL-CA-360, 3 mM DTT, 0.4 mM Pefabloc, Complete Mini protease inhibitors (Roche)). The lysates were centrifuged at 16000 g (15 sec, 4°C), and supernatants spun again for 5 min at 16000 g. The supernatants resulting from this centrifugation step were collected as cytoplas-mic extracts, while the pellets were resuspended in 50 µl nuclear

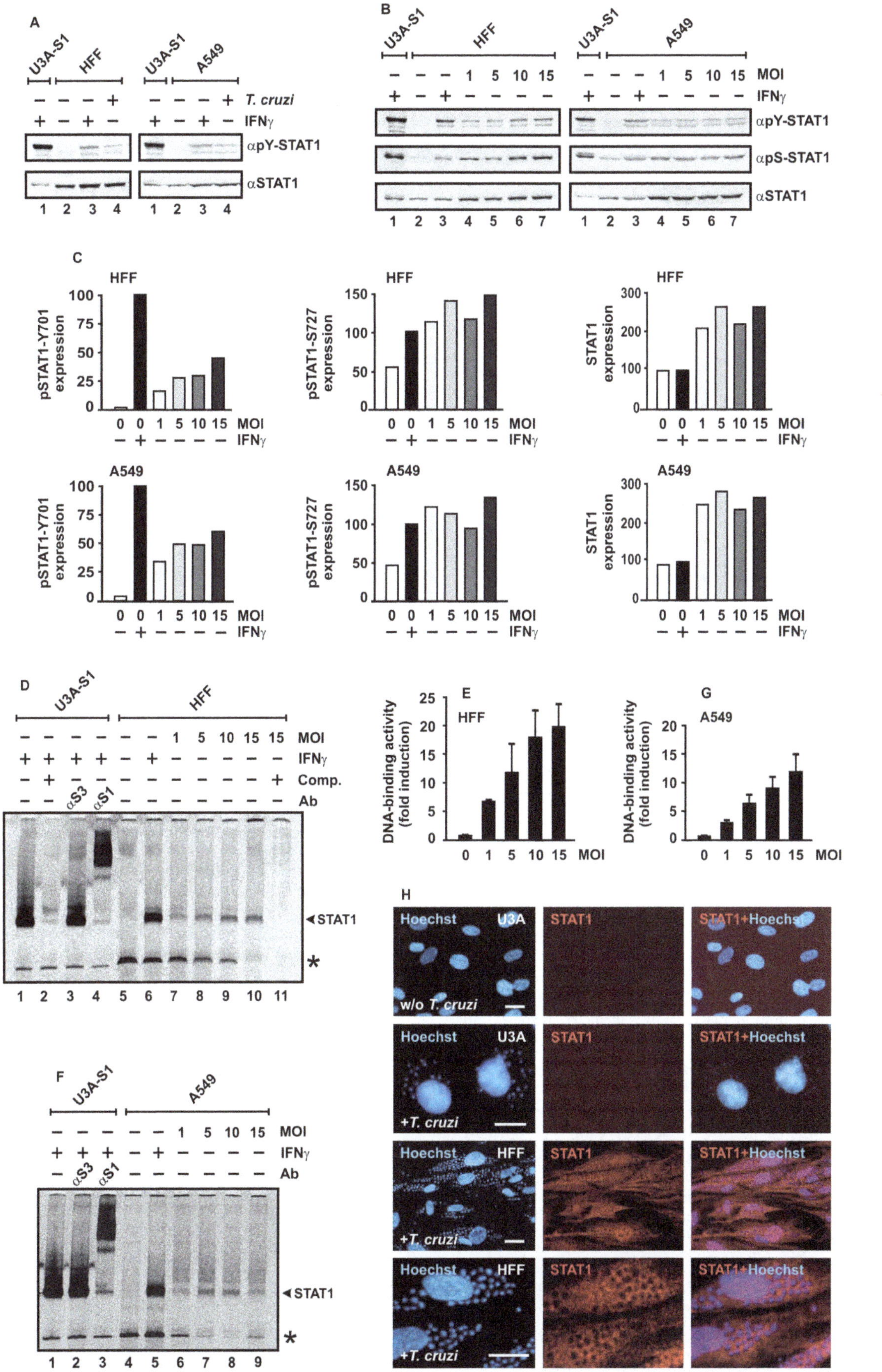
A
U3A-S1
HFF
U3A-S1
A549
T. cruzi
IFNγ
αpY-STAT1
αSTAT1
1 2 3 4 1 2 3 4
B
U3A-S1
HFF
U3A-S1
A549
MOI
IFNγ
αpY-STAT1
αpS-STAT1
αSTAT1
1 2 3 4 5 6 7 1 2 3 4 5 6 7
C
HFF
A549
pSTAT1-Y701 expression
pSTAT1-S727 expression
STAT1 expression
0 0 1 5 10 15 MOI
IFNγ
D
U3A-S1
HFF
MOI
IFNγ
Comp.
Ab
αS3
αS1
STAT1
*
1 2 3 4 5 6 7 8 9 10 11
E
HFF
DNA-binding activity (fold induction)
0 1 5 10 15 MOI
G
A549
DNA-binding activity (fold induction)
0 1 5 10 15 MOI
F
U3A-S1
A549
MOI
IFNγ
Ab
αS3
αS1
STAT1
*
1 2 3 4 5 6 7 8 9
H
Hoechst
U3A
HFF
STAT1
STAT1+Hoechst
w/o T. cruzi
+T. cruzi

**Figure 2. Parasitic infection with *T. cruzi* induces a sustained and cell type-independent activation of STAT1.** (A) Immunoblots demonstrating tyrosine phosphorylation and expression levels of STAT1 in differentially treated HFF (left) and A549 cells (right) using antibodies against tyrosine-phosphorylated STAT1 (αpY-STAT1) and, after stripping off of bound immunoreactivity from the membranes, pan-STAT1 antibody (αSTAT1). Cells were either left untreated (lane 2), stimulated for 18 h with 5 ng/ml IFNγ (lane 3) or challenged for 18 h with *T. cruzi* trypomastigotes at an MOI of 10 (lane 4). Cellular extracts from IFNγ-treated U3A cells (45 min) expressing recombinant STAT1 were used as control (lane 1) (n = 3). (B) Infectious dose-dependent increase in STAT1 tyrosine and serine phosphorylation in cytokine-unstimulated HFF (left) and A549 cells (right). Equal numbers of cells were exposed to parasites for 18 h at different MOIs. Representative immunoblotting experiments using antibodies specifically reacting with phospho-Y701- (αpY-STAT1), phospho-S727- (αpS-STAT1) and pan-STAT1 antibody (αSTAT1) are shown (n = 4). (C) Quantification of Western blot results for expression of phospho-Y701-, phospho-S727- and total STAT1 in extracts from cells infected with increasing doses of *T. cruzi*, as depicted in Figure 2B. Phosphorylation and expression levels were compared to the signal intensity in uninfected cells stimulated for 3 h with IFNγ set as 100. (D-G) *T. cruzi* infection elicits GAS-binding activity. Similar extracts as used for Western blotting (B) were incubated for 30 min with [$^{33}$P]-labeled DNA containing a single STAT binding site (M67) and then loaded onto a polyacrylamide gel for detection by EMSA. (D,F) The band corresponding to M67-bound STAT1 dimers was identified by supershift with a STAT1-, but not STAT3 antibody (Ab) and, in addition, by competition with a 750-fold molar excess of unlabeled M67-DNA (Comp.). STAT1-M67 complexes are marked with an arrowhead, asterisks indicate an unspecific band. (E,G) The histograms demonstrate the M67-binding activity in cytokine-untreated HFF cells (E) and A549 cells (G) plotted against the infection dose (MOI). (H) Immunocytochemical staining of STAT1 in *T. cruzi*-infected HFF cells using anti-STAT1 antibody C-24 and Cy3-labeled secondary antibody. The fluorescence micrographs show the intracellular distribution of endogenous STAT1 in methanol-fixed, Hoechst-stained HFF cells and, in contrast, the lack of STAT1 expression in uninfected (w/o *T. cruzi*) and infected (+*T. cruzi*) U3A cells (scale bar 20 µm). Note that there was no co-localization of cytoplasmic STAT1 and Hoechst-stained amastigotes and that some parasite-containing cells showed nuclear accumulation of STAT1.

extraction buffer (20 mM Hepes, pH 7.4, 420 mM KCl, 20% (v/v) glycerol, 1 mM EDTA, 0.1 mM Na$_3$VO$_4$, 3 mM DTT, 0.4 mM Pefabloc, and Complete Mini protease inhibitors) and left on ice for 30 min. The nuclear extracts were spun at 16000 g and 4°C for 15 min. The supernatants of this centrifugation step were mixed with cytoplasmic extracts to obtain whole cell extracts, which were used for Western blotting, electrophoretic mobility shift assays (EMSAs) and *in vitro* dephosphorylation assays.

## Western blotting

For immunoblotting experiments, confluent cells grown in 6-well tissue culture plates in DMEM containing 10% FCS were incubated with IFNγ (5 ng/ml) and/or infected with trypomastigotes at different MOIs or left untreated. Supernatants were removed after 18 h and cells washed thrice with PBS. After cell fractionation, the combined cytoplasmic and nuclear lysates were boiled in SDS sample buffer and resolved by 10% SDS-PAGE with subsequent transfer onto PVDF membranes. The membranes were incubated first with a phospho-Tyr701-specific STAT1 antibody (Cell Signaling) and then with a conjugated anti-rabbit secondary antibody (Li-Cor). To determine the amount of total STAT1, blots were stripped for 60 min at 60°C in a buffer containing 2% SDS, 0.7% β-mercaptoethanol, and 62.5 mM Tris-HCl, pH 6.8 and then successively re-probed with phospho-Ser727-antibody and the pan-STAT1 polyclonal antibody C-24 (both rabbit antibodies from Santa Cruz Biotechnology), the latter reacting with both phosphorylated and non-phosphorylated STAT1. Bound immunoreactivity was detected with secondary IRDye 800CW antibodies visualized on a Li-Cor Odyssey imaging machine.

## In vitro dephosphorylation assay

To assess whether *T. cruzi*-infected cells express phosphatase or protease activity specifically targeting endogenous STAT1, *in vitro* dephosphorylation assays using recombinant STAT1-GFP as substrate were performed. Briefly, equal amounts of cell extracts from uninfected STAT1-GFP-expressing U3A cells (20 µl) were mixed with extracts from *T. cruzi*-infected or mock-treated STAT1-negative U3A cells and incubated for either 0 min or 45 min at room temperature. Samples were boiled in SDS sample buffer and monitored for serine and tyrosine phosphorylation by means of Western blotting.

## Electrophoretic mobility shift assay

To probe for STAT1 DNA-binding activity in *T. cruzi*-infected cells, electrophoretic mobility shift assays (EMSAs) were performed. Five microliters of cellular extracts were incubated with 1 ng of [$^{33}$P]-labeled M67 duplex oligonucleotide [39]. The radioactively labeled probe, which was generated by an end-filling reaction using the Klenow fragment (New England Biolabs), contained a consensus STAT1-binding site (5'-CGA-CATTTCCCGTAAATCTG-'3; GAS sequence underlined, 4 bp overhangs at the 5' end and the respective antisense oligo are not shown). For competition experiments, cellular extracts were first incubated with [$^{33}$P]-labeled M67 in EMSA buffer for 15 min at RT, and then a 750-fold molar excess of unlabeled M67 DNA was added. In supershift assays, 20 ng of either the STAT1-specific antibody C-24 or a non-specific STAT3 antibody were present in the shift reaction. The reactions were loaded on a 4% 29:1 acrylamide:bisacrylamide gel at 4°C and separated at 400 V. DNA-binding activity of STAT1 was visualized on vacuum-dried gels with a phosphoimaging system (FLA-5100, Fuji) using the programs Aida Image Analyzer v.4.06 and TINA 2.0 (Raytest).

## Analysis of mRNA expression

Cells were incubated for 18 h with medium alone, with *T. cruzi* trypomastigotes, or IFNγ. Total RNA was extracted from the cells by using the PeqGold Total RNA kit (Peqlab), according to the manufacturer's instructions. The High Capacity cDNA Reverse Transcription Kit (Applied Biosystems) was used to convert mRNA into cDNA at 42°C for 120 min and followed by a denaturation step at 95°C for 2 min. Amplification was then performed by *p*olymerase *c*hain *r*eaction (PCR) in an I-cycler (Bio-Rad) using SsoAdvanced SYBR Green Supermix (Bio-Rad) and the primer pairs indicated in File S1. The following PCR program was applied: initial denaturation at 95°C for 30 sec followed by 40 cycles of denaturation at 95°C for 5 sec and annealing/extension at 55°C for 20 sec. A melting curve analysis was run after final amplification via a temperature gradient from 55 to 94°C in 0.5°C increment steps measuring fluorescence at each temperature for a period of 10 s. All reactions were carried out in duplicate for each sample. Using the Bio-Rad software, the threshold (C$_t$) at which the cycle numbers were measured was adjusted to areas of exponential amplification of the traces. The $\Delta\Delta$C$_t$ method was used to compare expression levels of two samples by applying the formula $2^{-(\Delta C_t \text{ target} - \Delta C_t \text{ reference sample})}$, with *gapdh* as control [41].

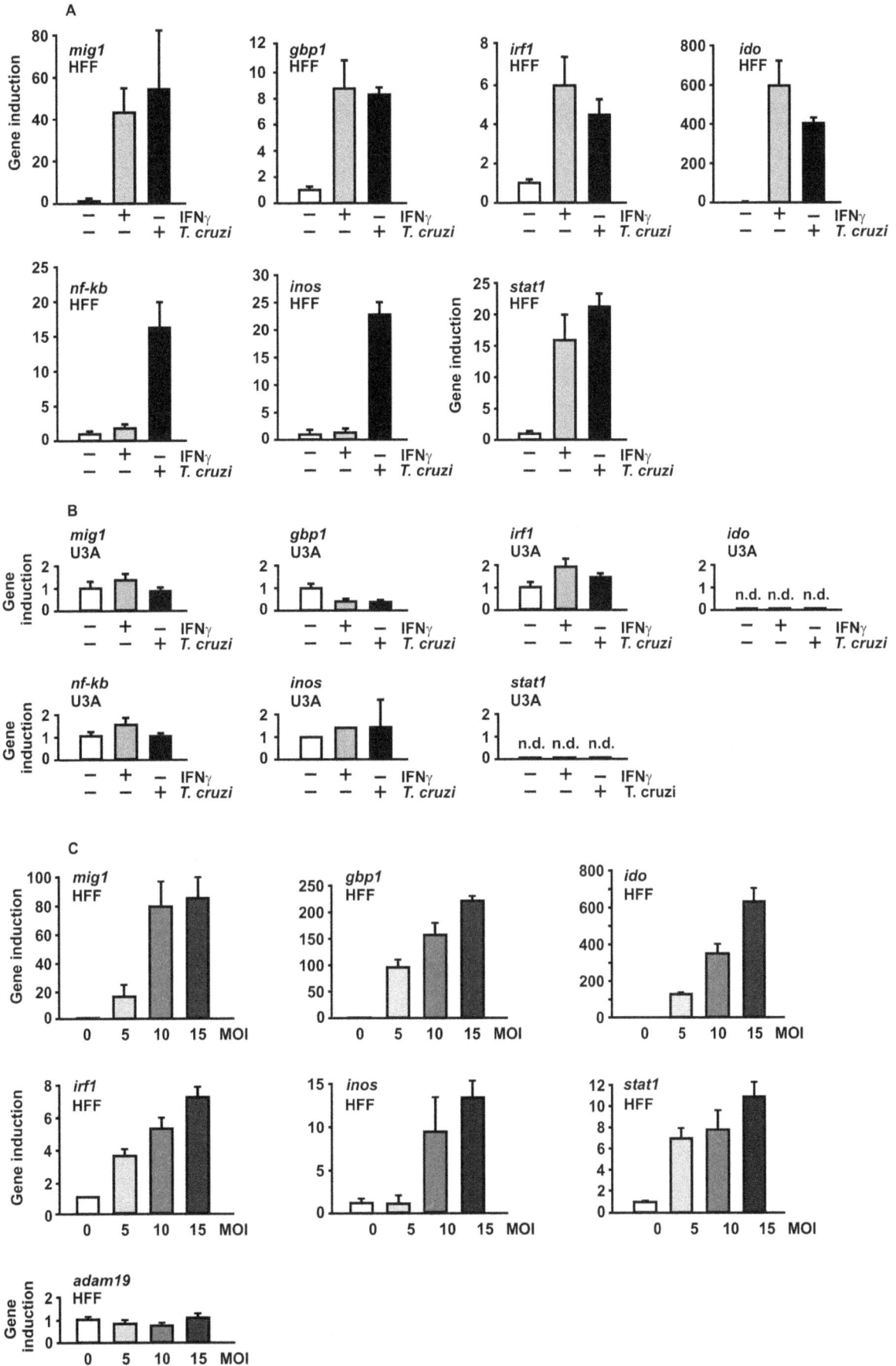

**Figure 3. Induction of STAT1-regulated genes in *T. cruzi*-infected cells.** (A) HFF cells were either left untreated, stimulated for 6 h with 5 ng/ml of IFNγ or infected for 18 h with *T. cruzi* parasites at an MOI of 10 in the absence of cytokine exposure, as determined by real-time PCR assays. Histograms depict expression levels of the *mig1*, *gbp1*, *irf1*, *ido*, *stat1*, *inos*, and *nf-kb* gene before (white columns) and after 6 h stimulation of cells with IFNγ (grey columns) as well as after parasite infection (black columns). Specific gene induction was normalized to the expression level of the housekeeping gene *gapdh*. The data are presented as means and standard deviations from at least three independent experiments. (B) U3A cells lacking STAT1 expression showed no induction of these genes under the same experimental conditions as used in (A) (n.d. = not detectable). (C) Infectious dose-dependent increase in STAT1-regulated gene expression in *T. cruzi*-infected HFF cells. The *adam19* gene was used as a negative control (n = 3).

## Statistical analyses

Means and standard deviations were calculated for each infection and stimulation mode. Differences in infection status as well as STAT1 phosphorylation level and DNA-binding activity were assessed using Student's *t* tests and Mann-Whitney-Wilcoxon tests, where appropriate. Data were analyzed using the Sigmastat (Systat Software) program. In all analyses, a p value ≤0.05 was used to indicate statistical significance.

## Results

### STAT1 is required for IFNγ-induced protection against *T. cruzi* infection

To assess the role of STAT1 transcription factor in IFNγ-dependent protection against infection by *T. cruzi*, we first performed invasion assays using trypomastigotes in HFF cells expressing endogenous STAT1 and in STAT1-negative U3A cells,

both in the absence and presence of IFNγ. The data showed that infection of HFF cells by both the pathogenic Y strain and the low virulent CL Brener strain of *T. cruzi* resulted in high invasion and multiplication indices over 72 h, as determined by microscopic examination using crystal violet and May-Grünwald stainings (Fig. 1A). We confirmed that both the percentages of infected cells and the numbers of intracellular amastigotes per cell were significantly higher in untreated HFF cells as compared to cells which had been pre-treated for 6 h with 5 ng/ml of IFNγ before exposure to parasites (Fig. 1B,C). Similar results were observed when the incubation time was extended to 12 h (Fig. S1A,B in File S1) or 24 h (Fig. 1D,E), except that the protective effect of IFNγ was more pronounced. The inhibitory effect of IFNγ on parasite replication required pre-treatment of cells with the cytokine (for 6 h) and was not observed when cells were exposed simultaneously to IFNγ and parasites (co-incubation, Fig. S1C,D in File S1). As demonstrated in Figure 1D and F, the mean number of *T. cruzi*-

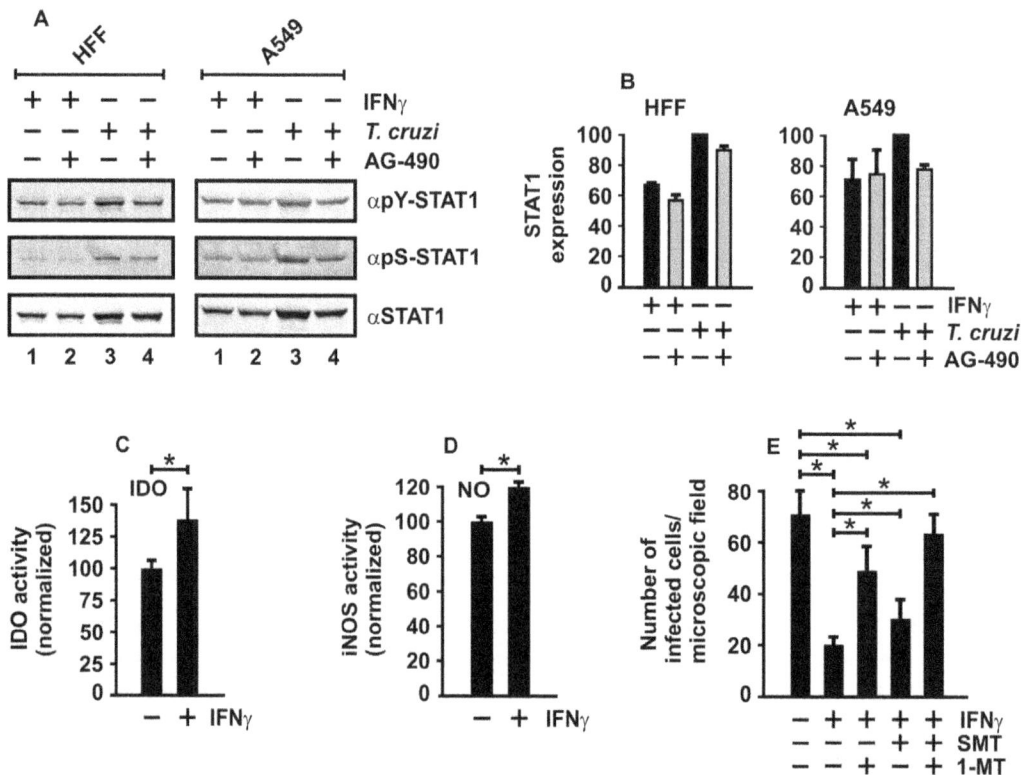

**Figure 4. STAT1-regulated expression of *ido* is involved in the control of *T. cruzi* infection.** (A,B) Exposure of cells to the JAK inhibitor AG-490 resulted in decreased STAT1 phosphorylation both at tyrosine residue 701 and serine residue 727 and is associated with a reduced intracellular STAT1 expression. Equal numbers of HFF and A549 cells were either pre-treated with 5 ng/ml of IFNγ or challenged for 18 h with parasites at an MOI of 15, in the absence or presence of AG-490 (50 µM). Representative Western blot results (A) and the corresponding quantification of STAT1 expression (B) are shown (n = 4). (C,D) Stimulation of HFF cells with IFNγ leads to increased enzymatic activity of indoleamine 2,3-dioxygenase (IDO, C) and elevated NO production (D), as measured with Ehrlich and Griess reagent, respectively (n = 3 in triplicate). (E) Inhibition of IDO by 1-methyltryptophan (1-MT, 1.5 mM) or iNOS by S-methylisothiourea sulfate (SMT, 1 mM) in IFNγ-pre-treated HFF cells resulted in significantly elevated numbers of *T. cruzi*-replicating cells (n = 4 in triplicate).

infected cells per microscopic field was higher in U3A cells than in HFF cells. Since U3A cells derived from a human fibrosarcoma typically displayed a smaller cytoplasmic volume than non-tumorigenic HFF cells, the average number of intracellular amastigotes was higher in the latter cell line. However, when U3A cells lacking endogenous STAT1 expression were substituted for HFF cells, we found that the added cytokine no longer elicited beneficial effects with respect to the resistance to *T. cruzi* infection. In particular, neither the number of infected cells (Fig. 1F) nor the number of replicating amastigotes (Fig. 1G) was significantly reduced upon pre-treatment with a high dose of IFNγ. Thus, treatment with IFNγ had no inhibitory effect on parasite growth most probably due to the STAT1 deficiency of this cell line.

To exclude the possibility that this negative finding resulted from defective IFNγ receptor activation, we reconstituted U3A cells with STAT1 and subsequently stimulated the transfected cells with 5 ng/ml of IFNγ. As expected, cytokine stimulation of STAT1-reconstituted U3A cells induced tyrosine phosphorylation of the recombinant STAT1 (Fig. 2A,B) and resulted in sequence-specific DNA binding (Fig. 2D,F), confirming that U3A cells consisted of all components of an intact IFNγ signal pathway with the exception of STAT1. Together, the infection experiments in two cultured human cell lines demonstrated that IFNγ is engaged in the protection from *T. cruzi* infection and, moreover, point to a role of STAT1 in this process.

### *T. cruzi* infection leads to phosphorylation and GAS binding of STAT1

Given the fact that IFNγ has lost its anti-microbial effects in *T. cruzi*-infected STAT1-negative U3A cells, we examined, in more detail, the vital impact of STAT1 in counteracting parasite invasion. As shown in Figure 2A, IFNγ stimulation of STAT1-reconstituted U3A cells induced tyrosine phosphorylation of the recombinant STAT1. In addition, immunoblotting experiments showed that 18-hours exposure of STAT1-expressing cells to high doses of IFNγ resulted in a sustained phosphorylation of the critical tyrosine residue 701, which is required for the formation of transcriptionally active homodimers (Fig. 2A). Tyrosine phosphorylation was observed in IFNγ-stimulated mesenchymal HFF cells as well as epithelial A549 cells. Moreover, tyrosine phosphorylation was also detected in *T. cruzi*-infected cells in the absence of any IFNγ added to the culture media. While cellular lysates from untreated HFF and A549 cells showed no detectable STAT1 tyrosine phosphorylation, there was notable phosphate incorporation at residue 701 in *T. cruzi*-infected cells, which was clearly above the detection threshold as demonstrated by Western blotting. However, in infected HFF and A549 cells, the amount of tyrosine-phosphorylated STAT1 was substantially lower as compared to STAT1-reconstituted U3A cells stimulated with IFNγ for 45 min to achieve maximal levels of tyrosine-phosphorylated STAT1. After 6 h of continuous exposure to IFNγ, the phospho-tyrosine signal had ceased in non-infected cells, whereas *T. cruzi*-infected HFF cells still exhibited a strong signal (Fig. S2 in File S1).

Next, we tested whether adding increasing parasite numbers leads to elevated levels of STAT1 tyrosine phosphorylation. For this purpose, we used cellular extracts from an equal number of cells infected with increasing parasite titers and probed for the amount of tyrosine- and serine-phosphorylated STAT1 using antibodies specifically recognizing phosphoY701 and phosphoS727, respectively. As expected, the results demonstrated that increasing MOIs were associated with elevated levels of tyrosine-phosphorylated STAT1. Furthermore, also the amount of serine-phosphorylated STAT1 increased in relation to the number of

added parasites (Fig. 2B,C). Given that in IFNγ-stimulated cells phospho-STAT1 induces the up-regulation of its own gene at a transcriptional level [42], we were not surprised that due to this positive feed-back loop also the amount of total cellular STAT1 was significantly increased in an infectious dose-dependent manner.

Although weaker than in the case with IFNγ treatment of STAT1-reconstituted U3A cells, electrophoretic mobility shift assays still demonstrated a detectable DNA-binding activity to a GAS element in lysates from *T. cruzi*-infected HFF cells (Fig. 2D,E) as well as A549 cells (Fig. 2F,G). Similar to the results from the Western blot experiments, infection with *T. cruzi* alone induced STAT1 activation as demonstrated by binding to high-affinity GAS elements and, in addition, this binding activity correlated positively with parasite burden.

To assess the intracellular distribution of STAT1 in infected cells, immunocytochemical stainings using a pan-STAT1 antibody were performed. While U3A cells used as a negative control showed no detectable immunofluorescence signals, HFF cells infected with *T. cruzi* displayed positive immunoreactivity both in the cytosol and the nucleus (Fig. 2H). In numerous cells containing amastigotes, there was evidence of a nuclear accumulation of STAT1, as defined by significantly higher immunofluorescence intensities in the nuclear as compared to the corresponding cytoplasmic compartment. In general, cytoplasmic STAT1 localization was confined to areas free of replicating parasites.

### Intracellular *T. cruzi* parasites activate IFNγ-driven STAT1 target genes

The experiments presented thus far have shown that incubation of two human cell lines of either fibroblast or epithelial origin with *T. cruzi* trypomastigotes was linked to an activation of STAT1, as shown by increased tyrosine and serine phosphorylation and binding to GAS elements. Taking into consideration that IFNγ stimulation of cells promotes STAT1-mediated gene expression, we sought to determine whether STAT1 functions as a transcription factor in the signaling pathway that is initiated by *T. cruzi* infection. To this end, we stimulated equal numbers of HFF cells for 6 h with 5 ng/ml IFNγ or infected them for 18 h with *T. cruzi* at an MOI of 10 in the absence of cytokine addition, before in cell lysates specific mRNA levels were measured by means of real-time PCR. As controls, we determined the transcript levels in HFF cells which had neither been treated with IFNγ nor infected with parasite and, in addition, used STAT1-negative U3A cells treated according to the protocol described above. In HFF cells, both IFNγ treatment and *T. cruzi* infection resulted in a robust activation of the STAT1-dependent target genes *mig1*, *gbp1*, *irf1*, *ido*, and *stat1*, whereas parasite infection alone, but not IFNγ, induced *inos* and *nf-kb* gene expression (Fig. 3A). However, no induction of these genes was observed in U3A cells, irrespective of whether they had been pre-treated with IFNγ or challenged with parasites (Fig. 3B).

To evaluate whether the induction of mRNA expression correlated with the parasite burden, we infected equal numbers of HFF cells with increasing MOIs ranging from 0 to 15 and subsequently measured transcript levels by RT-PCR. With the exception of *adam19* used as a negative control, all tested genes up-regulated by parasite infection showed a titer-dependent increase (Fig. 3C). Similar results were obtained in A549 cells (Fig. S3 in File S1). These data confirmed that STAT1 participates in the regulatory effects of IFNγ on *T. cruzi* parasitism.

Next, the action of the JAK inhibitor AG-490, also termed tyrphostin B42, on *T. cruzi*-mediated STAT1 activation was studied by measuring its effect on tyrosine and serine phosphor-

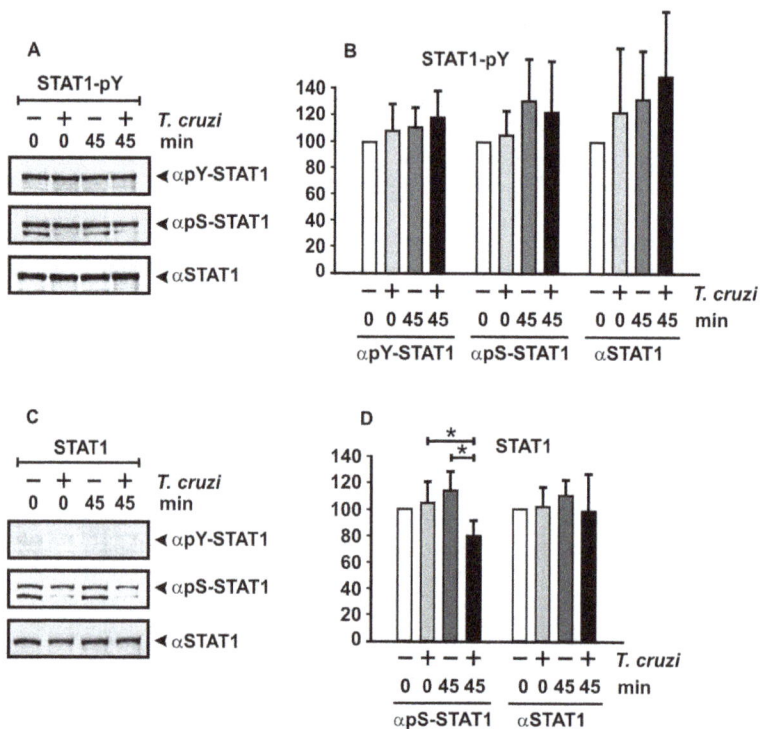

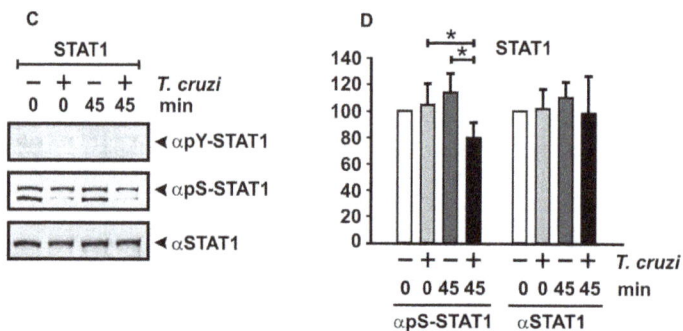

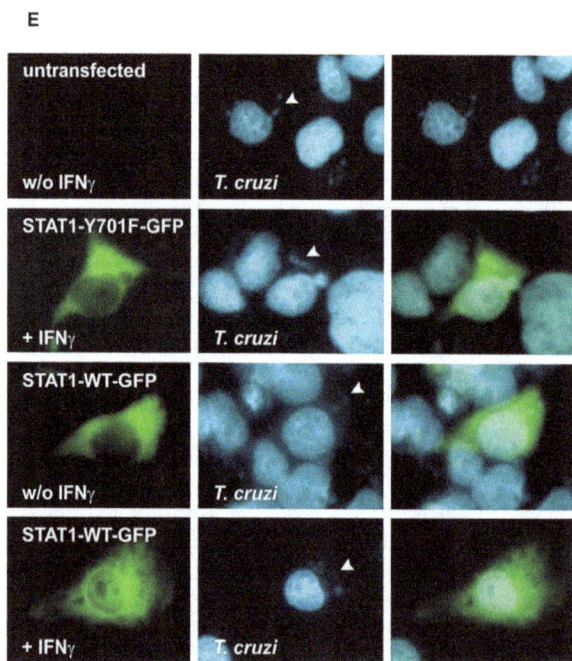

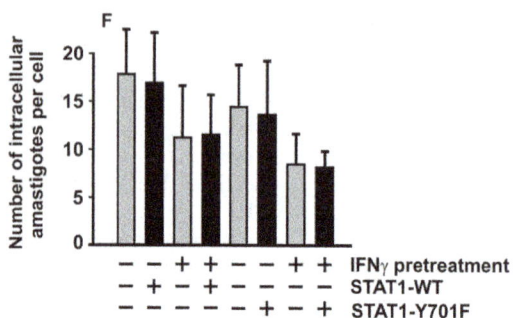

**Figure 5. STAT1 serine phosphorylation is impaired in *T. cruzi*-infected cells.** *In vitro* dephosphorylation assays were performed by incubating equal amounts of cell extracts from STAT1-GFP-expressing U3A cells (20 µl) with cell extracts from uninfected or parasite-infected STAT1-negative U3A cells. Co-incubation of the lysates lasted for 0 min and 45 min at room temperature, respectively. Before cell lysis, the U3A cells expressing the marker protein STAT1-GFP had either been treated with IFNγ (A,B) or left untreated (C,D). The reactions were then loaded on SDS polyacrylamide gels and tested for the amount of phosphorylated and total STAT1 by means of Western blotting using the antibodies indicated. Representative Western blots (A,C) and a quantification of these results (B,D) are shown. Arrowheads at the right-hand margin of the membranes mark bands corresponding to STAT1. Note that the faster-migrating band labeled with the anti-phospho-S727 antibody is an unknown non-STAT1 protein, whose expression is typically reduced in extracts from parasite-infected cells. The experiment was repeated at least four times with similar results. (E) Detection of intracellular parasites in STAT1-reconstituted U3A cells. Cells expressing a GFP fusion of wild-type STAT1 or the tyrosine-phosphorylation-deficient point mutant Y701F were treated for 6 h with 5 ng/ml IFNγ or left untreated, as indicated. Subsequently, cells were infected for 18 h with parasites at an MOI of 2, and parasitic and human nuclei were stained in fixed cells with Hoechst dye. Adjacent non-transfected cells were used as control. Arrowheads mark localization of parasites in the cytoplasm of human cells. The experiment was performed twice for each condition with similar results. (F) Quantification of results from (E) in at least n=14 cells per sample showing that STAT1-GFP expression did not prevent replication of amastigotes.

ylation. As shown in Figure 4, incubation of either HFF cells or A549 cells for 18 h with AG-490 at a concentration of 50 µM resulted in a significant reduction of both tyrosine- and serine-phosphorylated STAT1 as well as the total cellular amount of STAT1 (Fig. 4A,B). The impaired phosphorylation status was particularly observed in cells infected with parasites at an MOI of 15. This finding corroborated that pharmacological suppression of IFNγ-mediated signaling in cells challenged with parasites critically affected the activation of STAT1.

Based on our observation that infection with *T. cruzi* increased expression of the *ido* and *inos* gene titer-dependently (Fig. 3C), we next assessed the enzymatic activity of the two gene products in cytokine-stimulated HFF cells. The results showed that in cells stimulated with IFNγ for 72 h the synthesis of N-formyl-kynurenine from tryptophan catalyzed by IDO was increased as compared to untreated cells (Fig. 4C). Likewise, NO generation was enhanced as a consequence of IFNγ treatment (Fig. 4D).

In the light of the previous results, we also tested whether inhibition of either IDO or iNOS had any effect on parasite susceptibility. To this end, HFF cells were infected with *T. cruzi* trypomastigotes at an MOI of 1 for 72 h and the cells treated with the IDO blocker 1-methyltryptophan (1-MT, 1.5 mM) and the iNOS inhibitor S-methylisothiourea sulfate (SMT, 1 mM), respectively. Again, the invasion assay demonstrated that IFNγ pre-treatment of cells dramatically reduced the numbers of infected cells (Fig. 4E). However, exposure of IFNγ-treated cells to either 1-MT or SMT significantly blunted this effect, although no additive or synergistic effect was observed when the two inhibitors were added in combination to the cells. These results together suggested that the protective role of IFNγ treatment is in part executed by the action of the STAT1-regulated enzyme IDO, which is involved in parasite clearance.

## Evidence that *T. cruzi* infection directly impedes STAT1 signaling

Finally, we wondered whether the presence of intracellular *T. cruzi* counteracts the parasite-killing action of phosphorylated STAT1 by directly targeting the STAT1 molecule. Given our observation that STAT1-mediated IFNγ signaling represents a critical component in the protection against the parasite, we first assessed whether *T. cruzi*-infected cells circumvent IFNγ-stimulated anti-parasitic activity through abnormal dephosphorylation of STAT1. For this purpose, we incubated equal amounts of extracts from *T. cruzi*-infected or uninfected U3A cells with extracts from uninfected U3A cells expressing a recombinant fusion of green fluorescent protein with STAT1 (STAT1-GFP). The GFP fusion instead of untagged STAT1 was chosen as substrate, since from our previous experiments, as described above, it was known that parasite infection resulted in an increased

expression of endogenous STAT1. The GFP-tagged STAT1 was readily distinguishable from its untagged native counterpart due to the presence of a 27 kD GFP domain. Tyrosine-phosphorylated and non-tyrosine-phosphorylated STAT1-GFP was obtained separately from IFNγ-pre-treated and untreated U3A cells, respectively. The STAT1-GFP marker protein was then incubated *in vitro* for either 0 min or 45 min with cellular extracts from non-infected and infected cells. Immunoblotting results using tyrosine-phosphorylated STAT1-GFP showed that neither tyrosine 701 nor serine 727 phosphorylation was modulated when reacted with cellular extracts from *T. cruzi*-infected cells (Fig. 5A,B). Thus, we found no evidence that amastigotes replicating in the cytoplasm inhibit the establishment of an IFNγ-inducible anti-parasite state through secreting effector molecules that directly affect tyrosine-phosphorylated STAT1.

However, different results were obtained when we substituted non-phosphorylated STAT1-GFP for tyrosine-phosphorylated STAT1-GFP. By incubating non-phosphorylated STAT1-GFP with extracts from *T. cruzi*- or mock-uninfected U3A cells, we unexpectedly found significantly reduced amounts of serine-phosphorylated STAT1 in *T. cruzi*-infected cells as compared to those in uninfected cells (Fig. 5C,D). This finding suggested that a component in the lysates of parasite-infected cells had reduced exclusively the level of serine-phosphorylated, but not tyrosine-phosphorylated STAT1-GFP marker protein during the 45 min of *in vitro* incubation.

Finally, we wondered whether *T. cruzi* parasites replicate in STAT1-reconstituted U3A cells. For this purpose, we transfected STAT1-negative U3A cells with expression plasmids coding for either wild-type STAT1-GFP or its tyrosine-phosphorylation-deficient point mutant Y701F. Twenty hours after transfection, STAT1-GFP-expressing cells were treated for 6 h with IFNγ (5 ng/ml) or left untreated. Cells were then infected with parasites at an MOI of 2, and 18 h later in fixed cells intracellular parasites were stained with Hoechst dye. As shown in Fig. 5E and F, amastigotes were readily detectable in the cytoplasm of non-transfected cells as well as adjacent STAT1-GFP-expressing cells, irrespective of whether or not STAT1 was functional in IFNγ signaling.

## Discussion

Subversion of innate host immune responses, particularly those induced by the pro-inflammatory cytokine IFNγ, is increasingly recognized as a key feature that contributes to the success of obligate intracellular protozoan microorganisms which after invasion of warm-blooded animals cause a persistent life-long infection [43]. Anti-microbial effector mechanisms by which IFNγ confers resistance to intracellular parasitic pathogens are best studied in the model organism *Toxoplasma gondii* [29,43]. Among

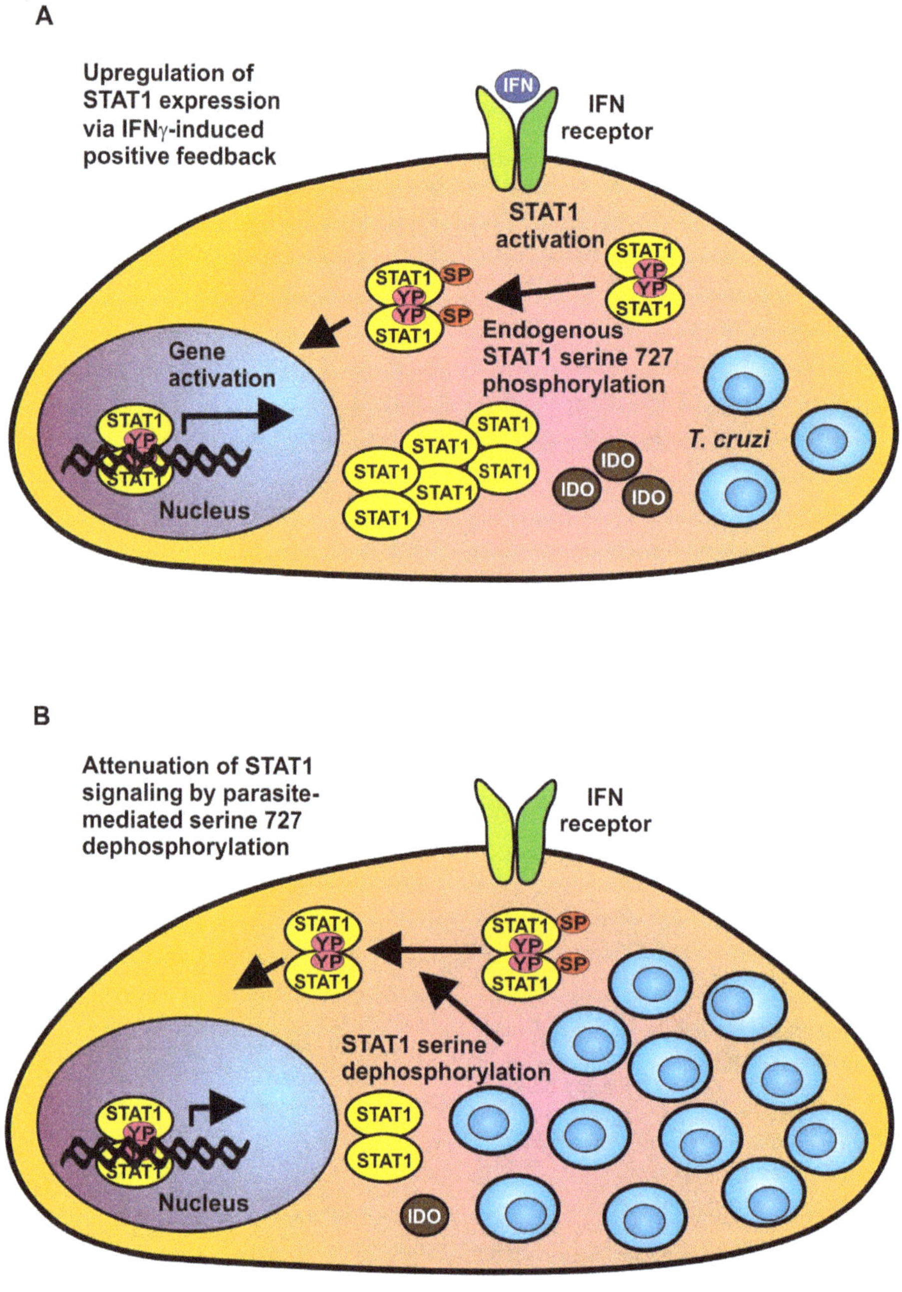

**Figure 6. STAT1-regulated suppression of *T. cruzi* replication and its inhibition by amastigotes.** The model depicts two phases in host-parasite interactions: (A) Induction of STAT1 expression by phosphorylated STAT1 via a positive feedback loop which resulted from the stimulation of cells with IFNγ. (B) Inhibition of STAT1 transcriptional activity by *T. cruzi*-mediated dephosphorylation of serine 727. Serine dephosphorylation of STAT1 leads to blunted indoleamine 2,3-dioxygenase (IDO) expression as part of a parasitic defense mechanism. The established autocrine production of interferons through a TLR-dependent pathway in the parasite-infected cells is not depicted in the figure.

other mechanisms, including killing by iNOS and IDO, the IFNγ-induced p47 GTPases have been placed at the centre of the initial defense against invading *Toxoplasma gondii* parasites, since shortly after infection they facilitate the disruption of the parasitophorous vacuole [44]. However, much less is known about the components of the IFNγ signal pathway that combat the etiologic agent of Chagas disease, and, conversely, how *T. cruzi* counteracts this attack [6,12,13,16,19,20].

In the present study, we revealed the crucial role of STAT1 in reducing the susceptibility to *T. cruzi* infection and, in addition,

identified the infection-regulated *ido* gene responsible for parasite killing to be under the transcriptional control of STAT1. Our data demonstrate that IFNγ pre-treatment causes *T. cruzi*-infected cells to enter an anti-parasitic state through the activation of the transcription factor STAT1. While the replication rate of parasites in STAT1-expressing cells was substantially reduced upon IFNγ stimulation, no such effect was observed in U3A cells deficient in STAT1 expression (Fig. 1). There are reports in the literature showing that cytokines synthesized during *T. cruzi* infection, including IFNγ, stimulate infected cells in an autocrine manner

[6,45]. In HFF cells, the duration of pre-treatment with IFN$\gamma$ is pivotal for resistance against the parasite, as measured by the numbers of infected cells and intracellular parasites. An 18-hour exposure to IFN$\gamma$ significantly increased the amount of non-phosphorylated STAT1 molecules in uninfected HFF and A549 cells (Fig. 2A), which was not surprising given the fact that tyrosine-phosphorylated STAT1 up-regulates its own gene in a positive feedback loop (Fig. 3C). Stark and colleagues reported that the newly synthesized, non-phosphorylated STAT1, which persists for days after IFN$\gamma$ priming, functions as a key transcriptional regulator for a subset of genes by mechanisms distinct from those used by tyrosine-phosphorylated STAT1 (for a review, see [42]).

The induction of non-phosphorylated STAT1 as a secondary transcription factor may be an element of a more general strategy to combat intracellular parasites by circumventing the potentially harmful action of cytokine-driven STAT1 activation. Best studied, albeit not in parasite infection, is the regulation of the IFN$\gamma$-inducible *lmp2* gene, which encodes a component of the 20S proteasome (*low molecular mass polypeptide 2*) and requires a complex consisting of non-phosphorylated STAT1 and interferon-regulatory factor 1 for promoter binding and basal transcription [46]. Thus, it will be interesting to know whether, also in *T. cruzi*-infected cells, non-phosphorylated STAT1 acts as a transcriptional regulator by interacting with other cofactors or *cis*-acting elements.

In our study we confirm that, even in the absence of any exogenous IFN$\gamma$ added to the cells, infection with the highly pathogenic Y strain of *T. cruzi* results in a robust increase in the amount of cellular STAT1 [26]. The induction of STAT1 was observed in HFF and A549 cells both at the mRNA and protein level (Fig. 3A and 2A-C) and was associated with increased binding of STAT1 homodimers to GAS elements (Fig. 2D-G). Moreover, we found that infection of HFF cells and A549 cells with the parasite leads to phosphorylation of the critical tyrosine 701 and serine 727 residue (Fig. 2B), both of which are required for maximal transcriptional response to IFN$\gamma$ [34–37]. These findings demonstrate a profound cellular response to infection with *T. cruzi*, albeit intracellular parasites could not be cleared and parasite multiplication successfully continued even in STAT1-expressing cells (Fig. 5E,F).

Our findings in HFF fibroblasts and A549 epithelial cells differ from a previous study in murine macrophages infected with *T. cruzi*, in which Bergeron and Olivier failed to detect tyrosine phosphorylation or GAS binding activity of STAT1 [27]. While these authors reported that in murine macrophages there is no evidence that the JAK2/STAT1 signaling pathway accounts for the observed *T. cruzi*-mediated NO production, we demonstrate here a significant up-regulation of *inos* mRNA in response to intracellular replicating *T. cruzi* parasites in two different cell lines, as determined by real-time PCR (Fig. 3A). However, infection with *T. cruzi* outranged the elevated *inos* mRNA synthesis as compared to IFN$\gamma$ stimulation of cells. Additionally, we found that the amount of *inos* mRNA rises with increased MOI (Fig. 3C). Similar observations were made for the *ido* and *stat1* transcripts but not for a control gene, suggesting that the transcriptional regulation of these three and the other known STAT1-regulated target genes tested (*mig1*, *gbp1*, and *irf1*) is dependent on tyrosine phosphorylation of STAT1.

Despite the activation of STAT1 and its downstream genes, we found signs of ongoing intracellular multiplication of the parasite and subsequent cell lysis. This observation suggests an evasion mechanism of *T. cruzi* from the cellular innate immune system. In our *in vitro* dephosphorylation assays using GFP-tagged STAT1 as an artificial substrate, we demonstrated that intracellular *T.*

*cruzi* amastigotes counteract the microbicidal effects of STAT1 by directly targeting serine-phosphorylated STAT1. The presence of a dephosphorylating activity in extracts from *T. cruzi*-infected cells was restricted to STAT1 molecules phosphorylated on serine 727, while there was no depletion of tyrosine-phosphorylated STAT1 triggered by the interaction between the pathogen and the host cell. These data support the hypothesis that *T. cruzi* parasites have evolved effective ways to repress STAT1 signaling by directing newly synthesized serine-phosphorylated STAT1 molecules to either dephosphorylation or proteosomal degradation. The fact that we observed no apoptotic cell death following *T. cruzi* infection suggests that the diminution of STAT1 serine phosphorylation is not a result of general cell stress or specific phagocytosis, but rather accounts for an effective strategy whereby amastigotes can subvert microbicidal activity, most probably by directly impairing the serine phosphorylation status (Fig. 6). Since it has been reported that serine residue 727 is essential for the induction of apoptosis [47,48], we hypothesize that *T. cruzi*-mediated serine dephosphorylation may prevent apoptotic cell death of infected cells and contributes to the long-term survival of the parasite.

Our observation is reminiscent of the action of *Leishmania donovani* in the proteasome-mediated degradation of STAT1 in macrophages [49,50] or of paramyxovirus-encoded proteins that degrade STAT1 by recruiting a cellular E3-ubiquitin-protein ligase for specific proteolysis [51–53]. Given that STAT1 serine dephosphorylation occurred in cell-free extracts within 45 min of incubation, it is unlikely that this inhibitory impact on the IFN$\gamma$/STAT1 signal pathway in *T. cruzi*-infected cells is a consequence of transcriptional responses such as an up-regulation of SOCS (*suppressor of cytokine signaling*) proteins.

In summary, our data demonstrate that the STAT1 signal pathway contributes to the IFN$\gamma$-induced anti-parasitic state in *T. cruzi*-infected human cells. The protective effect of STAT1 activation results in part from the induction of the *ido* gene, whose gene product catalyzes tryptophan depletion and inhibits growth of the intracellular amastigotes. However, once intracellular, the parasite keeps on replicating, despite the expression of activated STAT1 and its downstream genes. Obviously, the cellular response is insufficient to impair further parasite propagation, suggesting a parasitic evasion strategy. In line with this assumption, we show that *T. cruzi* amastigotes replicating intracellularly antagonize STAT1 signal transduction by promoting selective dephosphorylation of serine 727. Serine phosphorylation has been well established as a prerequisite for maximal transcriptional activation, and our observations underscore the notion that post-transcriptional modification of STAT1 is the target for the counter-attack by the parasites to override this protective function. These findings suggest that the control of the IFN$\gamma$/STAT1 pathway has evolved as an important issue in the fight between *Trypanosoma cruzi* and the host for supremacy.

## Supporting Information

**File S1** Contains the following files: **Figure S1**. The inhibitory effect of IFN$\gamma$ on parasite replication requires cytokine pre-treatment. (A,B) Twelve hours of pre-treatment with interferon-$\gamma$ significantly decreased parasite load, as determined by reduced numbers of both infected cells (A) and intracellular amastigotes per cell (B). HFF cells were either left untreated (-IFN$\gamma$, black columns) or treated with 5 ng/ml IFN$\gamma$ (+IFN$\gamma$, grey columns) before being infected with *T. cruzi* trypomastigotes at an MOI of 2 and 10, respectively (n = 2 in duplicate). (C,D) The inhibitory effect of IFN$\gamma$ on parasite replication requires cytokine pre-treatment of cells and is not observed when cells are exposed simultaneously to

IFNγ and parasites. Equal cell numbers of HFF cells were left untreated without IFNγ (black column), pre-treated for 6 h with IFNγ before parasite infection (dark grey columns) or simultaneously exposed to IFNγ and the indicated parasites strain (light grey columns). All cells were infected with *T. cruzi* at an MOI of 5, and 20 h post-infection numbers of infected cells (C) and intracellular amastigotes (D) per microscopic field were counted in each sample (n = 2 in duplicate). **Figure S2**. Long-time exposure of HFF cells to IFNγ resulted in up-regulation of endogenous STAT1 expression. Cells were stimulated for 0 h, 6 h or 12 h with 5 ng/ml IFNγ in the presence or absence of *T. cruzi* infection (MOI 20), as indicated. In lane 5, lysates from cells pre-treated for 4 h with IFNγ following by 8 h of simultaneous exposure to IFNγ and *T. cruzi* were loaded on the gel. A representative immunoblot result using phospho-tyrosine-specific STAT1 and pan-STAT1 antibodies is shown (n = 2). **Figure S3**. *Trypanosoma cruzi* infectious-dose-dependent increased levels of *gpb1*, *ido*, *irf1*, and *stat1* gene expression in A549 cells, as determined by real-time PCR assays. Histograms depict levels of gene activation after infection of cells with parasites at MOIs of 0, 5 and 10, respectively. *Adam19* was used as a negative control. Data are normalized to the expression level of the housekeeping gene *gapdh* and presented as means and standard deviations.

## Author Contributions

Conceived and designed the experiments: PS TM. Performed the experiments: PS VR TM. Analyzed the data: PS VR TM. Contributed reagents/materials/analysis tools: RS TM. Wrote the paper: PS TM.

## References

1. Boscardin SB, Torrecilhas AC, Manarin R, Revelli S, Rey EG, et al. (2010) Chagas' disease: an update on immune mechanisms and therapeutic strategies. J Cell Mol Med 14: 1373–1384.
2. Lescure FX, Le Loup G, Freilij H, Develoux M, Paris L, et al. (2010) Chagas disease: changes in knowledge and management. Lancet Infect Dis 10: 556–570.
3. Nagajyothi F, Machado FS, Burleigh BA, Jelicks LA, Scherer PE, et al. (2012) Mechanisms of *Trypanosoma cruzi* persistence in Chagas disease. Cell Microbiol 14: 634–643.
4. Epting CL, Coates BM, Engman DM (2010) Molecular mechanisms of host cell invasion by *Trypanosoma cruzi*. Exp Parasitol 126: 283–291.
5. Sánchez LV, Ramírez JD (2013) Congenital and oral transmission of American trypanosomiasis: an overview of physiopathogenic aspects. Parasitol 140: 147–159.
6. Machado FS, Dutra WO, Esper L, Gollob KJ, Teixeira MM, et al. (2012) Current understanding of immunity to *Trypanosoma cruzi* infection and pathogenesis of Chagas disease. Semin Immunopathol 34: 753–770.
7. Gonçalves VM, Matteucci KC, Buzzo CL, Miollo BH, Ferrante D, et al. (2013) NLRP3 controls *Trypanosoma cruzi* infection through a caspase-1-dependent IL-1R-independent NO production. PLoS Negl Trop Dis 7: e2469.
8. Truyens C, Angelo-Barrios A, Torrico F, van Damme J, Heremans H, et al. (1994) Interleukin-6 (IL-6) production in mice infected with *Trypanosoma cruzi*: effect of its paradoxical increase by anti-IL-6 monoclonal antibody treatment on infection and acute-phase and humoral immune responses. Infect Immun 62: 692–696.
9. Aliberti JC, Cardoso MA, Martins GA, Gazzinelli RT, Vieira LQ, et al. (1996) Interleukin-12 mediates resistance to *Trypanosoma cruzi* in mice and is produced by murine macrophages in response to live trypomastigotes. Infect Immun 64: 1961–1967.
10. Camargo MM, Almeida IC, Pereira ME, Ferguson MA, Travassos LR, et al. (1997) Glycosylphosphatidylinositol-anchored mucin-like glycoproteins isolated from *Trypanosoma cruzi* trypomastigotes initiate the synthesis of proinflammatory cytokines by macrophages. J Immunol 158: 5890–5901.
11. Almeida IC, Camargo MM, Procópio DO, Silva LS, Mehlert A, et al. (2000) Highly purified glycosylphosphatidylinositols from *Trypanosoma cruzi* are potent proinflammatory agents. EMBO J. 19: 1476–1485.
12. Antúnez MI, Cardoni RL (2000) IL-12 and IFNγ production, and NK cell activity, in acute and chronic experimental *Trypanosoma cruzi* infections. Immunol Lett 71: 103–109.
13. Michailowsky V, Silva NM, Rocha CD, Vieira LQ, Lannes-Vieira J, et al. (2001) Pivotal role of interleukin-12 and interferon-γ axis in controlling tissue parasitism and inflammation in the heart and central nervous system during *Trypanosoma cruzi* infection. Am J Pathol 159: 1723–1733.
14. Bastos KR, Barboza R, Sardinha L, Russo M, Alvarez JM, et al. (2007) Role of endogenous IFNγ in macrophage programming induced by IL-12 and IL-18. J Interferon Cytokine Res 27: 399–410.
15. Kierszenbaum F, Sonnenfeld G (1984) Beta-interferon inhibits cell infection by *Trypanosoma cruzi*. J Immunol 132: 905–908.
16. Plata F, Wietzerbin J, Pons FG, Falcoff E, Eisen H (1984) Synergistic protection by specific antibodies and interferon against infection by *Trypanosoma cruzi* in vitro. Eur J Immunol 14: 930–935.
17. Koga R, Hamano S, Kuwata H, Atarashi K, Ogawa M, et al. (2006) TLR-dependent induction of IFNβ mediates host defense against *Trypanosoma cruzi*. J Immunol 177: 7059–7066.
18. Chessler AD, Ferreira LR, Chang TH, Fitzgerald KA, Burleigh BA (2008) A novel IFN regulatory factor 3-dependent pathway activated by trypanosomes triggers IFNβ in macrophages and fibroblasts. J Immunol 181: 7917–7924.
19. Wirth JJ, Kierszenbaum F, Sonnenfeld G, Zlotnik A (1985) Enhancing effects of gamma interferon on phagocytic cell association with and killing of *Trypanosoma cruzi*. Infect Immun 49: 61–66.
20. Rodrigues AA, Saosa JS, da Silva GK, Martins FA, da Silva AA, et al. (2012) IFNγ plays a unique role in protection against low virulent *Trypanosoma cruzi* strain. PLoS Negl Trop Dis 6: e1598.
21. Campos MA, Closel M, Valente EP, Cardoso JE, Akira S, et al. (2004) Impaired production of proinflammatory cytokines and host resistance to acute infection with *Trypanosoma cruzi* in mice lacking functional myeloid differentiation factor 88. J Immunol 172: 1711–1718.
22. Huang H, Calderon TM, Berman JW, Braunstein VL, Weiss LM, et al. (1999) Infection of endothelial cells with *Trypanosoma cruzi* activates NF-κB and induces vascular adhesion molecule expression. Infect Immun 67: 5434–5440.
23. Hall BS, Tam W, Sen R, Pereira ME (2000) Cell-specific activation of nuclear factor-κB by the parasite *Trypanosoma cruzi* promotes resistance to intracellular infection. Mol Biol Cell 11: 153–160.
24. Hovsepian E, Penas F, Siffo S, Mirkin GA, Goren NB (2013) IL-10 inhibits the NF-κB and ERK/MAPK-mediated production of pro-inflammatory mediators by up-regulation of SOCS-3 in *Trypanosoma cruzi*-infected cardiomyocytes. PLoS One 8: e79445.
25. Hölscher C, Köhler G, Müller U, Mossmann H, Schaub GA, et al. (1998) Defective nitric oxide effector functions lead to extreme susceptibility of *Trypanosoma cruzi*-infected mice deficient in gamma interferon receptor or inducible nitric oxide synthase. Infect Immun 66: 1208–1215.
26. Vaena de Avalos S, Blader IJ, Fisher M, Boothroyd JC, Burleigh BA (2002) Immediate/early response to *Trypanosoma cruzi* infection involves minimal modulation of host cell transcription. J Biol Chem 277: 639–644.
27. Bergeron M, Olivier M (2006) *Trypanosoma cruzi*-mediated IFNγ-inducible nitric oxide output in macrophages is regulated by iNOS mRNA stability. J Immunol 177: 6271–6280.
28. Poncini CV, Giménez G, Pontillo CA, Alba-Soto CD, de Isola EL, et al. (2010) Central role of extracellular signal-regulated kinase and Toll-like receptor 4 in IL-10 production in regulatory dendritic cells induced by *Trypanosoma cruzi*. Mol Immunol 47: 1981–1988.
29. Cerávolo IP, Chaves AC, Bonjardim CA, Sibley D, Romanha AJ, et al. (1999) Replication of *Toxoplasma gondii*, but not *Trypanosoma cruzi*, is regulated in human fibroblasts activated with gamma interferon: requirement of a functional JAK/STAT pathway. Infect Immun 67: 2233–2240.
30. Stahl P, Ruppert V, Meyer T, Schmidt J, Campos MA, et al. (2013) Trypomastigotes and amastigotes of *Trypanosoma cruzi* induce apoptosis and STAT3 activation in cardiomyocytes *in vitro*. Apoptosis 18: 653–663.
31. Darnell JE Jr, Kerr IM, Stark GR (1994) Jak-STAT pathways and transcriptional activation in response to IFNs and other extracellular signaling proteins. Science 264: 1415–1421.
32. Vinkemeier U (2004) Getting the message across, STAT! Design principles of a molecular signaling circuit. J Cell Biol 167, 197–201.
33. Sehgal PB (2008) Paradigm shifts in the cell biology of STAT signaling. Semin Cell Dev Biol 19: 329–340.
34. Lödige I, Marg A, Wiesner B, Malecová B, Oelgeschläger T, et al. (2005) Nuclear export determines the cytokine sensitivity of STAT transcription factors. J Biol Chem 280: 43087–43099.
35. Wen Z, Zhong Z, Darnell JE Jr (1995) Maximal activation of transcription by Stat1 and Stat3 requires both tyrosine and serine phosphorylation. Cell 82: 241–250.
36. Kovarik P, Mangold M, Ramsauer K, Heidari H, Steinborn R, et al. (2001) Specificity of signaling by STAT1 depends on SH2 and C-terminal domains that regulate Ser727 phosphorylation, differentially affecting specific target gene expression. EMBO J 20: 91–100.
37. Varinou L, Ramsauer K, Karaghiosoff M, Kolbe T, Pfeffer K, et al. (2003) Phosphorylation of the Stat1 transactivation domain is required for full-fledged IFNγ-dependent innate immunity. Immunity 19: 793–802.
38. Müller M, Laxton C, Briscoe J, Schindler C, Improta T, et al. (1993) Complementation of a mutant cell line: central role of the 91 kDa polypeptide of

ISGF3 in the interferon-α and -γ signal transduction pathways. EMBO J 12: 4221–4228.

39. Begitt A, Meyer T, van Rossum M, Vinkemeier U (2000) Nucleocytoplasmic translocation of Stat1 is regulated by a leucine-rich export signal in the coiled-coil domain. Proc Natl Acad Sci USA 97: 10418–10423.

40. Däubener W, Wanagat N, Pilz K, Seghrouchni S, Fischer HG, et al. (1994) A new, simple, bioassay for human IFNγ. J Immunol Methods 168: 39–47.

41. Pfaffl MW (2001) A new mathematical model for relative quantification in real-time RT-PCR. Nucleic Acids Res 29: e45.

42. Cheon H, Yang J, Stark GR (2011) The functions of signal transducers and activators of transcriptions 1 and 3 as cytokine-inducible proteins. J Interferon Cytokine Res 31: 33–40.

43. Schneider AG, Abi Abdallah DS, Butcher BA, Denkers EY (2013) *Toxoplasma gondii* triggers phosphorylation and nuclear translocation of dendritic cell STAT1 while simultaneously blocking IFNγ-induced STAT1 transcriptional activity. PLoS One 8: e60215.

44. Martens S, Parvanova I, Zerrahn J, Griffiths G, Schell G, et al. (2005) Disruption of *Toxoplasma gondii* parasitophorous vacuoles by the mouse p47-resistance GTPases. PLoS Pathog 1: e24.

45. Machado FS, Martins GA, Aliberti JC, Mestriner FL, Cunha FQ, et al. (2000) *Trypanosoma cruzi*-infected cardiomyocytes produce chemokines and cytokines that trigger potent nitric oxide-dependent trypanocidal activity. Circulation 102: 3003–3008.

46. Chatterjee-Kishore M, Wright KL, Ting JP, Stark GR (2000) How Stat1 mediates constitutive gene expression: a complex of non-phosphorylated Stat1 and IRF1 supports transcription of the LMP2 gene. EMBO J 19, 4111–4122.

47. Kumar A, Commane M, Flickinger TW, Horvath CM, Stark GR (1997) Defective TNF-α-induced apoptosis in STAT1-null cells due to low constitutive levels of caspases. Science 278: 1630–1632.

48. Stephanou A, Scarabelli TM, Brar BK, Nakanishi Y, Matsumura M, et al. (2001) Induction of apoptosis and Fas receptor/Fas ligand expression by ischemia/reperfusion in cardiac myocytes requires serine 727 of the STAT-1 transcription factor but not tyrosine 701. J Biol Chem 276: 28340–28347.

49. Forget G, Gregory DJ, Olivier M (2005) Proteasome-mediated degradation of STAT1α following infection of macrophages with *Leishmania donovani*. J Biol Chem 280: 30542–30549.

50. Olivier M, Gregory DJ, Forget G (2005) Subversion mechanisms by which Leishmania parasites can escape the host immune response: a signaling point of view. Clin Microbiol Rev 18: 293–305.

51. Andrejeva J, Young DF, Goodbourn S, Randall RE (2002) Degradation of STAT1 and STAT2 by the V proteins of simian virus 5 and human parainfluenza virus type 2, respectively: consequences for virus replication in the presence of alpha/beta and gamma interferons. J Virol 76: 2159–2167.

52. Parisien JP, Lau JF, Rodriguez JJ, Ulane CM, Horvath CM (2002) Selective STAT protein degradation induced by paramyxoviruses requires both STAT1 and STAT2 but is independent of α/β interferon signal transduction. J Virol 76, 4190–4198.

53. Ramachandran A, Horvath CM (2009) Paramyxovirus disruption of interferon signal transduction: STATus report. J Interferon Cytokine Res 29: 531–537.

# 8

# Spatial Variation in Genetic Diversity and Natural Selection on the Thrombospondin-Related Adhesive Protein Locus of *Plasmodium vivax* (*PvTRAP*)

**Rattiporn Kosuwin[1], Chaturong Putaporntip[1], Hiroshi Tachibana[2], Somchai Jongwutiwes[1]***

1 Molecular Biology of Malaria and Opportunistic Parasites Research Unit, Department of Parasitology, Faculty of Medicine, Chulalongkorn University, Bangkok, Thailand,
2 Department of Infectious Diseases, Tokai University School of Medicine, Kanagawa, Japan

## Abstract

Thrombospondin-related adhesive protein (TRAP) of malaria parasites is essential for sporozoite motility and invasions into mosquito's salivary gland and vertebrate's hepatocyte; thereby, it is a promising target for pre-erythrocytic vaccine. TRAP of *Plasmodium vivax* (PvTRAP) exhibits sequence heterogeneity among isolates, an issue relevant to vaccine development. To gain insights into variation in the complete *PvTRAP* sequences of parasites in Thailand, 114 vivax malaria patients were recruited in 2006–2007 from 4 major endemic provinces bordering Myanmar (Tak in the northwest, n = 30 and Prachuap Khirikhan in the southwest, n = 25), Cambodia (Chanthaburi in the east, n = 29) and Malaysia (Yala and Narathiwat in the south, n = 30). In total, 26 amino acid substitutions were detected and 9 of which were novel, resulting in 44 distinct haplotypes. Haplotype and nucleotide diversities were lowest in southern *P. vivax* population while higher levels of diversities were observed in other populations. Evidences of positive selection on *PvTRAP* were demonstrated in domains II and IV and purifying selection in domains I, II and VI. Genetic differentiation was significant between each population except that between populations bordering Myanmar where transmigration was common. Regression analysis of pairwise linearized *Fst* and geographic distance suggests that *P. vivax* populations in Thailand have been isolated by distance. Sequence diversity of *PvTRAP* seems to be temporally stable over one decade in Tak province based on comparison of isolates collected in 1996 (n = 36) and 2006–2007. Besides natural selection, evidences of intragenic recombination have been supported in this study that could maintain and further generate diversity in this locus. It remains to be investigated whether amino acid substitutions in PvTRAP could influence host immune responses although several predicted variant T cell epitopes drastically altered the epitope scores. Knowledge on geographic diversity in PvTRAP constitutes an important basis for vaccine design provided that vaccination largely confers variant-specific immunity.

**Editor:** Georges Snounou, Université Pierre et Marie Curie, France

**Funding:** Ratchadapiseksompotch Fund, Faculty of Medicine, Chulalongkorn University (Grant No. RA56/059) to SJ. Financial support from The Thailand Research Fund through The Royal Golden Jubilee Ph.D. Program (Grant PHD/0278/2552) to RK and SJ. The funders had no role in study design, data collection and analysis, decision to publish, or preparation of the manuscript.

**Competing Interests:** The authors have declared that no competing interests exist.

* Email: jongwutiwes@gmail.com

## Introduction

In low- and middle-income countries in tropical areas, malaria remains one of the leading ten causes of morbidity and mortality, resulting in an estimated economic loss of nearly 40 million disability-adjusted life-years (DALYs) [1]. Although *Plasmodium falciparum* is the most pernicious and prevalent species, the significance of *P. vivax* should not be underappreciated because it can cause chronic relapsing illness due to reactivation of hypnozoites and it can potentially lead to severe complications similar to those caused by *P. falciparum* [2]. Furthermore, vivax malaria occupies a broader geographical range than falciparum malaria whilst the emergences of chloroquine- and primaquine-resistant *P. vivax* are of particular concern if they would be widespreading as chloroquine-resistant *P. falciparum* [3]. The progress of malaria control and elimination has been hampered by not only the emergence of drug resistant parasites but also insecticide resistant mosquito vectors; thereby, development of a malaria vaccine offers an important preventive measure [4]. Because *P. vivax* and *P. falciparum* co-circulate in several endemic areas outside Africa where co-infections of both species are not uncommon [5], effective malaria control requires vaccines against both species.

To date, malarial circumsporozoite (CS) protein is a prime candidate for pre-erythrocytic vaccine development [4]. Although CSP-derived immunogens could elicit immunity against sporozoites, the subunit vaccines derived from this molecule such as RTS, S recombinant vaccine against *P. falciparum* has resulted in limited clinical efficacy in field studies [6]. Because vaccines derived from irradiation-attenuated or live sporozoites consistently outperform vaccines incorporating single sporozoite proteins, a more effective pre-erythrocytic stage vaccine may require combination of multiple protective immunogens [7,8]. In a murine model, co-immunization of CSP with thrombospondin-related adhesive protein (TRAP), a protein mobilized from microneme to the surface of sporozoite [9], has conferred complete protection against parasite challenge whereas vaccination using each of these immunogens could elicit only partial protection [10]. It is important to note that TRAP-specific CD8+ T lymphocytes are prime mediators for protection against sporozoite challenge in mouse vaccination trials, resulting in significant reduction in liver stage parasites [11]. Furthermore, seroepidemiological study has shown that anti-TRAP antibodies were negatively correlated with parasite density among infected individuals in malaria endemic areas [12].

TRAP has been shown to mediate gliding motility and invasion processes of malarial sporozoites into vertebrate's hepatocyte and mosquito's salivary gland [13,14]. TRAP contains a hydrophobic N-terminal peptide (domain I), an integrin-like magnesium binding (or von Willebrand factor) A domain (domain II), thrombospondin type I repeats (domain III), an acidic proline/asparagine-rich region (domain IV), hydrophobic transmembrane domain (domain V) and a cytoplasmic tail (domain VI) [15,16]. The locomotion of sporozoites is mediated by the subpellicular actomyosin system that linked to the cytoplasmic tail of TRAP [17]. Despite functional importance of TRAP in parasite survival, analysis of the TRAP loci of *P. falciparum* (PfTRAP) and of *P. vivax* (PvTRAP) from clinical isolates revealed microheterogeneity of sequence that has been maintained by positive selective pressure [18–20]. Although it remains to be explored whether polymorphism in T cell epitopes of malarial TRAP could alter host cell immune recognition as that observed in CSP of *P. falciparum* [21], a rationale design of a malaria subunit vaccine targeting TRAP requires knowledge on the extent and pattern of sequence diversity in this molecule from diverse endemic areas.

Both *P. vivax* and *P. falciparum* have been circulating in Thailand with almost comparable prevalence since the past 2 decades. However, regional prevalence of *P. vivax* relative to *P. falciparum* seems to differ across major endemic areas of Thailand [22,23]. Our recent studies have shown that *P. vivax* populations in this country exhibited spatial variation in the extent of sequence diversity of major malaria vaccine candidate loci [24,25]. An effective malaria vaccine design undoubtedly requires knowledge on this issue to circumvent possible vaccine escape variants. Herein, we analyzed sequence diversity in the PvTRAP locus of clinical isolates from 4 major malaria endemic areas of Thailand.

## Materials and Methods

### Human Ethics Statement

The study protocol was approved by the Institutional Review Board on Human Research of Faculty of Medicine, Chulalongkorn University (IRB303/56). Written informed consent was obtained from all participants.

### Parasite populations

Blood samples were obtained from 114 *P. vivax*-infected patients in northwestern (Tak province, n = 30), southwestern (Prachuap Khirikhan province, n = 25), eastern (Chanthaburi province, n = 29) and southern (Yala and Narathiwat provinces, n = 30) Thailand during 2006 and 2007. Venous blood samples were taken from each patient and were preserved in EDTA anticoagulant. These isolates have been previously determined to contain single clone infection based on sequence analysis of the merozoite surface protein-1 gene (*Pvmsp-1*) as described previously [24,26].

## DNA extraction, amplification and sequencing

DNA was prepared from blood samples using QIAamp kit (Qiagen, Hilden, Germany). The complete *PvTRAP* sequence was amplified by nested PCR using a forward outer primer (PvTRAP-F0: 5′-ATGTGTGTACATTTGCGTATG-3′, nucleotides −442 to −422 before the start codon of the Salvador I sequence, GenBank accession number XM_001614097) and a reverse outer primer (PvTRAP-R0: 5′-TCATGAAGTGGC-GAAACAAAC-3′, nucleotides 1961 to 1968 from the start codon). The inner pairs of primers were PvTRAP-F1 (5′-ATGTCTGTGTGAGCGCGCGGT-3′, nucleotides −398 to −378) and PvTRAP-R1 (5′-TCATCAGAAGCAGTTCCAAG-3′, nucleotides 1791 to 1810). PCR amplification was performed in a total volume of 30 µl of the reaction mixture containing template DNA, 2.5 mM MgCl$_2$, 300 mM each deoxynucleoside triphosphate, 3 µl of 10x ExTaq PCR buffer, 0.3 µM of each primer and 1.25 units of ExTaq DNA polymerase (Takara, Seta, Japan). Thermal cycling profile for primary PCR included the preamplification denaturation at 94°C for 2 min followed by 35 cycles of 94°C for 30 s, 55°C for 30 s and 72°C for 2 min, and a final extension at 72°C for 5 min. Nested PCR was done using the same thermal profile except that amplification was done for 25 cycles. DNA amplification was carried out by using a GeneAmp 9700 PCR thermal cycler (Applied Biosystems, Foster City, CA). ExTaq DNA polymerase reportedly possesses efficient 5′→3′ exonuclease activity to increase fidelity and no strand displacement (Takara, Japan). The PCR product was examined by electrophoresis in a 1% agarose gel and was purified by using QIAquick PCR purification kit (QIAGEN, Germany). The PvTRAP sequences were determined directly and bi-directionally from PCR-purified templates and were performed on an ABI3100 Genetic Analyzer using the Big Dye Terminator v3.1 Cycle Sequencing Kit (Applied Biosystems, USA). Whenever singleton substitution occurred, sequence was re-determined using PCR products from two independent amplifications from the same DNA template. Sequences in this study have been deposited in the GenBank Database under the accession numbers KJ807186 to KJ807299.

## Statistical analyses

Sequence alignment with the Clustal X program [27] was done by using the *PvTRAP* sequence of the Salvador I strain as reference. The *PvTRAP* sequences of our previous Thai and Brazilian isolates previously published [19] and those reported by others [28] were included for analysis. For interspecific comparison, all *PvTRAP* sequences were aligned with the TRAP sequences of *P. knowlesi* (PkTRAP) and *P. cynomolgi* (PcyTRAP) (GenBank accession numbers XM_002259951 and XM_004223600, respectively). Haplotype diversity ($h$) and its sampling variance were computed according to equations 8.4 and 8.12 but replacing 2n by n [29]. Nucleotide diversity ($\pi$) was computed from the average number of pairwise sequence differences in the sample [29] using the MEGA version 6 program [30]. Evidence of genetic recombination was determined by 2 different approaches: (i) estimation of the minimum number of recombination events (Rm) that can be parsimoniously inferred from a sample of sequences with Monte Carlo simulations [31] using the DnaSP version 5 software [32] and (ii) searching for recombination breakpoints by phylogenetic approach by the Genetic Algorithm Recombination Detection (GARD) method. Goodness of fit for the GARD method was calculated by Akaike Information Criterion derived from a maximum likelihood model fit to each segment (AICc) using the HyPhy package [33,34].

The rates of synonymous substitutions per synonymous site ($d_S$) and nonsynonymous substitutions per nonsynonymous site ($d_N$)

were computed by using Nei and Gojobori's method [35] with Juke and Cantor correction [36] and their standard errors of these parameters were estimated by the bootstrap method with 1,000 pseudoreplicates as implemented in the MEGA 6.0 program [30]. A statistical significance level of differences in $d_S$ and $d_N$ was set at 5%.

Tests for departure from neutral evolution based on intraspecific comparisons were done by Tajima's $D$, Fu and Li's $D^*$ and $F^*$ statistics. Tajima's $D$ test determines the differences between the average number of nucleotide differences and an estimate of $\theta$ from the number of segregating sites where $\theta = 4Ne\mu$, in which $Ne$ and $\mu$ are the effective population size and the mutation rate, respectively [37]. Fu and Li's statistics measure the differences between the number of singletons and the total number of mutations ($D^*$ test) and the differences between the number of singletons and the average number of nucleotide differences between pairs of sequences ($F^*$ test) [38]. To examine whether deviation from neutrality occurs in particular fragments of *PvTRAP*, a sliding window analysis of Tajima's $D$ across the coding sequences was done using a window size of 250 bp and a step size of 10 bp [32].

Evidence of departure from neutrality based on interspecific comparison included Fu and Li's $D$ and $F$ tests, and McDonald-Kreitman tests. Fu and Li's $D$ statistics measure the differences between the number of mutations in external branches of the genealogy and the number of mutations while their $F$ statistic compares the differences between the number of mutations in external branches of the genealogy and the average number of nucleotide differences between pairs of sequences [38]. Coalescence simulation with 10,000 pseudoreplicates was done to estimate significance levels of these parameters as implemented in the DnaSP program [32]. The McDonald-Kreitman test contrasts intraspecific polymorphism with interspecific divergence of closely related species. Statistical departure from neutral expectation was computed by Fischer's exact test [39].

Tests of departure from neutrality at specific codons were performed based on estimation of the global ratio of the rate of non-synonymous to synonymous substitutions (dN/dS) across the *PvTRAP* gene. The single-likelihood ancestor counting (SLAC), fixed effects likelihood (FEL), random effects likelihood (REL) and mixed effects model of evolution (MEME) methods implemented in the HyPhy package [33] were used for analysis. SLAC model is highly conservative based on the maximum likelihood reconstruction of the ancestral sequences and the counts of synonymous and nonsynonymous changes at each codon position in a phylogeny under the assumption of neutral evolution. FEL model compares the ratio of nonsynonymous to synonymous substitution on a site-by-site basis, without assuming an *a priori* distribution of rates across sites whereas REL model first fits a distribution of rates across sites and then infers the substitution rate for individual sites. MEM algorithm detects codons under episodic positive selection unmasked by the abundance of purifying selection along the lineages [40]. Significance level settings for SLAC, FEL and MEME were $p$ values<0.1 and Bayes Factor>1000 for REL followed the default values available on the Datamonkey Web Server [41].

The population genetic structure was analyzed by molecular variance approach (AMOVA) using the Arlequin 3.5 software which is similar to the Weir and Cockerham's method but it takes into account the number of mutations between haplotypes [42]. The implemented algorithm calculates the fixation index $F_{ST}$ identical to the weighted average $F$-statistic over loci, $\theta w$ [43] and the significance levels of the fixation indices were estimated by a non-parametric permutation [42]. Pairwise Slakin's linearized *Fst*

was calculated based on $Fst/(1-Fst)$ to address the relationship between the magnitude of population differentiation and distance between endemic areas [44].

The genetic structure was also determined by STRUCTURE 2.3.4 program that deploys the Bayesian model–based clustering approach [45]. The most probable number of populations (K) was estimated using an admixture model. All sample data were run for values K = 1–10, each with a total of 15 iterations. We used 500,000 Markov Chain Monte Carlo generations for each run following a burn-in of 50,000 steps. The most likely number K in the data was estimated by calculating $\Delta K$ values [46] and identifying the K value that maximizes the log probability of data, lnP(D) [45] as implemented in STRUCTURE HARVESTER program [47].

The phylogenetic tree of the *PvTRAP* haplotypes was constructed by using the Maximum Likelihood method based on the model with the lowest Bayesian Information Criterion (BIC) scores [30]. All positions containing gaps and missing data were eliminated. Initial tree(s) for the heuristic search were obtained by applying Neighbor-Join and BioNJ algorithms to a matrix of pairwise distances estimated using the Maximum Composite Likelihood (MCL) approach. The final tree topology was selected based on superior log likelihood value as implemented in MEGA version 6 program [30].

Amino acid properties were characterized based on polarity and charge. Polarity of each residue is categorized as polar (S, Y, C, W, H, Q, T and N) or nonpolar (the remainder). Charge property includes positive (H, R and K), negative (D and E) or neutral (the remainder). Prediction of HLA-I binding peptides was based on the method taken into account proteasomal C terminal cleavage and transporter associated with antigen processing (TAP) transport efficiency. The predicted scores were determined by the NetCTL program [48].

## Results

### Genetic diversity in *PvTRAP*

Of 114 complete *PvTRAP* sequences, 109 isolates contained 1668 bp and 5 isolates from Tak province whose sequences were identical had 1677 bp. Sequence alignment with the Salvador I strain revealed 36 nucleotide substitutions within 33 codons, resulting in 26 nonsynonymous and 7 synonymous changes. Comparing with previous data from Brazilian, Thai [19], South Korean isolates [28], 9 new amino acid substitutions were identified (Table 1). Like our previous findings, the majority of amino acid changes occurred in domains II and IV [19]. Three codons contained more than two substituted nucleotides: codon 206 (after the Salvador 1 sequence) having 4 different substitutions: c.616C>T (P206S), c.617C>G (P206S), c.617C>T (P206L) and c.616_617delinsTT (P206L), codon 268 having double mutations: c.802_803delinsGG (K268G), and codon 424 containing 2 changes: c.1271A>C (N424T) and c.1272C>A (N424K). Among isolates examined herein, no discernible bias in frequency of nucleotide substitutions in each position of the codons occurred, i.e. 11, 11 and 13 substitutions in the first, second and third positions of the codons, respectively. In total, 44 distinct haplotypes were identified in this analysis, one of which had a duplication of 9 nucleotides encoding PDS in domain IV as previously noted [19] and responsible for size variation (Tables 1 and S1).

The distribution of *PvTRAP* haplotypes in southern Thailand (Yala and Narathiwat provinces) was rather skewed towards few haplotypes as shown by a relatively lower value of haplotype diversity ($h = 0.653 \pm 0.063$) whereas parasite populations from

**Table 1.** Distribution of PvTRAP haplotypes from 4 major malaria endemic areas of Thailand.

| Haplotype | Amino acid residue (Salvador-1 reference sequence across polymorphic sites) | Tak | Prachuap Khirikhan | Chanthaburi | Yala & Narathiwat | Total |
|---|---|---|---|---|---|---|
| Salvador-1 (ref) | L T D E Q R V N Q S V P R K T D G N N N G D R E | | | | | |
| #1 | S E . . . . T K . . G . R . . D . | 3 | 4 | – | – | 7 |
| #2 | S . . . . K . T K . G . R . . D . | 3 | – | – | – | 3 |
| #3 | S E . . . . T K . . K G . R . . G | 2 | – | – | – | 2 |
| #4 | S E . . . . T K A . . . K . R . . | 1 | – | – | – | 1 |
| #5 | S E . . . . T K . . P D S K . D . R | 5 | – | – | – | 5 |
| #6 | S . . . . . . K . G . K T . . R . G | 3 | 1 | – | – | 4 |
| #7 | S E . . . . T K . . G A . R . . G | 1 | 1 | – | – | 2 |
| #8 | S E . . . . T K . . K . . R . D . | 1 | – | – | – | 1 |
| #9 | S . . . E . . . S . G . . R . D . | 2 | – | – | – | 2 |
| #10 | S . . . . . T K . . . . R . . D . | 1 | 1 | – | – | 2 |
| #11 | S . . . . . . K . G . . R . . . | 1 | – | – | – | 1 |
| #12 | S E . . . . T K . . G . R . . D . | 1 | – | – | – | 1 |
| #13 | S . . . . . T K K . G . R . . D . | 1 | 4 | – | – | 5 |
| #14 | S E . . . . T K . . K . R . A . | 2 | – | – | – | 2 |
| #15 | S . . . . . T K . . G . R . . . | 1 | 1 | – | – | 2 |
| #16 | S . . . . . T K . . K . R . D . | 1 | 1 | – | – | 2 |
| #17 | S E . . . . T K . . K . R . A . | 2 | – | 1 | – | 3 |
| #18 | S . D . . . T K . . K K D . R D . | 1 | – | – | – | 1 |
| #19 | S . . . . R . . . . G . R . D . | – | 1 | – | – | 1 |
| #20 | S . . . . . T K . . K T G . A . | – | 1 | – | – | 1 |
| #21 | S . . . . . T K . . . T D . R . | – | 1 | – | – | 1 |
| #22 | S . . R . . . T K K . G . R . . D . | – | 1 | – | – | 1 |
| #23 | S . . . L . . T . . R . R T . . | – | 1 | – | – | 1 |
| #24 | F . S . I . . T K K . G . R . . D . | 2 | – | – | – | 2 |
| #25 | S . . R . . . T K K . G . R . . D . | – | 1 | – | – | 1 |
| #26 | S . . . . . T K . . . . R . . D . | – | 1 | – | – | 1 |
| #27 | S . . . . . T K . . G . R . . D A . | 1 | – | – | – | 1 |
| #28 | S . . . . . T K . . G . R . . G G | – | – | 10 | – | 10 |
| #29 | S . . . . . T K . . G . R . . G G | – | – | 1 | – | 1 |

**Table 1.** Cont.

| Amino acid residue → Haplotype | 145 | 150 | 124 | 135 | 143 | 146 | 162 | 172 | 180 | 186 | 206 | 254 | 256 | 258 | 263 | 284 | 290 | 293 | 303 | 304 | 305 | 313 | 342 | 344 | 346 | Tak | Prachuap Khirikhan | Chanthaburi | Yala & Narathiwat | Total |
|---|---|---|---|---|---|---|---|---|---|---|---|---|---|---|---|---|---|---|---|---|---|---|---|---|---|---|---|---|---|---|
| **Domain** | I | II | II | II | II | II | II | II | II | II | III | IV | IV | IV | IV | IV | IV | IV | IV | IV | IV | IV | IV | IV | IV | | | | | |
| **Salvador-1** | L | T | D | E | Q | R | V | R | N | Q | S | V | P | R | K | T | D | G | N | – | N | N | G | R | E | | | | | |
| #30 | . | S | . | . | . | . | . | K | . | . | – | . | . | G | . | R | . | . | . | . | . | . | . | . | **G** | – | – | 8 | – | 8 |
| #31 | . | S | E | . | . | . | . | K | T | . | – | . | . | G | . | R | . | . | . | . | K | . | . | **D** | . | – | – | 2 | – | 2 |
| #32 | . | S | . | **D** | . | . | . | . | T | . | – | . | . | G | . | R | . | . | . | . | K | . | . | . | . | – | – | 1 | – | 1 |
| #33 | . | S | . | . | . | . | . | K | . | . | – | . | . | G | . | R | **T** | . | . | . | . | . | . | . | **G** | – | – | 1 | – | 1 |
| #34 | . | S | . | . | . | . | **L** | . | T | . | – | . | . | G | . | R | . | . | . | . | K | . | . | **D** | . | – | – | 1 | – | 1 |
| #35 | . | S | E | . | **Q** | . | . | K | T | **K** | – | . | . | G | . | R | . | . | . | . | K | . | . | **D** | . | – | – | 1 | – | 1 |
| #36 | . | S | . | . | . | . | . | . | **T** | . | – | . | . | G | . | S | . | . | . | . | . | . | . | . | . | – | – | 1 | – | 1 |
| #37 | . | S | . | . | . | . | . | K | . | . | – | . | . | G | . | R | . | . | . | . | . | . | . | . | **G** | – | – | 1 | – | 1 |
| #38 | . | S | . | . | . | . | . | K | . | . | – | . | . | S | . | . | . | . | . | . | . | . | . | . | . | – | – | 1 | – | 1 |
| #39 | . | S | . | . | . | . | . | K | . | . | – | . | . | G | . | R | **T** | . | . | . | . | . | . | . | **G** | – | – | – | 15 | 15 |
| #40 | . | S | . | . | . | . | . | K | . | . | – | . | . | G | . | R | **T** | . | . | . | . | . | . | . | . | – | – | – | 2 | 2 |
| #41 | . | S | . | . | . | . | . | . | . | . | – | . | . | G | . | R | . | . | . | . | . | . | . | . | **G** | – | – | 1 | 1 |
| #42 | . | S | . | . | . | . | . | K | . | . | – | . | . | G | . | R | . | . | . | . | K | . | . | . | . | – | – | – | 10 | 10 |
| #43 | . | S | . | . | . | . | . | K | . | . | – | . | . | G | . | R | **T** | . | . | . | K | . | . | . | **G** | – | – | – | 1 | 1 |
| #44 | . | S | . | . | . | . | . | K | . | . | – | . | . | G | . | R | . | . | . | . | . | . | . | . | . | – | – | 1 | 1 |
| **Protein Structure\*** | β1 | α3 | α4 | α4 | α4 | α4 | | | | | | | | | | | | | | | | α5 | α5 | α5 | α5 | 31 | 24 | 29 | 30 | 114 |

Newly identified amino acid changes are in bold. \*Protein secondary structure and segment are after Song, et al. 2012 [51].

Tak, Prachuap Khirikhan and Chanthaburi provinces had significantly higher haplotype diversity values, suggesting a more even distribution of haplotype frequencies in these latter endemic regions (Table 2). It is noteworthy that the predominant haplotypes differed between endemic areas. For examples, haplotypes #39 (n = 15) and #42 (n = 10) were exclusively found in Yala and Narathiwat provinces whereas haplotypes #28 (n = 10) and #30 (n = 8) were unique for Chanthaburi province. In contrast, several haplotypes were shared between isolates from Tak and Prachuap Khirikhan provinces (Table 1). The nucleotide diversity ($\pi$) of *PvTRAP* varied from 0.00117±0.00014 (isolates from Yala and Narathiwat provinces) to 0.00272±0.00032 (isolates from Prachuap Khirikhan province). The magnitudes of nucleotide diversity in this locus did not significantly differ among isolates from Tak, Prachuap Khirikhan and Chanthaburi provinces but these were significantly greater than that from Yala and Narathiwat provinces ($p < 10^{-4}$) (Table 2).

## Population differentiation

We did not observe population differentiation between isolates from Tak and Prachuap Khirikhan provinces in which the $F_{ST}$ values did not deviate significantly from zero, implying that they shared a common origin or a high genetic admixture. However, *P. vivax* populations from Tak and Prachuap Khirikhan provinces displayed the $F_{ST}$ values that were significantly greater than zero when compared with populations from Chanthaburi province ($p < 10^{-5}$) as well as that from Yala and Narathiwat provinces ($p < 10^{-5}$). A significant departure from neutrality was also noted when isolates from Yala and Narathiwat provinces were compared with those from Chanthaburi province ($p < 10^{-5}$), implying genetic

differentiation between populations with minimum or absence of gene flow between these areas. Furthermore, the *Fst* values between Tak isolates in the present study (collected in 2006–2007) and those collected in 1996 [19] did not differ significantly (p = 0.23), suggesting genetic stability in *P. vivax* population in this area (Table 3). On the other hand, *P. vivax* populations from Brazil, South Korea and Thailand were significantly different from one another, indicating strong differentiation and reproductive isolation among populations. These results were in line with analysis using STRUCTURE version 2.3.4 software in which the subpopulation likelihood reached a plateau from K = 3 to K = 6. Consistently, a greatest delta K value was also obtained at K = 3 (Figure 1). The genetic structure at K = 3 could differentiate Brazilian isolates from Thai and South Korean parasite populations while a separation of South Korean and southern Thai *P. vivax* populations were not perceived. Similar findings were observed at K = 4. A more concordant results between those obtained by STRUCTURE program and the *Fst* parameters occurred when K = 5 in which isolates from Tak collected during different periods and from Prachuap Khirikhan were indistinguishable whereas South Korean and southern Thai populations were distinct (Figure 1). Regression analysis has shown positive correlation between pairwise linearized *Fst* and the natural logarithm of linear distance between endemic areas (Mantel test, r = 0.4619, p = 0.024). Likewise, positive correlation was found when geographic linear distance between endemic areas was considered (Mantel test, r = 0.4503, p = 0.028), implying that the majority of *P. vivax* populations in Thailand were genetically isolated by distance (Figure 2).

**Table 2.** Variation in the PvTRAP sequences of Thai isolates from diverse endemic areas.

| | Tak | Prachuap Khirikhan | Chanthaburi | Yala and Narathiwat | All |
|---|---|---|---|---|---|
| N | 30 | 25 | 29 | 30 | 114 |
| M | 14 | 24 | 18 | 7 | 32 |
| S | 14 | 23 | 18 | 7 | 31 |
| H | 17 | 18 | 12 | 6 | 43 |
| $h$±S.D. | 0.956±0.018 | 0.957±0.027 | 0.818±0.053 | 0.653±0.063 | 0.954±0.008 |
| $\pi$±S.D. | 0.00220±0.00017 | 0.00272±0.00032 | 0.00212±0.00032 | 0.00117±0.00014### | 0.00272±0.00009 |
| $d_S$±S.E. (all) | 0.00018±0.00017 | 0.00074±0.00043 | 0.00069±0.00052 | 0.00201±0.00142 | 0.00118±0.00081 |
| Domain I | 0.00000±0.00000 | 0.00397±0.00467 | 0.00000±0.00000 | 0.00000±0.00000 | 0.00087±0.00097 |
| Domain II | 0.00000±0.00000 | 0.00029±0.00027 | 0.00048±0.00046 | 0.00176±0.00130 | 0.00067±0.00038 |
| Domain III | 0.00000±0.00000 | 0.00000±0.00000 | 0.00000±0.00000 | 0.00000±0.00000 | 0.00000±0.00000 |
| Domain IV | 0.00000±0.00000 | 0.00002±0.00002 | 0.00000±0.00000 | 0.00000±0.00000 | 0.00000±0.00000 |
| Domain V | 0.00000±0.00000 | 0.00000±0.00000 | 0.00000±0.00000 | 0.00000±0.00000 | 0.00000±0.00000 |
| Domain VI | 0.00250±0.00281 | 0.00575±0.00593 | 0.00721±0.00781 | 0.01906±0.00000 | 0.01245±0.01187 |
| $d_N$±S.E. (all) | 0.00278±0.00090** | 0.00327±0.00086** | 0.00252±0.00077* | 0.00091±0.00052 | 0.00314±0.00090 |
| Domain I | 0.00000±0.00000 | 0.00233±0.00232 | 0.00000±0.00000 | 0.00000±0.00000 | 0.00053±0.00055 |
| Domain II | 0.00264±0.00134* | 0.00428±0.00161* | 0.00406±0.00161* | 0.00014±0.00014 | 0.00396±0.00173 |
| Domain III | 0.00423±0.00426 | 0.00442±0.00439 | 0.00067±0.00071 | 0.00000±0.00000 | 0.00261±0.00266 |
| Domain IV | 0.00381±0.00162* | 0.00314±0.00132* | 0.00248±0.00118* | 0.00227±0.00141 | 0.00373±0.00157* |
| Domain V | 0.00000±0.00000 | 0.00000±0.00000 | 0.00000±0.00000 | 0.00000±0.00000 | 0.00000±0.00000 |
| Domain VI | 0.00000±0.00000 | 0.00000±0.00000 | 0.00000±0.00000 | 0.00000±0.00000 | 0.00000±0.00000 |

N, number of sequences; M, number of mutations; S, number of segregating sites; H, number of haplotypes.
Z-tests of the hypothesis that $\pi$ equals the corresponding value for each population: ###p<0.0001.
Z-tests of the hypothesis that mean $d_S$ equals that of mean $d_N$ : *p<0.05; **p<0.01.

**Table 3.** Genetic differentiation ($F_{ST}$ indices) of *P. vivax* populations based on the PvTRAP locus.

| | Tak (2006–7) | Prachuap Khirikhan | Chanthaburi | Yala and Narathiwat | Tak (1996) | Brazil |
|---|---|---|---|---|---|---|
| Prachuap Khirikhan | 0.0001 | | | | | |
| Chanthaburi | 0.2324*** | 0.2319*** | | | | |
| Yala and Narathiwat | 0.5001*** | 0.4657*** | 0.3615*** | | | |
| Tak (1996) | 0.0098 | 0.0003 | 0.2147*** | 0.3867*** | | |
| Brazil | 0.5072*** | 0.4708*** | 0.4380*** | 0.4603*** | 0.4325*** | |
| Korea | 0.7057*** | 0.6907*** | 0.5514*** | 0.7305*** | 0.5742*** | 0.5630*** |

***$p < 0.00001$.

## Tests for departure from neutral evolution

Analysis of patterns of nucleotide substitutions in the *PvTRAP* sequences of isolates in this study reveals that the rate of nonsynonymous substitutions per nonsynonymous sites or $d_N$ significantly exceeds that of synonymous substitutions per synonymous sites or $d_S$ in all parasite populations ($p < 0.05$ or $p < 0.01$) except *P. vivax* isolates from Yala and Narathiwat provinces (Table 2). In contrast, $d_S$ was greater than $d_N$ for isolates from Yala and Narathiwat provinces although the difference was not statistically meaningful ($p = 0.467$). Closer looks into the rate of nucleotide substitutions in each domain of *PvTRAP* has revealed that $d_N$ significantly outnumbered $d_S$ in domains II and IV of

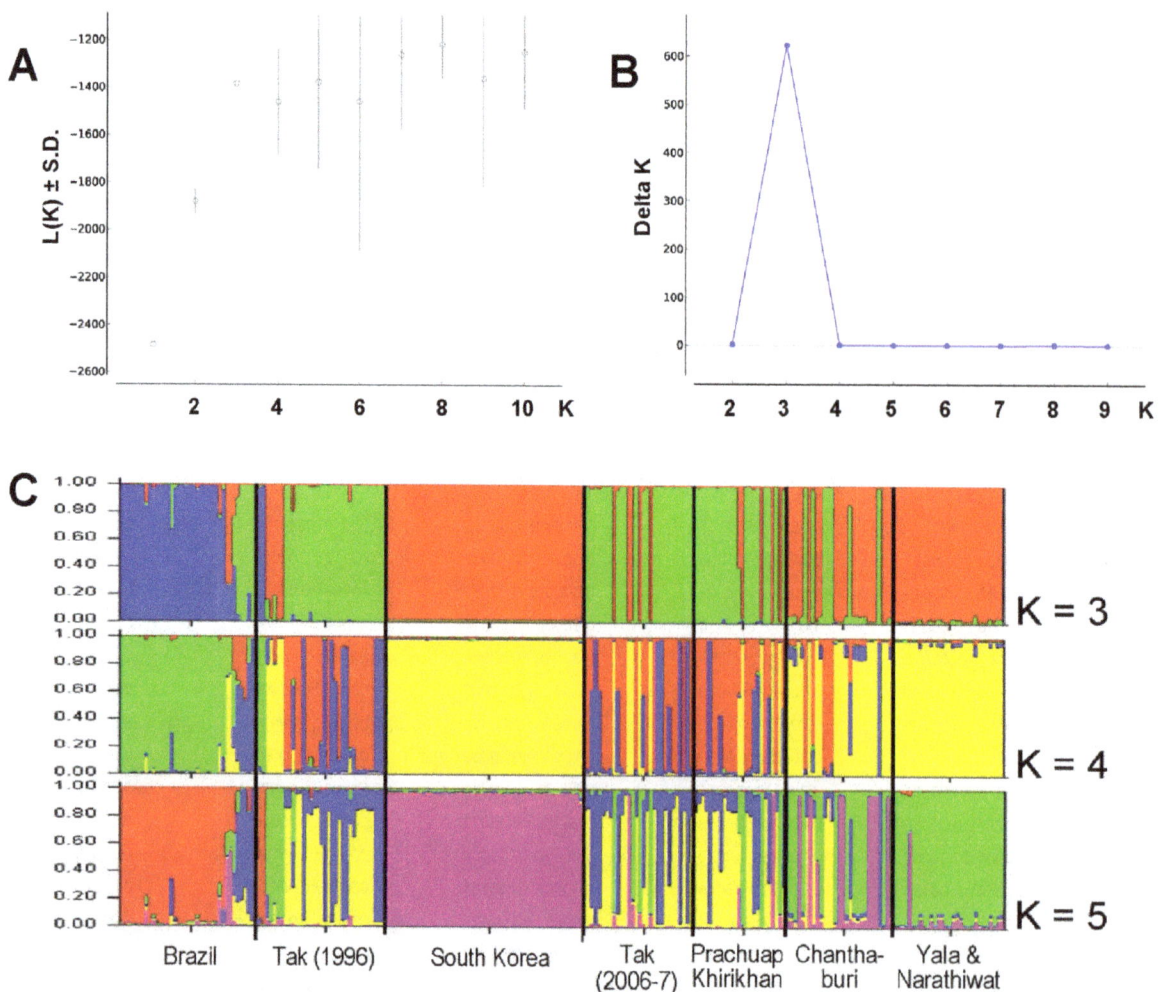

**Figure 1. Determination of the most likely number of clusters of *PvTRAP* haplotypes by *ad hoc* methods: (A) relationship between K and mean of estimated mean log likelihood of K and its standard deviation or L(K) ± S.D. [45], (B) relationship between K and delta K [46].** (C) Graphical representation of the genetic structure of populations from Brazil, South Korea and each endemic area of Thailand. Results are given for K3, K4 and K5.

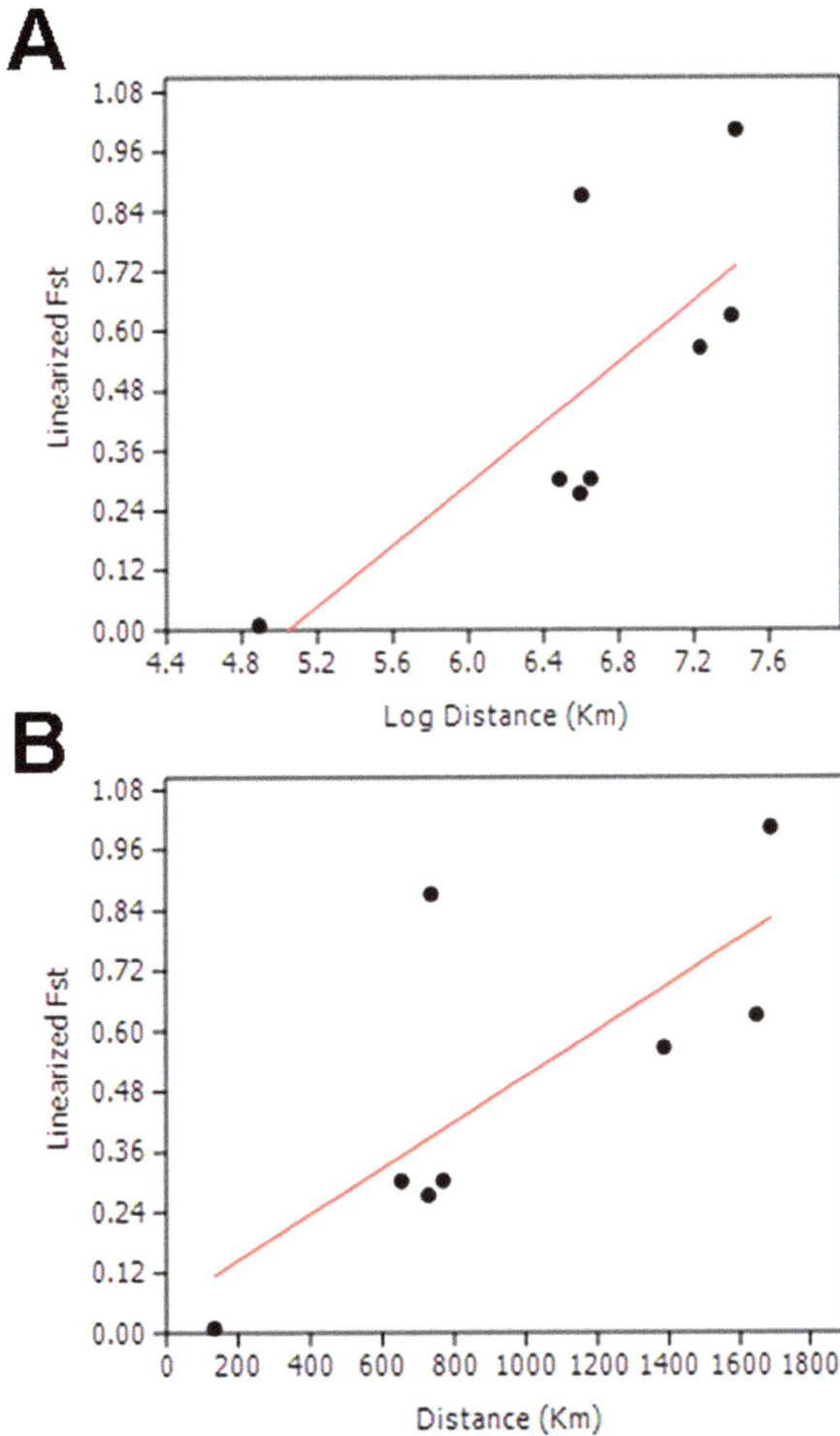

**Figure 2. Regression analysis of pairwise linearized *Fst* regressed on natural logarithm geographic distances between *P. vivax* populations in Thailand (A) and pairwise linearized *Fst* regressed on geographic distances between populations (B).**

isolates from Tak, Prachuap Khirikhan and Chanthaburi provinces. Although no significant difference in $d_S$ and $d_N$ for the entire coding region of the *PvTRAP* gene was observed when all parasite populations were considered, $d_N$ in domain IV was obviously greater than $d_S$ (Table 2). Meanwhile, statistics based on Tajima's *D* and Fu & Li's *D** and *F** did not show evidences of departure

from neutral expectation (Table 4). However, sliding window analysis reveals significant negative Tajima's *D* in domain II of *P. vivax* populations from Prachuap Khirikhan and Chanthaburi provinces, suggesting purifying selection at certain residues in this gene. In contrast, evidence of balancing selection was detected in domain IV of Yala and Narathiwat isolates as shown by significant

**Table 4.** Intraspecific and interspecific neutrality tests for the PvTRAP locus from each endemic area of Thailand.

| Province/Endemic area | Tajima's D | Fu & Li's D* | Fu & Li's F* | Out group: P. cynomolgi | | Out group: P. knowlesi | |
|---|---|---|---|---|---|---|---|
| | | | | Fu & Li's D | Fu & Li's F | Fu & Li's D | Fu & Li's F |
| Tak (n = 30) | 0.097 | −0.143 | −0.080 | 0.431 | 0.350 | 0.987 | 1.067 |
| Prachuap Khirikhan (n = 25) | −0.967 | −0.686 | −0.905 | −0.086 | −0.401 | 0.284 | 0.025 |
| Chanthaburi (n = 29) | −0.818 | −1.429 | −1.451 | −0.490 | −0.798 | −0.080 | −0.383 |
| Yala and Narathiwat (n = 30) | 0.278 | −0.120 | 0.000 | 0.010 | 0.227 | 0.427 | 0.591 |
| All (n = 114) | −0.698 | −1.714 | −1.572 | −1.055 | −1.120 | −0.534 | −0.553 |

All values did not have significance departure from zero.

positive $D$ values. Meanwhile, no departure from neutrality was detected in malaria population from Tak province (Figure 3).

Results from interspecific comparison between *PvTRAP* and *PcyTRAP* using the MacDonald-Kreitman test has shown that each parasite population except that from Yala and Narathiwat provinces had significant deviation in positive direction from neutral expectation ($p < 0.01$), reflecting a signature of positive or balancing selection. Consistent results were obtained when the TRAP gene of *P. knowlesi* was used as the outgroup sequence (Table 5). In contrast, application of the Fu & Li's $D$ and $F$ tests using either *PcyTRAP* or *PkTRAP* as outgroup sequences did not yield significant departure from neutrality. Meanwhile, positively selected sites were identified at 4 codons in domain II and 3 codons in domain IV by one or more of the methods implemented in the HyPhy package (Table 6). Likewise, negatively selected sites were detected in residue 21 in domain I, residues 30, 100, 133 in domain II and residues 519 and 539 in domain VI, suggesting functional constraint occurring at certain residues in this protein.

## Amino acid substitutions and predicted HLA binding

The amino acid substitutions in PvTRAP of Thai isolates had slightly higher percentage of conservative changes with respect to polarity and charge property as 67.7% and 58.1% of these substitutions, respectively, were unchanged. Similar percentages of these changes were observed among worldwide isolates, being 64.6% and 56.9%, respectively (Table S2). Closer look into amino acid substitutions and potential HLA-binding peptides as predicted by the high scores for the C-terminal cleavage and the transporter associated with antigen processing efficiency [48] have shown that a number of substituted residues have remarkably reduced predicted scores; thereby, the property of these epitopes could be altered. It is noteworthy that substituted residues in several of these potential epitopes occurred in domains II and IV of PvTRAP (Table 6).

## Intragenic recombination

Evidence of recombination in the *PvTRAP* locus analyzed by using the HyPhy package has identified evidence of 1 recombination breakpoint with significant topological incongruence between AICc score of the best fitting GARD model ($p < 0.05$) for *P. vivax* populations in Thailand except those from Yala and Narathiwat provinces ($0.05 < p < 0.1$). Although these recombination breakpoints occur at different positions, i.e. between nucleotides 1008 and 1009, 1170 and 1171, and 1173 and 1174 for Chanthaburi, Prachuap Khirikhan and Tak provinces, respectively, it is noteworthy that all were located in domain IV of the *PvTRAP* gene. Meanwhile, estimation of the parameter Rm reveals 3 and 5 recombination sites in parasite populations from Tak and Prachuap Khirikhan provinces whereas Chanthaburi isolates and southern Thai *P. vivax* populations had equal number of potential recombination sites (Rm = 2). Locations of potential recombination sites for each population are listed in Table S3, most of which span domain IV. These results suggest that intragenic recombination could shape diversity at the *PvTRAP* locus.

## Phylogenetic relationship

Model test reveals that the Tamura 3-parameter with the rate variation model that allowed for some sites to be evolutionarily invariable gave the lowest BIC value. The tree with the highest log likelihood (−3285.15) reveals that the majority of Brazilian isolates were placed in a separate lineage while the majority of Tak isolates collected in 1996, in 2006–2007 and most Prachuap Khirikhan isolates were clustered together. Several isolates from Chantha-

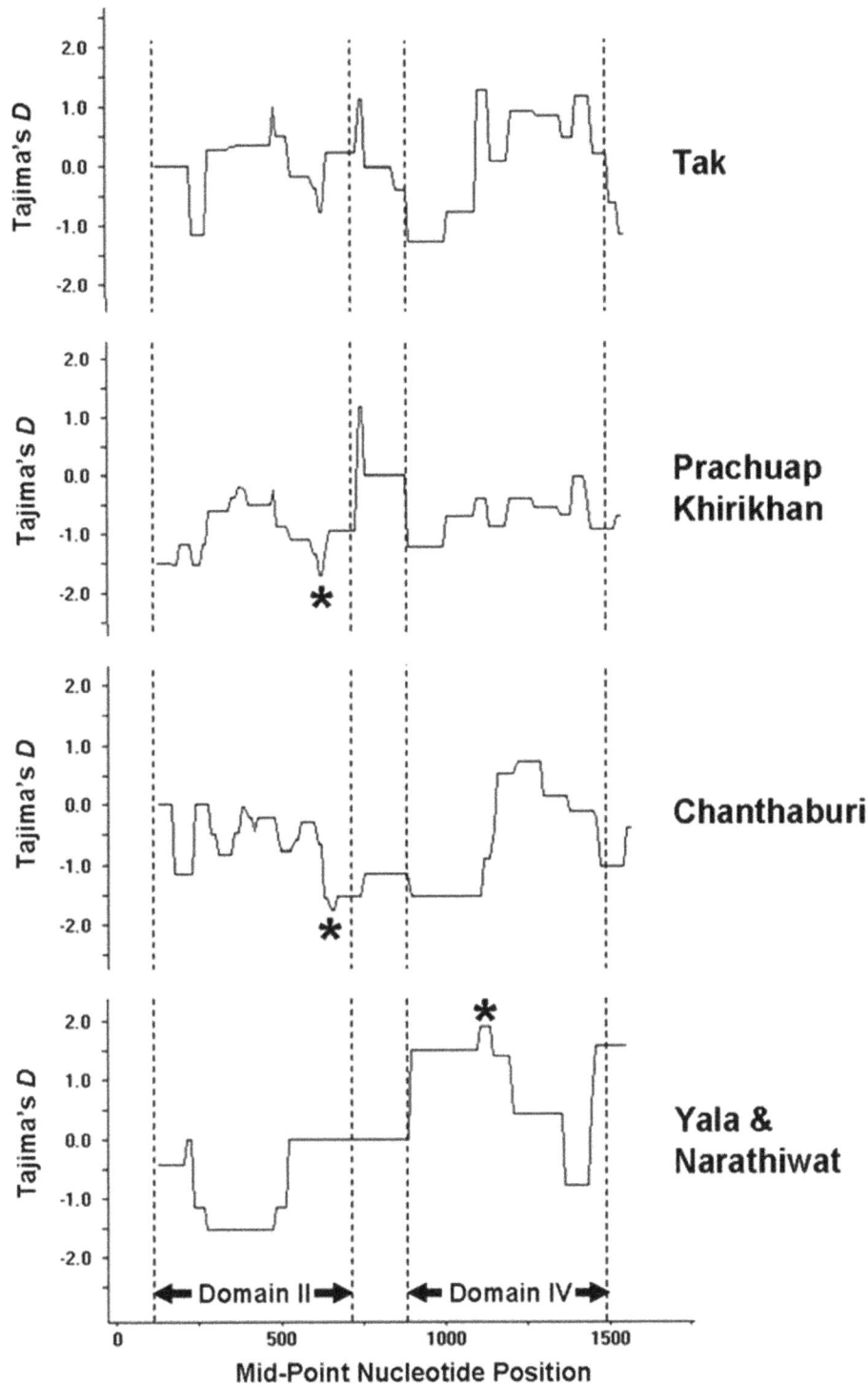

**Figure 3. Sliding window plots of Tajima's $D$ across *PvTRAP* sequences from each population in Thailand.** Asterisks denote values with $p<0.05$. Plots are based on a window size of 250 bp and a step size of 10 bp.

buri, Yala and Narathiwat provinces seem to be located in separate branches from the remaining Thai isolates. However, there was no clear lineage for each parasite population because several isolates from different endemic areas shared or was placed in closely related branches. More importantly, all branches in this phylogenetic inference received bootstrap values less than 80%, suggesting no distinct lineage of the *PvTRAP* haplotypes from worldwide origins (Figure 4).

## Discussion

Host adhesion and motility are fundamental properties of malarial sporozoites to establish infection that involves sporozoite-specific

**Table 5.** McDonald–Kreitman tests on TRAP of *Plasmodium vivax* from diverse geographic origins with *P. cynomolgi* orthologue as outgroup species.

| Population | Polymorphic changes within *P. vivax* | | Fixed differences between *species* | | Neutrality index | *p* value |
|---|---|---|---|---|---|---|
| | Synonymous | Nonsynonymous | Synonymous | Nonsynonymous | | |
| Tak | 1 | 13 | 116 | 140 | 10.77 | **0.0045** |
| Prachuap Khirikhan | 3 | 21 | 117 | 138 | 5.93 | **0.0011** |
| Chanthaburi | 2 | 16 | 117 | 140 | 6.69 | **0.0053** |
| Yala and Narathiwat | 3 | 4 | 115 | 145 | 1.06 | 1.0000 |
| All | 6 | 23 | 117 | 136 | 3.298 | **0.0096** |

Note: Repeats are excluded from all analyses.

proteins, one of which is TRAP. It has been shown that TRAP-deficient sporozoites were severely impaired in their host cell adherence property and gliding motility; thereby, invasion of mosquito's salivary glands and hepatocytes was interrupted [14,49]. Two extracellular portions of malarial TRAP, the von Willebrand A-domain and the thrombospondin repeats located in domains II and IV, respectively, are crucial for initial host cell adherence and stabilization of adhesion/deadhesion during gliding mobility of sporozoites [49,50]. Crystal structure analyses spanning the von Willebrand A-domain and the thrombospondin repeats of PvTRAP and PfTRAP reveal two conformational states, open and closed structures. Such structural phase transition is possibly responsible for 'stick-and-slip' or gliding motility of sporozoites through the actomyosin motility apparatus [51,52]. The sequence motif of metal-ion-dependent adhesion sites (MIDAS) in von Willebrand A-domain and the extensible β ribbon that implicates in ligand binding during the open high-affinity state [51] were perfectly conserved in all PvTRAP haplotypes as shown in this study. Although amino acid substitutions were observed in the β1 and α3 domains, charge and polarity of these substituted residues were retained, suggesting structural constraint in these portions of the protein (Table 1). Experimental studies have shown that artificially engineered mutations of the MIDAS-coordinating amino acid residues that were conserved across malaria species could impair gliding motility and sporozoite infectivity to both mosquito and mammalian hosts [49,50]. On the other hand, alteration in both charge and polarity profiles of some substituted amino acids (residues 143, 166, 172 and 176, Tables 1 and S2) were identified in α4 and α5 domains, implying that changes in these residues may not drastically affect functional conformation of PvTRAP. Nevertheless, further experimental studies are required to address this issue.

The extent of sequence diversity of the *PvTRAP* locus among parasite populations in Thailand seems to be comparable among endemic areas except that from southern region (Table 2). The results from this study are in line with our recent analyses of other major vaccine antigens, i.e. apical membrane antigen-1 (Pvama-1), merozoite surface protein-1 (Pvmsp-1), merozoite surface protein-4 (Pvmsp-4) and merozoite surface protein-5 (Pvmsp-5) of *P. vivax* showing that southern parasite population possesses significantly lower levels of nucleotide diversity, number of haplotypes and haplotype diversity than those of northwestern parasite population [24,25,53]. Heterogeneity in the *PvTRAP* sequences could have been arisen from selective pressures and intragenic recombination. Recombination confers evolutionary advantages because creation of adaptive traits or removal of deleterious mutants is enabled. Importantly, most recombination sites were identified within or spanning domain IV of the *PvTRAP* locus where signature of

positive selection was identified. The relatively higher levels of nucleotide diversity of *PvTRAP* from Thai-Mynmar border (Tak and Prachuap Khirikhan isolates) than that from Thai-Cambodia border (Chanthaburi isolates) seems to be in line with Rm in each population. It is likely that transmigration of people could maintain and spread of *P. vivax* harboring variant *PvTRAP* haplotypes while recombination could further generate novel alleles. Intriguingly, it seems that reduction in genetic diversity of southern parasite population may not be simply explained by clonal population structure of *P. vivax* in this region because recombination event could be traced despite the paucity of potential recombination sites observed (Rm = 2). In contrast, several decades of extensive malaria control in this region and a lack of remarkable re-introduction of malaria cases from Malaysia could have resulted in population bottleneck as previously suggested [24]. Therefore, after bottleneck effect and relaxation of malaria control in southern Thailand because of difficulty in implementation due to local political unrest, it could be possible that recombination between different *PvTRAP* haplotypes may gradually increase the level of genetic diversity in this population.

Our analyses have shown that domains II and IV of the *PvTRAP* locus had significant difference in the rate of nonsynonymous substitutions than that of synonymous substitutions. Evidence of positive or balancing selection was also reaffirmed by the McDonald-Kreitman test. Although departure from neutrality in the entire coding PvTRAP sequence was not apparent by using Tajima's *D* and related statistics, sliding window analysis has identified significant negative Tajima's *D* values in domains II of *P. vivax* population from Prachuap Khirikhan and Chanthaburi provinces while significant positive values occurred in domain IV among isolates from Yala and Narathiwat provinces. Differential patterns of departure from neutrality among populations could reflect different evolutionary forces exerted on PvTRAP of each population. Alternatively, intrinsic statistical property of Tajima's *D* statistics can be confounded by demographic processes [54]. Therefore, significant positive Tajima's *D* observed in southern Thai isolates could imply balancing selection or a recent bottleneck because an excess of intermediate frequency alleles as rare alleles are lost from the population due to genetic drift can be found immediately after the bottleneck. Meanwhile, significant negative Tajima's *D* value in domains II could imply purifying selection and selective sweep while migration and population growth may be the confounders [55]. Nevertheless, positively selected codons were found based on analyses using SLAC, FEL, REL and MEME methods while negatively selected residues were detected in domains I, II and VI (Table 6). Taken together, it seems that co-existence of both

**Table 6.** Codon-based analysis of departure from neutrality in the PvTRAP locus of Thai isolates.

| Domain | Codon | SLAC | | FEL | | REL | | MEME | | Consensus* | | | |
|---|---|---|---|---|---|---|---|---|---|---|---|---|---|
| | | $d_N - d_S$ | p value | $d_N - d_S$ | p value | $d_N - d_S$ | Bayes Factor | ω+ | p value | SLAC | FEL | REL | MEM |
| II | 120 | 93.598 | 0.213 | 2821.14 | 0.091 | 73.567 | 9897.75 | >100 | 0.115 | + | + | + | |
| II | 134 | 74.642 | 0.496 | 2675.58 | 0.208 | 73.381 | 195736 | >100 | 0.230 | | | + | |
| II | 206 | 84.790 | 0.280 | 3760.95 | 0.105 | 73.323 | 702283 | >100 | 0.051 | | | + | + |
| II | 255 | 149.777 | 0.112 | 3884.15 | 0.063 | 73.501 | $2.27475 \times 10^{10}$ | >100 | 0.084 | | + | + | + |
| IV | 410 | 95.051 | 0.401 | 3087.63 | 0.178 | 73.377 | 8780900 | >100 | 0.203 | | | + | |
| IV | 421 | 55.136 | 0.594 | 1821.01 | 0.252 | 73.389 | 2484.53 | >100 | 0.267 | | | + | |
| IV | 439 | 83.134 | 0.304 | 2728.63 | 0.104 | 73.541 | 267258 | >100 | 0.129 | | | + | |
| I | 21 | -102.062 | 0.151 | -2637.21 | 0.063 | -38.418 | 35812.5 | <100 | >0.100 | - | - | - | |
| II | 30 | -204.123 | 0.023 | -5228.32 | 0.011 | -38.410 | 29828.7 | <100 | >0.100 | | | - | |
| II | 100 | -46.299 | 0.333 | -1399.13 | 0.131 | -38.407 | 29051.4 | <100 | >0.100 | - | - | - | |
| II | 133 | -31.970 | 0.483 | -1019.35 | 0.210 | -38.100 | 8558.41 | <100 | >0.100 | | | - | |
| VI | 519 | -92.599 | 0.111 | -2873.15 | 0.032 | -38.407 | 29054.6 | <100 | >0.100 | - | - | - | |
| VI | 539 | -83.719 | 0.199 | -3571.07 | 0.051 | -38.391 | 27525.8 | <100 | >0.100 | | | - | |

*positively selected site, +; negatively selected site, -.
Default significance levels in Datamonkey program: SLAC, p<0.1; FEL, p<0.1; REL, Bayes Factor>1000; MEME, p<0.1.

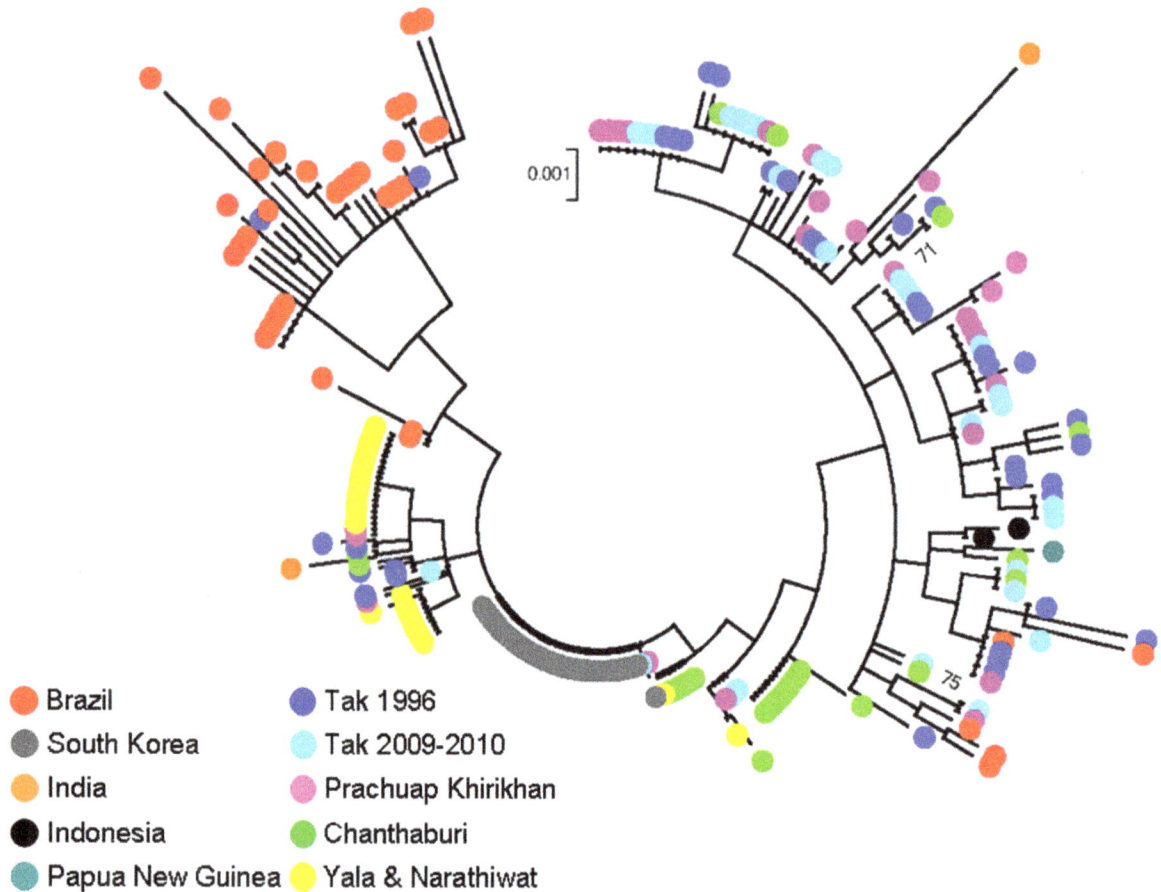

**Figure 4. Phylogenetic analysis inferred from a total of 246 PvTRAP sequences from this study and those previously reported [19,28] using the Maximum Likelihood method based on the Tamura 3-parameter model.** Gaps and missing data are excluded from analysis. Bootstrap values more than 70% are shown. Dots with different colors represent isolates and their geographic origins as listed on the lower left.

negatively selected amino acids and residues under positive selection could stem from functional constraint of this molecule and probably driven by host immune pressure, respectively. Meanwhile, it seems that intragenic recombinantion was pronounced in populations from Tak and Prachuap Khirikhan provinces while relatively few recombinations occurred in parasites from Chanthaburi province and southern isolates. These findings are in accord with our previous analysis on *PvMSP4* and *PvMSP5* in which frequency of recombination seems to correlate with endemicity of each area [25,53].

Despite low level of sequence diversity, genetic distances inferred from the *PvTRAP* locus among *P. vivax* populations could be deployed for analysis of population structure. Pairwise comparison of populations revealed that South Korean isolates displayed high and significant *Fst* indices when compared with malaria populations in other endemic areas. Likewise, slightly lower but significant *Fst* indices were observed when isolates from Brazil were compared with populations elsewhere (Table 3). Therefore, gene flow between *P. vivax* populations in Thailand, Brazil and Korea may not occur because geographic distance has precluded intra- and intercontinental spread of parasites. Although malaria can be transferred between remote endemic areas through infected travelers, this situation could be inefficient due to rarity in numbers of such cases to contribute to genetic admixture of parasites in these three countries. However, closer look into genetic differentiation between 4 major malaria endemic areas of

Thailand, significant genetic differentiation of *P. vivax* population between endemic areas was observed except parasites from Tak province collected over a decade apart (1996 and 2006–2007) having low and non-significant *Fst* indices. Therefore, diversity in the *PvTRAP* locus of parasites in Thailand seems to exhibit spatial but not temporal variation. It is noteworthy that regression analysis of linear geographic as well as normal logarithm of geographic distances and pairwise linearized *Fst* values yielded significant correlation (Mantel test, $p = 0.024$ and 0.028, respectively), indicating that *P. vivax* populations in these 4 endemic areas have been isolated by distance (Figure 2). Therefore, it seems likely that the flight range of mosquito vectors was the major determinant for magnitude of genetic differentiation of *P. vivax* populations between endemic areas. However, there remain some other factors that could influence and shape genetic structure of malaria in Thailand. Although geographic distance between Tak and Prachuap Khirikhan provinces is about 160 kilomerters longer than that between Tak and Chanthaburi provinces, significant genetic differentiation occurred in the latter but not the former. It should be noted that transmigration of people along Thai-Myanmar border where both Tak and Prachuap Khirikhan provinces are located is intense during the past decades whereas transmigration of gem miners who, not uncommonly, carried malaria parasites in their circulation between Tak and Chantha-buri provinces had ceased for over 3 decades ago [56]. Further-more, the abundance of malaria vectors along Thai-Myanmar

**Table 7.** Amino acid substitutions in potential T cell epitopes of PvTRAP and predicted scores*.

| HLA | Domain | Predicted epitopes | Predicted score | %Reduction | No. isolates |
|-----|--------|--------------------|-----------------|------------|--------------|
| A1 | I | YLLVVFLLY | 2.1645 | | 112 |
| | | YLLVVFFLY | 1.6240 | 24.97 | 2 |
| A1 | II | VCNESVDLY | 0.9376 | | 113 |
| | | VCNESVDLI | 0.7571 | 19.25 | 1 |
| A2 | I | FLLYVSIFA | 1.1395 | | 112 |
| | | FFLYVSIFA | 0.5416 | 52.47 | 2 |
| A2 | II | NMTAALEEV | 1.1851 | | 47 |
| | | NMTAALDEV | 1.1104 | 6.30 | 65 |
| | | NMTAALDDV | 0.7819 | 34.02 | 2 |
| A3 | II | YTALEVAKK | 0.8093 | | 63 |
| | | YRALEVAKK | 0.2512 | 68.96 | 51 |
| A3 | II | KVTELRKSY | 0.7907 | | 102 |
| | | KVTELRKTY | 0.5974 | 24.45 | 12 |
| A24 | I | VFLLYVSIF | 1.7214 | | 112 |
| | | VFFLYVSIF | 1.4858 | 13.69 | 2 |
| A26 | II | ELRKTYSPY | 2.0070 | | 12 |
| | | ELRKSYSPY | 1.9132 | 4.67 | 102 |
| | | ESVDLYLLV | 1.3830 | 31.09 | 1 |
| | | ESVDIYLLV | 1.3656 | 31.96 | 113 |
| B7 | II | KLKQRNVSL | 0.8786 | | 112 |
| | | KLKKRNVSL | 0.8400 | 4.39 | 1 |
| | | KLKQRNVTL | 0.7263 | 17.33 | 1 |
| B7 | II | KQRNVTLAV | 0.9688 | | 1 |
| | | KKRNVSLAV | 0.8745 | 9.73 | 1 |
| | | KQRNVSLAV | 0.7908 | 18.37 | 112 |
| B7 | IV | LPVPAPLPA | 1.3094 | | 2 |
| | | LPVPAPLPT | 1.1464 | 12.45 | 112 |
| B8 | II | RPRERNCKF | 1.9566 | | 1 |
| | | RPRELNCKF | 1.3892 | 29.00 | 2 |
| | | RPRESNCKF | 1.0950 | 44.04 | 2 |
| | | RPREPNCKF | 1.0440 | 46.64 | 109 |
| B8 | II | ELRKTYSPY | 0.9655 | | 12 |
| | | ELRKSYSPY | 0.8494 | 12.02 | 102 |
| B27 | II | KRNVSLAVI | 1.2095 | | 1 |
| | | QRNVSLAVI | 0.9591 | 20.70 | 112 |
| | | QRNVTLAVI | 0.9591 | 20.70 | 1 |
| B27 | II | KQRNVSLAV | 0.7781 | | 112 |
| | | KQRNVTLAV | 0.7781 | 0 | 1 |
| | | KKRNVSLAV | 0.5384 | 30.81 | 1 |
| B27 | III | GRGTHSRSR | 1.2775 | | 96 |
| | | GKGTHSRSR | 0.4904 | 61.61 | 18 |
| B39 | I | SYLLVVFLL | 0.7915 | | 112 |
| | | SYLLVVFFL | 0.6755 | 14.66 | 2 |
| B44 | IV | NEKVIPNPL | 1.4650 | | 110 |
| | | NEKVIPTPL | 1.4469 | 1.24 | 3 |
| | | NEKVIPKPL | 1.2157 | 17.02 | 1 |
| B58 | I | KSYLLVVFF | 1.4989 | | 2 |
| | | KSYLLVVFL | 0.8347 | 44.31 | 112 |
| B62 | II | KQRNVTLAV | 1.1504 | | 1 |
| | | KQRNVSLAV | 1.1443 | 0.53 | 112 |

**Table 7.** Cont.

| HLA | Domain | Predicted epitopes | Predicted score | %Reduction | No. isolates |
|-----|--------|-------------------|-----------------|------------|--------------|
|     |        | KKRNVSLAV | 0.2828 | 75.42 | 1 |

*Based on the C-terminal cleavage and the transporter associated with antigen processing efficiency [48].

border could facilitate gene flow of *P. vivax* in these areas. In contrast, ecological niches of *Anopheles* vectors between Tak and Chanthaburi provinces are largely interrupted by focal deforestation and urbanization, precluding natural spread of malaria between these areas.

Analysis of hierarchical splitting of gene pools by STRUCTURE has shown that $K = 3$ is an optimum value. Intriguingly, $K = 3$ could not subdivide South Korean and southern Thai isolates that in fact differed biologically and genetically. A very long incubation period of vivax malaria from 230 to 300 days has been recognized among South Korean isolates [28] but, to our knowledge, has never been recorded in Thai strains. Furthermore, phylogenetic analysis has placed all South Korean isolates to a separate cluster from southern Thai isolates, *albeit* without significant bootstrap support (Figure 3). In contrast, the partition at $K = 5$ is more consistent with results from analysis using *Fst* indices in which distinct subpopulation structure was perceived between isolates from southern Thailand and those from South Korea whilst at $K < 5$, no discernible difference was observed between these populations. Meanwhile, phylogenetic inference does not support distinct clusters of isolates relating to their geographic origins because no branch received remarkably high bootstrap values, consistent with our previous study [19]. Therefore, genetic diversity of the *PvTRAP* locus displays tendency towards geographic variation (Figure 4).

Protective immunity against malarial TRAP involves both humoral and cell mediated immune responses. Antibodies against TRAP targeting domains involved in gliding motility and hepatocyte binding ligands could prevent sporozoite invasion into hepatocytes [13] while TRAP-specific cytotoxic T cells could destroy sporozoite-infected hepatocytes [57]; thereby, exoerythrocytic development does not ensue. The roles of anti-TRAP antibodies in partial protection from or significantly reduced risk of falciparum malaria have been demonstrated in African endemic areas [12,58,59]. With analogy to the thrombospodin-related motif (TRM) of PfTRAP that binds heparin sulfate proteoglycan of hepatocyte, the TRM in domain III of PvTRAP is relatively conserved with two amino acid substitutions, i.e. WTACSVTCGR(or K)GTH (or Q) SRSR, while the MIDAS motif is perfectly conserved among clinical isolates. Therefore, limited diversity at the functional domains in PvTRAP has encouraged vaccine incorporation. Meanwhile, several CD4+ and CD8+ T cell epitopes have been mapped across PfTRAP, some of which have been associated with protection [57,60,61]. Importantly, T cell responses to PfTRAP were commonly allele-specific although certain epitopes

could induce or serve as targets of cross-reactive immunity [60]. Recent studies have highlighted the importance of protective roles of CD8+ T cell and memory T cell responses to PfTRAP from clinical malaria [61,62]. The significance of diversity in these T cell epitopes among natural malaria population remains to be elucidated in term of vaccine efficacy. To date, little is known about immunological responses to PvTRAP although vaccination study using synthetic peptide encompassing hepatocyte-binding ligand of the protein has conferred protection against parasite challenge in *Aotus* monkeys [63]. Our analysis suggests that polymorphisms in PvTRAP could affect T cell recognition because amino acid substitutions in several predicted HLA-binding peptides have remarkably altered the predicted epitope scores (Table 7).

In conclusion, our analysis has shown that genetic diversity in the *PvTRAP* locus of Thai *P. vivax* isolates exhibits geographic variation and population structure. Positive selection and intragenic recombination have shaped diversity at the PvTRAP locus. The low level of sequence diversity in PvTRAP among clinical isolates as shown in this study may not drastically compromise vaccine design.

## Supporting Information

**Table S1   Haplotypes of PvTRAP from Thai isolates.**

**Table S2   Characteristics of amino acid substitutions in PvTRAP of worldwide and Thai isolates.**

**Table S3   Minimum number of recombination events (Rm) and recombination sites in the PvTRAP gene of each parasite population in Thailand.**

## Acknowledgments

We are grateful to all patients who donated their blood samples for this study. We thank the staff of the Bureau of Vector Borne Disease, Department of Disease Control, Ministry of Public Health, Thailand, for assistance in field work.

## Author Contributions

Conceived and designed the experiments: SJ CP RK. Performed the experiments: RK CP SJ. Analyzed the data: SJ CP RK. Contributed reagents/materials/analysis tools: SJ CP HT. Contributed to the writing of the manuscript: SJ CP. Reviewed the paper: SJ CP RK HT.

## References

1. Lopez AD, Mathers CD, Ezzati M, Jamison DT, Murray CJ (2006) Global and regional burden of disease and risk factors, 2001: systematic analysis of population health data. Lancet 367: 1747–1757.
2. Price RN, Tjitra E, Guerra CA, Yeung S, White NJ, et al. (2007) Vivax malaria: neglected and not benign. Am J Trop Med Hyg 77(Suppl.): 79–87.
3. Hay SI, Guerra CA, Tatem AJ, Noor AM, Snow RW (2004) The global distribution and population at risk of malaria: past, present, and future. Lancet Infect Dis 4: 327–336.
4. Crompton PD, Pierce SK, Miller LH (2010) Advances and challenges in malaria vaccine development. J Clin Invest 120: 4168–4178.
5. Mayxay M, Pukrittayakamee S, Newton PN, White NJ (2004) Mixed-species malaria infections in humans. Trends Parasitol 20: 233–240.
6. Casares S, Brumeanu TD, Richie TL (2010) The RTS,S malaria vaccine. Vaccine 28: 4880–4894.
7. Hoffman SL, Goh LM, Luke TC, Schneider I, Le TP, et al. (2002) Protection of humans against malaria by immunization with radiation-attenuated *Plasmodium falciparum* sporozoites. J Infect Dis 185: 1155–1164.

8. Roestenberg M, McCall M, Hopman J, Wiersma J, Luty AJ, van Gemert GJ, et al. (2009) Protection against a malaria challenge by sporozoite inoculation. N Engl J Med 361: 468–477.

9. Rogers WO, Malik A, Mellouk S, Nakamura K, Rogers MD, et al. (1992) Characterization of *Plasmodium falciparum* sporozoite surface protein 2. Proc Natl Acad Sci USA 89: 9176–9180.

10. Khusmith S, Charoenvit Y, Kumar S, Sedegah M, Beaudoin RL, et al. (1991) Protection against malaria by vaccination with sporozoite surface protein 2 plus CS protein. Science 252: 745–718.

11. Hafalla JCR, Bauza K, Friesen J, Gonzalez-Aseguinolaza G, Hill AVS, et al. (2013) Identification of targets of CD8+ T cell responses to malaria liver stages by genome-wide epitope profiling. PLoS Pathog 9: e1003303.

12. Scarselli E, Tolle R, Koita O, Diallo M, Müller HM, et al. (1993) Analysis of the human antibody response to thrombospondin-related anonymous protein of *Plasmodium falciparum*. Infect Immun 61: 3490–3495.

13. Muller HM, Reckman I, Hollingdale MR, Bujard H, Robson KJH, et al. (1993) Thrombospondin related anonymous protein (TRAP) of *Plasmodium falciparum* binds specifically to sulfated glycoconjugates and to HepG2 hepatoma cells suggesting a role for this molecule in sporozoite invasion of hepatocytes. EMBO J 12: 2881–2889.

14. Sultan AA, Thathy V, Frevert U, Robson KJH, Crisanti A, et al. (1997) TRAP is necessary for gliding motility and infectivity of *Plasmodium* sporozoites. Cell 90: 511–522.

15. Robson KJ, Hall JR, Jennings MW, Harris TJ, Marsh K, et al. (1988) A highly conserved amino-acid sequence in thrombospondin, properdin and in proteins from sporozoites and blood stages of a human malaria parasite. Nature 335: 79–82.

16. Templeton TJ, Kaslow DJ. (1997) Cloning and cross-species comparison of the thrombospondin-related anonymous protein (TRAP) gene from *Plasmodium knowlesi*, *Plasmodium vivax* and *Plasmodium gallinaceum*. Mol Biochem Parasitol 84: 13–24.

17. Kappe SH, Buscaglia CA, Bergman LW, Coppens I, Nussenzweig V (2004) Apicomplexan gliding motility and host cell invasion: overhauling the motor model. Trends Parasitol 20: 13–16.

18. Jongwutiwes S, Putaporntip C, Kanbara H, Tanabe K (1998) Variation in the thrombospondin-related adhesive protein (TRAP) gene of *Plasmodium falciparum* from Thai field isolates. Mol Biochem Parasitol 92: 349–353.

19. Putaporntip C, Jongwutiwes S, Tia T, Ferreira MU, Kanbara H, et al. (2001) Diversity in the thrombospondin-related adhesive protein gene (TRAP) of *Plasmodium vivax*. Gene 268: 97–104.

20. Ohashi J, Suzuki Y, Naka I, Hananantachai H, Patarapotikul J (2014) Diversifying selection on the thrombospondin-related adhesive protein (TRAP) gene of *Plasmodium falciparum* in Thailand. PLoS One 9: e90522.

21. Gilbert SC, Plebanski M, Gupta S, Morris J, Cox M, et al. (1998) Association of malaria parasite population structure, HLA, and immunological antagonism. Science 279: 1173–1177.

22. Putaporntip C, Hongsrimuang T, Seethamchai S, Kobasa T, Limkittikul K, et al. (2009) Differential prevalence of *Plasmodium* infections and cryptic *Plasmodium knowlesi* malaria in humans in Thailand. J Infect Dis 199: 1143–1150.

23. Jongwutiwes S, Buppan P, Kosuvin R, Seethamchai S, Pattanawong U, et al. (2011) *Plasmodium knowlesi* malaria in humans and macaques, Thailand. Emerg Infect Dis 17: 1799–1806.

24. Jongwutiwes S, Putaporntip C, Hughes AL (2010) Bottleneck effects on vaccine-candidate antigen diversity of malaria parasites in Thailand. Vaccine 28: 3112–3117.

25. Putaporntip C, Udomsangpetch R, Pattanawong U, Cui L, Jongwutiwes S. (2010) Genetic diversity of the *Plasmodium vivax* merozoite surface protein-5 locus from diverse geographic origins. Gene 456: 24–35.

26. Putaporntip C, Jongwutiwes S, Sakihama N, Ferreira MU, Kho W-G, et al. (2002) Mosaic organization and heterogeneity in frequency of allelic recombination of the *Plasmodium vivax* merozoite surface protein-1 locus. Proc Nat Acad Sci USA 99: 16348–16353.

27. Thompson JD, Gibson TJ, Plewniak F, Jeanmougin F, Higgins DG (1997) The ClustalX windows interface: flexible strategies for multiple sequence alignment aided by quality analysis tools. Nucleic Acids Res 25: 4876–4882.

28. Nam MH, Jang JW, Kim H, Han ET, Lee WJ, et al. (2011) Conserved sequences of thrombospondin-related adhesive protein gene of *Plasmodium vivax* in clinical isolates from Korea. Trop Med Int Health 16: 923–928.

29. Nei M (1987) Molecular evolutionary genetics. New York: Columbia University Press.

30. Tamura K, Stecher G, Peterson D, Filipski A, Kumar S (2013) MEGA6: Molecular Evolutionary Genetics Analysis version 6.0. Mol Biol Evol 30: 2725–2729.

31. Hudson RR, Kaplan NL (1985) Statistical properties of the number of recombination events in the history of a sample of DNA sequences. Genetics 111: 147–164.

32. Librado P, Rozas J (2009) DnaSP v5: A software for comprehensive analysis of DNA polymorphism data. Bioinformatics 25: 1451–1452.

33. Kosakovsky Pond SL, Frost SDW, Muse SV (2005) HyPhy: hypothesis testing using phylogenies. Bioinformatics 21: 676–679.

34. Kosakovsky Pond SL, Posada D, Gravenor MB, Woelk CH, Frost SDW (2006) Automated phylogenetic detection of recombination using a genetic algorithm. Mol Biol Evol 23: 1891–1901.

35. Nei M, Gojobori T (1986) Simple methods for estimating the numbers of synonymous and nonsynonymous nucleotide substitutions. Mol Biol Evol 3: 418–426.

36. Jukes TH, Cantor CR (1969) Evolution of protein molecules. In Munro HN editor. Mammalian Protein Metabolism 21–132 Academic Press New York.

37. Tajima F (1983) Evolutionary relationship of DNA sequences in finite populations. Genetics 105: 437–460.

38. Fu YX, Li WH (1993) Statistical tests of neutrality of mutations. Genetics 133: 693–709.

39. McDonald JH, Kreitman M (1991) Adaptive protein evolution at the Adh locus in *Drosophila*. Nature 351: 652–654.

40. Murrell B, Wertheim JO, Moola S, Weighill T, Scheffler K, et al. (2012) Detecting individual sites subject to episodic diversifying selection. PLoS Genet 8: e1002764.

41. Kosakovsky Pond SL, Frost SDW (2005) Datamonkey: Rapid detection of selective pressure on individual sites of codon alignments. Bioinformatics 21: 2531–2533.

42. Excoffier L, Smouse PE, Quattro JM (1992) Analysis of molecular variance inferred from metric distances among DNA haplotypes: Application to human mitochondrial DNA restriction data. Genetics 131: 479–491.

43. Weir BS, Cockerham CC (1984) Estimating $F$ statistics for the analysis of population structure. Evolution 38: 1358–1370.

44. Slatkin M (1993) Isolation by distance in equilibrium and non equilibrium populations. Evolution 47: 264–279.

45. Pritchard JK, Stephens M, Donnelly P (2000) Inference of population structure using multilocus genotype data. Genetics 155: 945–959.

46. Evanno G, Regnaut S, Goudet J. (2005) Detecting the number of clusters of individuals using the software STRUCTURE: a simulation study. Mol Ecol 14: 2611–2620.

47. Earl DA, vonHoldt BM (2012) STRUCTURE HARVESTER: a website and program for visualizing STRUCTURE output and implementing the Evanno method. Conservation Genet Resour 4: 359–361.

48. Larsen MV, Lundegaard C, Lamberth K, Buus S, Brunak S, et al. (2005) An integrative approach to CTL epitope prediction: a combined algorithm integrating MHC class I binding, TAP transport efficiency, and proteasomal cleavage predictions. Eur J Immunol 35: 2295–2303.

49. Wengelnik K, Spaccapelo R, Naitza S, Robson KJH, Janse CJ, et al. (1999) The A-domain and the thrombospondin-related motif of *Plasmodium falciparum* TRAP are implicated in the invasion process of mosquito salivary glands. EMBO J 18: 5195–5204.

50. Matuschewski K, Nunes AC, Nussenzweig V, Menard R (2002) *Plasmodium* sporozoite invasion into insect and mammalian cells is directed by the same dual binding system. EMBO J 21: 1597–1606.

51. Song G, Koksal AC, Lu C, Springer TA (2012) Shape change in the receptor for gliding motility in *Plasmodium* sporozoites. Proc Nat Acad Sci USA 109: 21420–21425.

52. Pihlajamaa T, Kajander T, Knuuti J, Horkka K, Sharma A, et al. (2013) Structure of *Plasmodium falciparum* TRAP (thrombospondin-related anonymous protein) A domain highlights distinct features in apicomplexan von Willebrand factor A homologues. Biochem J 450: 469–476.

53. Putaporntip C, Jongwutiwes S, Ferreira MU, Kanbara H, Udomsangpetch R, et al. (2009) Limited global diversity of the *Plasmodium vivax* merozoite surface protein 4 gene. Infect Genet Evol 9: 821–826.

54. Aris-Brosou S, Excoffier L (1996) The impact of population expansion and mutation rate heterogeneity on DNA sequence polymorphism. Mol Biol Evol 13: 494–504.

55. Maruyama T, Fuerst PA (1984) Population bottlenecks and non-equilibrium models in population genetics. I. Allele numbers when populations evolve from zero variability. Genetics 108: 745–763.

56. Thimasarn K, Jatapadma S, Vijaykadga S, Sirichaisinthop J, Wongsrichanalai C (1995) Epidemiology of malaria in Thailand. J Travel Med 2: 59–65.

57. Schneider J, Gilbert SC, Blanchard TJ, Hanke T, Robson KJH, et al. (1998) Enhanced immunogenicity for CD8+ T cell induction and complete protective efficacy of malaria DNA vaccination by boosting with modified vaccinia virus Ankara. Nat Med 4: 397–402.

58. John CC, Zickafoose JS, Sumba PO, King CL, Kazura JW (2003) Antibodies to the *Plasmodium falciparum* antigens circumsporozoite protein, thrombospondin-related adhesive protein, and liver-stage antigen 1 vary by ages of subjects and by season in a highland area of Kenya. Infect Immun 71: 4320–4325.

59. John CC, Tande AJ, Moormann AM, Sumba PO, Lanar DE, et al. (2008) Antibodies to pre-erythrocytic *Plasmodium falciparum* antigens and risk of clinical malaria in Kenyan children. J Infect Dis 197: 519–526.

60. Flanagan KL, Plebanski M, Odhiambo K, Sheu E, Mwangi T, et al. (2006) Cellular reactivity to the p. *Falciparum* protein trap in adult kenyans: novel epitopes, complex cytokine patterns, and the impact of natural antigenic variation. Am J Trop Med Hyg 74: 367–375.

61. Ewer KJ, O'Hara GA, Duncan CJ, Collins KA, Sheehy SH, et al. (2013) Protective CD8+ T-cell immunity to human malaria induced by chimpanzee adenovirus-MVA immunisation. Nat Commun 4: 2836.

62. Todryk SM, Bejon P, Mwangi T, Plebanski M, Urban B, et al. (2008) Correlation of memory T cell responses against TRAP with protection from clinical malaria, and CD4+ CD25 high T Cells with susceptibility in Kenyans. PLoS ONE 3: e2027.

63. Castellanos A, Arévalo-Herrera M, Restrepo N, Gulloso L, Corradin G, et al. (2007) *Plasmodium vivax* thrombospondin related adhesion protein: Immunogenicity and protective efficacy in rodents and Aotus monkeys. Mem Inst Oswaldo Cruz 102: 411–416.

# 9

# Identification of MMV Malaria Box Inhibitors of *Perkinsus marinus* Using an ATP-Based Bioluminescence Assay

**Yesmalie Alemán Resto[2], José A. Fernández Robledo[1]***

**1** Bigelow Laboratory for Ocean Sciences, Boothbay, Maine, United States of America, **2** Research Experiences for Undergraduates (REU) NSF Program - 2013 - Bigelow Laboratory for Ocean Sciences, Boothbay, Maine, United States of America

## Abstract

"Dermo" disease caused by the protozoan parasite *Perkinsus marinus* (Perkinsozoa) is one of the main obstacles to the restoration of oyster populations in the USA. *Perkinsus* spp. are also a concern worldwide because there are limited approaches to intervention against the disease. Based on the phylogenetic affinity between the Perkinsozoa and Apicomplexa, we exposed *Perkinsus* trophozoites to the Medicines for Malaria Venture Malaria Box, an open access compound library comprised of 200 drug-like and 200 probe-like compounds that are highly active against the erythrocyte stage of *Plasmodium falciparum*. Using a final concentration of 20 µM, we found that 4 days after exposure 46% of the compounds were active against *P. marinus* trophozoites. Six compounds with $IC_{50}$ in the µM range were used to compare the degree of susceptibility *in vitro* of eight *P. marinus* strains from the USA and five *Perkinsus* species from around the world. The three compounds, MMV666021, MMV665807 and MMV666102, displayed a uniform effect across *Perkinsus* strains and species. Both *Perkinsus marinus* isolates and *Perkinsus* spp. presented different patterns of response to the panel of compounds tested, supporting the concept of strain/species variability. Here, we expanded the range of compounds available for inhibiting *Perkinsus* proliferation *in vitro* and characterized *Perkinsus* phenotypes based on their resistance to six compounds. We also discuss the implications of these findings in the context of oyster management. The *Perkinsus* system offers the potential for investigating the mechanism of action of the compounds of interest.

**Editor:** Tobias Spielmann, Bernhard Nocht Institute for Tropical Medicine, Germany

**Funding:** This study was supported by institutional funds from Bigelow Laboratory for Ocean Sciences, grant OCE0755142 (REU Program) from the National Science Foundation, and grant 1R21AI076797-01A2 from the National Institute of Health. The funders had no role in study design, data collection and analysis, decision to publish, or preparation of the manuscript.

**Competing Interests:** José A. Fernández Robledo is current Academic Editor.

* Email: jfernandez-robledo@bigelow.org

## Introduction

*Perkinsus marinus* and *Perkinsus chesapeaki* cause Dermo disease in oysters and clams in the USA. Described in the early 1950s, Dermo disease is associated with mass mortalities of eastern oysters (*Crassostrea virginica*) in the Gulf Coast [1]; now it is under surveillance by the World Organization for Animal Health (OIE; http://www.oie.int/; Aquatic Animal Health Code, Section 11: Diseases of Molluscs). The Chesapeake Bay (Maryland, Virginia, USA) is a clear example of where *P. marinus* has contributed to the decimation of the oyster industry (today's production in Maryland is 4.2% of the production in the mid-1960s). The expansion of the *P. marinus* distribution range in the USA has been associated with global warming and the shellfish trade [2,3]. Dermo remains an important obstacle to the restoration of oyster populations in numerous eastern states [3,4]. Interestingly, *P. marinus* has also been reported with high prevalence in oysters from eastern states with no noticeable mollusk mortality [5], and recent records of *P. marinus* in oysters from the West Coast of North America were not associated with mortalities [6]. The presence of *P. marinus* phenotypes and genotypes might account for differences in virulence [7–9]. In the USA, *P. chesapeaki*

displays a high preference for infecting clams and it appears to be better adapted to lower salinities and temperatures than *P. marinus* [5] and recently it has been detected in cockles (*Cerastoderma edule*) in Europe [10]. Worldwide, seven *Perkinsus* spp. have been described, most of them in the last decade with five of them available in *in vitro* culture (reviewed in [11]).

Compared to parasites of human and veterinary relevance, the pharmacopoeia for marine protozoan parasites is still very limited, and some of these compounds are toxic in the marine environment [12,13]. *Perkinsus* and other non-photosynthetic relatives of both dinoflagellates and apicomplexans lineages have lost the ability to perform photosynthesis; still, they have retained a cryptic plastid and its pathways (Chromalveolata hypothesis), which are recognized as promising drug targets [14,15]. Government agencies, drug companies, and non-profit organizations have screened multiple compound libraries against *Plasmodium falciparum* resulting in the Medicines for Malaria Venture (MMV) Malaria Box (http://www.mmv.org/malariabox) [16]. This compound library is being used to find inhibitors of defined parasite life stages [17,18], to describe mechanisms of action [19], and to find active compounds against other protozoan parasites [20,21]. Here, we followed similar approach and tested the MMV Malaria Box for the discovery of novel hits against *Perkinsus* using an adenosine tri-phosphate (ATP) content-based assay to test *P. marinus* proliferation growth [13].

## Materials and Methods

### Materials

The MMV Malaria Box constitutes 200 drug-like and 200 probe-like compounds with activity against the blood-stage of *Plasmodium falciparum* 3D7 (http://www.mmv.org/research-development/malaria-box-supporting-information). Stock solutions (20 mM) (Batch April2013; Table S1) in dimethyl sulfoxide (DMSO) were diluted in water and tested in the primary screening at a final concentration of 20 µM. The compounds were not repurchased nor re-synthetized; consequently, the results should be considered as primary unconfirmed hits until the identification of these compounds is followed up by a proper confirmation.

### Parasite strains and *in vitro* culture

Experiments were carried out with eight *P. marinus* strains and five *Perkinsus* species (Table 1). Cultures were maintained in Dulbecco modified Eagle's: Ham's F12 (1:2) supplemented with 5% fetal bovine serum in 25 cm$^2$ (5 ml) vented flasks in a 26–28°C microbiology incubator as reported elsewhere [22]. For the compound library screening, *P. marinus* PRA240 [13,23] cultures were expanded in a 75 cm$^2$ (30–50 ml) vented flask in a microbiology incubator fitted with orbital shaking (70–80 rpm).

### 96-Well format *Perkinsus* growth-inhibition primary screen

*Perkinsus marinus* PRA240 (100 µl, 2.0–5.0×10$^6$ cells/ml or 2000–4000 relative fluorescence units, RFU) were prepared in sterile 96-well plates (white OptiPlateTM-96, PerkinElmer Life Sciences, Boston, MA). *Perkinsus marinus* cells were exposed once to the MMV Malaria Box (20 µM; final concentration of DMSO was 0.1%; concentrations above 0.1% are toxic to *P. marinus* trophozoites) in triplicate. Control wells (×3) included DMSO

with *Perkinsus* cells, culture medium with cells and culture medium with no cells. The effect of the compounds on *P. marinus* proliferation was evaluated using the ATPlite assay at day 4 post-exposure, as reported elsewhere [13]. Readings for each well were normalized to the control wells with cells and DMSO (100% activity).

### Secondary *Perkinsus* growth-inhibition screen (IC$_{50}$) and *Perkinsus* strain and species sensitivity

Six of the best hits from the MMV Malaria Box (Table 2) were retested on *P. marinus* PRA240; the IC$_{50}$ was calculated in an 8-point dose-response curve (10 µM to 0.156 µM) using Prim6 (sigmoidal) (Graphpad Software, Inc.). Eight *P. marinus* strains and five *Perkinsus* spp. (Table 1) were tested to compare their relative sensitivity using 2 µM day 2 post-exposure. *In vitro* cultures *Perkinsus olseni*, *P. chesapeaki*, and *P. mediterraneus* are characterized by either sporulating, making the culture medium acid or remaining in clumps, or having very large trophozoites [24–27]. Consequently, to standardize the assay, aliquots from the cultures in the exponential phase were used for ATP measurement and then the experimental-well plates seeded with cells- ATP activity equivalent to *P. marinus* PRA240 2.0–5.0×10$^6$ cells (2000–4000 RLU) [13]. The effect of the compounds on *P. marinus* strains and *Perkinsus* spp. proliferation was evaluated as above.

## Results and Discussion

### MMV Malaria Box screen

In this study, we screened the MMV Malaria Box for compounds that might inhibit *P. marinus* proliferation *in vitro*, an approach that has been successfully used to identify compounds against other protozoan parasites [19–21]. In our previous study, the effect of the drugs on *P. marinus* proliferation was evaluated at days 2, 4, and 8 post-exposure; however, it was at day 4 post-exposure when the inhibitory effect(s) of most drugs tested became apparent [13]. Consequently, for the MMV Malaria Box

**Table 1.** *Perkinsus* spp. and *Perkinsus marinus* strains used in the study.

| Perkinsus sp.* | Strain | ATCC # | Location/year of isolation | Host | Reference |
|---|---|---|---|---|---|
| P. marinus | C13-11 [MA-2-11] | 50896 | Cotuit, MA (USA)/1998 | Crassostrea virginica | |
| | LICT-1 [CT-1] | 50508 | Long Island Sound, CT (USA)/1998 | Crassostrea virginica | |
| | DBNJ-1 [NJ-1] | 50509 | Delaware Bay, NJ (USA)/1993 | Crassostrea virginica | |
| | CB5D4 | PRA240 | Bennett Point, MD (USA)/2008 | Crassostrea virginica | [23] |
| | CB5D4 | PRA393 | GFP mutant derived from PRA240 | Crassostrea virginica | [13,23] |
| | HCedar2 | 50757 | Cedar Keys, FL (USA)/1998 | Crassostrea virginica | |
| | HTtP14 [FL-6] | 50763 | Fort Pierce, FL (USA) | Crassostrea virginica | |
| | TXsc | 50983 | Galveston Bay, TX (USA)/1993 | Crassostrea virginica | [28] |
| P. chesapeaki (= andrewsi) | A8-4a | 50807 | Fox Point, MD (USA)/2001 | Macoma balthica | [25] |
| P. olseni (= atlanticus) | ALG1 | 50984 | Ria Formosa, Algarve (Portugal)/2002 | Tapes decussatus | [24] |
| P. mediterraneus | G2 | PRA238 | Menorca (Spain)/2003 | Ostrea edulis | [47] |
| P. honshuensis | Mie-3G/H8 | PRA177 | Gokasho Bay, Mie Pref. (Japan)/2002 | Venerupis philippinarum | [56] |

*Perkinsus marinus* PRA240 was used for the primary screen. A total of eight *P. marinus* strains isolated from oysters from the East and Gulf Coast of the USA and five *Perkinsus* spp. from around the world were used for the secondary screen. In all the cases cultures were maintained in Dulbecco modified Eagle's: Ham's F12 (1:2) supplemented with 5% fetal bovine serum.
*Perkinsus qugwadi* and *Perkinsus beihaiensis* have never been available in culture [57,58].

**Table 2.** List of compounds active against *Plasmodium falciparum* selected for the MMV Malaria Box (http://www.mmv.org/malariabox) for secondary *Perkinsus marinus* growth-inhibition screen (IC$_{50}$) and *Perkinsus marinus* strain and *Perkinsus* species sensitivity.

| HEOS Compound ID | Target | Smiles | EC$_{50}$ (µM)* | Set | MW (KDa) | EC$_{50}$ (µM)** |
|---|---|---|---|---|---|---|
| MMV665941 | Unknown | CN(C)c1ccc(cc1)C(O)(c2ccc(cc2)N(C)C)c3ccc(cc3)N(C)C | 0.255 | Probe-like | 389.53 | 5.35 |
| MMV666021 | Yes, 29 | Cc1ccc(cc1)c2cc3C(=O)c4ccccc4c3nn2 | 0.094 | Probe-like | 272.30 | 1.05 |
| MMV665807 | TM protease serine 4 | Oc1ccc(Cl)cc1C(=O)Nc2cccc(c2)C(F)(F)F | ND | Drug-like | 315.67 | 2.00 |
| MMV666102 | Functional 17 | CN(C)c1ccc(cc1)c2nc3cc(N)ccc3[nH]2 | ND | Drug-like | 252.31 | 1.77 |
| MMV396719 | Functional 11 | n1(c2c(cccc2)n3)c3c4c(cccc4)NC1(C)c5cccc(OC)c5 | 1.150 | Drug-like | 341.40 | 2.08 |
| MMV006522 | Functional 19, Cytotoxic | CCOc1ccc2nc(C)cc(Nc3ccc(Br)cc3)c2c1 | 0.480 | Probe-like | 357.24 | 35.61 |

*Plasmodium falciparum* 3D7;
**Perkinsus marinus* PRA240 primary screen.

screening we measured cell viability at day 4 post-exposure. We found that 46% of the compounds active against the *P. falciparum* erythrocyte life stage were also active against *P. marinus* trophozoites (Table S2). A total of 58 compounds (31.8%) resulted in at least 50% inhibition; from these compounds, 13 (7.1%) resulted in at least 90% inhibition (Figure 1). The repertoire of available anti-*Perkinsus* drugs has gradually increased over the past two decades thanks to the establishment of the culture methodologies for *Perkinsus* spp. [28–30] (Figure 2A). Still, prior to this study, the number of available compounds against *Perkinsus* spp. was very limited (Figure 2B) compared to compounds against protozoan parasites of medical and veterinary relevance [31–34]. Previous screenings for compounds inhibiting *Perkinsus* proliferation have been based on the strong line of evidence for the presence in *Perkinsus*, like those in apicomplexan parasites, of pathways linked to a relic plastid [12,13,35,36]. Here we have shown that the MMV Malaria Box offers a promising alternative way of finding compounds effective against *Perkinsus* spp.

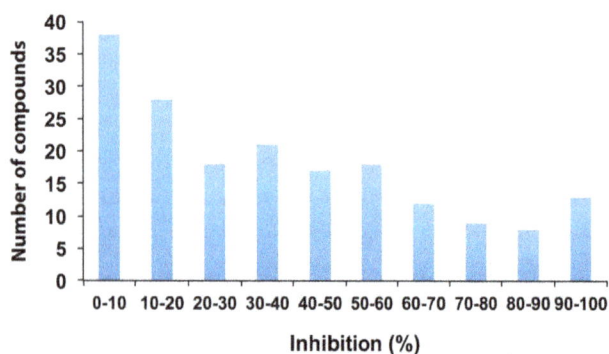

**Figure 1. Percentage of inhibition of *Perkinsus marinus* using the MMV Malaria Box.** Biological triplicate cultures were grown in sterile 96-well plates (100 µl; 2.0×10$^6$ cells/ml) and cells were exposed to the MMV Malaria Box (20 µM). The effect of the drugs on *P. marinus* proliferation was evaluated using the ATPlite assay at day 4 post-exposure to the selected drugs. Readings for each concentration were normalized to the control wells with each solvent (100% activity). A total of 122 (67.0%) compounds resulted in at least 50% inhibition; from these compounds, 13 (7.1%) resulted in at least 90% inhibition (Figure 2).

## Secondary *Perkinsus* growth-inhibition screen (IC$_{50}$)

Three drug-like and three probe-like of the 13 compounds with the highest inhibitory effect on *P. marinus* (Table 2) were randomly selected for calculating the IC$_{50}$ in an 8-point dose-response curve (10 µM to 0.156 µM). We found that the IC$_{50}$ varied between 1.05 µM for MV66602 and 5.35 µM for MMV665941; for MV006522 the IC$_{50}$ was 35.6 µM a high concentration or leaving the compound longer time would have resulted in a fitted sigmoidal curve. In this study the IC$_{50}$ for the selected compounds (Figure 3) was in the lower µM range and much lower than for the compounds tested in our previous study [13], still it was higher than the corresponding *P. falciparum* IC$_{50}$ values (Table 2); consequently, without knowing the mechanism of action of the compounds, we cannot rule out off-target effects due to non-specific cytotoxic agents, including detergent effects, multi-targeting and oxidative effects. The nature of the assays (*Plasmodium falciparum* relies on infected erythrocytes and the *P. marinus* screen is performed in the absence of the host cells) and culture medium can also account for the differences in the IC$_{50}$ values. With a direct life cycle, *P. marinus* trophozoites are phagocytized by the oyster hemocytes [37,38] where they resist oxidative killing [39]. Interestingly, MMV666021 has been involved in the inhibition of glutathione-S-transferase (GST) activity of prostaglandin D2 synthase (PGDS) [40]. GST are involved in parasite survival by protecting them against oxidative stress from the host or from products derived from their own metabolism [41], and in *P. falciparum* it has been associated with chloroquine-resistance [42]. We grow *Perkinsus* in a host cell-free culture medium; hence, if the MMV666021 is indeed affecting the oxidative stress, it is most likely dealing with the ROS derived from the parasites' own metabolism. *Perkinsus marinus* trophozoites have an expanded transporter repertoire, which is useful not only for transporting nutrients but also for secreting extracellular products (ECP) intended to inactivate the host defense and to break down host tissues [43,44]. Protease activity variations significantly decrease the migration of hemocytes [44] and have been associated with differences in the average cell size and growth rate [45]. MMV665807 is believed to target transmembrane serine proteases. Interestingly, the *P. marinus* genome encodes multiple putative serine protease genes (*e.g.* XM_002788359, XM_002786609, XM_002766692); numerous studies have identified serine protease activities in the spent medium of cultured *P. marinus* and *P. mediterraneus* [46,47] and mutations in the promoter region of serine protease inhibitors (SPIs) in *C. virginica*

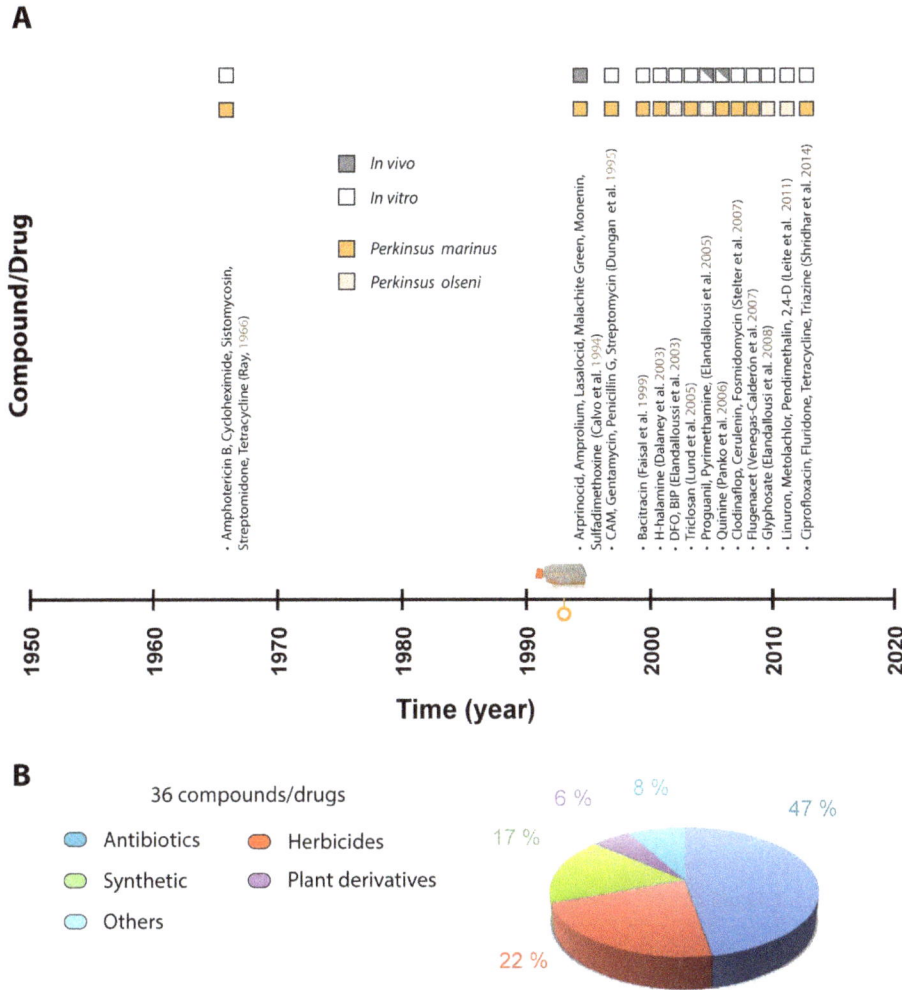

**Figure 2. Drug discovery against Dermo disease.** (A) Time line for the discovery of drugs against Dermo disease, starting when the etiological agent was described until this study. Most of the discoveries did happen after the development of the culture methodologies for *Perkinsus* spp. in 1993 and most studies have been carried out in *in vitro* cultures. (B) Percentage of the compounds active against *Perkinsus* based on their chemical nature. (C) Percentage of available compounds against Dermo tested in *Perkinsus marinus* and *Perkinsus olseni*.

which confer resistance to Dermo disease [48,49]. The parasite proteases could be the target of MMV665807; however, to prove this hypothesis would require further experiments outside the scope of this study.

## Perkinsus marinus strains and Perkinsus species sensitivity

Our results comparing the effect of a single compound concentration (2 µM) against seven *P. marinus* strains indicate that for MMV666021, MMV665807 and MMV666102, the inhibition of the different strains was within an equivalent range (Table 3). Interestingly, for MMV665941, MMV396719 and MMV006522, there was a high degree of variability among the *P. marinus* strains. The presence of *Perkinsus* in low salinity estuaries and the sudden spread of the parasite between oyster beds is often seen as indicating the presence of strains adapted to low salinity and strains of variable virulence; parasites isolated from the Atlantic coast are more virulent than Gulf isolates [50,51]. Indeed, *P. marinus* "races" and genetic strains have been documented along the Atlantic and Gulf coasts of the USA [7–9]. *Perkinsus marinus* strains from Maryland and Virginia appeared to be more susceptible to treatment with the antimalarial drug

Quinine (130 µM) [52]. We also found strain variability to the compounds tested; *P. marinus* HCedar 2 from Florida appears to be much more sensitive to MMV006522 than the other *P. marinus* strains. On the other hand, MMV665941 appears to be less effective against *P. marinus* LICT-1 [CT-1] and DBNJ-1 [NJ-1], isolated from oysters from Connecticut and Delaware respectively. *Perkinsus marinus* LICT-1 [CT-1] also appears to be the strain less sensitive to MMV396719. There is a genetic base beneath *P. marinus* strains [8,9] and recent microsatellite analyses suggest that *P. marinus* utilizes both sexual and asexual reproduction and, over the short-term, selection acts upon independent parasite lineages rather than upon individual *loci* in a cohesive, interbreeding population [53]. Drawing a parallel to other protozoan parasites with markedly clonal population structure and variable degree of virulence [54], it is possible that the observed variability respond to a clonal population structure (strains derived from one original single clone) with variable virulence. Indeed, Reece et al. [8] grouped 76 *P. marinus* isolates in 12 different composite genotypes with >88% of isolates possessing one of three predominant genotypes.

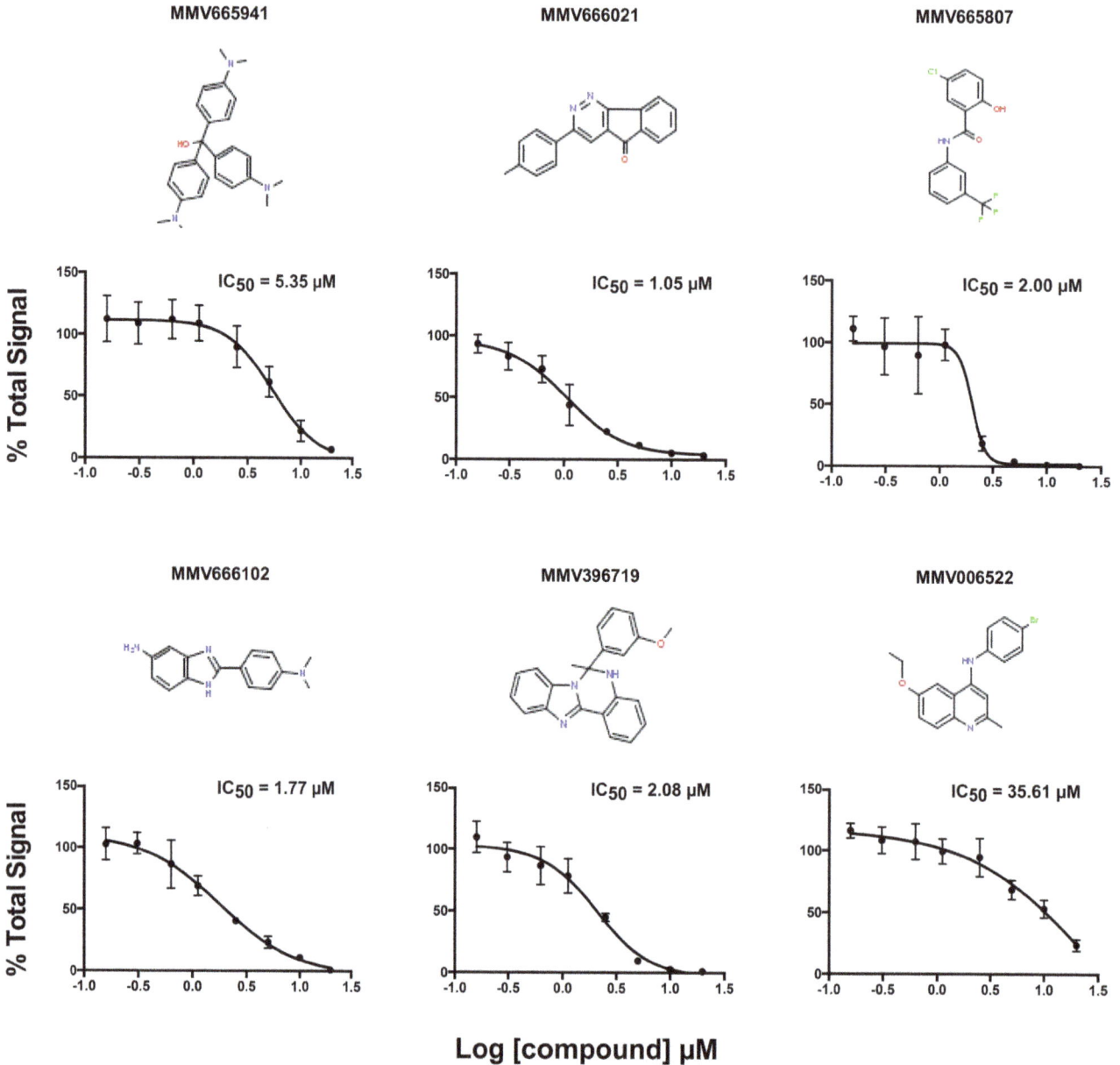

**Figure 3. Secondary** *Perkinsus marinus* **growth-inhibition screen (IC$_{50}$).** Biological triplicate cultures were exposed to an 8 -point dose-response curve (10 μM to 0.156 μM). The effect of the drugs on *P. marinus* proliferation was evaluated as above.

We also compared all five *Perkinsus* spp. in culture using the panel of six drug-like and probe-like compounds at a concentration within the range of the determined IC$_{50}$ (2 μM). We found that MMV666021, MMV665807 and MMV666102 were active against all five *Perkinsus* spp. Interestingly, some compounds were not equally effective against all the *Perkinsus* spp. *P. chesapeaki* was not sensitive to MMV665941, MMV396719 and MMV006522 at the concentration tested. *Perkinsus chesapeaki* affects both oysters and clams along the East Coast of the USA [5] and recently it was detected in cockles from Europe [10]. Compared to other *Perkinsus* spp., *P. chesapeaki* A8-4a is characterized by making the culture medium acidic [25,26], which could affect the potency or the uptake of the compounds tested. Both *P. mediterraneus* G2 and *P. honshuensis* Mie-3G/H8 were not sensitive to MMV665941 at the concentration tested;

interestingly, they were also less sensitive to MMV666102, a compound that showed a high degree of inhibition in most *P. marinus* strains and in *P. chesapeaki*. This study highlights an unexpected degree of variability between *Perkinsus* spp. A plausible explanation could be variations in the propagation rates and strategies in the *in vitro* culture affecting the uptake of the compounds. For example, some *Perkinsus* spp. in culture appear to be "locked" at the trophozoite stage while other *Perkinsus* spp. continuously zoosporulate [26,55]. *P. mediterraneus* cell density only increases two- to sixfold over a 6-week period compared to ten- to thirtyfold in *P. marinus* [47]. With the culture medium indicated above, both *P. mediterraneus* G2 and *P. honshuensis* Mie-3G/H8 are characterized by forming large in clumps in culture; whether this phenotype conditions the uptake of the compounds remains to be demonstrated. To answer these

**Table 3.** Activity of antimalarial drug-like and probe-like (2 μM) compounds on *P. marinus* strains and *Perkinsus* spp. determined at day 2 post-exposure.

| Species | Strain | HEOS Compound ID | | | | | | | | | | | |
|---|---|---|---|---|---|---|---|---|---|---|---|---|---|
| | | MMV665941 | | MMV666021 | | MMV665807 | | MMV666102 | | MMV396719 | | MMV006522 | |
| | | Mean | SDV | Mean | SDV | Mean | SDV | Mean | SDV | Mean | SDV | Mean | SDV |
| *P. marinus* | C13-11 [MA-2-11] | 34.9 | 1.9 | 86.3 | 1.5 | 97.1 | 0.8 | 74.5 | 1.6 | 39.7 | 7.7 | −10.3 | 0.2 |
| | LICT-1 [CT-1] | −20.7 | 10.2 | 79.6 | 1.6 | 95.6 | 0.2 | 64.6 | 1.9 | 36.7 | 10.2 | −4.6 | 7.7 |
| | DBNJ-1 [NJ-1] | −6.8 | 11.0 | 71.6 | 3.2 | 92.4 | 1.1 | 72.6 | 1.7 | −1.3 | 8.6 | −47.4 | 3.3 |
| | CB5D4 | 34.3 | 2.3 | 84.8 | 2.4 | 96.4 | 0.7 | 64.4 | 1.7 | 44.4 | 3.6 | 4.5 | 2.9 |
| | CB5D4-GFP | 41.8 | 3.0 | 85.8 | 1.7 | 95.2 | 1.6 | 64.7 | 3.7 | 48.1 | 2.8 | −14.3 | 6.9 |
| | HCedar2 | 39.7 | 5.6 | 86.5 | 1.8 | 97.5 | 0.1 | 75.0 | 0.3 | 54.4 | 7.6 | 20.9 | 3.0 |
| | HTtP14 [FL-6] | 44.0 | 3.1 | 82.3 | 5.1 | 95.0 | 0.9 | 59.1 | 4.1 | 53.9 | 2.2 | −1.7 | 5.0 |
| | TXsc | 10.7 | 6.8 | 72.5 | 3.8 | 91.5 | 0.6 | 68.1 | 2.3 | 7.9 | 10.5 | −22.4 | 8.3 |
| *P. chesapeaki* | A8-4a | −47.8 | 15.6 | 55.7 | 6.8 | 84.1 | 2.0 | 88.0 | 3.3 | −12.0 | 8.2 | −96.0 | 9.0 |
| *P. olseni* | ALG1 | 14.3 | 14.7 | 80.4 | 1.4 | 94.9 | 0.7 | 51.1 | 1.3 | 22.5 | 1.4 | 10.5 | 3.2 |
| *P. mediterraneus* | G2 | −41.9 | 27.4 | 78.2 | 1.3 | 92.6 | 0.4 | 57.0 | 7.5 | 36.5 | 10.9 | 1.2 | 21.7 |
| *P. honshuensis* | Mie-3G/H8 | −38.5 | 15.4 | 86.2 | 3.3 | 95.4 | 1.5 | 8.8 | 14.1 | 64.6 | 4.5 | 38.0 | 5.9 |

Data expressed as inhibition mean (%).

questions would require fine-tuning cultures and dedicated experiments outside the scope of this preliminary large screening.

## Conclusions

By taking full advantage of both the open access Malaria Box and having *Perkinsus* spp. in culture, we have identified numerous compounds that affect the *in vitro* proliferation of the parasite. These primary "hits", if confirmed, would expand the number of available compounds against *P. marinus* fivefold. To determine whether the drugs tested in this study will be effective in treating or preventing *P. marinus* infections in bivalves, we must first find a delivery method that administers an effective dose to the oyster tissue and toxicity to the bivalve hosts and other organisms in the environment [13]. In this study, we have taken an indirect approach for identifying and characterizing geographic pheno-types on the basis of resistance to selected compounds from the Malaria Box. At this point, we do not have an explanation for this variability. Most of the compounds tested have a very low molecular weight and are most likely taken up non-specifically by the parasite transporters [*P. marinus* genome encodes excess in secondary active transporters (41 out of 66 super families) including Major Facilitator Superfamily, Amino Acid/Auxin Permease, and Drug/Metabolite, unpublished data]. This would require more targeted research outside of the scope of this work. Moreover, these findings should also make the scientific commu-nity aware that the conclusions may be limited or can change depending on the particular strain used in the study. What the targets are, and the biological functions affected by these compounds in *Perkinsus* spp., remains an open question. Still, *Perkinsus* can be use as a model to ascertain the mechanism of action of the probe-like compounds.

## Supporting Information

**Table S1    List of Medicines for Malaria Venture Malaria Box compounds developed against *Plasmodium falci-parum* 3D7 and used in this study** (http://www.mmv.org/research-development/malaria-box-supporting-information).

**Table S2    Percentage of inhibition of *Perkinsus marinus* trophozoites growth using drug-like compounds and probe-like compounds (20 μM) at day 4 post-exposure.**

## Acknowledgments

This study was supported institutional funds from Bigelow Laboratory for Ocean Sciences as well as by grants OCE0755142 (REU Program) awarded to David M. Fields from the National Science Foundation and 1R21AI076797-01A2 from the National Institute of Health to JAFR.

## Author Contributions

Conceived and designed the experiments: JAFR. Performed the experi-ments: YAR JAFR. Analyzed the data: YAR JAFR. Contributed reagents/materials/analysis tools: JAFR. Wrote the paper: JAFR.

## References

1. Mackin JG, Owen HM, Collier A (1950) Preliminary note on the occurrence of a new protistan parasite, *Dermocystidium marinum* n. sp., in *Crassostrea virginica* (Gmelin). Science 111: 328–329.
2. Ford SE, Chintala MM (2006) Northward expansion of a marine parasite: Testing the role of temperature adaptation. J Exp Mar Bio Ecol 339: 226–235.
3. Ford SE, Smolowitz R (2007) Infection dynamics of an oyster parasite in its newly expanded range. Mar Biol 151: 119–133.
4. Smolowitz R (2013) A review of current state of knowledge concerning *Perkinsus marinus* effects on *Crassostrea virginica* (Gmelin) (the eastern oyster). Vet Pathol 50: 404–411.
5. Pecher WT, Alavi MR, Schott EJ, Fernández-Robledo JA, Roth L, et al. (2008) Assessment of the northern distribution range of selected *Perkinsus* species in eastern oysters (*Crassostrea virginica*) and hard clams (*Mercenaria mercenaria*) with the use of PCR-based detection assays. J Parasitol 94: 410–422.
6. Cáceres-Martínez J, Vasquez-Yeomans R, Padilla-Lardizabal G, del Rio Portilla MA (2008) *Perkinsus marinus* in pleasure oyster *Crassostrea corteziensis* from Nayarit, Pacific coast of Mexico. J Invertebr Pathol 99: 66–73.
7. Bushek D, Allen SK Jr. (1996) Races of *Perkinsus marinus*. J Shellfish Res 15: 103–107.
8. Reece K, Bushek D, Hudson K, Graves J (2001) Geographic distribution of *Perkinsus marinus* genetic strains along the Atlantic and Gulf coasts of the USA. Mar Biol 139: 1047–1055.
9. Robledo JAF, Wright AC, Marsh AG, Vasta GR (1999) Nucleotide sequence variability in the nontranscribed spacer of the rRNA locus in the oyster parasite *Perkinsus marinus*. J Parasitol 85: 650–656.
10. Carrasco N, Rojas M, Aceituno P, Andree KB, Lacuesta B, et al. (2014) *Perkinsus chesapeaki* observed in a new host, the European common edible cockle *Cerastoderma edule*, in the Spanish Mediterranean coast. J Invertebr Pathol 117: 56–60.
11. Fernández Robledo JA, Vasta GR, Record NR (2014) Protozoan parasites of bivalve molluscs: Literature follows culture. PLoS One 9: e100872.
12. Leite MA, Alfonso R, Cancela ML (2011) Herbicides and Protozoan Parasite Growth Control: Implications for New Drug Development. In: M . Larramendy, Soloneski, S, editor editors. Herbicides, Theory and Applications. Rijeka, Croatia: InTech. pp. 567–580.
13. Shridhar S, Hassan K, Sullivan DJ, Vasta GR, Fernández Robledo JA (2013) Quantitative assessment of the proliferation of the protozoan parasite *Perkinsus marinus* using a bioluminescence assay for ATP content. Int J Parsitol: Drug Drug Resist 3: 85–92.
14. Keeling PJ (2010) The endosymbiotic origin, diversification and fate of plastids. Philos Trans R Soc Lond B Biol Sci 365: 729–748.
15. Fichera ME, Roos DS (1997) A plastid organelle as a drug target in apicomplexan parasites. Nature 390: 407–409.
16. Spangenberg T, Burrows JN, Kowalczyk P, McDonald S, Wells TN, et al. (2013) The open access malaria box: a drug discovery catalyst for neglected diseases. PLoS One 8: e62906.
17. Duffy S, Avery VM (2013) Identification of inhibitors of *Plasmodium falciparum* gametocyte development. Malar J 12: 408.
18. Lucantoni L, Duffy S, Adjalley SH, Fidock DA, Avery VM (2013) Identification of MMV malaria box inhibitors of *Plasmodium falciparum* early-stage gametocytes using a luciferase-based high-throughput assay. Antimicrob Agents Chemother 57: 6050–6062.
19. Bowman JD, Merino EF, Brooks CF, Striepen B, Carlier PR, et al. (2014) Antiapicoplast and gametocytocidal screening to identify the mechanisms of action of compounds within the Malaria Box. Antimicrob Agents Chemother 58: 811–819.
20. Bessoff K, Spangenberg T, Foderaro JE, Jumani RS, Ward GE, et al. (2014) Identification of *Cryptosporidium parvum* active chemical series by Repurposing the open access malaria box. Antimicrob Agents Chemother 58: 2731–2739.
21. Njuguna JT, von Koschitzky I, Gerhardt H, Lammerhofer M, Choucry A, et al. (2014) Target evaluation of deoxyhypusine synthase from *Theileria parva* the neglected animal parasite and its relationship to *Plasmodium*. Bioorg Med Chem pii: S0968-0896(14)00347-2. doi: 10.1016/j.bmc.2014.05.007. [Epub ahead of print].
22. Gauthier JD, Feig B, Vasta GR (1995) Effect of fetal bovine serum glycoproteins on the *in vitro* proliferation of the oyster parasite *Perkinsus marinus*: Development of a fully defined medium. J Eukaryot Microbiol 42: 307–313.
23. Fernández-Robledo JA, Lin Z, Vasta GR (2008) Transfection of the protozoan parasite *Perkinsus marinus*. Mol Biochem Parasitol 157: 44–53.
24. Robledo JA, Nunes PA, Cancela ML, Vasta GR (2002) Development of an *in vitro* clonal culture and characterization of the rDNA locus of *Perkinsus atlanticus*, a protistan parasite of the clam *Tapes decussatus*. J Eukaryot Microbiol 49: 414–422.
25. Coss CA, Robledo JA, Ruiz GM, Vasta GR (2001) Description of *Perkinsus andrewsi* n. sp. isolated from the Baltic clam (*Macoma balthica*) by characterization of the ribosomal RNA locus, and development of a species-specific PCR-based diagnostic assay. J Eukaryot Microbiol 48: 52–61.
26. Coss CA, Robledo JA, Vasta GR (2001) Fine structure of clonally propagated *in vitro* life stages of a *Perkinsus* sp. isolated from the Baltic clam *Macoma balthica*. J Eukaryot Microbiol 48: 38–51.
27. Casas SM, Grau A, Reece KS, Apakupakul K, Azevedo C, et al. (2004) *Perkinsus mediterraneus* n. sp., a protistan parasite of the European flat oyster *Ostrea edulis* from the Balearic Islands, Mediterranean Sea. Dis Aquat Organ 58: 231–244.
28. Gauthier JD, Vasta GR (1993) Continuous *in vitro* culture of the eastern oyster parasite *Perkinsus marinus*. J Invertebr Pathol 62: 321–323.

29. La Peyre JF, Faisal M, Burreson EM (1993) *In vitro* propagation of the protozoan *Perkinsus marinus*, a pathogen of the eastern oyster, *Crassostrea virginica*. J Eukariot Microbiol 40: 304–310.

30. Kleinschuster SJ, Swink SL (1993) A simple method for the *in vitro* culture of *Perkinsus marinus*. Nautilus 107: 76–78.

31. Chen CZ, Kulakova L, Southall N, Marugan JJ, Galkin A, et al. (2011) High-throughput *Giardia lamblia* viability assay using bioluminescent ATP content measurements. Antimicrob Agents Chemother 55: 667–675.

32. Goodman CD, Su V, McFadden GI (2007) The effects of anti-bacterials on the malaria parasite *Plasmodium falciparum*. Mol Biochem Parasitol 152: 181–191.

33. Grimberg BT, Mehlotra RK (2011) Expanding the antimalarial drug drsenal-now, but how? Pharmaceuticals (Basel) 4: 681–712.

34. Monzote L, Siddiq A (2011) Drug development to protozoan diseases. Open Med Chem J 5: 1–3.

35. Stelter K, El-Sayed NM, Seeber F (2007) The expression of a plant-type ferredoxin redox system provides molecular evidence for a plastid in the early dinoflagellate *Perkinsus marinus*. Protist 158: 119–130.

36. Fernández Robledo JA, Caler E, Matsuzaki M, Keeling PJ, Shanmugam D, et al. (2011) The search for the missing link: A relic plastid in *Perkinsus*? Int J Parasitol 41: 1217–1229.

37. Tasumi S, Vasta GR (2007) A galectin of unique domain organization from hemocytes of the eastern oyster (*Crassostrea virginica*) is a receptor for the protistan parasite *Perkinsus marinus*. J Immunol 179: 3086–3098.

38. Feng C, Ghosh A, Amin MN, Giomarelli B, Shridhar S, et al. (2013) The Galectin CvGal1 from the Eastern oyster (*Crassostrea virginica*) binds to blood group A oligosaccharides on the hemocyte surface. J Biol Chem 288: 24394–24409.

39. Schott EJ, Pecher WT, Okafor F, Vasta GR (2003) The protistan parasite *Perkinsus marinus* is resistant to selected reactive oxygen species. Exp Parasitol 105: 232–240.

40. Hohwy M, Spadola L, Lundquist B, Hawtin P, Dahmen J, et al. (2008) Novel prostaglandin D synthase inhibitors generated by fragment-based drug design. J Med Chem 51: 2178–2186.

41. Na BK, Kang JM, Kim TS, Sohn WM (2007) *Plasmodium vivax*: molecular cloning, expression and characterization of glutathione S-transferase. Exp Parasitol 116: 414–418.

42. Rojpibulstit P, Kangsadalampai S, Ratanavalachai T, Denduangboripant J, Chavalitshewinkoon-Petmitr P (2004) Glutathione-S-transferases from chloro-quine-resistant and -sensitive strains of *Plasmodium falciparum*: what are their differences? Southeast Asian J Trop Med Public Health 35: 292–299.

43. Joseph SJ, Fernández-Robledo JA, Gardner MJ, El-Sayed NM, Kuo CH, et al. (2010) The Alveolate *Perkinsus marinus*: biological insights from EST gene discovery. BMC Genomics 11: 228.

44. Garreis KA, La Peyre JF, Faisal M (1996) The effects of *Perkinsus marinus* extracellular products and purified proteases on oyster defence parameters *in vitro*. Fish Shellfish Immunol 6: 581–597.

45. Brown GD, Reece KS (2003) Isolation and characterization of serine protease gene(s) from *Perkinsus marinus*. Dis Aquat Organ 57: 117–126.

46. Faisal M, Schafhauser DY, Garreis KA, Elsayed E, La Peyre JF (1999) Isolation and characterization of *Perkinsus marinus* proteases using bacitracin-sepharose affinity chromatography. Comp Biochem Physiol, B 123B: 417–426.

47. Casas SM, Reece KS, Li Y, Moss JA, Villalba A, et al. (2008) Continuous culture of *Perkinsus mediterraneus*, a parasite of the European flat oyster *Ostrea edulis*, and characterization of its morphology, propagation, and extracellular proteins *in vitro*. J Eukaryot Microbiol 55: 34–43.

48. He Y, Yu H, Bao Z, Zhang Q, Guo X (2012) Mutation in promoter region of a serine protease inhibitor confers *Perkinsus marinus* resistance in the eastern oyster (*Crassostrea virginica*). Fish Shellfish Immunol 33: 411–417.

49. Yu H, He Y, Wang X, Zhang Q, Bao Z, et al. (2011) Polymorphism in a serine protease inhibitor gene and its association with disease resistance in the eastern oyster (*Crassostrea virginica* Gmelin). Fish Shellfish Immunol 30: 757–762.

50. Perkins FO, Menzel RW (1966) Morphological and cultural stages in the life cycle of *Dermocystidium marinum*. Proc Natl Shellfish Assoc 56: 23–30.

51. Bushek D, Allen SK Jr. (1996) Host-parasite interactions among broadly distributed populations of the eastern oyster *Crassostrea virginica* and the protozoan *Perkinsus marinus*. Mar Ecol Prog Ser 139: 127–141.

52. Panko C, Volety A, Encomio V, Barreto J (2006) Evaluation of the antimalarial drug quinine as a potential chemotherapeutic agent for the eastern oyster parasite, *Perkinsus marinus*. J Shellfish Res 25: 760.

53. Thompson PC, Rosenthal BM, Hare MP (2011) An evolutionary legacy of sex and clonal reproduction in the protistan oyster parasite *Perkinsus marinus*. Infect Genet Evol 11: 598–609.

54. Sibley LD, Mordue DG, Su C, Robben PM, Howe DK (2002) Genetic approaches to studying virulence and pathogenesis in *Toxoplasma gondii*. Philos Trans R Soc Lond B Biol Sci 357: 81–88.

55. Casas SM, La Peyre JF (2013) Identifying factors inducing trophozoite differentiation into hypnospores in *Perkinsus* species. Eur J Protistol 49: 201–209.

56. Dungan CF, Reece KS (2006) *In vitro* propagation of two *Perkinsus* spp. parasites from Japanese Manila clams *Venerupis philippinarum* and description of *Perkinsus honshuensis* n. sp. J Eukaryot Microbiol 53: 316–326.

57. Blackbourn J, Bower SM, Meyer GR (1998) *Perkinsus qugwadi* sp.nov. (*incertae sedis*), a pathogenic protozoan parasite of Japanese scallops, *Patinopecten yessoensis*, cultured in British Columbia, Canada. Can J Zool Rev Can Zool 76: 942–953.

58. Moss JA, Xiao J, Dungan CF, Reece KS (2008) Description of *Perkinsus beihaiensis* n. sp., a new *Perkinsus* sp. parasite in oysters of Southern China. J Eukaryot Microbiol 55: 117–130.

# 10

# Assessment of Helminth Biodiversity in Wild Rats Using 18S rDNA Based Metagenomics

**Ryusei Tanaka**[1,⬥], **Akina Hino**[1,⬥], **Isheng J. Tsai**[2], **Juan Emilio Palomares-Rius**[1], **Ayako Yoshida**[1], **Yoshitoshi Ogura**[3], **Tetsuya Hayashi**[3], **Haruhiko Maruyama**[1], **Taisei Kikuchi**[1*]

**1** Division of Parasitology, Faculty of Medicine, University of Miyazaki, Miyazaki, Japan, **2** Biodiversity Research Center, Academia Sinica, Taipei, Taiwan, **3** Division of Microbial Genomics, Department of Genomics and Bioenviromental Science, Frontier Science Research Center, University of Miyazaki, Miyazaki, Japan

## Abstract

Parasite diversity has important implications in several research fields including ecology, evolutionary biology and epidemiology. Wide-ranging analysis has been restricted because of the difficult, highly specialised and time-consuming processes involved in parasite identification. In this study, we assessed parasite diversity in wild rats using 18S rDNA-based metagenomics. 18S rDNA PCR products were sequenced using an Illumina MiSeq sequencer and the analysis of the sequences using the QIIME software successfully classified them into several parasite groups. The comparison of the results with those obtained using standard methods including microscopic observation of helminth parasites in the rat intestines and PCR amplification/sequencing of 18S rDNA from isolated single worms suggests that this new technique is reliable and useful to investigate parasite diversity.

**Editor:** Geoffrey N. Gobert, Queensland Institute of Medical Research, Australia

**Funding:** This work was supported in part by JSPS KAKENHI (grants numbers 24780044, 24659190, 26460510 and 26292178), the Project of Establishing Medical Research Base Networks against Infectious Diseases in Okinawa (grants number CBB13001), and the Integrated Research Project for Human and Veterinary Medicine of the University of Miyazaki. The funders had no role in study design, data collection and analysis, decision to publish, or preparation of the manuscript.

**Competing Interests:** The authors have declared that no competing interests exist.

\* Email: taisei_kikuchi@med.miyazaki-u.ac.jp

⬥ These authors contributed equally to this work.

## Introduction

Parasitism is one of the most common and successful lifestyles on the Earth [1] and has evolved independently at least 60 times during the evolutionary history of animal life [1] [2]. Several parasite lineages have diversified greatly over geological time. As a result, parasites outnumber their free-living relatives in some taxonomic groups in animal kingdom [1].

Studying parasite diversity is important for at least 3 major reasons. First, parasites are now recognised as playing important roles in ecosystem fractions [3] by influencing the populations and communities of their hosts [4]. Second, many parasite species are medically and agriculturally important. Although little is known about their evolutionary origins, several human parasites may have evolved by switching to humans from wild or domestic animals [5,6]. Additionally, species interactions involving parasites are a key to understanding many biological invasions and emerging infectious diseases [3]. Finally, because of the many independent transitions to parasitism within taxonomic groups, researchers can study the processes of evolution as the phenomena is related to speciation rates and diversification [1]. Therefore, the number of studies investigating the patterns of parasite diversity among/within host species and among geographical regions has been increasing in recent years [4]. However, the traditional approach of identifying all individual parasitic worms based on microscopic observation and PCR amplification/sequencing of

18S rDNA from isolated single parasites is time consuming, and requires highly specialised experience in morphology. In addition, morphological identification is simply impossible in some cases. As a result, parasite communities are not well classified, leaving diversity analysis ambiguous and non-holistic. Recent advances in high-throughput massively parallel sequencing, also called 'next generation sequencing' (NGS), are revolutionising the description of microbial diversity within and across complex biomes from the human body to the Earth's biosphere [7,8]. The greater sequence coverage and lower per-base sequence cost offered by NGS instruments including Illumina sequencers and 454 pyrosequencers have been greatly contributing to this progress.

Most of the metagenomic studies performed to date have targeted the biodiversity of prokaryotic communities using 16S ribosomal RNA gene (rDNA) sequences [9,10,11]. Attempts to assess eukaryotic diversity using NGS techniques have just begun for fungi [12], nematodes [13,14] and marine microbes [15]. In this study, we performed eukaryotic 18S rDNA-based metagenomics to assess biodiversity of helminth parasites (i.e. Nematoda, Cestoda and Trematoda) in the alimentary tract of wild rats. We analysed massive numbers of sequence reads obtained by 18S rDNA-PCR amplification followed by Illumina sequencing. To evaluate accuracy, sensitivity and resolution power of the method, we compared these results with those from the standard methods including extraction of helminth parasites from the intestine, microscopic observation and single-worm PCR amplification/

sequencing. Our results suggest this new technique is useful for the identification of animal parasites and the assessment of parasite diversity.

## Materials and Methods

### Collecting wild rats

Nine wild rats (7 *Rattus norvegicus* and 2 *R. rattus*), which were captured under the rodent extermination programmes at 2 locations of Miyazaki City, Japan, from November 2013 to January 2014 (Table 1) were received from Miyazaki City Phoenix Zoo or Miyazaki Pest Control Association. The rats were transported to the laboratory and maintained in clean cages for 12 hr before sacrificing by ether inhalation. Faecal pellets they excreted *ad libitum* were collected for metagenomic analysis.

### Ethical statement

Animal experiments were performed in accordance with the procedures approved by the Animal Experiment Committee of the University of Miyazaki under an approval no. 2009-506-6, as specified in the Fundamental Guidelines for Proper Conduct of Animal Experiment and Related Activities in Academic Research Institutions under the jurisdiction of the Ministry of Education, Culture, Sports, Science and Technology, Japan, 2006.

### Isolation of parasitic worms from rat intestines

Parasites were isolated from rat intestines as previously described with some modification [16]. Briefly, whole intestines were extirpated from freshly sacrificed rats and separated into 2 parts (20 cm from the pylorus ring and the remainder). They were dispreaded, washed and incubated separately in PBS (phosphate buffered saline) at 37°C for 2 h to let worms emerge. PBS was then centrifuged at $3000 \times g$ for 5 min to concentrate the isolated parasites, which were observed using light microscopy.

### DNA extraction, sequencing from individual parasites and phylogenetic analyses

DNA from nematodes was extracted into DirectPCR Lysis Reagent (Viagen) containing 20 mM dithiothreitol (Wako) and 0.5 mg/mL proteinase K (Qiagen). Individual nematodes were transferred to a 10-μL aliquot of the lysis buffer and incubated at 60°C for 20 min followed by 95°C for 10 min. DNA extraction from cestodes was performed using the QIAamp DNA Mini Kit according to the manufacturer's instructions (Qiagen). One microlitre of the extract was used for PCR amplification of the 18S ribosomal RNA gene. These PCR reactions contained primers 988F and 1912R [17] for nematodes and wormA and wormB for cestodes [18], along with GoTaq Green Master Mix (Promega). PCR products were purified using the MinElute 96 UF Kit (Qiagen) and sequenced using BigDye Terminator v3.1 and an ABI 3130 sequencer (Applied Biosystems).

Phylogenetic analyses were performed with obtained and published 18S rDNA sequences. Sequences were aligned with MAFFT v6.864b [19] using '–auto' option, and the alignments were cleaned with Gblocks v.0.91b [20] using flags '−b4 = 10 − b5 = n −b6 = y −s = y'. Phylogenetic analyses were performed with RAxML v.7.2.8 [21]. The trees were bootstrapped 100 times for support.

### Illumina library construction and sequencing

DNA extraction from individual faecal pellets (approximately 0.1 g) obtained from wild rats as described above was performed using the PowerSoil DNA extraction kit (MoBio) as recommended in the Earth Microbiome Project (http://www.earthmicrobiome.org/). Barcoded PCR products were generated according to the protocol of the Earth Microbiome Project [22] (http://www.earthmicrobiome.org/emp-standard-protocols/18s/). Briefly, the V9 region of the eukaryotic 18S rRNA gene was amplified in triplicates for each sample with 1391f and EukBr primers [23] containing Illumina adaptors and a unique 12 bp Golay barcode using Ex Taq Polymerase (Takara). PCR amplification was performed in the presence of the mammal blocking oligo [24] on a 30-μL scale under the following conditions: 94°C for 3 min; 22 cycles of 94°C for 45 s; 65°C for 15 s; 57°C for 30 s; 72°C for 90 s and 72°C for 10 min. PCR products from the 3 reactions were combined and purified using the MinElute 96 UF PCR Purification Kit (Qiagen). The final sequencing library was prepared by mixing equal amounts of PCR products and purifying the mixture using agarose gel electrophoresis and the QIAquick Gel Extraction Kit (Qiagen).

Libraries were sequenced with Illumina MiSeq using the MiSeq Reagent Kit v2 (500cycles) (Illumina) and custom sequencing primers (http://www.earthmicrobiome.org/emp-standard-protocols/18s/) according to the manufacturer's recommended protocol (https://icom.illumina.com/). The linearization, blocking and hybridization step was repeated in situ to regenerate clusters, release the second strand for sequencing and hybridise the R2 sequencing primer. This was followed by another 250 cycles of sequencing to produce paired-end reads. The sequencing data have been deposited to DDBJ sequence read archive (DRA) under the BioProject PRJDB3050.

### Illumina data analysis

Illumina sequence data was processed using QIIME version 1.8.0 [25]. Paired-end reads were joined using the 'fastq-join' method in QIIME (join_paired_ends.py). After QIIME quality filtering and library splitting according to the Golay barcode sequences (split_libraries_fastq.py: –store_qual_scores -q 9–max_-barcode_errors 2–sequence_max_n 1–max_bad_run_length 2 −p 0.75), 18S rRNA OTUs were picked from the reads using a closed-reference OTU picking protocol against the SILVA 108 database (Eukarya_only) [26] at 95% identity with 'uclust' (pick_otus.py: –max_accepts 1–max_rejects 8–stepwords 8–word_length 8).

## Results

### Identification of parasites from rat intestines

Seven and two wild rats were collected at 2 contrasting locations in Miyazaki City, Japan: a restaurant downtown in the middle of the city (TR) and a zoo in the suburbs (ZR), respectively (Table 1). The rat IDs were named with the header TR or ZR according to the location from where they were collected. The sample was composed of 3 males and 6 females from 2 species (*R. norvegicus* and *R. rattus*) with varying body sizes (body weight ranging from 95 g to over 300 g) (Table 1). The parasite isolation protocol from the intestines detected nematodes in 8 out of the 9 rats (Table 2). On the other hand, cestodes were identified in only 2 *R. rattus* rats collected in the zoo (*R. rattus*). They were more than 20 cm in length and difficult to distinguish from each other by their morphology (Table 2, Figure 1). The sequences of their 18S rRNA genes were highly similar (99.9% identity) to those of *Hymenolepis diminuta* (JX310720) (Table 3, 4, Figure S1E). We found no flukes in the wild rat intestines.

**Table 1.** Wild rats used in this study.

| Rat ID* | Species | Gender** | Body length (cm) | Tail length (cm) | Body weight (g) | Location | Collection Date |
|---------|---------|----------|------------------|------------------|-----------------|----------|-----------------|
| TR2 | *Rattus norvegicus* | F | 18 | NA | NA | NishiTachibana St., Miyazaki City | 6-Nov-13 |
| TR3 | *Rattus norvegicus* | F | 19 | NA | NA | NishiTachibana St., Miyazaki City | 6-Nov-13 |
| TR4 | *Rattus norvegicus* | F | 19 | 16 | 151.6 | NishiTachibana St., Miyazaki City | 8-Nov-13 |
| TR5 | *Rattus norvegicus* | M | 23 | 21 | 311.2 | NishiTachibana St., Miyazaki City | 8-Nov-13 |
| TR6 | *Rattus norvegicus* | M | 16.5 | 14.5 | 97.5 | NishiTachibana St., Miyazaki City | 8-Nov-13 |
| TR7 | *Rattus norvegicus* | F | 20.5 | 18 | 191.2 | NishiTachibana St., Miyazaki City | 8-Nov-13 |
| TR8 | *Rattus norvegicus* | F | 18 | 17 | 159 | NishiTachibana St., Miyazaki City | 9-Nov-13 |
| ZR1 | *Rattus rattus* | F | 14 | 18 | 100.8 | Phoenix zoo, Miyazaki City | 28-Jan-14 |
| ZR2 | *Rattus rattus* | M | 14 | 15 | 96.3 | Phoenix zoo, Miyazaki City | 28-Jan-14 |

*Rat IDs were assigned based on collection locations (TR or ZR).

**F: female, M: male.

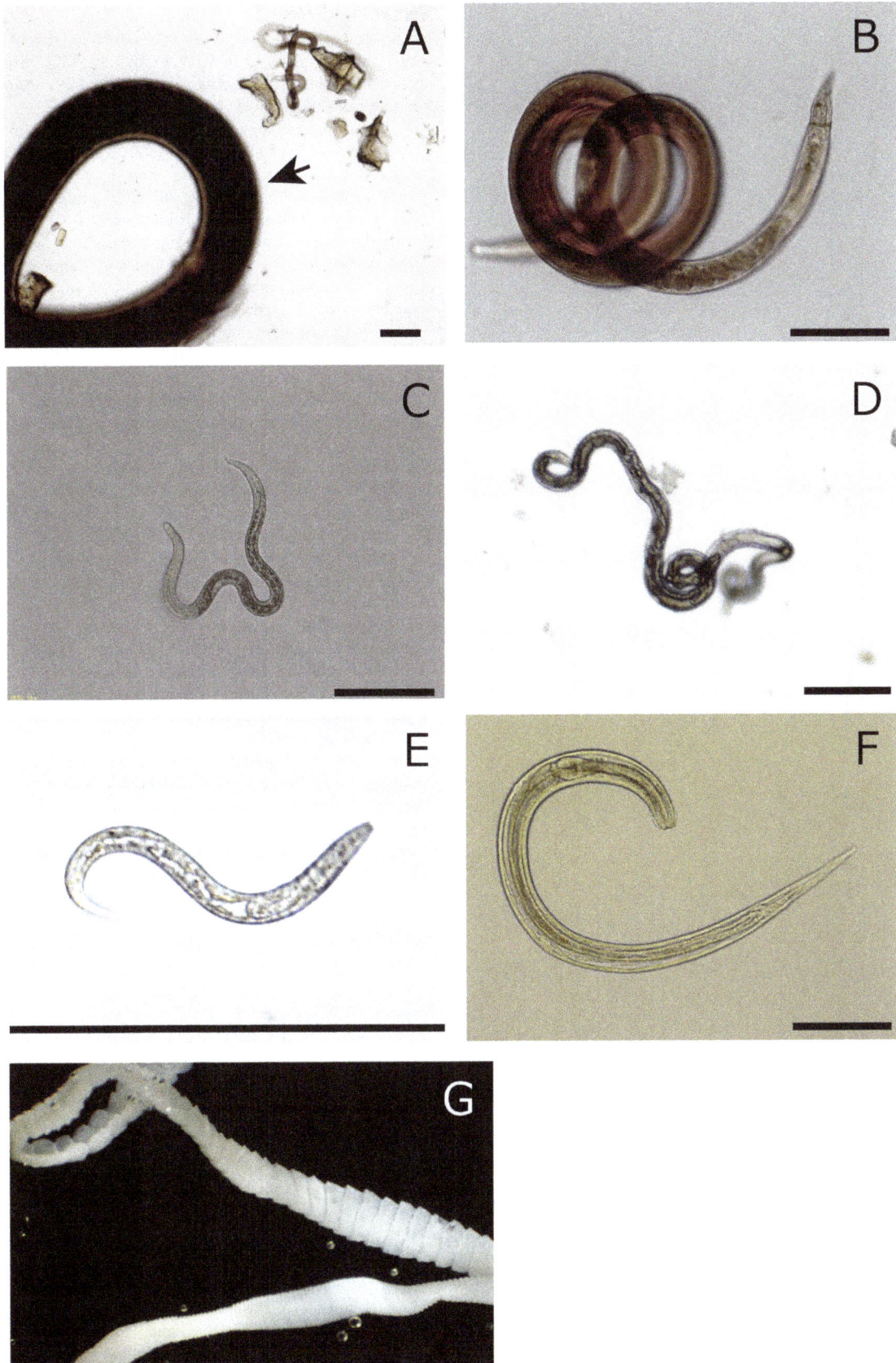

**Figure 1. Seven morphological types identified in wild rat intestines.** (A–F) Nematoda, (G) Cestoda (bar = 200 μm).

**Table 2.** Numbers of helminth parasites identified in rat intestines.

| Rat ID | Nematodes | | | | | | Cestodes |
|---|---|---|---|---|---|---|---|
| | morphA | morphB | morphC | morphD | morphE | morphF | |
| TR2 | 0 | 75 (35) | 1 (1) | 85 (40) | 11 (5) | 0 | 0 |
| TR3 | 2 (0) | 19 (19) | 6 (3) | 142 (84) | 7 (2) | 0 | 0 |
| TR4 | 0 | 134 (117) | 2 (0) | 42 (33) | 0 | 0 | 0 |
| TR5 | 0 | 177 (177) | 0 | 128 (125) | 0 | 0 | 0 |
| TR6 | 0 | 37 (37) | 2 (1) | 165 (160) | 0 | 0 | 0 |
| TR7 | 0 | 344 (333) | 6 (4) | 273 (234) | 0 | 0 | 0 |
| TR8 | 0 | 130 (117) | 0 | 94 (74) | 0 | 0 | 0 |
| ZR1 | 0 | 0 | 0 | 0 | 0 | 0 | 3 |
| ZR2 | 0 | 0 | 0 | 0 | 0 | 2 (0) | 3 |

Helminth parasites observed in the rat intestines were classified into 7 groups based on their morphological traits (Figure 1). The total number of parasites is shown in each cell. Values in parentheses are the number of parasites identified in the first 20 cm from the pylorus ring.

## Nematode diversity

The total numbers of nematodes observed in each of the 8 rats varied, ranging from 2 in ZR2 to 623 in TR7. Most of the nematodes isolated were obtained from the first 20 cm of the intestine from the pylorus ring (Table 2).

Based on morphology, the nematodes were classified into 6 groups (Figure 1A–1F, Table 2). Briefly, morph A had a very large body size (more than 1 cm long) with a cylindrical shape and creamy-white colour, morph B had a differentiable red spiralled body, morph C had an medium body size (~1.0 mm in length) among the observed nematodes and a rhabditiform morphology, morph D was thin, 2–3 mm in length and characterised by the presence of a long pharynx, morph E was a small rhabditiform, probably first- or second-stage larvae and morph F was 1.4–2.0 mm long and differentiable by prominent and broad cervical alae (Figure 1A–1F). Those morphological characteristics are summarised in Table 3.

The 18S rDNA sequencing analysis of individual nematode suggested that these morphological groups consisted of at least 6 distinct species including those belonging to the genera *Ascaridia*, *Nippostrongylus*, *Heligmosomoides*, *Strongyloides* and *Aspiculuris* (Table 4, Figure S1). Morphological descriptions of the species are mostly consistent with the morphological characteristics observed. The 18S rDNA sequences of *N. brasiliensis* and *H. polygyrus* were very similar to each other. In morph B and morph C samples, we found 2 distinct sequence groups showing high similarity to the *N. brasiliensis* 18S rDNA sequence (99.2% and 97.5% identity, respectively). But one of them also showed comparable high similarity to *H. polygyrus* (97.5% identity) (Table 4). In the maximum-likelihood tree, those two sequences were classified together into a cluster which consisted of *N. brasiliensis* and *H. polygyrus* sequences, but they were sub-clustered separately from each other (Fig. S1B).

These results suggested the rats collected in the restaurant downtown were heavily infected by multiple parasitic nematode species including *Ascaridia*, *Nippostrongylus* (or *Heligmosomoides*) and *Strongyloides* nematodes. Of those, the majority of the infections were from *N. brasiliensis* and *Strongyloides* species. In contrast, rats collected in the zoo showed infrequent nematode infection; only a pin worm species was detected (Table 2, 3, 4).

## 18S rDNA Illumina sequencing

DNA was extracted from faecal samples collected from individual rats. The variable regions (V9) of the eukaryotic 18S ribosomal RNA genes present in each faecal community were amplified by PCR, and the resulting amplicons were sequenced on an Illumina MiSeq using 500 cycles.

Of the 6 million Illumina reads from the V9 regions of the eukaryotic 18S rRNA genes that passed the QIIME quality filters and were correctly assigned to each sample on the basis of the barcode sequences, 80.6% matched reference sequences in the SILVA 108 database at 95% sequence identity. They clustered into a total of 391 Operational Taxonomic Units (OTUs).

The QIIME level-2 OTU classifications (phylum level) are shown in Figure 2. In TR samples, 90–99% of the reads were assigned to Nematoda sequences. In TR6 samples, approximately 10% of reads were assigned to Apicomplexa (Figure 2).

Less than 10% of the reads in ZR samples were assigned to nematodes. Reads that were assigned to Streptophyta (Planta), Chordata (Animalia) or Dikarya (Fungi) were more represented. Reads assigned to Platyhelminthes were also found in ZR samples (approximately 2% in ZR1 samples and 3–49% in ZR2 samples) (Figure 2). Deeper classification revealed that most of the Streptophyta reads in ZR samples were assigned to corn (*Zea*),

**Table 3.** Morphological characters of isolated helminths.

| Morphology Type | No. measured | Stage | Body length (mm)[a] | Body width (μm)[a] | Length of esophagus (μm)[a] | Descriptive characters |
|---|---|---|---|---|---|---|
| morph A | 1 | ND | 210 | 294 | ND | large body size, cylindrical shape, creamy-white colour |
| morph B | 4 | adult male | 2.26 (1.74–2.76) | 74.6 (61.1–92.7) | 247 (192–294) | red spiralled body, a prominent unbrella-like bursa and two spicules at the posterior end |
| | 4 | adult female | 2.81 (2.37–3.26) | 68.71 (63.9–72.2) | 269 (241–330) | red spiralled body, ellipsoidal eggs inside of the body, vulva opens at the posterior end |
| morph C | 4 | larva | 0.83 (0.77–0.88) | 29.0 (19.9–40.5) | 178 (140–226) | middle size (~1.0 mm) rhabditiform |
| morph D | 3 | adult female | 1.98 (1.83–2.10) | 30.0 (28.7–31.0) | 147 (146–147) | thin body, long pharynx, only females found, ellipsoidal eggs and vulva in the middle of the body |
| morph E | 3 | larva | 0.19 (0.17–0.20) | 14.4 (14.2–14.7) | 66.2 (53.6–72.6) | small size (~0.20 mm) rhabditiform |
| morph F | 1 | adult male | 1.57 | 74.5 | 258 | braod cervical alae, oval esophageal bulb, slightly hooked tail with no clear spiclues |
| | 1 | unmatured female[b] | 1.32 | 60.1 | 254 | braod cervical alae, oval esophageal bulb, pharynx plainly visible, conical shaped tail |
| morph G | 2 | adult | >200 | 1181 (1153–1231) | NA | flat segemented body, 4 suckers at the scolex, proglottids with both male/female sexual organs |

[a]mean; range of the size in parentheses.
[b]probably 4th stage lavae.
ND; Not determined, NA; Not applicable.

Chordata reads were to assigned to pig (*Sus*) and Dikarya reads were assigned to yeast (*Saccharomyces* or *Candida*) (Table S1). Samples from each rat (different pieces of faecal pellets from the same day) showed similar contents although their ratios were different (Figure 2). Other taxa to which more than 0.1% of the reads were assigned in any sample included Hexamitidae (approximately 0.04%–2.3% in ZR1 samples), Arthropoda (> 0.1% in ZR1 and ZR2, <0.1% in TR samples) and Trichomo-

**Table 4.** Species identification of isolated helminths based on 18S rDNA sequencing.

| Morphology type | Number of sequences | Top hit in nematode 18S database | Sequence similarity (%) | Alignment length (bp) | sequence ID in Fig. S1 |
|---|---|---|---|---|---|
| morph A | 1 | *Ascaridia galli* [EF180058] | 98.2 | 895 | A1 |
| morph B | 26 | *Nippostrongylus brasiliensis* [AJ920356] | 99.2 | 906 | B1 |
| | 12 | [AJ920356] or *Heligmosomoides polygyrus* [AJ920355] | 97.5 | 906 | B2 |
| morph C | 6 | *Nippostrongylus brasiliensis* [AJ920356] | 99.2 | 906 | C1 |
| | 3 | [AJ920356] or *Heligmosomoides polygyrus* [AJ920355] | 97.5 | 906 | C2 |
| | 1 | *Ascaridia galli* [EF180058] | 98.1 | 906 | C3 |
| morph D | 12 | *Strongyloides venezuelensis* [AB923887] | 99.7 | 902 | D1 |
| | 14 | *Strongyloides ratti* [AB923889] | 99.8 | 895 | D2 |
| morph E | 8 | *Strongyloides ratti* [AB923889] | 99.8 | 895 | E1 |
| morph F | 2 | *Aspiculuris tetraptera* [EF464551] | 100 | 135 | F1 |
| morph G | 2 | *Hymenolepis diminuta* [JX310720] | 99.9 | 2005 | G1 |

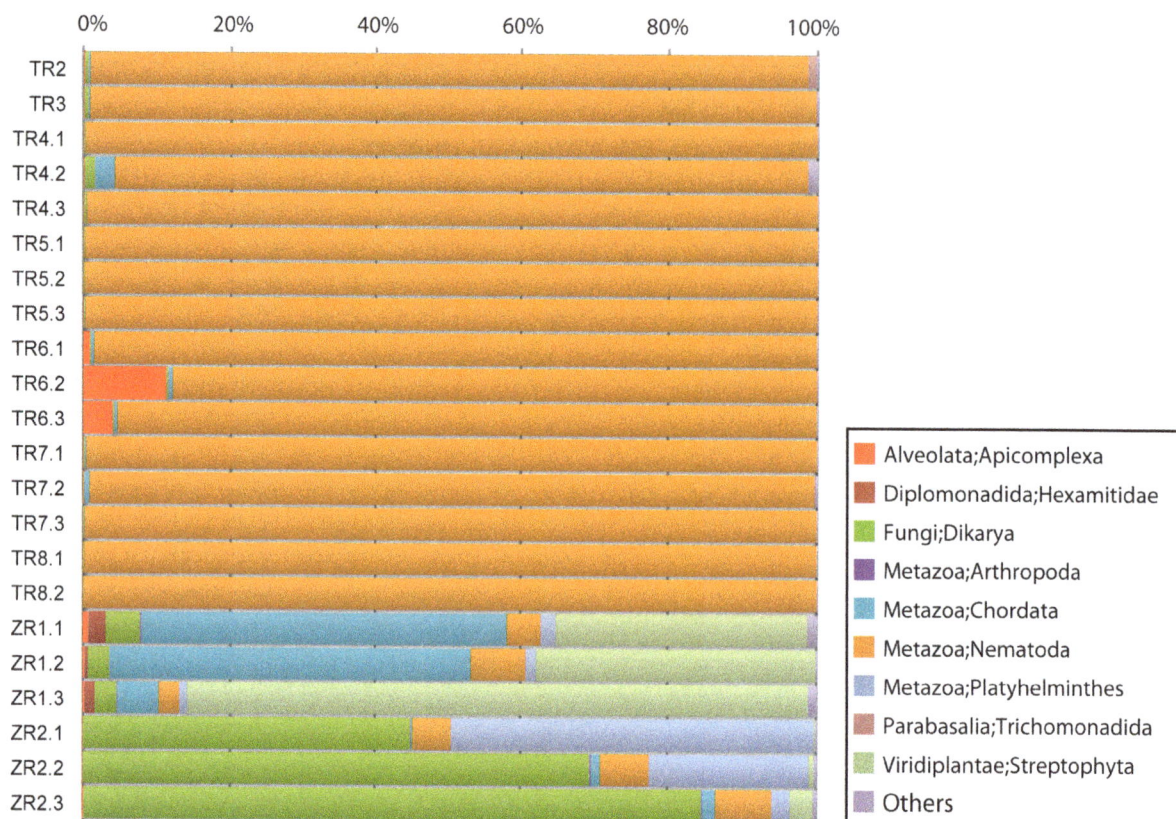

**Figure 2. QIIME phylum-level classification of the 18S rDNA Illumina sequencing data.**

nadida (>0.1% TR2 and TR7 samples, <0.1% in other TR samples) (Figure 2).

The QIIME level-6 classifications (mostly superfamily level) clustered OTUs into 135 taxonomic groups (Table S1) and revealed that more than 90% of the total assigned reads in TR samples were classified into Panagrolaimoidea or Strongylida, which include *Strongyloides* species and *Nippostrongylus* species, respectively (Figure 3). Other nematode taxonomic groups in TR samples to which reads were assigned (with a minimum number of read filter of 4) include Ascaridiea, Heterakoidea, Oxyuroidea, Rhabditoidea and Acuarioidea although they were all rare OTUs (<0.1%, Figure 3, Table S1). In ZR samples, only 3 nematode superfamilies (Panagrolaimoidea, Strongylida and Oxyuroidea) had assigned reads (Figure 3, Table S1).

Because the SILVA 108 database contained only an essential number of nematodes reads, we sought to reclassify the reads assigned to Nematoda in the QIIME classification by BLAST similarity analysis using an in-house nematode 18S database. Although it was still difficult to classify them into deeper levels due to the short read length (~150 bp), the database enabled us to get a better insight into the nematode community. The reads assigned to Panagrolaimoidea or Strongylida in the QIIME classification showed the highest similarity to *Strongyloides* species and *N. brasiliensis* sequences in the nematode 18S database, respectively (Table 5). The reads assigned to Ascaridiea, Heterakoidea or Acuarioidea were highly similar to each other and showed the greatest similarity to *Ascaridia galli* or *Heterakis gallinarum*. The reads assigned to Oxyuroidea were most similar to *Aspiculuris tetraptera* (Table 5). These results were mostly consistent with the results of the standard method (direct microscopic observation)

(Table 2, 6). A trend was observed in the *Strongyloides/Nippostrongylus* ratio; samples with a lower ratio in the direct observation showed a lower reads ratio although *Strongyloides* reads ratios were always higher than the observed ratios (Table 2, 6, Figure 3). In addition to the species that we identified in the standard method, we found 18S rDNA reads which matched sequences from a free-living species *Oscheius spp*.

The Platyhelminthes reads both in TR and ZR samples were all classified into Hymenolepididae at the superfamily-level classification. The ZR2 rat had a higher percentage of Platyhelminthes reads (28%) than the ZR1 rat while they were rare in TR samples (<0.1%).

Reads assigned to the taxa that included parasitic Protozoa species were also identified in the QIIME classifications. Reads which were classified into Eimeriorina and Trichomonadidae were found in several samples (Figure 3). Notably, as much as 6.34% of the reads were classified into Eimeriorina in TR6 (Figure 3). Reads that were assigned to Giardiinae were found in TR8 and ZR1 samples (<0.01% and 1.30%, respectively), Trypanosomatidae in TR5 and TR8 and Acanthamoebidae in TR samples although they were rare (<0.1%) (Figure 3).

## Discussion

Although 'metagenomics-based' studies of bacterial communities using 16S rDNA sequences have been extensively performed recently, there have been only few metagenomics reports targeting eukaryotic communities. In this study, we showed the power and usefulness of 18S rDNA Illumina sequencing for population studies of eukaryote parasites. Compared with traditional methods (isolation from rat intestines, microscopic observations and single-

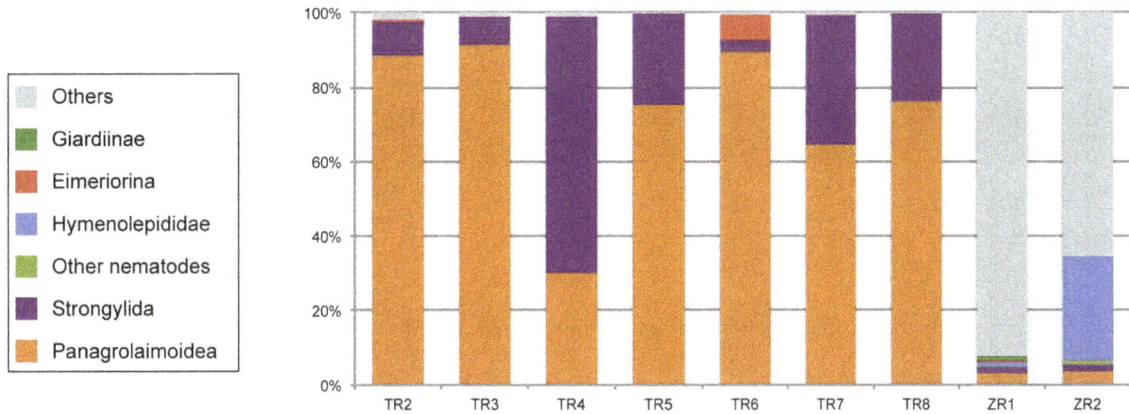

| Phylum/Subphylum | Superfamily | TR2 | TR3 | TR4 | TR5 | TR6 | TR7 | TR8 | ZR1 | ZR2 |
|---|---|---|---|---|---|---|---|---|---|---|
| Nematoda | Ascaridiea | <0.01% | <0.01% | nd | nd | nd | <0.01% | <0.01% | nd | nd |
| | Heterakoidea | 0.06% | 0.02% | 0.01% | <0.01% | <0.01% | <0.01% | 0.03% | nd | <0.01% |
| | Oxyuroidea | nd | nd | nd | <0.01% | nd | nd | nd | nd | 0.97% |
| | Panagrolaimoidea | 88.63% | 91.52% | 30.24% | 75.13% | 89.52% | 64.35% | 76.09% | 3.37% | 3.77% |
| | Rhabditoidea | <0.01% | <0.01% | <0.01% | nd | <0.01% | <0.01% | <0.01% | 0.03% | nd |
| | Strongylida | 9.14% | 7.49% | 68.78% | 24.68% | 3.31% | 35.11% | 23.67% | 1.40% | 1.68% |
| | Acuarioidea | <0.01% | <0.01% | <0.01% | nd | nd | <0.01% | <0.01% | nd | nd |
| Cestoda | Hymenolepididae | <0.01% | <0.01% | <0.01% | <0.01% | <0.01% | 0.01% | <0.01% | 1.45% | 28.18% |
| Apicomplexa | Eimeriorina | 0.07% | 0.04% | <0.01% | <0.01% | 6.34% | <0.01% | <0.01% | 0.38% | 0.04% |
| Mastigophora | Trichomonadidae | 0.10% | 0.06% | <0.01% | <0.01% | 0.03% | 0.03% | nd | nd | nd |
| | Trypanosomatidae | nd | nd | nd | <0.01% | nd | nd | <0.01% | nd | nd |
| | Giardiinae | nd | nd | nd | nd | nd | nd | <0.01% | 1.30% | nd |
| Sarcordina | Acanthamoebidae | 0.02% | 0.01% | <0.01% | nd | <0.01% | nd | nd | nd | nd |

**Figure 3. Parasite sequences in the QIIME superfamily-level classification.**

worm sequencing of 18S rDNA), the 18S metagenomics-based method is easy, quick to apply and sensitive. The 18S metagenomics-based method requires only faecal samples (no need to sacrifice hosts) and requires neither special techniques nor knowledge of parasite morphologies. Most importantly, 18S Illumina sequencing identified more varieties of parasite than the traditional methods in addition to all of the parasites identified using the traditional method.

The amount of parasite DNA in a faecal sample is expected to vary widely depending on their life cycles or conditions. The higher prevalence of *Strongyloides* species in the Illumina sequencing results than that in the result of the standard technique may have arisen because they produced and excreted more eggs/larvae into faeces than those by the other parasite species. We found a large Ascaridia-like nematode in the intestine of one of the rats (Figure 1). However, the number of reads that were assigned to this nematode group (Ascaridiae) or a closely related group (Heterakoidea) in level 6 QIIME was very small (Figure 3). This suggests that the nematode was not very active in the rat body and did not produce eggs, possibly because it was sexually immature. Ascaridia/Heterakoidea nematodes are known as parasites of birds [27], and as far as we know, no Ascaridia/Heterakoidea nematodes have been reported from rodents. Therefore, the presence of Ascaridia/Heterakoidea nematode might reflect a recent accidental swallow of the parasite by the rat.

DNA extraction efficiencies from eggs or larvae of each species can also affect the number of sequence reads obtained. To reduce this kind of effects we used a well-established DNA extraction protocol using MoBio PowerSoil kit, which was used as a standard method in the Earth Microbiome Project (http://www.earthmicrobiome.org/) and the Human Microbiome project [28]. The protocol, combining mechanical and chemical sample disruptions, has been widely used for various types of samples including soils and faeces, and it was shown that has been shown capable to extract DNA even from tough organisms such as spores or fungal mats (http://www.mobio.com/). The bias due to extraction efficiency in this study, therefore, can be low, but tests and optimisations of DNA extraction methods from parasites are still needed.

It is an interesting challenge to quantify parasites using 18S Illumina reads. To achieve the estimate, at least 2 normalisation steps seem to be needed. First, absolute read numbers need to be normalised using control DNA. Because DNA extracted from faeces was amplified by PCR and then appropriate amounts of the products for each sample were mixed and used for sequencing, we should not simply use the read numbers to estimate the amount of each parasite's DNA. Control DNA can be endogenous or exogenous (i.e. artificially added to each sample). For example, a large number of reads were assigned to corn (*Zea*) or pig (*Sus*) (Figure 3) in ZR samples. These probably resulted from the DNA of foods that the rats ingested. These reads were also present in TR samples although the ratios of read numbers were very small (< 0.01%). These kinds of reads, which can be assumed to always exist in relatively fixed amounts, could be a candidate for endogenous control DNA. Second, we need to use 'species factors' that are calculated on the basis of DNA emission in faeces per

**Table 5.** Nematode reads re-assigned with in-house nematode 18S database.

| Order; Family | TR2 | TR3 | TR4 | TR5 | TR6 | TR7 | TR8 | ZR1 | ZR2 | Possible nematode species |
|---|---|---|---|---|---|---|---|---|---|---|
| Rhabditida; Strongyloididae | 119998 | 500677 | 208730 | 1085734 | 862840 | 292994 | 477358 | 129 | 526 | Strongyloides spp. |
| Strongylida; Heligmonellidae | 12381 | 40979 | 474806 | 356714 | 31943 | 159843 | 148495 | 55 | 238 | N. brasiliensis, H. polygyrus |
| Ascaridida; Ascarididae, Heterakidae | 87 | 134 | 78 | 5 | 0 | 41 | 208 | 0 | 0 | A. galli, H. gallinarum |
| Oxyurida; Heteroxynematidae | 0 | 0 | 0 | 0 | 0 | 0 | 0 | 0 | 137 | A. tetraptera |
| Rhabditida; Rhabditidae | 0 | 0 | 27 | 0 | 0 | 0 | 0 | 0 | 0 | Oscheius sp |

Minimum 4-read filter was applied.

**Table 6.** Comparison of the detection results from the standard method and the 18S Illumina method.

| Standard method species | Illumina best classification | TR2 Std | TR2 Ilm | TR3 Std | TR3 Ilm | TR4 Std | TR4 Ilm | TR5 Std | TR5 Ilm | TR6 Std | TR6 Ilm | TR7 Std | TR7 Ilm | TR8 Std | TR8 Ilm | ZR1 Std | ZR1 Ilm | ZR2 Std | ZR2 Ilm |
|---|---|---|---|---|---|---|---|---|---|---|---|---|---|---|---|---|---|---|---|
| A. galli, H. gallinarum | A. galli, H. gallinarum | | | | | | | | | | | | | | | | | + | |
| A. tetraptera | A. tetraptera | | | | + | | | | | | | | | | | | | + | + |
| N. brasiliensis[a] | N. brasiliensis, H. polygyrus | | + | + | + | + | + | + | + | + | + | + | + | + | + | | + | | + |
| S. ratti | Strongyloides sp. | + | + | + | + | + | + | + | + | + | + | + | + | + | + | | + | | + |
| S. venezuelensis | Strongyloides sp. | + | + | + | + | - | + | + | + | + | + | + | + | + | + | | | | |
| | Oscheius sp. | | | | | | + | | | | | | | | | | | | |
| H. diminuta | Hymenolepis sp. | | | | | | | | | | | | | | | + | + | + | + |

"+" indicates worm/sequence detected. Std; Standard method, Ilm; 18S Illumina method.
[a]these samples can be divided into two groups based on the 18S sequences.

individual parasite to estimate the number of parasites in a host body. In addition, copy number differences of rRNA genes between each parasite genome are also a subject to be considered [29].

We used the SILVA 108 database as a reference to classify 18S Illumina reads in QIIME. This study showed the database is very useful to provide a rough estimate of the structure of the helminth community in host bodies. However, the nematode and cestode sequences in the SILVA database were limited to only essential sequences and were not broad enough to cover nematode and cestode diversities. For example, the 18S rDNA sequences of rat parasites like *S. ratti*, *N. brasiliensis* and *H. polygyrus* do not exist in the SILVA 108 database. Therefore, we used an in-house nematode 18S rDNA database to obtain a better insight into the community structures. Additionally, because Illumina sequencing read lengths have been getting longer (up to 300 bp paired-end with the v3 kit), it will be able to use a region of the 18S rRNA gene longer than the V9 region (~150 bp) used in this study to obtain a better resolution during the classification step. Attempts to develop better primers to amplify and classify eukaryote organisms by metagenomics have already started [30,31]. Because parasites have considerably diverged within some small taxonomic groups and parasite species that researchers want to distinguish are often closely related to each other, superfamily-level classifications achievable with the QIIME software are sometimes not sufficient enough in depth to make fine distinctions. Therefore, developing a more complete database and using longer 18S sequencing reads are the next key steps to improve this technique as a more powerful tool to study parasite community structures.

Another important finding is that we identified a number of Eimeriorina sequences in the 18S Illumina data from TR rats (6.3% of total reads in TR6). Eimeriorina contains *Eimeria* species, which are protozoan parasites of animals [32]. These sequences were likely from parasites that have infected the rats. We also identified sequences that were assigned to taxa which include parasitic Protozoa species including *Trichomonas*, *Giardiia*, *Trypanosoma* and *Acanthamoeba* spp. This suggests 18S Illunima sequencing is also useful to study protozoan parasite community structures. We also identified some fungal (yeast) sequences in a ZR rat. It is not clear whether they are parasitic or endosymbiotic species, but this suggests that 18S Illumina sequencing may enable us to investigate associations between helminths and protozoan parasites or parasites and other eukaryotic micro-organisms.

## Conclusions

Studying the diversity of parasites has been recognised as an interesting and useful approach in several research fields, such as those investigating the evolution of life, ecosystem fractions and invasion and migration of emerging diseases. However, the difficult and time-consuming processes required to identify parasites have restricted analysis covering a wide range of parasite species or dealing many samples. In this study, we showed the power and usefulness of 18S rDNA-based metagenomics in the investigation of parasite diversity. We also showed this approach still needs improvements in database completeness and read length in order to classify various parasites into a sufficient level. This approach with those improvements will enable us to analyse a large number of samples in a high throughput manner and can be the next standard to investigate parasite diversity.

## Supporting Information

**Figure S1   Maximum likelihood phylogenies based on 18S rDNA sequences to show relationships of the isolated worms with other nematode or cestode species/isolates.** Selected nematode or cestode species were used in each of five trees; [A] with the sequences from morph A (A1) and morph C (C3), [B] morph B (B1, B2) and morph C (C1, C2), [C] morph D (D1, D2) and morph E (E1), [D] morph F (F1), [E] morph G (G1). Sequences obtained in this study were boxed in red, in which sequence group IDs (shown in Table 4) and individual sequence IDs (in parentheses) were given. Bootstrap values greater than 60% are shown on appropriate nodes. Genbank accession numbers were shown after the species names in each tip label.

## Acknowledgments

The authors thank the Miyazaki City Phoenix Zoo and the Miyazaki Pest Control Association for providing the wild rats, Mika Kuroki, Mie Nagata and Yasunobu Maeda for their technical assistances.

## Author Contributions

Conceived and designed the experiments: HM TH TK. Performed the experiments: RT AH JEPR. Analyzed the data: IJT TK. Contributed reagents/materials/analysis tools: AY YO. Contributed to the writing of the manuscript: AH TK.

## References

1. Poulin R, Morand S (2000) The diversity of parasites. Quarterly Review of Biology 75: 277–293.
2. Sukhdeo MV, Bansemir AD (1996) Critical resources that influence habitat selection decisions by gastrointestinal helminth parasites. International Journal for Parasitology 26: 483–498.
3. Hatcher MJ, Dunn AM (2011) Parasites in ecological communities: from interactions to ecosystems: Cambridge University Press.
4. Korallo NP, Vinarski MV, Krasnov BR, Shenbrot GI, Mouillot D, et al. (2007) Are there general rules governing parasite diversity? Small mammalian hosts and gamasid mite assemblages. Diversity and Distributions 13: 353–360.
5. Combes C (1990) Where do human schistosomes come from? An evolutionary approach. Trends Ecol Evol 5: 334–337.
6. Waters AP, Higgins DG, McCutchan TF (1991) Plasmodium falciparum appears to have arisen as a result of lateral transfer between avian and human hosts. Proceedings of the National Academy of Sciences of the United States of America 88: 3140–3144.
7. Desai N, Antonopoulos D, Gilbert JA, Glass EM, Meyer F (2012) From genomics to metagenomics. Current Opinion in Biotechnology 23: 72–76.
8. Weinstock GM (2012) Genomic approaches to studying the human microbiota. Nature 489: 250–256.
9. Yatsunenko T, Rey FE, Manary MJ, Trehan I, Dominguez-Bello MG, et al. (2012) Human gut microbiome viewed across age and geography. Nature 486: 222–227.
10. Sogin ML, Morrison HG, Huber JA, Mark Welch D, Huse SM, et al. (2006) Microbial diversity in the deep sea and the underexplored "rare biosphere". Proceedings of the National Academy of Sciences of the United States of America 103: 12115–12120.
11. Amir A, Zeisel A, Zuk O, Elgart M, Stern S, et al. (2013) High-resolution microbial community reconstruction by integrating short reads from multiple 16S rRNA regions. Nucleic Acids Research 41: e205.
12. Jumpponen A, Jones K (2009) Massively parallel 454 sequencing indicates hyperdiverse fungal communities in temperate Quercus macrocarpa phyllosphere. New Phytologist 184: 438–448.
13. Porazinska DL, Giblin-Davis RM, Powers TO, Thomas WK (2012) Nematode spatial and ecological patterns from tropical and temperate rainforests. PLoS ONE 7: e44641.

14. Porazinska DL, Giblin-Davis RM, Faller L, Farmerie W, Kanzaki N, et al. (2009) Evaluating high-throughput sequencing as a method for metagenomic analysis of nematode diversity. Molecular ecology resources 9: 1439–1450.

15. Logares R, Audic S, Bass D, Bittner L, Boutte C, et al. (2014) Patterns of rare and abundant marine microbial eukaryotes. Current Biology 24: 813–821.

16. Webster JP, Macdonald DW (1995) Parasites of wild brown rats (Rattus norvegicus) on UK farms. Parasitology 111 (Pt 3): 247–255.

17. Holterman M, van der Wurff A, van den Elsen S, van Megen H, Bongers T, et al. (2006) Phylum-wide analysis of SSU rDNA reveals deep phylogenetic relationships among nematodes and accelerated evolution toward crown Clades. Molecular Biology and Evolution 23: 1792–1800.

18. Waeschenbach A, Webster BL, Bray RA, Littlewood DT (2007) Added resolution among ordinal level relationships of tapeworms (Platyhelminthes: Cestoda) with complete small and large subunit nuclear ribosomal RNA genes. Molecular Phylogenetics and Evolution 45: 311–325.

19. Katoh K, Misawa K, Kuma K, Miyata T (2002) MAFFT: a novel method for rapid multiple sequence alignment based on fast Fourier transform. Nucleic Acids Res 30: 3059–3066.

20. Castresana J (2000) Selection of conserved blocks from multiple alignments for their use in phylogenetic analysis. Molecular Biology and Evolution 17: 540–552.

21. Stamatakis A (2006) RAxML-VI-HPC: maximum likelihood-based phylogenetic analyses with thousands of taxa and mixed models. Bioinformatics 22: 2688–2690.

22. Caporaso JG, Lauber CL, Walters WA, Berg-Lyons D, Huntley J, et al. (2012) Ultra-high-throughput microbial community analysis on the Illumina HiSeq and MiSeq platforms. ISME J 6: 1621–1624.

23. Amaral-Zettler LA, McCliment EA, Ducklow HW, Huse SM (2009) A Method for Studying Protistan Diversity Using Massively Parallel Sequencing of V9 Hypervariable Regions of Small-Subunit Ribosomal RNA Genes. PLoS ONE 4: e6372.

24. Vestheim H, Jarman S (2008) Blocking primers to enhance PCR amplification of rare sequences in mixed samples - a case study on prey DNA in Antarctic krill stomachs. Frontiers in Zoology 5: 12.

25. Caporaso JG, Kuczynski J, Stombaugh J, Bittinger K, Bushman FD, et al. (2010) QIIME allows analysis of high-throughput community sequencing data. Nat Meth 7: 335–336.

26. Quast C, Pruesse E, Yilmaz P, Gerken J, Schweer T, et al. (2013) The SILVA ribosomal RNA gene database project: improved data processing and web-based tools. Nucleic Acids Res 41: D590–596.

27. Yamaguti S (1961) Systema helminthum. Vol. 3. The nematodes of vertebrates. New York and London: Interscience Publishers.

28. Turnbaugh PJ, Ley RE, Hamady M, Fraser-Liggett CM, Knight R, et al. (2007) The Human Microbiome Project. Nature 449: 804–810.

29. Bik HM, Fournier D, Sung W, Bergeron RD, Thomas WK (2013) Intra-genomic variation in the ribosomal repeats of nematodes. PLoS ONE 8: e78230.

30. Hugerth LW, Muller EE, Hu YO, Lebrun LA, Roume H, et al. (2014) Systematic Design of 18S rRNA Gene Primers for Determining Eukaryotic Diversity in Microbial Consortia. PLoS ONE 9: e95567.

31. Hadziavdic K, Lekang K, Lanzen A, Jonassen I, Thompson EM, et al. (2014) Characterization of the 18S rRNA gene for designing universal eukaryote specific primers. PLoS ONE 9: e87624.

32. Zhao X, Duszynski DW (2001) Phylogenetic relationships among rodent Eimeria species determined by plastid ORF470 and nuclear 18S rDNA sequences. International Journal for Parasitology 31: 715–719.

# 11

# Getting What Is Served? Feeding Ecology Influencing Parasite-Host Interactions in Invasive Round Goby *Neogobius melanostomus*

**Sebastian Emde[1], Judith Kochmann[2]\*, Thomas Kuhn[1], Martin Plath[3], Sven Klimpel[1,2]**

1 Institute for Ecology, Evolution and Diversity, Goethe-University, Frankfurt am Main, Hesse, Germany, 2 Senckenberg Gesellschaft für Naturforschung, Biodiversity and Climate Research Centre, Frankfurt am Main, Hesse, Germany, 3 College of Animal Science and Technology, Northwest Agriculture & Forestry University, Yangling, Shaanxi Province, P. R. China

## Abstract

Freshwater ecosystems are increasingly impacted by alien invasive species which have the potential to alter various ecological interactions like predator-prey and host-parasite relationships. Here, we simultaneously examined predator-prey interactions and parasitization patterns of the highly invasive round goby (*Neogobius melanostomus*) in the rivers Rhine and Main in Germany. A total of 350 *N. melanostomus* were sampled between June and October 2011. Gut content analysis revealed a broad prey spectrum, partly reflecting temporal and local differences in prey availability. For the major food type (amphipods), species compositions were determined. Amphipod fauna consisted entirely of non-native species and was dominated by *Dikerogammarus villosus* in the Main and *Echinogammarus trichiatus* in the Rhine. However, the availability of amphipod species in the field did not reflect their relative abundance in gut contents of *N. melanostomus*. Only two metazoan parasites, the nematode *Raphidascaris acus* and the acanthocephalan *Pomphorhynchus* sp., were isolated from *N. melanostomus* in all months, whereas unionid glochidia were only detected in June and October in fish from the Main. To analyse infection pathways, we examined 17,356 amphipods and found *Pomphorhynchus* sp. larvae only in *D. villosus* in the river Rhine at a prevalence of 0.15%. *Dikerogammarus villosus* represented the most important amphipod prey for *N. melanostomus* in both rivers but parasite intensities differed between rivers, suggesting that final hosts (large predatory fishes) may influence host-parasite dynamics of *N. melanostomus* in its introduced range.

**Editor:** Raul Narciso C. Guedes, Federal University of Viçosa, Brazil

**Funding:** The present study was financially supported by the research funding programme "LOEWE –Landes-Offensive zur Entwicklung Wissenschaftlichökonomischer Exzellenz" of Hesse's Ministry of Higher Education, Research, and the Arts. The funders had no role in study design, data collection and analysis, decision to publish, or preparation of the manuscript.

**Competing Interests:** The authors have declared that no competing interests exist.

\* Email: judith.kochmann@senckenberg.de

## Introduction

Biological invasions have increased exponentially in recent years due to human activities, especially shipping, along with the adverse effects of environmental changes such as global warming [1–3]. Although brackish waters have the highest risk for species introductions, freshwater ecosystems are also strongly affected, especially by the introduction of non-indigenous fishes [4,5]. Once established in their new environment, invasive non-indigenous species can have tremendous effects on local populations of indigenous species, e.g., through competitive [6], predator-prey [7–9], or host-parasite interactions [10,11], all of which have the potential to result in altered ecosystem functioning (see review by Strayer [12]).

To date, several studies in aquatic ecosystems have considered the question of how invasive predators can affect native prey populations [13–15], or how invasive prey populations can alter indigenous prey communities [16,17], and whether or not non-indigenous prey species become integrated into the prey spectrum of indigenous predators [18]. Furthermore, studies have started to concentrate on parasitization patterns of native and invasive species, and several different scenarios are possible: (i) invasive hosts may lose their original parasite load ('enemy release hypothesis'), providing invasive species with an initial benefit in their novel range [10,19,20]. (ii) Introduced hosts may carry new parasite species (parasite spill-over), which may adversely affect native host species [21]. (iii) Invasive hosts may serve as intermediate hosts or vectors for local parasites or diseases (parasite spillback) [21]. (iv) Finally, shift and/or loss of local parasite species would be predicted if the invader is replacing local host species but cannot function as intermediate or definitive host in the parasite life cycles (dilution effect) [22,23]. Few studies, however, have simultaneously considered predator-prey interactions and parasitization patterns of different trophic levels in ecosystems that are heavily influenced by invasive species [24–26]. This is surprising, given that many parasites with indirect life-cycles rely on the ingestion of their intermediate hosts by further (intermediate or final) host species to successfully complete their life cycles [27,28]. Biological invasions could provide large numbers of host specimens within a very short time-span (e.g.

[29]) that could affect parasite transmission patterns in entire fish communities.

The round goby *Neogobius melanostomus* (Pallas, 1814) is a frequent invader of brackish and freshwater habitats worldwide, reaching enormous population densities and causing changes of food web dynamics at different trophic levels, e.g., in the North American Great Lakes [30] and in large European rivers, e.g. the Danube [29]. Round gobies nowadays make up app. 80% of fish catches in the Rhine [31], and so an alteration of ecological interactions is also expected for the Rhine. For example, it is known that round gobies act as competitors of spawning or foraging sites with native species [30]. Feeding patterns of *N. melanostomus* vary in different distribution areas. While dreissenid mussels play an important role in the feeding ecology of *N. melanostomus* in the Great Lakes and in the Baltic Sea [32,33], amphipods seem to be their main forage in German rivers [24,25,31]. In the Rhine, the Ponto-Caspian amphipod *Dikerogammarus villosus* (Sowinsky, 1894) has been described as dominating communities of macroinvertebrates and as an important prey species of *N. melanostomus* [24,25,31,34,35]. Both species, *D. villosus* and *N. melanostomus* function as intermediate hosts for different parasites (e.g., *Pomphorhynchus* spp. and *Raphidascaris* spp.) and may be responsible for the spread of these parasites, which could increasingly affect native vertebrate and invertebrate hosts as well [24,36].

Studies on *N. melanostomus* that combine the analysis of their feeding habits with parasitological analyses are rare and have focused on the Danube [25,29] and Rhine [24,31]. To analyse the role of different amphipod species for metazoan fish parasite transmission as well as temporal variation of diet compositions in invasive *N. melanostomus*, samples from the rivers Main and Rhine were compared in this study. We hypothesized that (a) *N. melanostomus* will mainly feed on amphipods throughout the course of our repeated monthly sampling and in both rivers, and accordingly, (b) the availability of amphipod species in a given river will reflect their relative contribution to gut contents of *N. melanostomus*. Moreover, we expected that (c) monthly infestation rates of amphipods with parasite species and monthly feeding rates of amphipods by *N. melanostomus* should reflect parasite infestation rates in *N. melanostomus*. Finally, a detailed description of parasite fauna for two sampling locations in the rivers Main and Rhine was intended to complement current parasite diversity estimates of *N. melanostomus* in its introduced range.

## Materials and Methods

### Sampling

A total of $n = 350$ *N. melanostomus* were collected from June to October 2011 in the rivers Rhine (49°51′54.7″N 8°21′40.2″E) and Main (50°04′48.9″N 8°31′19.6″E) in Germany. Both sites were similar in habitat structure with rip-rap embanked shorelines (technolithal) that led into bottom substrate of sand and gravel. In contrast to the Rhine, the river bank of the Main had little more vegetation with roots partly reaching into the water.

35 *N. melanostomus* specimens per site were caught randomly on top of and around rip-raps (depths of ~40–200 cm) during one day at the end of each month (between ~9 am–2 pm) using a hook and line technique. Since standardized angling is known to yield an equilibrated sex ratio and homogenously distributed, relatively large-sized specimens in *N. melanostomus* [37], a fishing rod equipped with an anti-tangle bottom rig consisting of a special sinker (Tiroler Hölzl, 80 g) was used to avoid entanglement between rip-rap interstices. A small, round hook (Owner, barb special, size 14, FRL-044) was baited with 1–3 fly maggots. All

hooked fish were used for subsequent examination in the laboratory without any size or sex selection. Each fish was carefully hooked off with a special hook removal tool and was humanely killed inside a plastic bag in order to avoid losing gut contents or parasites. To prevent further digestion or migration of parasites to other organs, fishes were kept separately in plastic bags in a cooling box filled with ice and stored afterwards at −20°C for later examination.

Amphipods were also collected monthly at the same sampling sites turning around large stones and using the 'kick-sampling' method after Storey et al. [38]. A small fishing net (15×20 cm, mesh size ~1 mm) was used to catch as many amphipods as possible within 30 minutes along a 10 m stretch at a depth of up to 50 cm. Amphipods were kept together with organic material and some stones in plastic bags. Entire samples were frozen at −20°C and later separated from sediment to identify amphipods to species level.

### Parasitological examination and feeding ecology of *N. melanostomus*

Gobies were measured for total length (cm) and weight (g), condition factors (CF) were calculated according to Schäperclaus [39]. These measures are key parameters in studies on fish biology and were reported in (Text S1, Table S1) to facilitate comparisons with other studies.

Fish were then examined for their metazoan parasite fauna and stomach content using a stereomicroscope (Olympus SZ 61, magnification x 6.7–45). At first, skin, fins and gills were inspected for ectoparasites. Afterwards, the body cavity was opened to separate the inner organs. Body cavity, rinsed with 0.9% NaCl, gastrointestinal tract, gonads, kidney, liver, mesenteries, spleen and eyes were dissected and examined for endoparasites. Isolated parasites were freed from host tissue and preserved in 70% ethanol (with 4% glycerol) for morphological identification. To this end, glycerine preparations were made according to Riemann [40]. Determination under a microscope (Leitz Dialux 22, magnification x 15.75–630) was aided by original descriptions and descriptions of Golvan [41] and Špakulová et al. [42] for acanthocephalans, and Moravec [43] for nematodes. Subsamples were stored in 100% ethanol for genetic analysis (see Text S2).

Since gobies have no clearly separated stomach and a very short gut, the entire gastrointestinal tract was carefully cut lengthwise with a small pair of scissors. The weights of full and empty stomachs and the weights of each food item were recorded to the nearest 0.001 g after pat-drying on absorbent paper. Very small, as well as almost digested and defragmented parts of one prey group that could not be identified to species level were referred to as 'not determined' (indet.) and weighted as a pooled subsample. Only specimens that could clearly be identified, e.g. using assignable parts like eyes or telson, were identified and counted. Other components, mainly mucus and sand, but also undeterminable items were neglected. Prey organisms were sorted and identified to the lowest possible taxon and grouped into the following categories: amphipods, molluscs, insects and 'others' (plants, vertebrates, Acari). Isolated food organisms and parasites were preserved in 70% ethanol (with 4% glycerol) for morphological identification.

Amphipods were identified to species-level following Eggers & Martens [44,45] and preserved in 70% ethanol. For parasitological examination, all amphipods were dissected and carefully screened under a stereomicroscope. Isolated parasites were stored in 100% ethanol. From each monthly sampling, fifty amphipods of each species were randomly taken to determine sex, body size and weight using an ocular micrometer and a micro-balance. Size was measured from the anterior rostrum to the base of the telson while animals were stretched in a straight position [46]. Data are reported in (Text S3, Figure S1).

## Statistical analyses

We first tested if the relative abundance of amphipods on site (covariate, arcsine(square root)-transformed percentages relative to the highest monthly abundance value observed for the respective site) determines the proportion of amphipods in *N. melanostomus* gut contents (monthly mean values were treated as the dependent variable) using analysis of covariance (ANCOVA using SPSS vs. 22), in which 'site' was a fixed factor. A Chi$^2$ goodness-of-fit test (using R; R Development Core Team [47]) was then applied to test whether amphipod species compositions as encountered on site are reflected in gut contents.

Gut content analyses comprised calculations of the numerical percentage of prey (N%), the weight percentage of prey (W%), and the frequency of occurrence of prey (F%) [48,49]. On the basis of these three indices, the index of relative importance (IRI) of different food items was calculated [50]. Differences in gut content assemblage structure between months and rivers were also assessed using two-factorial permutation ANOVA (PERMANOVA; 999 permutations) on Bray-Curtis dissimilarities of 4$^{th}$-root transformed weights (mg) of the different species in each fish gut using the PRIMER v6 and PERMANOVA+ add-on package (PRIMER-e, Plymouth, UK). The SIMPER procedure [51] was used for post hoc identification of the source of variation.

Parasitological analyses comprised calculations of standard parameters: the prevalence (P), mean intensity (mI), intensity (I) and mean abundance (mA) for each parasite species according to Bush et al. [52]. High mean intensities of *Pomphorhynchus* sp. infections were found (see results), and previous studies suggested transmission pathways into *N. melanostomus* via amphipods, especially *D. villosus* [24]. Therefore, we used a repeated-measures General Linear Model (rmGLM using SPSS vs. 22) to test if mean intensities of *Pomphorhynchus* sp. in round gobies (dependent variable) differed between sexes (rm) and sites (fixed factor), and if the proportion of amphipods in the gut contents (arcsine(square root)-transformed numerical percentages, covariate) had an effect. The nematode *R. acus* was also relatively abundant in fish samples, but we restricted our analysis to nonparametric Wilcoxon signed-rank test (using SPSS vs. 22) to test whether differences in infection rates existed between the two rivers.

## Results

### Amphipod communities

717 to 3,758 amphipods were collected during the monthly samplings, with a total of $n = 9,820$ in the Rhine and $n = 7,536$ in the Main (see Table S2). Five invasive but no native amphipod species were found in both rivers, namely *D. villosus*, *Echinogammarus trichiatus* (Martynov, 1932), *Echinogammarus ischnus* (Stebbing, 1899), *Chelicorophium curvispinum* (Sars, 1895) and *Chelicorophium robustum* (Sars, 1895). *Cryptorchestia cavimana* (Heller, 1865) occurred only in samples from the Main. *Dikerogammarus villosus* was dominating in all samples from the Main (total $n = 5,346$; 69%), except for September (Figure 1). In contrast, *E. trichiatus* was the dominant species in all samples from the Rhine (total $n = 8,463$; 86%; Figure 1). In both rivers a more balanced sex ratio was found for *D. villosus* (males:females, Rhine: 1:1.03; Main: 1:1.29) than for *E. trichiatus* (Rhine: 1:2.36; Main: 1:3.10).

## General feeding ecology of *N. melanostomus*

18 (Rhine) and 16 (Main) different prey items were identified in *N. melanostomus* guts (Table S3, Table S4). The index of relative importance (IRI) found amphipods to be the main diet component of *N. melanostomus*, with an overall contribution of 71% in the Rhine and 46% in the Main (Figure 2). In the Rhine, amphipods contributed with at least 30% in each monthly sample (Figure 2). The second most important group was molluscs, which contributed with 7–38% to the overall gut content. The widespread and common species *Bithynia tentaculata*, *Potamopyrgus antipodarum* and *P. antipodarum f. carinata* were distinguishable, but, due to a high degree of fragmentation, were combined into 'Gastropoda indet.'. Insects were rarely consumed, except for July where the IRI for Chironomidae rose to 2,288.83 (Table S3) when very little gut content was found overall. In the Main, highest proportions of amphipods (over 80%) occurred in September and October (Figure 2). Insects were consumed more often than in the Rhine, especially in June (79%) and August (36%). Fish diet was based on molluscs with 50% and 45% in July and August, respectively. Fishes, plants and Acari were rarely consumed in both rivers.

Gut content assemblage structure showed strong fluctuations between months and rivers. They differed significantly between June and July and June and August in the Rhine, whereas June and August were different from all other months in the Main (PERMANOVA: pseudo-$F = 8.64$, $df = 4$, $p = 0.001$ for the interaction 'river × month'; for post hoc results see Table 1). Amphipods mostly accounted for the highest average dissimilarity between different monthly samples in the Rhine, whereas amphipods and insects accounted for the highest average dissimilarity between months in the Main (SIMPER procedure).

## Amphipod prey preference of *N. melanostomus*

Few individuals of *C. curvispinum* were found in *N. melanostomus* guts, and the dominating amphipod species was *D. villosus*, especially in the Main, but to a lesser degree also in the Rhine. This was reflected in the ANCOVA, which detected a significant interaction between 'site' and 'relative abundance of amphipods on site' (Table 2).

*Dikerogammarus villosus* was disproportionally frequent in gut contents given its availability relative to that of other amphipod species on site (Chi$^2$ goodness-of-fit tests, $p<0.001$; except for the July sampling in the Main when *D. villosus* overall was highly abundant in the field; Figure 1). Therefore, an additional ANCOVA with similar model structure was run using percentages of *D. villosus* in the gut content of *N. melanostomus* as the dependent variable (Table 3). Whereas a decrease (not increase) of numerical percentages of *D. villosus* in the gut content of *N. melanostomus* with increasing availability of *D. villosus* on site was found in the Main (driving a significant main effect of the covariate; Table 3), this pattern was not observed in the river Rhine (see significant interaction effect in Table 3; Figure 3).

## Fish parasites: species identity and general biology

In total, three metazoan parasite species, two in the Rhine and three in the Main, could be isolated from *N. melanostomus*. The following taxa were identified morphologically: *Pomphorhynchus* sp., *Raphidascaris acus*, and Glochidia indet. (Table 4). As noted by Špakulová et al. [42] and Emde et al. [24], morphological identification of species within the acanthocephalan genus *Pomphorhynchus* can be difficult. Therefore, molecular barcoding was conducted on a subset of $n = 3$ specimens that were morphologically identified as *P. tereticollis*. Sequence data for ITS-1/5.8S/ITS-2 (Genbank accession numbers KJ756498–KJ756500) were almost identical (99.0% similarity, e-value: 0.00)

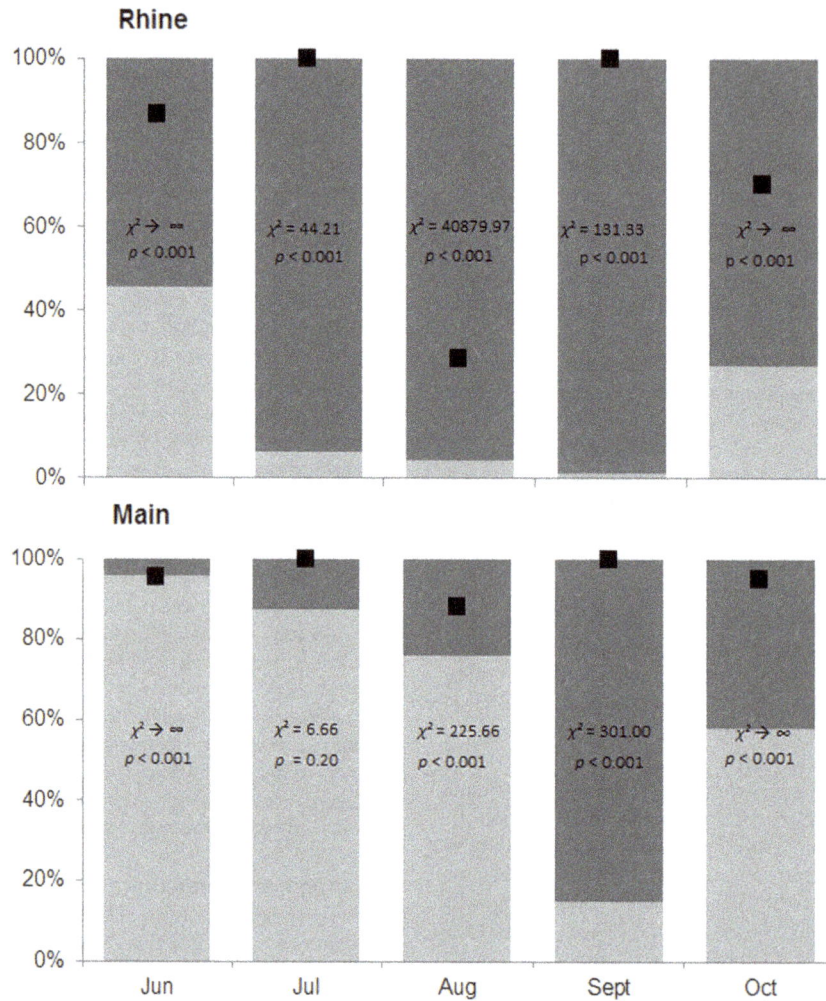

**Figure 1. Dominant amphipod species.** Fraction of the two dominant amphipod species (*D. villosus* = light grey, *E. trichiatus* = dark grey) in samples collected at our two study sites and numerical percentages of *D. villosus* in gut contents of *N. melanostomus* (black squares). Chi$^2$ goodness-of-fit tests were used to compare the availability of different amphipod species on site (expected values) with observed compositions in gut contents. For total numbers of individuals and amphipod species see Table S2.

to a sequence from *P. laevis* isolated from the cyprinid *Leuciscus cephalus* from the Czech Republic (Genbank accession number AY135415), suggesting that all acanthocephalan individuals in this study may belong to the same species. Due to a mismatch between the morphological identification characteristics and genetic information, acanthocephalan specimens were referred to as *Pomphorhynchus* sp. in this study.

All parasites were larval stages (Table 4). *Pomphorhynchus* sp. occurred only in the cystacanth stage. In the Rhine 91% of specimens were encysted in the mesenteries and liver and 9% were living freely in the body cavity. A similar pattern was found in the Main with 96% encysted in mesenteries and liver and 4% freely in the body cavity. The body cavity also harboured encysted *R. acus*, which occurred predominantly as L$_2$-larvae (91% in the Main, 88% in the Rhine), and L$_3$-larvae.

### Fish parasites: faunal composition

The most prevalent metazoan parasite type was *Pomphorhynchus* sp. with 100% prevalence in August and September in fish caught in the Rhine (Table 4). Maximum intensity reached 118 specimens per fish. Highest prevalence of *Pomphorhynchus* sp.

in the Main was recorded in June with 74.3%. Mean intensity of *Pomphorhynchus* sp. was an order of magnitude larger in fishes sampled from the Rhine (maximum mI = 34.6) than from the Main (maximum mI = 3.48) and always greater in female than in male *N. melanostomus* (rmGLM, significant interaction of 'sex × site'; Table 5; Figure 4). The nematode *R. acus* occurred with significantly lower prevalence in the Rhine (min. 28.57%, max. 57.14%) than in the Main (74.29% and 91.43%; Wilcoxon signed-rank test, $z = -2.023$, $p = 0.043$; Table 4). A maximum intensity of specimens of *R. acus* per fish was detected. Undetermined glochidia, i.e., parasitic larvae of unionid bivalves were detected on fish gills only in June (P = 54.3%) and October (P = 38.1%) in the Main.

### Parasites retrieved from amphipods

*Pomphorhynchus* sp. was the only parasite species that could be detected in amphipod samples. Two individuals were retrieved from *D. villosus* in the Rhine; the first was detected in samples from August (157 amphipods screened, P = 0.64%), the second in samples from October (671 amphipods screened, P = 0.15%). Overall, *Pomphorhynchus* sp. occurred at a prevalence of 0.15% in

## Rhine

## Main

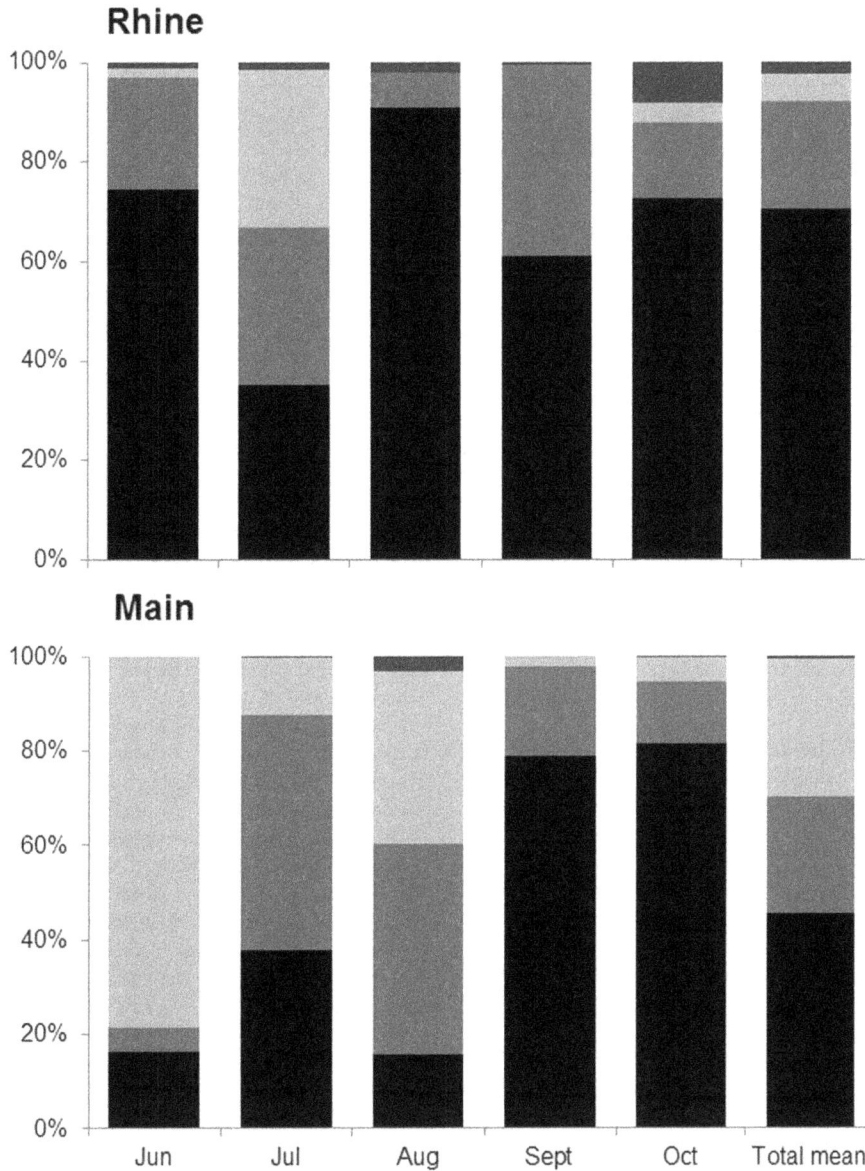

**Figure 2. Gut contents of** *Neogobius melanostomus.* Relative compositions (index of relative importance, IRI) of gut contents of *N. melanostomus* in two rivers from June until October 2011 as well as the total mean. Bar plot, from bottom to top: Amphipoda (black), Mollusca (medium grey), Insecta (light grey), others (dark grey).

**Table 1.** PERMANOVA results (post hoc procedure).

| River Rhine | t | p (perm) | River Main | T | p (perm) |
|---|---|---|---|---|---|
| **Jun, Jul** | 2.842 | 0.001 | Jun, Jul | 5.006 | 0.001 |
| **Jul, Aug** | 3.310 | 0.001 | Jun, Aug | 3.155 | 0.001 |
| | | | Jun, Sept | 5.943 | 0.001 |
| | | | Jun, Oct | 3.758 | 0.001 |
| | | | Jul, Aug | 2.835 | 0.001 |
| | | | Aug, Sept | 4.269 | 0.001 |
| | | | Aug, Oct | 2.447 | 0.003 |

Differences in gut contents assemblage structure (based on species' weights) of *N. melanostomus* between months. Significant results (permutation $p<0.05$) after Bonferroni correction for multiple comparisons are shown.

**Table 2.** ANCOVA results – all amphipods.

| | Source | df | MS | F | p | Partial Eta squared |
|---|---|---|---|---|---|---|
| All amphipods | Site | 1 | 0.317 | 9.330 | **0.022** | 0.609 |
| | Rel. abundance | 1 | 0.003 | 0.091 | 0.773 | 0.015 |
| | Site × rel. abundance | 1 | 0.549 | 16.183 | **0.007** | 0.730 |
| | Residuals | 6 | 0.034 | | | |

Numerical percentages of all amphipods in the gut content of *N. melanostomus* in relation to the relative abundance of amphipods on site. Significant effects are in bold.

*D. villosus* in the river Rhine (two out of 1,350 specimens). The total number of *D. villosus* was four times larger in the Main than in the Rhine (i.e., $n = 5,346$), still, no parasites were detected. Low overall abundance precluded an analysis of potential temporal fluctuation in parasite infections of amphipods. Numerical percentages of amphipods in fish gut contents did not predict mean intensities of acanthocephalan parasites in *N. melanostomus* (Table 5).

## Discussion

### Feeding ecology of *N. melanostomus*

Co-evolved trophic relationships can facilitate biological invasions, as exemplified by communities of coexisting invasive *N. melanostomus*, dreissenid mussels and *E. ischnus* in the North American Great Lakes [53,54]. Presence of co-evolved prey, however, appears not to be a prerequisite for *N. melanostomus* in German rivers, since *N. melanostomus* was characterized by an opportunistic and broad feeding strategy [see also 30,31]. Opportunistic feeding might also provide a plausible explanation for why we detected no positive correlation between the abundance of *D. villosus* in the field (generally a preferred type among amphipod prey) and their proportional contribution to gut contents. This was obvious especially during early summer, when prey species other than amphipods became more relevant (higher index of relative importance), especially in the Main, where insects and molluscs became the main food sources. Similarly, the importance of amphipod prey (*D. villosus* and others) for *N. melanostomus* in the Danube increased from early to late summer while the importance of chironomid larvae decreased [25]. Ingested insects in our present study were mostly nematoceran larvae, which are generally abundant in slow-flowing waterways like the Main. Non-biting midges (Chironomidae) no longer dominate the invertebrate community of the navigable main

channel of the upper Rhine [55], which may explain why insects, overall, were barely ingested. While *N. melanostomus* is commonly regarded as a predator of fish eggs and fry (e.g. [56]), these were only rarely retrieved from gut contents.

An ontogenetic size dependent diet shift from amphipods and insects to a diet dominated mainly by molluscs is known for round gobies (e.g. [25]), however, fish lengths where shifts seem to occur vary substantially between study regions and most likely depend on availability and abundance of prey organisms [57,58] as well as on time since invasion [29]. In our present study, the genus *Dreissena* seems to play a subordinate role compared to the Great Lakes and the Baltic Sea, which may be attributable to more readily available food sources, like insect larvae and amphipods. Generally, a tendency of increasing absolute numbers with increasing fish size was observed for *D. villosus* and nematocerans. In this context, Emde et al. [24] already demonstrated a size-dependent increase in acanthocephalan infections, which was inter alia explained by a correlation between goby and amphipod (*D. villosus*) prey body size, as it seems likely that the development of acanthocephalan larvae might only grow in amphipods above a certain size threshold. Thus, smaller gobies, feeding on smaller *D. villosus*, are less infected by acanthocephalans.

All amphipods found during monthly sampling were non-indigenous species from the Ponto-Caspian region (i.e., Black and Caspian Seas), corroborating studies in several European watersheds [24,44,59]. The most common non-indigenous amphipod species were *D. villosus* und *E. trichiatus*. *Dikerogammarus villosus* was dominant in samples from the Main, whereas *E. trichiatus* was dominant in Rhine samples, suggesting that faunal compositions of invasive amphipods may be more stable temporally and to a lesser degree spatially within the Rhine drainage (see also [24,60]). *Dikerogammarus villosus* was detected six years earlier than *E. trichiatus* in the Rhine and is known for its strong predation on other gammarids [7,61]. However, the total number

**Table 3.** ANCOVA results – *D. villosus*.

| | Source | df | MS | F | p | Partial Eta squared |
|---|---|---|---|---|---|---|
| *D. villosus* | Site | 1 | 0.155 | 31.261 | **0.001** | 0.839 |
| | Rel. abundance | 1 | 0.053 | 10.644 | **0.017** | 0.640 |
| | Site × rel. abundance | 1 | 0.127 | 25.487 | **0.002** | 0.809 |
| | Residuals | 6 | 0.005 | | | |

Numerical percentages of *D. villosus* in the gut content of *N. melanostomus* in relation to the relative abundance of *D. villosus* on site. Significant effects are in bold.

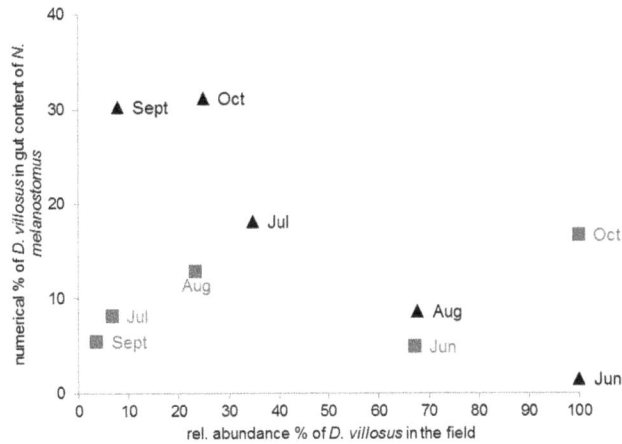

**Figure 3.** *Dikerogammarus villosus* **in fish guts and in the field.**
Numerical percentage of *D. villosus* in gut contents of *N. melanostomus* in relation to the relative abundance of *D. villosus* at the Main (black) and Rhine (grey) between June and October 2011.

of individuals caught in the Rhine was an order of magnitude lower than that of *E. trichiatus*. Whether higher predation on *D. villosus* by *N. melanostomus* in the Rhine compared to the Main could explain this pattern remains uncertain, since no fish densities at both sites were recorded herein.

Regardless of the high numbers of *E. trichiatus* in the Rhine, *N. melanostomus* primarily fed on *D. villosus*. How can this pattern be explained? Sih and Christensen [62] argued that variation in prey behaviour is more likely to affect the direction of predator-prey interactions than active prey choice of predators. Qualitatively, we noted that *E. trichiatus* at our study sites occurred closer to riverbanks, while *D. villosus* were found in both shallow and deeper waters, and so *E. trichiatus* could avoid fish predation in shallow habitats or by hiding between rip-rap interstices. Spatial niche segregation between *E. trichiatus* and *D. villosus* was previously observed in the Netherlands where the former seems to occur on soft substrates whereas the latter is most abundant on hard substrates [63]. Thus, different microhabitat use or different activity patterns in *D. villosus* are likely explanations for their dominance among amphipod prey in *N. melanostomus*.

Parasites can manipulate the predator avoidance of freshwater amphipods, rendering them more vulnerable to their fish predators (for *Gammarus pulex* see [64,65]). Whether infections by *Pomphorhynchus* sp. affect the predator avoidance of *D. villosus* is currently not known, but if infected individuals were indeed more prone to predation, this would provide a striking explanation for our finding that gobies were highly infected by *Pomphorhynchus* sp., yet infectious stages were barely found in their amphipod prey (i.e., *D. villosus*), and were even completely absent in the Main. It seems reasonable to argue that infected *D. villosus* were ingested at an accelerated rate compared to uninfected specimens. Generally, infection rates of invertebrate intermediate hosts, especially crustaceans, tend to be low, often ranging between 0.01 and 0.1% prevalence [23,28]. A possible reason for the higher infestation rates of *D. villosus* in the Rhine might be the presence of more final hosts (like common barbel *Barbus barbus* and European chub *Squalius cephalus,* however this assumption is not based on quantitative data but on personal observations only (S. Emde, personal observation).

*Pomphorhynchus* sp. is known to include a variety of different first intermediate hosts in its life cycle, such as *D. villosus* [24], *G. pulex* [64] and *C. curvispinum* [65]. *Gammarus pulex* seems to be

completely displaced by invasive species in the Rhine and Main [24] and was not part of the gobies' diet at both sampling sites. Following a massive decrease since 1995, *C. curvispinum* currently also plays a negligible role in the gobies' diet [65]. In the light of the decrease of other amphipod species and the observed dominance of *D. villosus* in the gobies' diet, we suggest that *D. villosus* currently represents the most relevant intermediate host for *Pomphorhynchus* sp. Still, future studies could investigate additional invertebrate groups and might uncover additional first intermediate hosts for the opportunistic parasites of the genus *Pomphorhynchus*.

## Parasite fauna of *N. melanostomus*

More than 94 parasites of *N. melanostomus* have been recorded worldwide [66], and in its introduced range in Europe, 35 metazoan parasite species have been detected so far (e.g., [66–69]). *Neogobius melanostomus* usually carries more than ten different parasite species per population in its native range [70]. Herein, only three parasite species could be detected in 350 round gobies examined, suggesting that the diversity of *N. melanostomus* parasites in the Rhine did not change over the past four years ([24], S. Emde personal observation). In other regions, the parasite fauna of invasive *N. melanostomus* increased rapidly, e.g., in the Gulf of Dansk, where numbers rose from 4 to 12 parasite species within two years [68]. Only 6 to 10 years have passed since round gobies were first recorded in German inland waterways, while the first report of round gobies at our sampling sites was in 2007 [71,72]. Our results support the 'enemy release hypothesis' [19], and release from natural parasites could be one reason promoting the fast spread of round gobies worldwide. This advantage over indigenous fishes, however, will likely be lost if the diversity of the parasite fauna of *N. melanostomus* increases with time [73]. Whether or not such an increase of parasite diversity will occur in the future requires further monitoring.

All parasites detected in *N. melanostomus* were larval stages, and so we tentatively argue that currently no native parasite species is able to use *N. melanostomus* as its final but only as a paratenic host. A higher parasitization of *N. melanostomus* was observed in the Rhine, where fishes were also smaller and had a lower condition factor than in the Main (Text S1, Table S1). A high parasite load can lead to decreased growth in their fish hosts [74], however, infection studies in controlled environments would be needed to further address this hypothesis.

*Pomphorhynchus* sp. (Acanthocephala) and *Raphidascaris acus* (Nematoda) have been detected before in *N. melanostomus* caught in the Rhine, with similar infection rates for *Pomphorhynchus* sp. [24]. Latest data of the Danube River also described high abundances of this parasite but detected highest abundances in more recently invaded areas [29]. Similarly high prevalences of *R. acus* as found in our current study (up to 91.43%) are known from studies in other sections of the Rhine (56%) [75] and the Danube (P = 57%) [67]. Generally, differences in infection rates (prevalence/intensities) among studies could be related to the presence/absence as well as abundance of the parasites' final hosts. For adult *R. acus* the European pike (*Esox lucius*) and brown trout (*Salmo trutta fario*) are known as principal final hosts [43], whereas it is barbel (*Barbus barbus*) and chub (*Squalius cephalus*) for *Pomphorhynchus* sp. [24]. However, *N. melanostomus* seems to represent a new, additional intermediate host for these parasites and thus, bridges the trophic level towards new potential, predatory final hosts. Other potential definitive hosts in the rivers Rhine and Main are trout (*Salmo trutta*) and catfish (*Silurus glanis*) for *Pomphorhynchus* [76,77] and the European eel (*Anguilla anguilla*), European perch (*Perca fluviatilis*) and pike-

**Table 4.** Parasitological parameters for the parasite fauna of *N. melanostomus*.

| | P (%) | | | | | I_min–I_max/Mi | | | | | mA | | | | |
|---|---|---|---|---|---|---|---|---|---|---|---|---|---|---|---|
| | Jun | Jul | Aug | Sept | Oct | Jun | Jul | Aug | Sept | Oct | Jun | Jul | Aug | Sept | Oct |
| **Rhine** | | | | | | | | | | | | | | | |
| **Nematoda** | | | | | | | | | | | | | | | |
| *Raphidascaris acus* (Cyst, L) in body cavity/mesentery/liver | 50.00 | 28.57 | 57.14 | 55.56 | 34.29 | 1–6/2.44 | 1–5/2.10 | 1–18/3.80 | 1–17/3.67 | 1–7/10.25 | 1.22 | 0.60 | 2.17 | 1.89 | 3.51 |
| **Acanthocephala** | | | | | | | | | | | | | | | |
| *Pomphorhynchus* sp. (L) in body cavity/mesentery/liver | 86.11 | 71.43 | 100 | 100 | 74.29 | 1–118/22.65 | 1–35/9.04 | 1–101/24.26 | 2–95/32.71 | 1–20/8.46 | 19.50 | 6.46 | 24.26 | 32.71 | 6.29 |
| **Total** | 100 | 77.14 | 100 | 100 | 88.57 | 1–118/20.72 | 1–39/9.15 | 1–105/26.43 | 2–95/34.60 | 1–23/11.06 | 20.72 | 7.06 | 26.43 | 34.60 | 9.80 |
| **Main** | | | | | | | | | | | | | | | |
| **Nematoda** | | | | | | | | | | | | | | | |
| *Raphidascaris acus* (Cyst, L) in body cavity/mesentery/liver | 74.29 | 85.71 | 82.86 | 88.57 | 91.43 | 1–20/3.85 | 1–29/6.17 | 1–24/6.55 | 1–9/3.39 | 1–15/5.63 | 2.86 | 5.29 | 5.43 | 3.00 | 5.14 |
| **Acanthocephala** | | | | | | | | | | | | | | | |
| *Pomphorhynchus* sp. (L) in body cavity/mesentery/liver | 74.29 | 60.00 | 37.14 | 37.14 | 25.71 | 1–16/2.77 | 1–15/3.48 | 1–6/2.62 | 1–9/3.00 | 1–16/3.22 | 2.06 | 2.09 | 0.97 | 1.11 | 0.83 |
| **Bivalvia** | | | | | | | | | | | | | | | |
| Glochidia indet. (L) in gills | 54.29 | – | – | – | 38.10 | 1–8/3.68 | – | – | – | 1–23/9.88 | 2.00 | – | – | – | 3.76 |
| **Total** | 97.14 | 85.71 | 94.29 | 94.29 | 94.29 | 1–30/7.12 | 1–44/8.60 | 1–29/7.72 | 1–17/4.36 | 1–35/8.73 | 6.91 | 7.37 | 6.40 | 4.11 | 8.23 |
| **Wilcoxon signed-rank test** | | | | | | | | | | | | | | | |
| *Raphidascaris acus* | $z=-2.023$ $p=$**0.043** (Rhine/Main) | | | | | $z=-0.405$ $p=0.686$ (Rhine/Main) | | | | | $z=-2.023$ $p=$**0.043** (Rhine/Main) | | | | |
| *Pomphorhynchus* sp. | $z=-2.023$ $p=$**0.042** (Rhine/Main) | | | | | $z=-2.023$ $p=$**0.043** (Rhine/Main) | | | | | $z=-2.023$ $p=$**0.043** (Rhine/Main) | | | | |

I = Intensity, L = larvae, mA = mean abundance, ml = mean intensity and P = prevalence.

**Table 5.** Repeated–measures GLM results.

|  | Source | df | MS | F | p | Partial Eta squared |
|---|---|---|---|---|---|---|
| **Within–subjects effects** | Sex | 1 | 35.935 | 3.689 | 0.096 | 0.345 |
|  | Sex × amphipods in gut | 1 | 3.973 | 0.408 | 0.543 | 0.055 |
|  | Sex × site | 1 | 129.212 | 13.263 | **0.008** | 0.655 |
|  | Residuals (Sex) | 7 | 9.742 |  |  |  |
| **Between–subjects effects** | Intercept | 1 | 510.007 | 3.202 | 0.117 | 0.314 |
|  | Amphipods in gut | 1 | 46.853 | 0.294 | 0.604 | 0.040 |
|  | Site | 1 | 1459.710 | 9.164 | **0.019** | 0.567 |
|  | Residuals | 7 | 159.282 |  |  |  |

Repeated–measures GLM on mean intensities of *Pomphorhynchus* sp. in round gobies in relation to fish sex, numerical percentages of amphipods (*D. villosus* and Amphipoda indet.) in the gut content and site. Significant effects are in bold.

perch (*Sander lucioperca*) for *R. acus* [78]. Infection studies need to show whether the female parasite attains gravidity in the potential definitive host or whether these predatory fishes may only act as para-definitive hosts in which the parasite matures but is unable to produce eggs [78]. If they do not act as definitive hosts, the large number of parasite larvae in *N. melanostomus* will

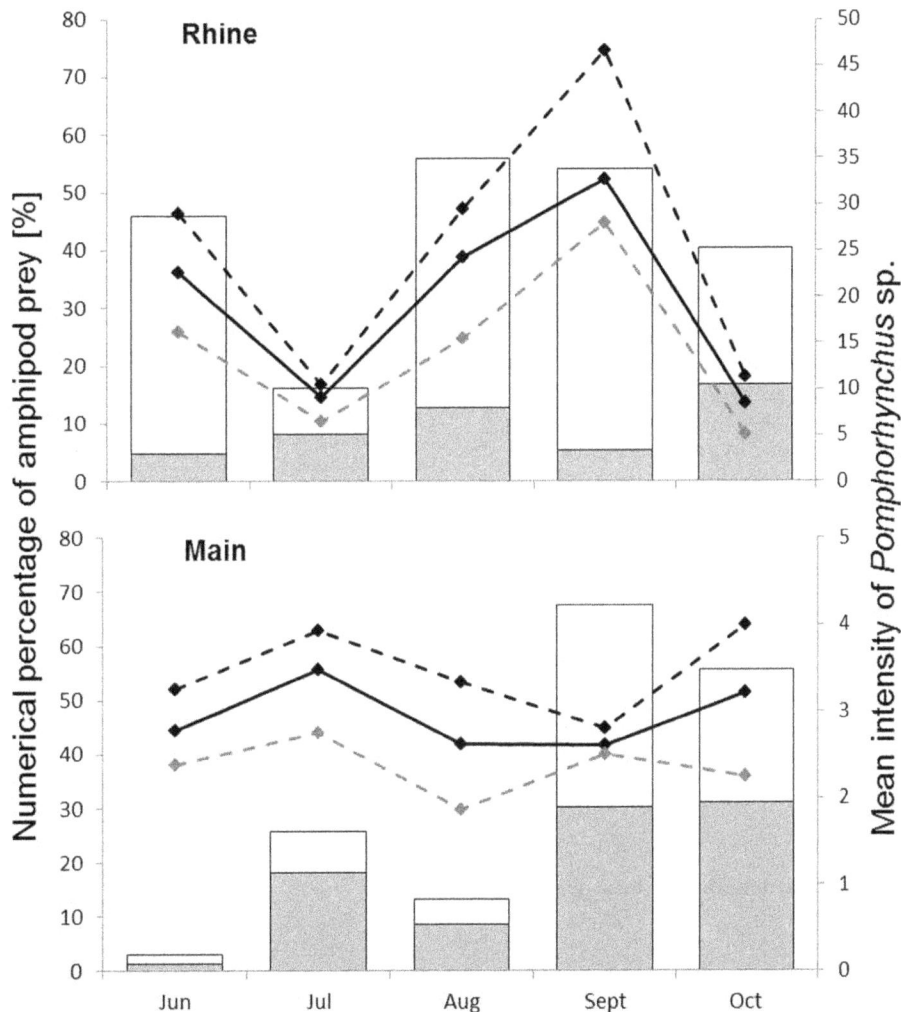

**Figure 4. Amphipod prey and infections with *Pomphorhynchus* sp.** Relationship between numerical percentages of *D. villosus* (grey) and Amphipoda indet. (white) in the gut content of *N. melanostomus* and mean intensities (ml, black line) of *Pomphorhynchus* sp. in male (grey dashed line) and female (black dashed line) *N. melanostomus*. For numbers of individuals please refer to Table S3 and Table S4.

be transmitted to these predatory fishes, however, not be able to complete their life cycle. This would lead to a dilution effect, resulting in a continued loss of infection within the system as has been described for different parasite-host communities before [79,80] and would therefore be an alternative plausible explanation for the lower infection rates in the Main than in the Rhine.

Parasitic larval stages (Glochidia) of freshwater mussels of the family Unionidae were found in samples from the river Main, which confirms a former report of *N. melanostomus* serving as a host for unionid glochidia in the Danube [67]. Glochidia could be detected only during some months, because river mussels (*Unio* sp.) spawn in early summer and swan mussels (*Anodonta* sp.) in late summer, and glochidia attach to fish gills for only a few weeks [81]. Although unionid mussels are known to occur in the Rhine [82], no glochidia were detected on the gills of *N. melanostomus*, which could suggest an abundance-correlated effect. Alternatively, *N. melanostomus* might be a bad host for unionids [83]. Authors infected gobies with Glochidia of which 98% were lost within 16 days. Based on that study, our findings of Glochidia attached to gills of *N. melanostomus* could therefore be a finding that was the result of a very recent infection.

We initially hypothesized monthly infestation rates of *D. villosus* with *Pomphorhynchus* sp. potentially reflecting infestation rates in *N. melanostomus*. Due to overall low abundances of *Pomphorhynchus* sp. in *D. villosus* a statistical analysis in this direction was not possible. We also tested whether the numerical percentage of *D. villosus* in gut contents predicts mean intensities of *Pomphorhynchus* sp. but found no such effect. The timing of the parasite's life cycle, however, has not yet been examined, and so our analysis (that was based on monthly sampling) may not have been appropriate to capture such potential effect.

Sex-related differences in parasite infections are common and can be ascribed to sex-specific behavioural, physiological or morphological differences [84,85]. In this study, mean intensity of *Pomphorhynchus* sp. was significantly higher in females than males in the Rhine, supporting the finding of Brandner et al. [29] from the Danube River. No significant sex differences were observed in the Main, but *Pomphorhynchus* sp. mean intensities were low in the Main overall. Males can allocate much less time to feeding than females (for poeciliid fishes see [86,87]) lowering their risk to take up parasites from food. Indeed, Charlebois et al. [88] found *N. melanostomus* males to cease feeding during brood care, while females producing eggs should have increased energy demands.

Our study confirmed that *D. villosus* functions as the main amphipod prey species for *N. melanostomus* in German rivers, however, parasite intensities in *N. melanostomus* differed between sampling locations of Rhine and Main independently of amphipod abundances. We suggest that a characterization of new final fish hosts, especially for *Pomphorhynchus* sp., at the sites investigated herein could provide important new insight into the ecological causes of variation in parasitization patterns of *N. melanostomus* in its introduced range.

## Supporting Information

**Figure S1   Box–plots of total length and total weight of two amphipod species.**

## References

1. Carlton JT, Geller JB (1993) Ecological roulette: the global transport of nonindigenous marine organisms. Science 261: 78–82.
2. Walther G-R, Post E, Convey P, Menzel A, Parmesan C, et al. (2002) Ecological responses to recent climate change. Nature 416: 389–395.

**Table S1   Biological parameters of *Neogobius melanostomus*.**

**Table S2   Amphipod fauna.**

**Table S3   Gut contents and parameters of *Neogobius melanostomus* for the river Rhine.**

**Table S4   Gut contents and parameters of *Neogobius melanostomus* for the river Main.**

**Text S1   Size measurements and condition factors of *N. melanostomus*.**

**Text S2   Genetic identification of parasites.**

**Text S3   Size measurements of *D. villosus* and *E. trichiatus*.**

## Acknowledgments

We thank J. Schneider (Office for fish ecological studies – BFS, Frankfurt), S. Gallus and S. Schierz (Goethe University, Frankfurt) for their support with data assessment. We further wish to thank C. Koehler (Dezernat V 51.1 Landwirtschaft-Landschaftspflege-Fischerei, Regierungspräsidium Darmstadt) for providing a fishing license to catch gobies. We are grateful to D. Green who helped with statistics. Finally, we thank the reviewers of this article for their helpful suggestions. The authors declare no conflict of interest.

### Ethics Statement

Approval of our present study by a review board institution or ethics committee was not necessary because all fish were caught by a person (S. Emde) holding a valid local fishing license (No. 06258) for the river Main, issued by the '*Höchster Fischereigenossenschaft*', 65830 Kriftel, Germany. For the river Rhine a special permit (F4/Di-Zi) was issued by the '*Hessische Landesgesellschaft mbH*', 34121 Kassel, Germany. No living animals were used. In Germany, the fishing license permits the holder to capture and sacrifice the fish, which can be used not only for consumption but also for research purposes. All fish were stunned by a blow on the head and expertly killed immediately by cervical dislocation and a cardiac stab according to the German Animal Protection Law (§ 4) and the ordinance of slaughter and killing of animals (*Tierschlachtverordnung* § 13). Because of public accessibility no permissions were required to enter the sampling sites.

## Author Contributions

Conceived and designed the experiments: SE SK. Performed the experiments: SE TK. Analyzed the data: SE JK MP SK. Contributed reagents/materials/analysis tools: SK MP. Contributed to the writing of the manuscript: SE JK TK MP SK.

3. Leprieur F, Beauchard O, Blanchet S, Oberdorff T, Brosse S (2008) Fish invasions in the world's river systems: when natural processes are blurred by human activities. Plos Biol 6: e28.
4. Vitule JRS, Freire CA, Simberloff D (2009) Introduction of non-native freshwater fish can certainly be bad. Fish Fish 10: 98–108.

5. Ricciardi A, MacIsaac HJ (2011) Impacts of biological invasions on freshwater ecosystems. In: Fifty Years of Invasion Ecology: The Legacy of Charles Elton (Ed. Richardson D. M.), 211–224, Blackwell Publishing.

6. Martin CW, Valentine MM, Valentine JF (2010) Competitive interactions between invasive Nile Tilapia and native fish: the potential for altered trophic exchange and modification of food webs. Plos One 5: e14395. doi:10.1371/journal.pone.0014395.

7. Dick JTA, Platvoet D (2000) Invading predatory crustacean *Dikerogammarus villosus* eliminates both native and exotic species. P Roy Soc Lond B Bio 267: 977–983.

8. Salo P, Korpimaki E, Banks PB, Nordstrom M, Dickman CR (2007) Alien predators are more dangerous than native predators to prey populations. P Roy Soc B-Biol Sci 274: 1237–1243.

9. Paolucci EM, MacIsaac HJ, Ricciardi A (2013) Origin matters: alien consumers inflict greater damage on prey populations than do native consumers. Divers and Distrib 19: 988–995.

10. Prenter J, MacNeil C, Dick JTA, Dunn AM (2004) Roles of parasites in animal invasions. Trends Ecol Evol 19: 385–390.

11. Douda K, Lopes-Lima M, Hinzmann M, Machado J, Varandas S, et al. (2013) Biotic homogenization as a threat to native affiliate species: fish introductions dilute freshwater mussel's host resources. Divers Distrib 19: 933–943.

12. Strayer DL (2012) Eight questions about invasions and ecosystem functioning. Ecol Lett 15: 1199–1210.

13. Witte F, Goldschmidt T, Wanink J, van Oijen M, Goudswaard K, et al. (1992) The destruction of an endemic species flock: quantitative data on the decline of the haplochromine cichlids of Lake Victoria. Environ Biol Fish 34: 1–28.

14. Kats LB, Ferrer RP (2003) Alien predators and amphibian declines Review of two decades of science and the transition to conservation. Divers and Distrib 9: 99–110.

15. Machida Y, Akiyama YB (2013) Impacts of invasive crayfish (*Pacifastacus leniusculus*) on endangered freshwater pearl mussels (*Margaritifera laevis* and *M. togakushiensis*) in Japan. Hydrobiologia 720: 145–151.

16. Ricciardi A, Whoriskey FG, Rasmussen JB (1996) Impact of *Dreissena polymorpha* on native unionid bivalves in the upper St. Lawrence River. Can J Fish Aquat Sci 53: 1434–1444.

17. Dick JTA, Platvoet D, Kelly DW (2002) Predatory impact of the freshwater invader *Dikerogammarus villosus* (Crustacea: Amphipoda). Can J Fish Aquat Sci 59: 1078–1084.

18. Carlsson NOL, Sarnelle O, Strayer DL (2009) Native predators and exotic prey – an acquired taste? Front Ecol Environ 7: 525–532.

19. Crawley MJ (1987) What makes a community invasible? In: Colonization, succession, and stability (Eds. Gray A.J., Crawley M.J., Edwards P.J.) 429–453, Blackwell, Oxford.

20. Torchin ME, Lafferty KD, Dobson AP, McKenzie VJ, Kuris AM (2003) Introduced species and their missing parasites. Nature 421: 628–630.

21. Kelly DW, Paterson RA, Townsend CR, Poulin R, Tompkins DM (2009) Parasite spillback: A neglected concept in invasion ecology? Ecology 90: 2047–2056.

22. Kopp K, Jokela J (2007) Resistant invaders can convey benefits to native species. Oikos 116: 295–301.

23. Paterson RA, Townsend CR, Poulin R, Tompkins DM (2011) Introduced brown trout alternative acanthocephalan infections in native fish. J Anim Ecol 80: 990–998.

24. Emde S, Rueckert S, Palm HW, Klimpel S (2012) Invasive Ponto-Caspian amphipods and fish increase the distribution range of the acanthocephalan *Pomphorhynchus tereticollis* in the river Rhine. Plos One 7: e53218 doi:10.1371/journal.pone.0053218.

25. Brandner J, Auerswald K, Cerwenka AF, Schliewen UK, Geist J (2013) Comparative feeding ecology of invasive Ponto-Caspian gobies. Hydrobiologia 703: 113–131.

26. Locke SA, Bulté G, Marcogliese DJ, Forbes MR (2014) Altered trophic pathway and parasitism in a native predator (*Lepomis gibbosus*) feeding on introduced prey (*Dreissena polymorpha*). Oecologia 175: 315–24.

27. Rohde K (2005) Marine parasitology. CSIRO Publishing.

28. Busch MW, Kuhn T, Münster J, Klimpel S (2012) Marine crustaceans as potential hosts and vectors for metazoan parasites. In: Arthropods as vectors of emerging diseases (Ed. H. Mehlhorn), 329–360, Parasitol Res Monographs 3, Springer, Berlin Heidelberg.

29. Brandner J, Cerwenka AF, Schliewen UK, Geist J (2013) Bigger is better: Characteristics of round gobies forming an invasion front in the Danube River. PLoS ONE 8(9): e73036.

30. Kornis MS, Mercado-Silva N, van der Zanden MJ (2012) Twenty years of invasion: a review of round goby *Neogobius melanostomus* biology, spread and ecological implications. J Fish Biol 80: 235–285.

31. Borcherding J, Dolina M, Heermann L, Knutzen P, Krüger S, et al. (2013) Feeding and niche differentiation in three invasive gobies in the Lower Rhine, Germany. Limnologica 43: 49–58.

32. Skóra KE, Rzeznik J (2001) Observations on diet composition of *Neogobius melanostomus* Pallas 1811 (Gobiidae, Pisces) in the Gulf of Gdansk (Baltic Sea). J Great Lakes Res 27: 290–299.

33. Rakauskas V, Bacevičius E, Pūtys Ž, Ložys L, Arbačiauskas K (2008) Expansion, feeding and parasites of the round goby, *Neogobius melanostomus* (Pallas, 1811), a recent invader in the Curonian Lagoon, Lithuania. Acta Zoologica Lituanica 18, 3: 180–190.

34. Haas G, Brunke M, Streit B (2002) Fast turnover in dominance of exotic species in the Rhine river determines biodiversity and ecosystem function: an affair between amphipods and mussels. In: Invasive aquatic species of Europe: distribution, impacts, and management (eds. Leppäkoski E, Gollasch S, Olenin S), 426–432, Dordrecht.

35. Bernauer D, Jansen W (2006) Recent invasions of alien macroinvertebrates and loss of native species in the upper Rhine River, Germany. Aquatic Invasions 1, 2: 55–71. Available: http://aquaticinvasions.net/2006/AI_2006_1_2_Bernauer_Jansen.pdf. Accessed 2012 June 5.

36. Ondračková M, Francová K, Dávidová M, Polačik M, Jurajda P (2010) Condition status and parasite infection of *Neogobius kessleri* and *N. melanostomus* (Gobiidae) in their native and non-native area of distribution of the Danube River. Ecol Res 25: 857–866.

37. Brandner J, Pander J, Mueller M, Cerwenka AF, Geist J (2013) Effects of sampling techniques on population assessment of invasive round goby *Neogobius melanostomus*. J Fish Biol 82: 2063–2079.

38. Storey AW, Edward DHD, Gazey P (1991) Surber and kick sampling: a comparison for the assessment of macroinvertebrate community structure in streams of south-western Australia. Hydrobiologia 211: 111–121.

39. Schäperclaus W (1991) Fish Diseases. Volume 1. (Ed. Kothekar V.S.), Akademie-Verlag, Berlin.

40. Riemann F (1988) Introduction to the study of meiofauna. Higgins RP und Thiel H (Eds.). Smithsonian Institution Press: 293–301.

41. Golvan YJ (1969) Systematique des acanthocephales (Acanthocephala, Rudolphi 1801). L'ordre des Palaeacanthocephala Meyer 1931. La superfamille des Echinorhynchoidea (Cobbold 1876) Golvan et Houin, 1963. Mémoires du Museum National d'Histoire Naturelle, Série A, Zoologie Band 57, Paris: 373 p.

42. Špakulová M, Perrot-Minnot M-J, Neuhaus B (2011) Resurrection of *Pomphorhynchus tereticollis* (Rudolphi, 1809) (Acanthocephala: Pomphorhynch-idae) based on new morphological and molecular data. Helminthologia 48, 3: 268–277.

43. Moravec F (1994) Parasitic nematodes of freshwater fishes of Europe. Academy of Sciences of the Czech Republic, Academia.

44. Eggers TO, Martens A (2001) Bestimmungsschlüssel der Süßwasser–Amphipoda (Crustacea) Deutschlands. Lauterbornia 42: 1–68.

45. Eggers TO, Martens A (2004) Ergänzungen und Korrekturen zum "Bestim-mungsschlüssel der Süßwasser–Amphipoda (Crustacea) Deutschlands". Lauterbornia 50: 1–13.

46. Quigley MA, Lang GA (1989) Measurement of amphipod body length using a digitizer. Hydrobiologia 171: 255–258.

47. R Development Core Team (2010) R: A language and environment for statistical computing. Foundation for Statistical Computing, Vienna, Austria.

48. Hyslop EJ (1980) Stomach content analysis - a review of methods and their application. J Fish Biol 17: 411–429.

49. Amundsen PA, Gabler HM, Staldvik FJ (1996) A new approach to graphical analysis of feeding strategy from stomach contents data – modification of the Costello (1990) method. J Fish Biol 48: 607–614.

50. Pinkas L, Oliphant MD, Iverson ILK (1971) Food habits of albacore, bluefin tuna and bonito in Californian waters. Calif Fish Game 152: 1–105.

51. Clarke KR (1993) Non-parametric multivariate analyses of changes in community structure. Aust J Ecol 18: 117–143.

52. Bush O, Lafferty AD, Lotz JM, Shostak AW (1997) Parasitology meets ecology on his own terms: Margolis, et al. revisited. J Parasitol 83: 575–583.

53. McKinney ML, Lockwood JL (1999) Biotic homogenizaton: a few winners replacing many losers in the next mass extinction. Trends Ecol Evol 14: 450–453.

54. Vanderploeg HA, Nalepa TF, Jude DJ, Mills EL, Holeck KT, et al. (2002) Dispersal and emerging ecological impacts of Ponto-Caspian species in the Laurentian Great Lakes. Can J Fish Aquat Sci 59: 1209–1228.

55. IKSR (2002) Das Makrozoobenthos des Rheins 2000, Internationale Kommis-sion zum Schutz des Rheins (IKSR), Bericht Nr. 128-d.doc; Koblenz.

56. Corkum LD, Sapota MR, Skora KE (2004) The round goby, *Neogobius melanostomus*, a fish invader on both sides of the Atlantic Ocean. Biol Invasions 6: 173–181.

57. Karlson AML, Almqvist G, Skóra KE, Appelberg M (2007) Indications of competition between non-indigenous round goby and native flounder in the Baltic Sea. ICES J Mar Sci 64: 479–486.

58. Campbell LM, Thacker R, Barton D, Muir DCG, Greenwood D, et al. (2009) Re-engineering the eastern Lake Erie littoral food web: the trophic function of non-indigenous Ponto-Caspian species. J Great Lakes Res 35: 224–231.

59. Grabowski M, Jaždžewski K, Konopacka A (2007) Alien Crustacea in Polish waters. Aquatic Invasions 2: 25–38.

60. Chen W, Bierbach D, Plath M, Streit B, Klaus S (2012) Distribution of amphipod communities in the Middle to Upper Rhine and five tributaries. BioInvasions Rec 1: 263–271.

61. Podraza P, Ehlert T, Roos P (2001) Erstnachweis von *Echinogammarus trichiatus* (Crustacea: Amphipoda) im Rhein. Lauterbornia: 41: 129–133.

62. Sih A, Christensen B (2001) Optimal diet theory: when does it work, and when and why does it fail? Anim Behav 61: 379–390.

63. Boets P, Lock K, Tempelman D, van Haaren T, Platvoet D, et al. (2012) First occurrence of the Ponto-Caspian amphipod *Echinogammarus trichiatus* (Martynov, 1932) (Crustacea: Gammaridae) in Belgium. BioInvasions Rec 1: 115–120.

64. Baldauf SA, Thünken T, Frommen JG, Bakker TC, Heupel O, et al. (2007) Infection with an acanthocephalan manipulates an amphipod's reaction to a fish predator's odours. Int J Parasitol 37: 61–65.

65. Van Riel MC, Van der Velde G, and Bij de Vaate A (2003): *Pomphorhynchus* spec. (Acanthocephala) uses the invasive amphipod *Chelicorophium curvispinum* (G.O. Sars, 1895) as an intermediate host in the river Rhine. Crustaceana 7: 241–246.

66. Kvach J, Stepien CA (2008) Metazoan parasites of introduced Round and Tubenose Gobies in the Great Lakes: Support for the "Enemy Release Hypothesis". J Great Lakes Res 34: 23–35.

67. Ondračková M, Dávidová M, Pečínková M, Blažek R, Gelnar M, et al. (2005) Metazoan parasites of *Neogobius* fishes in the Slovak section of the River Danube. J Appl Ichthyol 21: 345–349.

68. Kvach J, Skóra KE (2007) Metazoa parasites of the invasive round goby *Apollonia melanostoma* (*Neogobius melanostomus*) (Pallas) (Gobiidae: Osteichthyes) in the Gulf of Gdańsk, Baltic Sea, Poland: a comparison with the Black Sea. Parasitol Res 100: 767–774.

69. Francová K, Ondračová M, Polačik M, Jurajda P (2011) Parasite fauna of native and non-native populations of *Neogobius melanostomus* (Pallas, 1814) (Gobiidae) in the longitudinal profile of the Danube River. J Appl Ichthyol 27: 879–886.

70. Kvach Y (2005) A comparative analysis of helminth faunas and infection of ten species of gobiid fishes (Actinopterigii: Gobiidae) from the North-Western Black Sea. Acta Ichthyol Piscat 35: 103–110.

71. Borcherding J, Staas S, Krüger S, Ondračková M, Ślapansky L, et al. (2011) Non-native Gobiid species in the lower River Rhine (Germany): recent range extensions and densities. A review of Gobiid expansion along the Danube-Rhine corridor – geopolitical change as a driver for invasion. J Appl Ichthyol 27: 153–155.

72. Roche KF, Janač M, Jurajda P (2013) A review of Gobiid expansion along the Danube-Rhine corridor – geopolitical change as a driver for invasion. Knowl Manag Aquat Ec 411: 01.

73. Gendron AD, Marcogliese DJ, Thomas M (2012) Invasive species are less parasitized than native competitors, but for how long? The case of the round goby in the Great Lakes-St. Lawrence Basin. Biol Invasions 14: 367–384.

74. Woo PTK, Buchmann K (2012). Fish Parasites. Pathobiology and Protection. CABI.

75. Nachev M, Ondračková M, Severin S, Ercan F, Sures B (2010) The impact of invasive gobies on the local parasite fauna of the family percidae and the gudgeon (*Gobio gobio*) in the Rhine River. In: Tagungsband der Deutschen Gesellschaft für Protozoologie und Parasitologie 2010.

76. Hine PM, Kennedy CR (1974) Observations on the distribution, specificity and pathogenicity of the acanthocephalan *Pomphorhynchus laevis* (Müller). J Fish Biol 6: 521–535.

77. Dezfuli BS, Castaldelly G, Bo T, Lorenzoni M, Giari L (2011) Intestinal immune response of *Silurus glanis* and *Barbus barbus* naturally infected with *Pomphorhynchus laevis* (Acanthocephala). Parasite Immunol 33: 116–123.

78. Moravec F (2013) Parasitic Nematodes of Freshwater fishes of Europe. Academia Praha, 264–284.

79. Kopp K, Jokela J (2007) Resistant invaders can convey benefits to native species. Oikos 116: 295–301.

80. Telfer S, Bown KJ, Sekules R, Begon M, Hayden T, et al. (2005) Disruption of a host-parasite system following the introduction of an exotic host species. Parasitology 130: 661–668.

81. Brodniewicz I (1968) On glochidia of the genera Unio and Anodonta from the quaternary fresh-water sediments of Poland. Acta Palaeontol Pol XIII: 619–631.

82. Zieritz A, Gum B, Kuehn R, Geist J (2012) Identifying freshwater mussels (Unionoida) and parasitic glochidia larvae from host fish gills: a molecular key to the North and Central European species. Ecol Evol 2: 740–750.

83. Taeubert JE, Gum B, Geist J (2012) Host-specificity of the endangered thick-shelled river mussel (*Unio crassus*, Philipsson 1788) and implications for conservation. Aquat Conserv 22: 36–46.

84. Poulin R. (1996) Helminth growth in vertebrate hosts: Does host sex matter? Int J Parasitol, 2: 1311–1315.

85. Robinson SA, Forbes MR, Hebert CE, McLauglin JD (2010) Male biased parasitism in cormorants and relationships with foraging ecology on Lake Erie, Canada. Waterbirds, 33: 307–313.

86. Koehler A, Hildenbrand P, Schleucher E, Riesch R, Arias-Rodriguez L, et al. (2011) Effects of male sexual harassment on female time budgets, feeding behavior, and metabolic rates in a tropical livebearing fish (*Poecilia mexicana*). Behav Ecol Sociobiol 65: 1513–1523.

87. Scharnweber K, Plath M, Tobler M (2011) Trophic niche segregation between the sexes in two species of livebearing fishes (Poeciliidae). Bull Fish Biol 13: 11–20.

88. Charlebois PM, Marsden JE, Goettel RG, Wolfe RK, Jude DJ, et al. (1997) The round goby, *Neogobius melanostomus* (Pallas): a review of European and North American literature. Illinois-Indiana Sea Grant Program and Illinois Natural History Survey. INHS Special Publication No.20.

# 12

# A Nuclear Factor of High Mobility Group Box Protein in *Toxoplasma gondii*

**Hui Wang, Tao Lei, Jing Liu, Muzi Li, Huizhu Nan, Qun Liu***

Key Laboratory of Animal Epidemiology and Zoonosis, Ministry of Agriculture, National Animal Protozoa Laboratory, College of Veterinary Medicine, China Agricultural University, Beijing, China

## Abstract

High mobility group box 1 (HMGB1) is a nuclear factor that usually binds DNA and modulates gene expression in multicellular organisms. Three HMGB1 orthologs were predicted in the genome of *Toxoplasma gondii*, an obligate intracellular protozoan pathogen, termed TgHMGB1a, b and c. Phylogenetic and bioinformatic analyses indicated that these proteins all contain a single HMG box and which shared in three genotypes. We cloned TgHMGB1a, a 33.9 kDa protein that can stimulates macrophages to release TNF-α, and, we demonstrated that the TgHMGB1a binds distorted DNA structures such as cruciform DNA in electrophoretic mobility shift assays (EMSA). Immunofluorescence assay indicated TgHMGB1a concentrated in the nucleus of intracellular tachyzoites but translocated into the cytoplasm while the parasites release to extracellular. There were no significant phenotypic changes when the TgHMGB1a B box was deleted, while transgenic parasites that overexpressed TgHMGB1a showed slower intracellular growth and caused delayed death in mouse, further quantitative RT-PCR analyses showed that the expression levels of many important genes, including virulence factors, increased when TgHMGB1a was overexpressed, but no significant changes were observed in TgHMGB1a B box-deficient parasites. Our findings demonstrated that TgHMGB1a is indeed a nuclear protein that maintains HMG box architectural functions and is a potential proinflammatory factor during the *T.gondii* infection. Further studies that clarify the functions of TgHMGB1s will increase our knowledge of transcriptional regulation and parasite virulence, and might provide new insight into host–parasite interactions for *T. gondii* infection.

**Editor:** Stanislas Tomavo, Pasteur Institute Lille, France

**Funding:** This study was supported by the National Natural Science Foundation of China (Grant No. 31372424), the National Key Basic Research Program (973 program) of China (Grant No. 2015CB150300) and Research Fund for the Doctoral Program of Higher Education of China (Grant No. 20110008110006). The funders had no role in study design, data collection and analysis, decision to publish, or preparation of the manuscript.

**Competing Interests:** The authors have declared that no competing interests exist.

* Email: qunliu@cau.edu.cn

## Introduction

High mobility group box 1 (HMGB1) was first discovered in calf thymi as a nuclear protein that contained a unique DNA-binding domain and showed rapid migration in polyacrylamide gels, a property of the HMG superfamily. The two HMG-box containing HMGB proteins are only present in multicellular animals, and the HMGB gene apparently arose through the fusion of two different genes, each coding for one of the boxes [1]. Mammalian HMGB1 encodes 219 amino acids (aa) and contains two DNA-binding motifs (A-box and B-box) that are arranged in tandem, following a long negatively charged C-terminus that is rich in aspartic and glutamic acids, which differ in length (HMGB1–3) or are absent (HMGB4) [2]. By contrast, yeast HMGBs (Nhp6-a and -b) have only one HMG box domain and no acidic tail [3]. HMG proteins can bind to cruciform, double- and single-stranded DNA with high affinity through HMG-box and acidic C-terminus [4,5]. Interactions between DNA and HMGs are mediated by basic amino acid residues of the protein. Structural studies using nuclear magnetic resonance spectroscopy established that the DNA binding domain of HMG has an L-shaped structure made of three α-helices that provide surfaces for potential interactions with both DNA and protein [2,6].

In higher organisms, HMGBs are ubiquitously expressed in cell nuclei and act as DNA chaperones that influence multiple processes in chromatin, such as transcription, replication, recombination, DNA repair and genomic stability [7]. Variety of post-translational modifications (PTMs) such as acetylation, phosphorylation and methylation to HMGB can modulate not only HMGB1 protein function but also its subcellular location and eventual secretion [8]. HMGB1 is a prototypical damage-associated molecular pattern and the only one of these proteins that can be passively and actively secreted into the extracellular milieu, where it acts as cytokines, chemokines and growth factors that promotes cell migration and inflammation [9,10,11].

HMGB1 is highly conserved with >98% amino acid identity between humans and rodents [12], but appears to be more polymorphic among parasite species. HMGB proteins have been reported in many parasites including *Wuchereria bancrofti*, *Brugia malayi* [13], *Trypanosoma cruzi* [14,15], *Schistosoma mansoni* [16], *Schistosoma japonicum* [17], *Plasmodium falciparum* [18,19,20], *Entamoeba histolytica* [21] and *Babesia bovis* [22]. Mammalian HMGB1 bears two DNA binding motifs followed by an acidic C-terminus, whereas most HMGB1 family members of parasites have only one or more HMG box and no acidic tail.

*T. gondii* is an obligate intracellular protozoan that can actively invade almost all nucleated cells and can cause opportunistic disease in various animals and humans [23]. The pathological basis for Toxoplasmosis is tissue destruction and inflammation, which are a direct result of the parasite's cell lytic growth cycle of attachment, invasion, growth, and egress. There are three *T. gondii* genotypes, types I, II, and III, which have different growth characteristics [24] and cause variable levels of virulence in mice [25,26]. Type I strains are uniformly lethal at all doses (high virulence) in all strains of laboratory mice, whereas type II (intermediate virulence) and type III (low virulence) show much lower levels of pathogenicity. Virulence in type I strains is a result of rhoptry effector proteins (ROP18, ROP5 and ROP16) effectively eliminating critical host immune responses which leads to uncontrolled proliferation of the tachyzoite, and host survival is compromised due to excessive parasite burden. Type II strains induce stronger proinflammatory responses, including very high levels of IL-12 in comparison with either type I or III and the susceptible animals always die of severe inflammation. Like type I, Type III strains limit the initial production of pro-inflammatory cytokines, whereas be unable to avoid intracellular killing mediated by IRGs and late production of IL-12 by DCs triggers a Th1-type response that is sufficient to control parasite burden and induce cyst formation, leading to a chronic infection (reviewed by [27]). The transitions between the different stages of the *T. gondii* life cycle allow the parasite to be virulent and survive. These developmental transitions are accompanied by major changes in gene expression [28], and the control mechanisms for parasite proliferation (replication and differentiation) may be regulated by the cell cycle [29,30] and the micro-environments around the parasites [31]. Regulation of *T. gondii* gene expression is, in part, promoted by epigenetic events, such as histone modifications and interactions between histones and other nuclear factors [32].

In ToxoDB earlier version (version 9.0, http://toxodb.org/toxo/), several genes were predicted to encode HMG proteins in *T. gondii* genome, though there are many information even including the coding sequence were updated in the latest *T.gondii* genomic database (ToxoDB version 11), the characteristics of these protein, such as subcellular locations and biological functions are poorly understood to date. Therefore, we aimed in this work to identify and characterize HMGB homologues in *T. gondii* and also to assess whether HMGB proteins relate to gene regulation as an architectural factor and contribute to parasite growth or virulence to mice.

## Material and Methods

### Ethics statement

All experiments with animals in this study were performed in strict accordance with the recommendations in the Guide for the Care and Use of Laboratory Animals of the Ministry of Science and Technology of China. Formal animal ethics of all experimental procedures were approved by the Institutional Animal Care and Use Committee of China Agricultural University (The certificate of Beijing Laboratory Animal employee, Approval No: 18049). All efforts were made to minimize animal suffering.

*Toxoplasma* culture using HFF (human foreskin fibroblasts) monolayers are generally appeared in many papers about *Toxoplasma* (for example: John C. Boothroyd, Pernas L, Adomako-Ankomah Y, Shastri AJ, Ewald SE, Treeck M, et al. (2014) *Toxoplasma* effector MAF1 mediates recruitment of host mitochondria and impacts the host response. PLoS Biol 12:

e1001845.). In this study, we used the HFF cell to maintain *T.gondii* only, which is a normal system for culture the parasites in vitro. HFF cell line was bought from the ATCC company (USA) and serial passaged in our lab. From the company websites, products reference: Hovatta O, et al. A culture system using human foreskin fibroblasts as feeder cells allows production of human embryonic stem cells. Hum. Reprod. 18: 1404–1408, 2003. PubMed: 12832363.

### Parasites and cell culture

HFF (human foreskin fibroblasts) were cultured in complete DMEM, as described previously [33]. *T. gondii* tachyzoites were maintained in vitro by serial passage on confluent HFF monolayers in DMEM containing 25 mM glucose and 4 mM glutamine supplemented with 10% fetal bovine serum (FBS, Gibco, USA), and incubated at 37°C with 5% $CO_2$ in a humidified incubator. The medium was changed 6 hours after inoculation.

### Sequence alignment and molecular phylogenetic analyses of HMGB1s

The sequences used in this study were obtained from the EuPathDB database and National Center for Biotechnology Information (NCBI) GenBank (**Table S1**). Phylogenetic analyses were performed based on either the Neighbor-Joining method [34] using the substitution model with partial deletion of gap containing sites (1000 bootstrap replicates were performed) or the Maximum Likelihood (ML) method with an approximate likelihood-ratio test and JTT selected for amino acid substitution model analysis with a CNI search to start the initial tree [35]. All alignments were performed using DNAMAN version 5.2.2 and ClustalX2, and phylogenetic trees were visualized with Mega 5.22 [36]. Bootstraps values obtained from the phylogenetic analyses are indicated on trees presented in **Figure S1**. HMGB1 proteins from different organisms were considered to be significant, allowing clusters to form that are represented by colored clades.

### Bioinformatic analyses of the TgHMGB1s sequence

A BlastP search using the complete sequence of *Mus musculus* High mobility group box 1 deposited in GenBank (accession no. AAH83067.1) was performed using ToxoDB (http://toxodb.org/toxo/, ver.9.0) within the *Toxoplasma* GT1 database. This search identified three putative transcripts with significant similarity to HMGB1 protein, termed TgHMGB1a, TgHMGB1b and TgHMGB1c. The molecular weight, isoelectric point, amino acid composition and putative protein functions were also predicted using BlastP. The presence of a HMG-box was confirmed using the Conserved Domain Search Service (CD Search) from NCBI. Also, characteristics of the primary sequence of TgHMGB1a were analyzed using the following online servers: (1) signal peptide cleavage sites in the TgHMGB1a sequence were predicted using Signal IP4.1 (http://www.cbs.dtu.dk/services/SignalP/); (2) transmembrane regions and the orientation of TgHMGB1a were analyzed using TMHMM Server v. 2.0 (http://www.cbs.dtu.dk/services/TMHMM-2.0/); (3) subcellular localization (http://wolfpsort.org/) and nuclear localization signal prediction (https://rostlab.org/owiki/index.php/PredictNLS); (4) to identify conserved domains, online tools (http://myhits.isb-sib.ch/cgi-bin/motif_scan; http://www.expasy.org/tools/scanprosite/) were used to search for motifs and conduct additional bioinformatics analyses; (5) sequence alignment analysis and three-dimensional (3D) structure modeling performed using ESPript 2.2 (http://esprit.ibcp.fr/ESPript/cgi-bin/ESPript.cgi) and homology remodeling tools (http://zhanglab.ccmb.med.umich.edu/I-TASSER/; http://swissmodel.expasy.org/workspace/index; and http://www.cbs.dtu.dk/services/CPHmodels/). Software programs such as DNAstar, DNAMAN and Primer Premier 5 were also used in the analyses.

## Cloning and identification of TgHMGB1a

Based on the gene sequence of TgHMGB1s, overlapping primers (**Table S2**) were designed to amplify the complete coding sequence of TgHMGB1a from the RH strain cDNA prepared from total RNA using oligo (dT) primers. Overlapping PCR reactions and touchdown PCR programs were applied to gain high fusion rates throughout the process, and the resultant products were cloned into a pEASY-T-Blunt vector (Beijing TransGen Biotech Co., Ltd.). Finally, the DNA insert was sequenced and analyzed by Blast to confirm the authenticity of the cloned sequence. Then a truncated TgHMGB1a fragment (including the B box but not the transmembrane domains) was amplified using specific primers (TgHMGB1a 4E-F and -R listed in **Table S2**). The products were digested (*Bam*HI and *Xho*I, NEB, Beijing), cloned into the pET-28a vector (Novagen, Germany), and transformed into *Escherichia coli* (*E.coli*) for expression. Only the soluble rTgTHMGB1a 4E were purified by affinity chromatography using Ni-IDA agarose (QIAGEN, Germany) following the manufacturer's instructions. Protein purity was assessed using SDS-PAGE and identified by western blot using the *T.gondii* positive pig serum.

Normal immune procedure of the Freund's Adjuvant (Sigma) were used to raise antibodies against rTgTHMGB1a 4E in Balb/c mice. Anti-rTgH1a4E sera were collected 2 weeks after the last immunization. The specificities and titers of polyclonal antibodies were examined using immunoblotting and enzyme-linked immunosorbent assays (ELISA), respectively, using conventional protocols. All sera samples were sterilized by filtration through 0.22 μm filters (Millipore, USA) and stored at −80°C.

## Immunofluorescence assays and western blotting

Immunofluorescence assays (IFA) for TgHMGB1a subcellular localization were performed as described previously [37]. Briefly, freshly released RH stain *T. gondii* were passed through 5 μm filters, spun down by centrifugation at $300 \times g$ for 10 minutes, and resuspended using PBS. Parasites were seeded at appropriate amounts onto confluent monolayers of HFF cells grown on previously prepared glass coverslips in 12-well plates. Infected cells were incubated at 37°C in 5% $CO_2$ for 16–20 hours and the fixed for 15 min in 3.7% formaldehyde. Cells were next permeabilized with 0.3% Triton X-100 for 10–15 min, and blocked with 3% bovine serum albumin (BSA) in phosphate-buffered saline (PBS) for 30 minutes prior to incubation with the primary antibody (diluted 1:100 in 1% BSA-PBS) for 1 hour. Coverslips were washed three times in PBS and then incubated with FITC-conjugated goat anti-mouse IgG (H+L) that was diluted 1:100 in PBS with 1% BSA (Proteintech, USA) for 1 hour at 37°C. Nuclei was stained with Hoechst 33258 (Sigma) and coverslips were subsequently mounted onto microscope slides. Fluorescence images were obtained with a Leica microsystem (Leica TCS SP5 II, Germany) and epifluorescence optics using an oil immersion lens with 63× magnification. Collected images were processed using LAS AF lite 2.2.0 software (Leica). Rabbit anti-MIC3 antibodies and preimmune mouse sera were used as controls. To analyze TgHMGB1a in extracellular parasites, filtered parasites were allowed to adhere onto the coverslips precoated with poly-lysine (poly-L-Lys) for 20 minutes at room temperature and were then processed similarly to intracellular parasites.

Purified parasites were lysed using RIPA buffer (Beyotime, Beijing) with a cocktail of protease inhibitors, and 5 μg of lysate was loaded per lane for fractionation by 12% (w/v) SDS-PAGE. After electrophoresis, separated strips were transferred onto polyvinylidene fluoride (PVDF) membranes (Millipore, USA) together with a visible prestained protein marker. Membranes were blocked with 5% (w/v) skim milk in PBS for 1 h at 37°C, and then incubated with the mouse anti-rTgHMGB1a 4E antibody (diluted 1:600 in PBS with 3% BSA). After three rounds of extensive washing in PBST (1% Tween-20), membranes were incubated for 1 h with goat anti-mouse IgG (H+L) horseradish peroxidase (HRP)-labeled secondary antibody (Sigma, USA) diluted 1:10,000 in PBS with 3% BSA. Finally, reactive bands were visualized using enhanced chemiluminescence reagents (Co Win Biotech Co., Ltd., Beijing).

## Examination of TNF-α release

Peptides derived from TgHMGB1a B box, mouse HMGB1, mouse HMGB1 A box, mouse HMGB1 B box or a chimeric protein of mouse HMGB1 A box fused to TgHMGB1a B box were expressed in *E. coli*. Following sonication, recombinant proteins were purified as previously mentioned. Then, gel retardation experiments were carried out to test whether or not HMG peptides bind the DNA from *E. coli* genome. As the manual of molecular cloning, the natural and denatured (0.1% SDS) purified complexes of protein and DNA were resolved by electrophoresis on 1.5% agarose gels in 1× TBE buffer at 120 V for 15 minutes. DNA was stained with 0.5 μg mL$^{-1}$ gold views (TransGen Biotech Co., Ltd., Beijing) and photographed using a UV transilluminator system. To evaluate the ability of TgHMGB1s stimulates to TNF-α release, murine peritoneal macrophages ana1 cell line were culture and stimulated with the recombinant peptides at indicated concentrations, and the culture medium was assayed for TNF-α at 24 h after stimulation. All of the recombinants were treated with Polymyxin B agarose (Sigma-Aldrich) twice and endotoxin contents were examined using the Limulus amebocyte lysate (LAL) kit (Lonza, USA), and only if the endotoxin content of proteins less than 0.06 EU/mg, it can be used to do the stimuli tests. The levels of TNF in the culture medium were determined and calculated using a commercial ELISA kit (Dakewe Biotech Co., Ltd., Beijing) following the instructions of manufacturer.

## Electrophoretic mobility shift assay (EMSA) with synthetic cruciform DNA

The partially complementary oligonucleotides 4H-a to 4H-d for EMSA corresponded to previous reports [14,38] were synthesized to create cruciform DNA (4H) as well as "hairpin-like" structures (2H), by annealing the required oligonucleotides of equal amount in each case according to the instruction of EMSA Probe Biotin Labeling Kit (Beyotime, Beijing). Double-stranded (ds) linear DNA was obtained by annealing oligonucleotides 4H-a and 4H-c with corresponding fully complementary ones (Compl 4H-a and Compl 4H-c). Biotin-labeled DNA structures were obtained by annealing the biotin-labeled oligonucleotides labeled through the Terminal Deoxynucleotidyl Transferase (EMSA Probe Biotin Labeling Kit, Beyotime, Beijing). EMSA reactions were set up as follows (10 μL): 2.0 μL 5×DNA-binding buffer, increasing amounts of recombinant TgHMGB1a 4E protein (0.01–1.5 mg/mL), 5 nmol of the labeled cruciform DNA and add ddH$_2$O to final volume. The reaction volumes were incubated at 25°C for 30 min. For competition assays, 100-fold or 500-fold molar excess of cold complete cruciform DNA (4H), incomplete cruciform DNA (2H) or linear dsDNA were added to the reaction and incubated for an additional 20 min at 25°C. Protein–DNA complexes were separated in 6% non-denaturing polyacrylamide gels in 0.5× Tris-borate-EDTA buffer (pH 8.0), run at 50 V at 4°C for 1.5 h. After electrophoresis, the bound and free probes in the gel were transferred to a positive charged nylon membrane (Millipore, USA) at 60 V at 4°C for 2 h. The probes on the membrane were detected using Chemiluminescent EMSA Kit (Beyotime, Beijing). Purified TgGRA1 proteins were used as control under the same conditions.

## Generation of TgHMGB1a B box-deficient parasites

A stable B box-deficient clone was established to further establish the function of TgHMGB1a. We constructed a gene targeting plasmid based on the single-step, O-type insertion strategy, as described in gene targeting protocols [39]. Briefly, 2200 bp of homologous sequence from a region upstream of TgHMGB1a B box and eGFP were fused using overlapping PCR (all primers used are shown in **Table S3**). The products flanked with *Avr*II and *Pac*I sites were introduced to the equivalent sites of pTCR (modified from pTCY), which we named the pTCR eGFP KO TgH1a B box plasmid. After digestion with *Bsi*WI (NEB, Beijing), linearized vectors were purified using ethanol and then resuspended using cytomix to transfect into the recipient RHΔKU80 strain. Parasites cultures and genetic manipulations were carried out according to the descriptions above, and chloramphenicol (CAT) and RFP were used as selection markers. A positive clone (i.e., a strain in which the B box was replaced by eGFP) should exhibit triple-positivity for chloramphenicol resistance, RFP and eGFP, and the targeting event in triple-positive clones was confirmed using PCR and western blotting. A stable mutant was cloned after selection by FCM sorting.

## Overexpression of TgHMGB1a

To further characterize the role of TgHMGB1a, we generated a transgenic RH parasite strain that overexpressed TgHMGB1a, and TgHMGB1a lacking the B box transgenic strain was also constructed as control. A plasmid for transfection was constructed as follows: the complete sequence of TgHMGB1a was amplified from a pEASY-T-TgHMGB1a plasmid using primers (**Table S3**) that appended the flanking *Eco*RV and *Nsi*I restriction endonuclease sites. Amplification products were introduced into the equivalent sites of a pDMG plasmid, and the resulting vector was used as an overexpression vector for transfection that we designated as pDMG-TgHMGB1a. The fused TgHMGB1a-GFP gene was under control of the *T. gondii* GRA1 promoter. The TgHMGB1a overexpression strain was generated using electroporation transfection [40] and selected based on pyrimethamine resistance as described previously [41]. Finally, flow cytometry (FCM) was used to acquire a monoclone. B box lacked TgHMGB1a overexpress strain (TgHMGB1a$^{-\text{B box}}$) was generated as the same methods.

## Plaque assay and intracellular replication test

Plaque assays to compare the transgenic parasites with their parental strains were performed on HFFs cells in 6-well tissue culture plates (Corning costar, Beijing). Briefly, 500 parasites per well were seeded into confluent monolayers and infected cells were maintained in fresh DMEM containing 10% FBS and incubated undisturbed at 37°C in 5% $CO_2$ for 7 days. To stain the monolayers, media was aspirated and disassociated parasites were washed off using PBS. Cell monolayers were then fixed for 10 minutes in PBS with 4% formaldehyde and stained with crystal violet solution (12.5 g crystal violet dissolved in 125 mL ethanol and mixed with 500 mL 1% ammonium oxalate in water) at room temperature for 10 minutes, washed with deionized water, air dried and visualized by microscopy using image acquisition and plaque area measurement as previously described [42].

To analyze the intracellular growth rate of the transgenic parasites compared with their parental strains, about $1 \times 10^5$ parasites were inoculated on confluent HFFs in 24-well plates. The infected cells continuously incubated for 24 hours. Thereafter, cells were fixed in PBS with 4% formaldehyde, and RH and RHΔKU80 parasites were stained for IFA using rabbit anti-SAG1 polyclonal antibody as described above. TgHMGB1a overexpression and B box-deficient parasites were directly examined for fluorescence (GFP and eGFP, respectively). The numbers of parasites per vacuole for a minimum of 150 randomly chosen vacuoles were counted for each strain using a fluorescence microscope (IX71, Olympus, Japan) at 400× magnification.

## RNA isolation and real-time PCR

Total RNA was isolated from HFF cells that were uninfected or infected by either transgenic or parental parasites using the RNA isolation kit version 2.11 (Takara Biotechnology, Dalian, Co., Ltd) and treated with DNase I (Takara) to remove any residual genomic DNA. We synthesized cDNA using Oligo (dT)18 and random 6-mers according to the manufacturer's protocol for cDNA Synthesis Kit (Takara). Specific primers were designed using Primer Premier 5.0, including primers for TgHMGB1s (TgHMGB1b and TgHMGB1c), rhoptrys (ROP18, ROP16 and Toxofilin), micronemes (MIC3, PLP1), dense granules (GRA7), profilin and *T. gondii* actin (**Table S4**). The specificity of these primers was evaluated using conventional real-time (RT)-PCR. Quantitative RT-PCR (ΔΔCt method) was performed using the ABI Prism 7500 System (Applied Biosystems Inc. USA) with Tli RNaseH Plus kit (Takara). Resulting concentrations of RNA were normalized by TgActin and the relative expression levels of target genes were analyzed using ABI Prism 7500 software v2.0.5.

## Chromatin immunoprecipitation (ChIP) -qPCR analysis

ChIP assays was applied to investigate whether TgHMGB1a directly participate transcriptional regulation of the indicated genes in the wild type and the TgHMGB1a overexpress strain. ChIP assays were carried out according to the manufacturer's protocol of the ChIP Assay Kit (Beyotime, Beijing) with slight modifications. Briefly, the chromatin solutions from equal number of purified RH and TgHMGB1a overexpress strain were sonicated and immunoprecipitated with anti-TgHMGB1a 4E sera in rotating overnight at 4°C. Parallel controls for each experiment included samples with the normal mice IgG (negative control) and the input (positive control), and TgHMGB1a B box$^{-/\text{eGFP}}$ parasites were also used as control. After elution and purification, the recovered DNA samples were used as template for regular PCR or real-time PCR performed using the ABI Prism 7500 System with Tli RNaseH Plus kit (Takara). The promoter-specific primers of indicated genes were listed at **Table S5,** and the coding region primers (**Table S4**) were used as controls. Pull-down level of the target promoter sequences were normalized to the corresponding abundance in the input chromatin. After amplification, regular PCR products were resolved on 1.5% agarose gel and visualized by gold view staining, while the quantitative-PCR were analyzed using the ABI Prism 7500 Software v2.0.5.

## Mouse infectivity studies

All the 8- to 10-wk-old female Balb/c mice were purchased from the Center for Experimental Animals (Beijing). Rodent laboratory chow and tap water were provided ad libitum and maintained under specific pathogen-free conditions for 7 days before manipulation. To assess the virulence of TgHMGB1a overexpress strain compared with the parental line, TgHMGB1a overexpress parasites, RH and RH-GFP tarchyzoites were injected intraperitoneally (i.p.) into mice at doses of $10^2$, $10^3$ and $10^4$ (5 mice per dose), and mice were monitored until there were no remaining survivors. Similarly, to examine the virulence of the TgHMGB1a B box deficient, TgHMGB1a B box$^{-/eGFP}$ parasites were injected i.p. into mice at doses of $10^2$, $10^3$ and $10^4$ (5 mice per dose). In comparison, RHΔKU80 was injected into mice at the same doses (5 mice per dose). Mice were also monitored until there were no remaining survivors or for 30 d. All the infected mice were monitored three times a day (every 8 hours) for clinical signs and mortality until there were no remaining survivors. The mice were humanely euthanized when they were unable to reach food or water for more than 24 h and lost 20% normal body weight. The mice were humanely euthanized by cervical dislocation after anesthetization. The mice were anesthetized by subcutaneous injection of Atropine (0.02 mg/kg) before euthanasia. All procedures were in strict accordance with the PR China legislation on use. All efforts were made to minimize animal suffering. And a completed ARRIVE checklist for the animal experiments can be found in **Data S1**.

## Analysis of data and statistics

All data sets are presented as the mean ± SD and all statistical analyses were performed using SPSS 18.0 (USA). Unpaired two-tailed Student's $t$-tests with unequal variance were performed.

## Results

### Only one HMG box in *T. gondii* "high mobility group box" proteins

HMG box-containing proteins were identified in the three genotypes of *T. gondii* using the Blast sequence homology algorithm (**Table S6**). Here, TgHMGB1a, b, c were named to represent three HMGB1 proteins in the GT1 strain annotated in the ToxoDB database (corresponding to TGGT1_053220, TGGT1_030840 and TGGT1_043980 in ToxoDB ver.9.0, but renamed to TGGT1_210412, TGGT1_219832, TGGT1_263720 in ToxoDB ver.11, **Figure S1 and S2**), which encode 33.9, 74.1 and 20.1 kDa proteins, respectively. Phylogenetic analysis showed that TgHMGB1a, b and c could be classified into three significant clusters through both the neighbor-joining (NJ) and maximum likelihood (ML) algorithms (**Figure S1**), which most because for different conservation of HMG domain in their C-terminal. However, TGGT1_217500 may be impossible to assign specific homologues to this gene because of low bootstraps values. There were genes that were clearly orthologous between *T. gondii* and *N. caninum*. TgHMGB1a corresponds to NcLIV_043670, while TgHMGB1b and TgHMGB1c correspond to NcLIV_060790 and NcLIV_024230, respectively. Five HMG proteins were also predicted in *Eimeria. spp.* (**Figure S1**) in ToxoDB ver.11. The two HMGB genes of *P. falciparum* (PfHMGB1/2) mapped to the TgHMGB1b cluster, while *T. cruzi*, *T. brucei*, *S. mansoni*, and *L. major* HMGB1s all mapped to branches that were far away from the TgHMGBs. *S. cerevisiae* HMGB1s formed an independent branch that was near to the TgHMGB1s. The *A. thaliana*, *M. musculus* and *H. sapiens* HMG genes, HMGB1, HMGB2, HMGB3 and HMGB4, were also analyzed in this study. The HMGBs of vertebrate animals and plants were clearly branched in two significant clades, supporting the conserved evolution of these proteins. Unlike mammalian HMGB1, TgHMGB1 proteins have only one HMG box domain (~71 aa) that is similar to the B box

of *H. sapiens*, as box A and an acidic C-terminus are both absent (**Figure 1A and S2**). The isoelectric points (pI) of these three proteins were all close to 9.6, which is characteristic of many DNA binding proteins. The three-dimensional (3D) structures of the HMG box domains of TgHMGB1a, b and c were homology modeled with the SWISS-MODEL server using *M. musculus* 2gzkA as a template structure. The three α-helices in the C-terminal domain of TgHMGB1a, b and c fold into an L shape that formed the HMG box (**Figure 1B, C, and D**, respectively).

### TgHMGB1a is a typical one HMG box protein and concentrated in nuclear in intracellular tachyzoites but translocated into cytoplasm while it's in extracellular

The ternary structure of full length TgHMGB1a (**Figure 2**A) was modeled using the structure template of 1ckt A chain. TgHMGB1a encodes 302 aa and contains only one 71 aa HMG box (228–298 aa) that is independently folded at the C-terminus, and 4 upstream transmembrane helices (31–50, 60–82, 119–141 and 151–173 aa) (**Figure 2B and S3B**). As expected, the HMG box domains showed the typical three α-helices in an L-shaped fold (**Figure 1B** and **2A**). And eight residues (Ser233, Phe235, Leu257, Ala258, Gly261, Lys262, Gly265 and Trp268) were predicted to play a pivotal role in DNA binding ability. A model of a TgHMGB1a-DNA complex is shown in **Figure 2C**. All sequence analyses and structure predictions indicated that the TgHMGB1a is a DNA binding protein. Furthermore, classic signal peptide cleavage sites and nuclear localization signals were not found in TgHMGB1a (**Figure S3A**), but many residues involved in post-translational modifications (PTMs) were predicted (**Figure S3C**). The genomic DNA sequence annotation of TgHMGB1a in ToxoDB was 3391 bp, and the gene was interrupted by 3 introns. Using the fuse PCR program, 909 bp PCR products of TgHMGB1a coding sequence were obtained, and then confirmed by sequencing and alignment. Truncated TgHMGB1a (193–302 aa, termed rTgHMGB1a 4E, ~14 kDa) was prokaryotic expressed as a His-tagged fusion protein and was strongly recognized by sera from swine infected with *T. gondii* (data not shown). Anti-rTgHMGB1a 4E sera was prepared from immunized mice that showed a high titer ($10^6$) against rTgHMGB1a 4E. Immunoblotting showed that the preimmune sera did not react against *T. gondii* tachyzoite lysates, whereas mouse anti-rTgHMGB1a 4E antibodies elicited a 34 kDa protein band in *T. gondii* tachyzoite lysates, but not in HFF cell lysates (**Figure 2D**). Interestingly, the amount of TgHMGB1a was dramatically decreased in extracellular parasites, even almost undetectable while extending the time of extracellular (**Figure S4**).

Intracellular and extracellular RH strain tachyzoites were fixed and subjected to IFA using anti-rTgHMGB1a 4E antibodies to determine the subcellular location of TgHMGB1a (**Figure 2E**). In the up panel, TgHMGB1a (green) co-localized with Hoechst (blue), indicating that most of this protein concentrated into the nucleus in the intracellular RH strain tachyzoites (**Figure 2E** top panel), and even in the endodyogeny process, TgHMGB1a always locate in the nucleus (**Figure S5** top panels), but interestingly, while the parasites released to extracellular, the TgHMGB1a proteins do not remain in the nucleus, and have translocated into cytoplasm (**Figure 2E** bottom panel and **Figure S6** top panel), and dispersed around the cytoplasma membrane. For comparison, rabbit anti-TgMIC3 serum was used to stain microneme (denotes the cytoplasm) of parasites, and control experiments carried out with preimmune sera or secondary antibody alone showed no signal (data not shown). Additionally, the similar results were obtained when use the anti-SAG1 sera as the control (intracellular

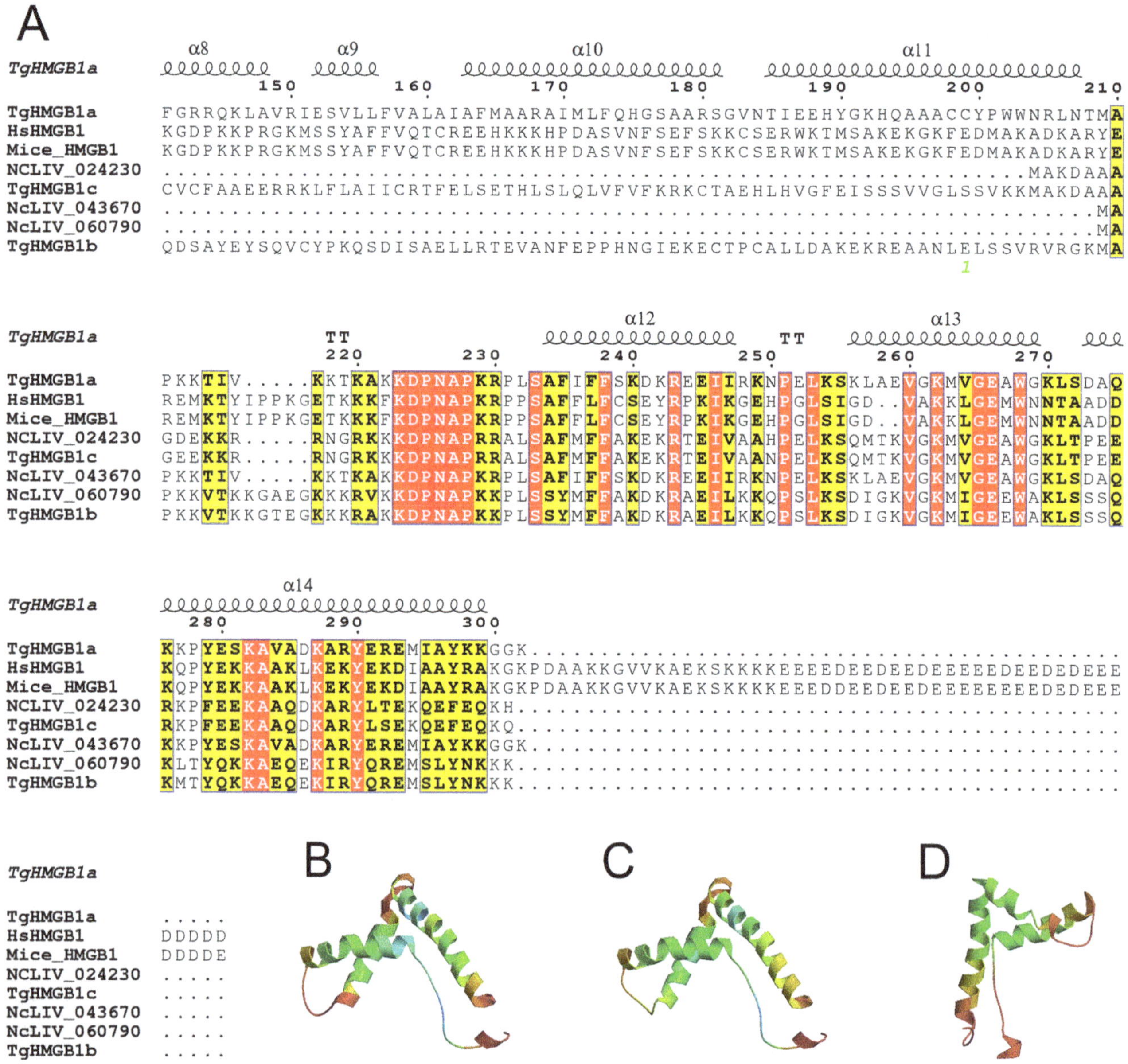

**A**

|  | α8 | α9 | α10 | α11 |
| --- | --- | --- | --- | --- |
| *TgHMGB1a* | 150 | 160 | 170 180 | 190 200 210 |

TgHMGB1a      FGRRQKLAVRIESVLLFVALAIAFMAARAIMLFQHGSAARSGVNTIEEHYGKHQAAACCYPWWNRLNTMA
HsHMGB1       KGDPKKPRGKMSSYAFFVQTCREEHKKKHPDASVNFSEFSKKCSERWKTMSAKEKGKFEDMAKADKARYE
Mice_HMGB1    KGDPKKPRGKMSSYAFFVQTCREEHKKKHPDASVNFSEFSKKCSERWKTMSAKEKGKFEDMAKADKARYE
NCLIV_024230  .............................................................MAKDAAA
TgHMGB1c      CVCFAAEERRKLFLAIICRTFELSETHLSLQLVFVFKRKCTAEHLHVGFEISSSVVGLSSVKKMAKDAAA
NcLIV_043670  ...............................................................MA
NcLIV_060790  ...............................................................MA
TgHMGB1b      QDSAYEYSQVCYPKQSDISAELLRTEVANFEPPHNGIEKECTPCALLDAKEKREAANLELSSVRVRGKMA

|  | | TT | α12 | TT α13 |
| --- | --- | --- | --- | --- |
| *TgHMGB1a* | 220 | 230 | 240 250 | 260 270 |

TgHMGB1a      PKKTIV.....KKTKAKKDPNAPKRPLSAFIFFSKDKREEIIRKNPELKSKLAEVGKMVGEAWGKLSDAQ
HsHMGB1       REMKTYIPPKGETKKKFKDPNAPKRPPSAFFLFCSEYRPKIKGEHPGLSIGD..VAKKLGEMWNNTAADD
Mice_HMGB1    REMKTYIPPKGETKKKFKDPNAPKRPPSAFFLFCSEYRPKIKGEHPGLSIGD..VAKKLGEMWNNTAADD
NCLIV_024230  GDEKKR.....RNGRKKKDPNAPRRALSAFMFFAKEKRTEIVAAHPELKSQMTKVGKMVGEAWGKLTPEE
TgHMGB1c      GEEKKR.....RNGRKKKDPNAPRRALSAFMFFAKEKRTEIVAANPELKSQMTKVGKMVGEAWGKLTPEE
NcLIV_043670  PKKTIV.....KKTKAKKDPNAPKRPLSAFIFFSKDKREEIIRKNPELKSKLAEVGKMVGEAWGKLSDAQ
NcLIV_060790  PKKVTKKGAEGKKKRVKKDPNAPKKPLSSYMFFAKDKRAEILKKQPSLKSDIGKVGKMIGEEWAKLSSSQ
TgHMGB1b      PKKVTKKGTEGKKKRAKKDPNAPKKPLSSYMFFAKDKRAEILKKQPSLKSDIGKVGKMIGEEWAKLSSSQ

|  | α14 |
| --- | --- |
| *TgHMGB1a* | 280 290 300 |

TgHMGB1a      KKPYESKAVADKARYEREMIAYKKGGK...........
HsHMGB1       KQPYEKKAAKLKEKYEKDIAAYRAKGKPDAAKKGVVKAEKSKKKKEEEEDEEDEEDEEEEEDEEDEDEEE
Mice_HMGB1    KQPYEKKAAKLKEKYEKDIAAYRAKGKPDAAKKGVVKAEKSKKKKEEEDDEEDEEDEEEEEEEDEEDEEE
NCLIV_024230  RKPFEEKAAQDKARYLTEKQEFEQKH...........
TgHMGB1c      RKPFEEKAAQDKARYLSEKQEFEQKQ...........
NcLIV_043670  KKPYESKAVADKARYEREMIAYKKGGK...........
NcLIV_060790  KLTYQKKAEQEKIRYQREMSLYNKKK...........
TgHMGB1b      KMTYQKKAEQEKIRYQREMSLYNKKK...........

|  | B | C | D |
| --- | --- | --- | --- |
| *TgHMGB1a* | | | |

TgHMGB1a      .....
HsHMGB1       DDDDD
Mice_HMGB1    DDDDE
NCLIV_024230  .....
TgHMGB1c      .....
NcLIV_043670  .....
NcLIV_060790  .....
TgHMGB1b      .....

**Figure 1. Sequence-structure alignment of HMGB1s and spatial structures prediction of HMG box in TgHMGB1a, b and c.** A. *T. gondii* and *N. caninum* HMG box domains were identified using Motif Scan and are marked by a box and compared with sequences of human and mice. The positions of identical and conserved residues contained in the HMG box are indicated by red- and yellow-filled rectangular frames, respectively. Dots indicate gaps or missing residues. The secondary structure of TgHMGB1a was also predicted and α-helices 12, 13 and 14 represent the three α-helices of the TgHMGB1a HMG box. B, C, and D. A three-dimensional model of the HMG box in TgHMGB1a (B), TgHMGB1b (C) and TgHMGB1c (D) were built based on homology modeling. As shown here, the three α-helices fold into a characteristic L-shaped domain expected for a HMG box.

parasites showed in **Figure S5** bottom panel, while **Figure S6** bottom panels indicated the extracellular tachyzoites).

## TgHMGB1a is a DNA binding protein and induces TNF-α secretion in macrophages

To comparative study of the proinflammatory role of TgHMGB1a, peptides (**Figure 3A**) of TgHMGB1a B box (TgH1a B), mouse HMGB1 (mH1), mouse HMGB1 A box (mH1A) and B box (mH1B), and the chimeric protein mH1A+ TgH1a B (c (A+B)) were expressed as His-tagged fusion proteins in *E. coli* and purified from soluble form. The purity and molecular weight of the recombinants were examined by SDS-PAGE

(**Figure 3B**). Agarose gel electrophoresis assays showed that mH1, TgH1a B, c(A+B), mH1A and mH1Bbound significant amounts of specific DNA fragments between 200 to 1000 bps (**Figure 3C** lanes 1–5), whereas natural recombinant peptides formed complexes with DNA that resulted in retarded bands in the sample loading wells (**Figure 3C** lanes 1′–5′). TgGRA1 did not bind to a specific DNA band as control (**Figure 3C** lanes 6 and 6′), demonstrating that DNA binding is due to the HMG box, but not dependent on the His-tag.

The endotoxin removed recombinants TgHMGB1a 4E stimulates anal macrophages to release TNF-α in a dose-dependent manner similar to miceHMGB1 and LPS (positive control)

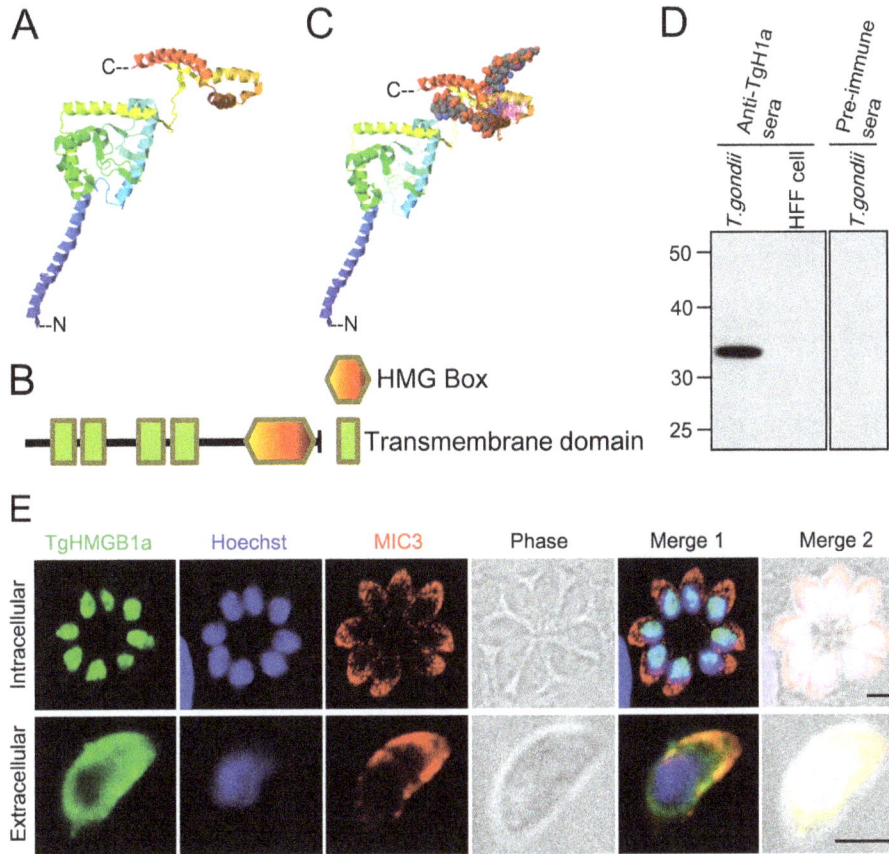

**Figure 2. Characterization of TgHMGB1a by structure modeling and cellular localization.** A. Ribbon cartoon representations of the predicted 3D structure of TgHMGB1a generated by homology modeling. Three α-helices independently fold into the HMG box at the C-terminal, while another four helices are clustered as the transmembrane domains. C indicated the C-terminal and N showed the N-terminal. B. Schematic representation of the primary structure of TgHMGB1a highlighting its different domains. TgHMGB1a has only one HMG box at the C-terminus and four transmembrane domains at the N- terminal; there is no acidic C-tail. C. A model of TgHMGB1a bound to DNA. TgHMGB1a can bind DNA through an interaction between the HMG box and the minor groove of DNA. D. Identification of native TgHMGB1a by western blotting. *T. gondii* represents the total antigens from cell cultured *T. gondii* tachyzoites; uninfected HFF cell lysates were used as a control. A specific band was elicited when anti-rTgHMGB1a 4E polyclonal antibodies were used as a primary antibody, but not when preimmune serum was used. E. TgHMGB1a is concentrated in the nucleus of parasite. Immunofluorescent localization of TgHMGB1a showed that TgHMGB1a (green) not only localized in the nucleus (blue) during cytokinesis (C) (top panel), but also during mitosis (M) (middle panels). Additionally, TgHMGB1a continued to accumulate in the nucleus of tachyzoites freshly released into extracellular media (bottom panel). MIC3 (red), nuclei stained with Hochest (blue). Scale bars, 3 μm. All immunofluorescent labeling was performed on HFF cells infected with tachyzoites of the RH strain.

(**Figure 3D**). As the A box of miceHMGB1 usually to be an antagonist of HMGB1, we made a chimeric peptide fused of miceHMGB1 A box and TgHMGB1a B box (c (A+B)) to verify whether or not the A box can also block the TgHMGB1a functions. To test this possibility, we examined the TNF-α secretion in anal culture supernatant after c (A+B) stimulation. At higher concentrations (1 and 10 μg/mL), c(A+B) induced TNF-α release significantly less than TgHMGB1a 4E and miceHMGB1 but still higher than the A box, GRA1 and even the B box, demonstrating that miceHMGB1 A box can also be used as an antagonist of TgHMGB1a. To determine whether it is the TgHMG protein alone or in complex with DNA that activates TNF secretion, protein kinase (PK) and DNase I were used to digest the recombinant TgHMGB1a B box respectively, and then we examined the TNF-α levels in the supernatant after stimulated by the treated products. The products after digested by PK almost completely lost the function of induce to TNF-α release, while a slight loss of TNF-α secretion after treat with DNase I. All the results indicated that TgHMGB1a is able to stimulate the TNF-α

release in the anal macrophages, and which partially but not completely dependents on bound DNA.

## TgHMGB1a specifically binds to cruciform DNA

DNA–protein complexes can be analyzed by EMSAs with four-way junction (4H) DNA, preferentially bound by HMGB proteins and a synthetic substrate commonly used to study proteins involved in recognizing and resolving Holliday-type junctions formed during in vivo genetic recombination events (**Figure 4A**). At least two different 4H DNA–TgHMGB1a 4E complexes formed with increasing of the protein concentration, resulting in two discrete retarded bands (**Figure 4B**). These two bands presumably correspond to complexes formed upon binding of one or more TgHMGB1a 4E molecules per 4H DNA. Binding competition with 2H "hairpin like" structures (**Figure 4A and C**, lanes 4–7) demonstrated that TgHMGB1a 4E preferentially binds to complete 4H DNA, since 500-fold excess of these 2H structures does not affect the 4H DNA–TgHMGB1a 4E complex migration. Simultaneously, competition experiments with linear dsDNA

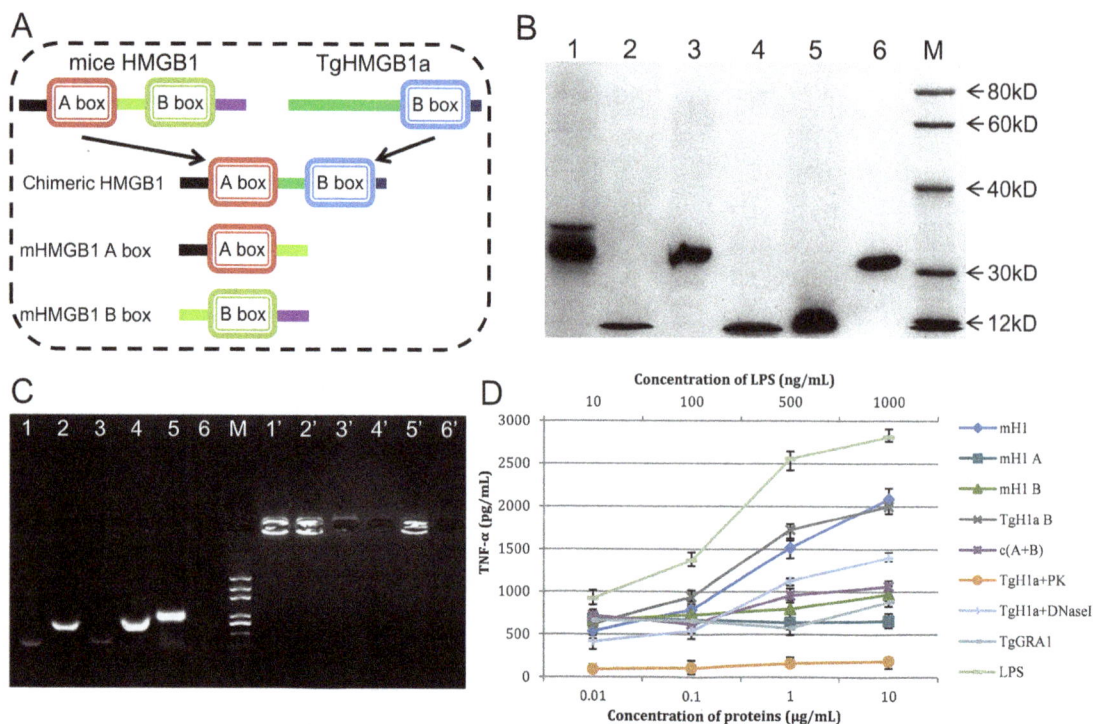

**Figure 3. The TgHMGB1a box bound DNA fragments from *E. coli* and stimulated macrophages to release TNF-α.** A. To examine TgHMGB1a DNA binding ability and to compare that with mouse HMGB1, full length mouse HMGB1 (mH1), TgHMGB1a 4E (TgH1a B), independent mH1 A and B box domains and a chimeric mH1 A box-TgH1a B box (c (A+B)) were expressed in the prokaryote *E. coli*. B. Purified recombinant peptides were detected by SDS-PAGE. Lane 1, mH1; lane 2, TgH1a B; lane 3, chimeric c(A+B); lane 4, mH1 A; lane 5, mH1 A; and lane 6, TgGRA1 (control). M indicates the protein marker. C. Gel retardation experiments were carried out to identify bound DNA fragments. Lanes 1–6: denatured DNA-protein complexes from lanes 1–6 of **Figure 3B**. Lanes 1′–6′: natural complexes of lanes 1–6 of **Figure 3B**. M, DNA marker. All specific bands ranged from 200 to 1000 bp. D. Endotoxin removed peptides stimulates macrophages to release TNF-α at indicated concentrations. Ana1 macrophages cells were stimulated with HMG peptides at the indicated concentrations, and the culture medium was assayed for TNF-α at 24 h after stimulation. LPS and TgGRA1 were used as controls at the indicated concentrations. Each bar indicates the mean ± SD of three independent experiments. TNF-α secretion induced by TgHMGB1a is significant higher than controls (p<0.01).

derived from the 4H arms (**Figure 4A**) indicated that TgHMGB1a 4E preferentially binds to the DNA cross-over of the cruciform DNA, since 500-fold excess of these duplex DNA molecules were not able to inhibit the formation of the cruciform DNA–TgHMGB1a 4E complexes. As expected, the retarded bands almost disappeared when 100-fold cold 4H was added (**Figure 4C**, lane 3), suggested that there is competition between labeled 4H DNA and excess cold 4H DNA. Purified TgGRA1 was used in similar EMSAs, but it didn't show mobility shift to the cruciform DNA, demonstrating that binding of TgHMGB1a 4E protein is specific and independent of the His-tag fusion (**Figure S7**).

## Disruption of TgHMGB1a did not cause obvious phenotype changes

To further characterize the biological role of TgHMGB1a, the B box of TgHMGB1a was targeted and replaced by eGFP in the RHΔKU80 strain using homologous recombination (**Figure 5A**). PCR was used to confirm both replacement of the B box and the insertion of a CAT-RFP cassette in the parasite genome (**Figure 5B**). Replacement of the B box of TgHMGB1a by eGFP was also confirmed using western blotting (**Figure 5C**). We also attempted to generate a complete knockout strain of TgHMGB1a, yet so far have been unsuccessful. Confocal microscopy showed that eGFP was distributed throughout parasite cells, but was not concentrated into the nucleus like integral TgHMGB1a (**Fig-**

**ure 5D**). Nevertheless, the absence of TgHMGB1a B box did not affect parasite growth in vitro, as TgHMGB1a B box$^{-/eGFP}$ parasites formed an equivalent number and size of plaques (**Figure S8A**) and had a replication rate similar to the RHΔKU80 strain (**Figure S8B**). These results suggested that the function of TgHMGB1a may be at least partially redundant, and other homologous proteins may play overlapping roles in parasites.

## Overexpression of TgHMGB1a affects intracellular parasite replication

To make sure that a position effect of the transformation was not involved, we duplicated the experiments with parasite clones isolated from two independent pDMG-TgHMGB1a (**Figure S9A**) transfections. Meanwhile, to determine the phenotype changes were due to the HMG domain of TgHMGB1a specifically, a strain overexpressing a construct lacking the HMG domain was also generated as control. TgHMGB1a overexpression was confirmed by western blotting (**Figure S9B**) and more importantly, exogenous TgHMGB1a-GFP was concentrated in the nucleus throughout the development of intracellular *T. gondii*, which colocalized with endogenous TgHMGB1a (**Figure 6A**), and B box lacked TgHMGB1a hasn't concentrated in nucleus but likely dispersed in cytoplasm (**Figure S9C**), which suggested the B box domain might effects on the localization of TgHMGB1a. Plaque assays showed that the RH and RH-GFP strains did not form significantly different plaques sizes, but the transgenic

## A

| | |
|---|---|
| 4H-a | 5' AACAGTAGCTTATTCGAGCTCGCGCCCTATCACGACGACTA 3' |
| 4H-b | 5' GTAGTCGTCGTGATAGGTGCAGGGGTTATAGGG 3' |
| 4H-c | 5' CCCTATAACCCCTGCATTGAATTCCAGTCTGATAA 3' |
| 4H-d | 5' TTTATCAGACTGGAATTCAAGCGCGAGCTCGAATAAGCTACTGT 3' |
| Compl 4H-a | 5' GTAGTCGTCGTGATAGGGCGCGAGCTCGAATAAGCTACTGTT 3' |
| Compl 4H-c | 5' TTTATCAGACTGGAATTCAATGCAGGGGTTATAGGG 3' |

## B

## C

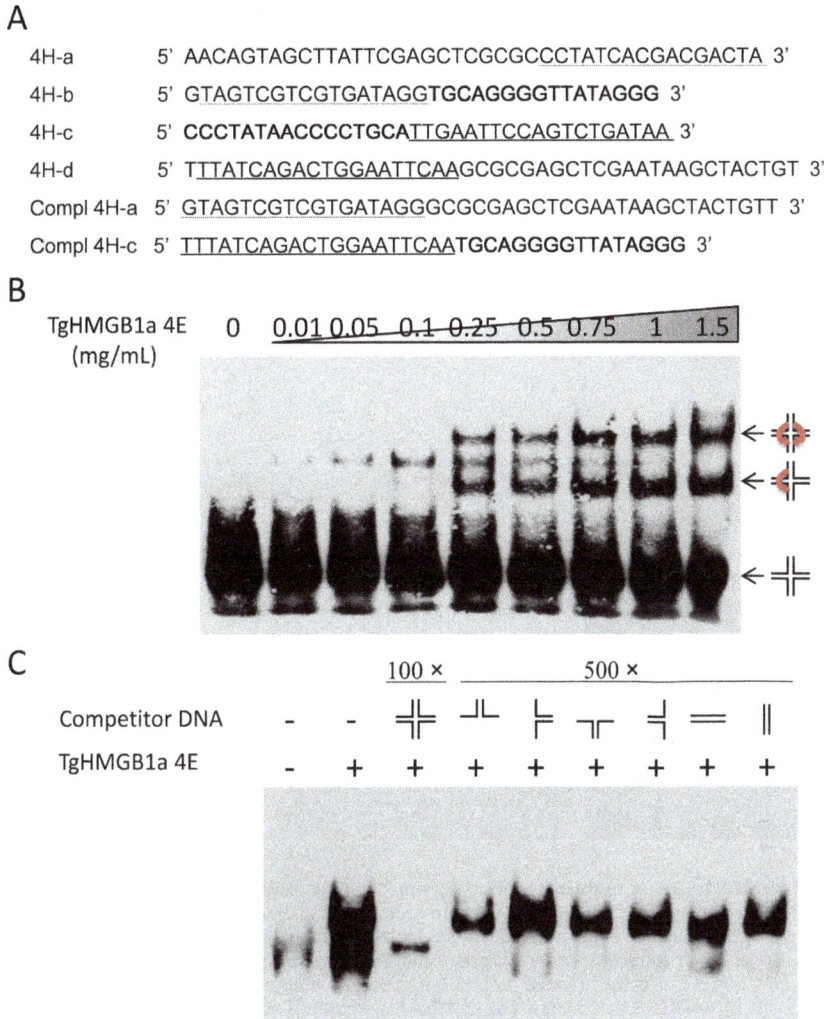

**Figure 4. TgHMGB1a binds to cruciform DNA.** A. Oligonucleotides used to form the different synthetic DNA structures. Identical text formats (italic, bold, dotted/solid underlines) represent complementary sequences. All of the primers were labeled using EMSA Probe Biotin Labeling Kit as described in methods. Cruciform 4H structures were obtained by annealing equal amounts of 4H-a, -b, -c and -d primers; 2H hairpin-like structures, by annealing of 4H-a and -b, -b and -c, -c and -d, -d and -a, respectively; double stranded (ds) linear DNA structures were obtained by annealing of 4H-a or 4H-c with its fully complementary sequence (compl 4H-a and compl 4H-c). B. EMSA with cruciform DNA (4H) and recombinant TgHMGB1a 4E. Increasing concentrations of TgHMGB1a 4E protein (0.01–1.5 mg/mL) were incubated with biotinlabeled 4H DNA for 30 min at 25°C. DNA–protein complexes formed were resolved on 6% non-denaturing polyacrylamide gel and transferred to nylon membrane and then detected by Chemiluminescent EMSA Kit. Lane 1, free labeled 4H; lanes 2–8, 4H DNA–protein complexes formed with increasing concentrations of TgHMGB1a 4E. C. Competition EMSA experiments were performed with different cold DNA structures. After incubation of TgHMGB1a 4E (1 mg/mL) with biotinlabeled 4H, cold complete 4H (in a 100-fold excess), hairpin-like 2H structures (in a 500- fold excess) or ds linear DNA (in a 500-fold excess) were added, and incubated for an additional 20 min. Complex formation was analyzed as in B. Lane 1, free labeled 4H; lane 2, 4H DNA–protein complexes formed with TgHMGB1a 4E without competitor DNA; lane 3, competition assay with 100-fold excess cold 4H; lanes 4–7 competition assay with 500-fold excess cold 2H hairpin-like structures; lanes 8–9, competition assay with 500-fold excess cold linear ds DNA.

parasites generated plaques that were smaller than the ones generated by its parental strain (**Figure 6B**), and almost there were no significant difference between the TgHMGB1a$^{-\text{B box}}$ parasites and RH or RH-GFP strains (**Figure S9D**). In another words, TgHMGB1a overexpression inhibited the formation of large plaques. Furthermore, it's most due to the HMG domain of TgHMGB1a. Quantification of plaque areas suggested that the size of plaques formed by RH and RH-GFP strains was almost two-fold greater than the ones formed by the transgenic parasites (**Figure 6C**). Furthermore, comparison of the replication rates showed a significant difference in the distribution of parasites per vacuole between the transgenic parasites and its parental strain

(**Figure 6D**); the transgenic parasites more often contained 4 and 8 tachyzoites vacuoles than its parental strain, while the parental strain more often contained 8 and 16 parasites vacuoles. Thus, these data suggested that overexpression of TgHMGB1a led to slower growth during intracellular replication in vitro.

## TgHMGB1a involved in the regulation of gene transcription

Quantitative RT-PCR analysis of $1 \times 10^7$ transgenic and their parental parasites were conducted to determine the potential involvement of TgHMGB1a in transcriptional regulation. Our quantitative RT-PCR analysis indicated that ROP18, toxofilin,

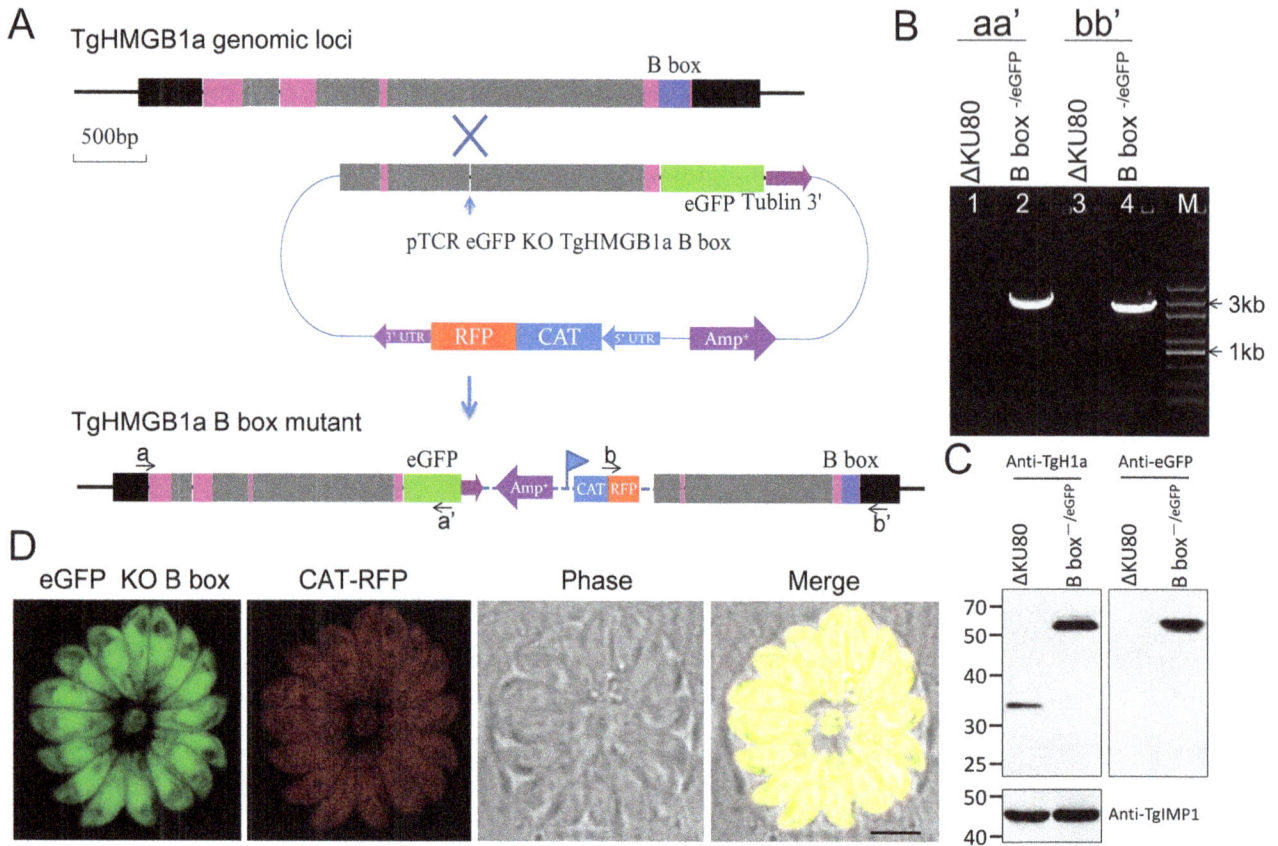

**Figure 5. Generation of a TgHMGB1a B box mutant strain.** A. Schematic of the experimental design. TgHMGB1a is encoded by 4 exons, and the B box is located in the fourth exon (shown for the TgHMGB1a genomic locus). An "O-type" knock-in/knock-out vector (pTCR eGFP KO TgHMGB1a B box) was constructed to target and replace the B box domain. The vector included an approximately 2200 bp homologous arm upstream of the TgHMGB1a B box, and a monoclonal site of restriction endonuclease (BsiWI) throughout the whole vector sequence was not only needed, but also necessary in the homologous arm, and it is was best ranged around the middle of homologous frame [55]. If the targeted gene was successfully disrupted, eGFP should replace the TgHMGB1a B box domain and create a fusion protein of TgHMGB1a B box$^{-/eGFP}$ expressed under the control of its endogenous promoter. CAT-RFP was used as a selection marker. B. Genomic PCR analysis of the ΔKU80-TgHMGB1a B box$^{-/eGFP}$ strain. The position of the primers and the expected sizes are shown in panel A. C. Western blot performed with anti-TgHMGB1a 4E antibodies on total extracts from RHΔKU80 and ΔKU80-TgHMGB1a B box$^{-/eGFP}$ strains. TgIMP1 was used as an internal control. D. Observation of the TgHMGB1a B box$^{-/eGFP}$ and CAT-RFP strains by fluorescent confocal microscopy, and the result indicated that B box mutant TgHMGB1a was dispersed throughout the parasite cells. Scale bar, 3 μm.

PLP, MIC3, profilin, and GRA7, but not ROP16, were up regulated at least 1.5-fold in TgHMGB1a overexpression parasites (**Figure 7A**), yet there were no obvious changes observed in the TgHMGB1a B box$^{-/eGFP}$ strain. However, a dramatic increase of ROP16 expression was detected in the TgHMGB1a B box$^{-/eGFP}$ strain (**Figure 7B**), even though profilin expression was reduced. Expression of the TgHMGB1a homologues, TgHMGB1b and TgHMGB1c, were also measured. TgHMGB1b and TgHMGB1c were almost undetectable and similarly expressed in transgenic and parental strains, while TgHMGB1a was overexpressed (**Figure 7A**). However, in the TgHMGB1a B box$^{-/eGFP}$ parasites, TgHMGB1b showed a significant increase, whereas TgHMGB1c showed a nonsignificant trend towards reduced expression (**Figure 7B**). Furthermore, ChIP-qPCR analysis showed that promoter levels of PLP1, profilin, ROP16, ROP18 and Toxofilin in the pull down of TgHMGB1a overexpress strain are significant higher than RH strain, although MIC3 and GRA7 are not obvious (**Figure 7C and D**). Meanwhile, there are almost no signals in TgHMGB1a B box$^{-/eGFP}$ parasites and in the negative controls (**Figure 7C**). Moreover, we also tested the coding regions through the regular and real-time PCR, and results

showed no significant difference between the wild type and TgHMGB1a overexpress strain (**Figure 7C bottom panel** and data not shown). These results suggested that TgHMGB1a overexpress enhanced the promoters binding ability, that is generally consistent with the transcription assays of these indicated genes (**Figure 7A**), demonstrating that TgHMGB1a plays role in regulating gene expression, and which is most like an activator of transcription and its roles might be somewhat redundant among HMG family members.

## Overexpress but not disrupt TgHMGB1a T.gondii delayed the death of mice

To evaluate the contribution of TgHMGB1a to parasite virulence, we measured the survival of mice infected with the transgenic T. gondii strains or their parental strains. Consistent with the in vitro assays, the survival of Balb/c mice infected with TgHMGB1a B box$^{-/eGFP}$ or RHΔKU80 strains were not significantly different (**Figure 7F**). However, mice infected with the TgHMGB1a overexpression showed a significantly delayed time to death (3 to 5 days) in a low doses ($10^2$ and $10^3$) infection

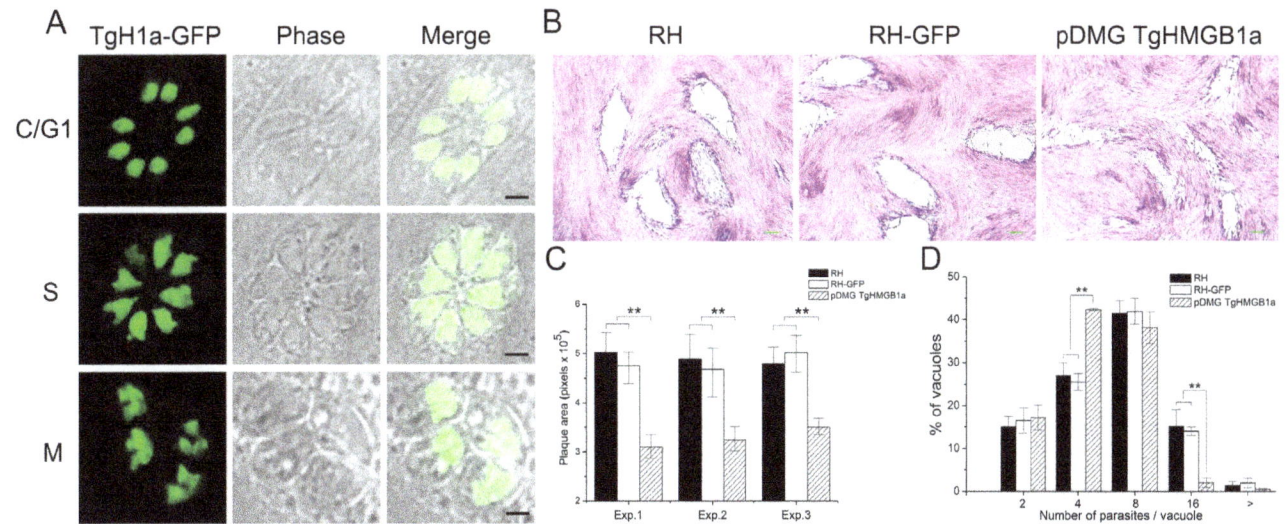

**Figure 6. Overexpression of TgHMGB1a affects intracellular parasite replication.** A. Transgenic TgHMGB1a localized to the nucleus, similar to endogenous TgHMGB1a. Plasmids of pDMG TgHMGB1a (**Figure S9A**) were introduced into the RH strain to overexpress TgHMGB1a. Through screening, clones were obtained in which the TgHMGB1a-GFP was expressed and concentrated in the nucleus throughout the entire cell cycle (cytokinesis, C; gap phase, G1; DNA synthesis, S; and mitosis, M), which was identical to endogenous TgHMGB1a, as shown by IFA (**Figure 2E**). Western blot were applied to confirm the overexpression of TgHMGB1a (**Figure S9B**). B. Plaque assays for TgHMGB1a overexpression parasites compared to its parental strains. Infected HFF-monolayers were fixed and stained (see Materials and methods). Scale bar, 0.2 mm. C. Plaque assays were carried out three times independently, the plaque area was quantified from three independent experiments. At least 30 plaques were quantified per strain in every experiment. Each bar indicates the mean ± SD; "exp." indicated the each experiment. D. An intracellular growth rate assay was carried out through counting the number of parasites per vacuole. Parasites of the TgHMGB1a overexpressed and parental strains (RH and RH-GFP) were allowed to invade and divide for 24 hours. IFA was performed with TgSAG1 abs and the numbers of parasites per vacuole were counted. At least 150 vacuoles were examined per strain. Each bar indicates the mean ± SD from three independent experiments; $n \geq 150$ vacuoles. Statistical significance was determined using Student's $t$-test (**$P < 0.05$).

compared to its parental RH or RH-GFP strains ($P < 0.0001$ was considered as statistically significant difference, see the **Data S1**), while the RH and RH-GFP strains showed almost equivalent virulence (**Figure 7E**). Collectively, these results established that the expression of TgHMGB1a is involved in parasitic virulence, however, it function may be partially redundant.

## Discussion

Homologous HMG proteins have been reported in many parasites, and almost all of these proteins share the characteristics of DNA binding ability and nuclear localization. However, the endogenous functions of these proteins are poorly understood in parasites. Previous studies indicated that most parasite HMGB1s are expressed throughout the life cycle of a parasite. In *S. mansoni*, HMGB1 appears to be a stage-specific protein that is abundantly expressed in the skin-stage schistosomula, adult female and egg stages of the parasite, whereas its expression is low or absent in the adult male and cercarial stages [16]. In *E. histolytica*, overexpression of EhHMGB1 caused the modulation of 33 transcripts, including four virulence factors that are involved in various cellular functions [21]. In *P. falciparum*, two HMGB factors were identified, and in vitro analyses showed that they were able to interact with distorted DNA structures and bend linear DNA with different affinities and to play roles in transcriptional regulation of *Plasmodium* development in erythrocytes [19,20]. In this study, many members of the HMGB1 family have been predicted in *T. gondii* (**Figure S1 and Table S6**). We cloned and characterized *T. gondii* HMGB proteins, and our results suggested that structure and functions of TgHMGB1a are similar to HMGB1 family proteins in their DNA binding ability and nuclear localization. Our findings confirmed that TgHMGB1a is expressed in *T.*

*gondii*, and which were conserved in three genotypes (**Table S1**). However, region of TGGT1_053220 (ToxoDB ver.9.0) has been renamed two independent genes (TGGT1_210412 and TGGT1_210408 in ToxoDB ver.11), the western blot showed a protein with molecular weight (about 33.9 kDa) in line with the TGGT1_053220. So our results suggested that TGGT1_210408 may be the last exon of TGGT1_210412, at least, there is a combined transcript of this two fragments (TGGT1_210408 and TGGT1_219828 in ToxoDB ver.11) in RH strain.

Both our sequence and structure analyses converged, indicating that only one *T. gondii* HMG box domain closely resembles the box B of mammalian HMGB (**Figure 1**). For human HMGB1, it was reported that structure-specific binding to cruciform DNA was mediated by box A and that box B, flanked by a basic region, displayed a marked DNA recognition activity [38,43]. In this study, we verified that TgHMGB1a B box can bind the DNA fragments of *E. coli* (**Figure 3**), however, the differences of the bound DNA might have resulted from the varying degrees of sonication. Nevertheless, many HMGB1 proteins in other organisms, for example *Plasmodium* [19], showed no specific binding ability to the DNA that interacted specifically with typical SOX and SRY transcription factors (TFs) in an EMSA. We approved the TgHMGB1a B box specifically binds the cruciform DNA in a high affinity (Fig 4) like the *T.cruzi* HMGB proteins [14], which may be valuable to understand the interaction between TgHMGB1s and DNA. Like the mammals HMGB1, parasites HMGB1 in general characterized by DNA binding ability and proinflammatory responses [20]. Comparison of the ability of induce to TNF-α release by the indicated recombinants (**Figure 3D**) showed that TgHMGB1a is able to stimulate the macrophages to release TNF-α as miceHMGB1, and the results of PK and DNase I treatments suggested that TgHMGB1a induced

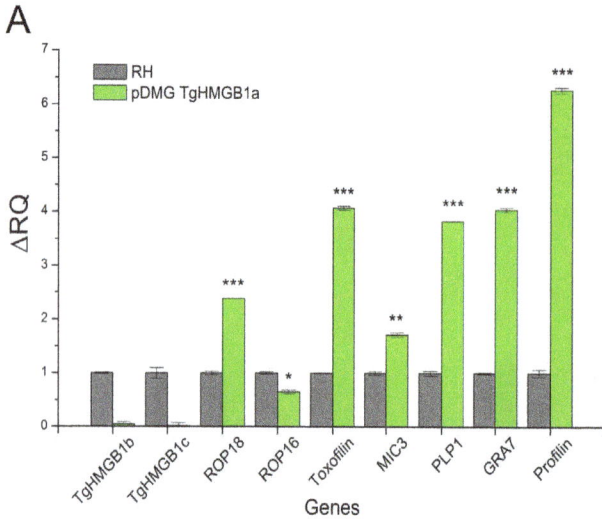

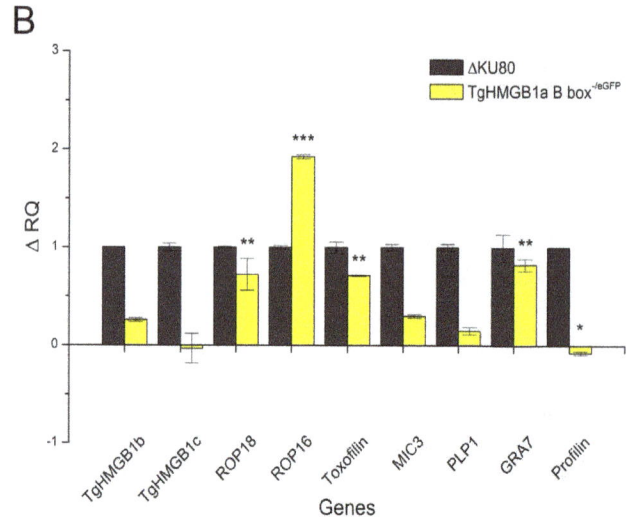

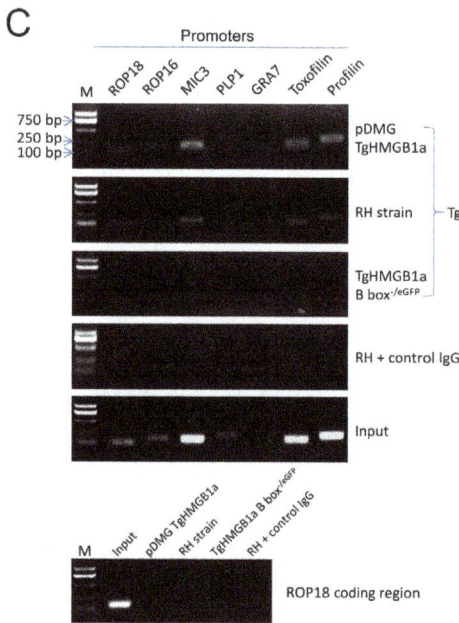

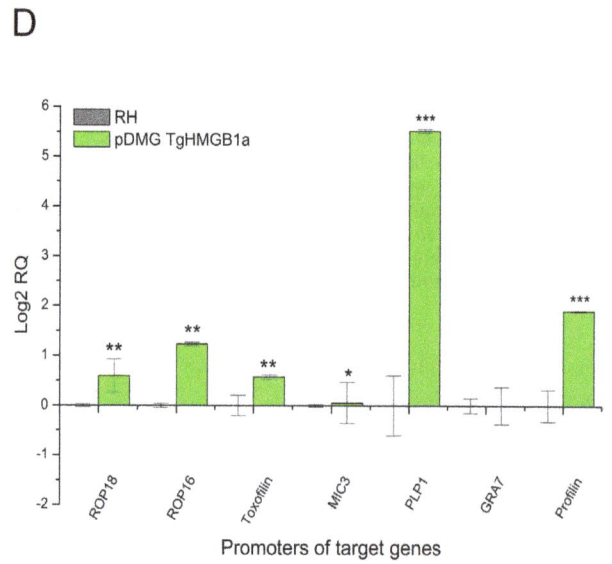

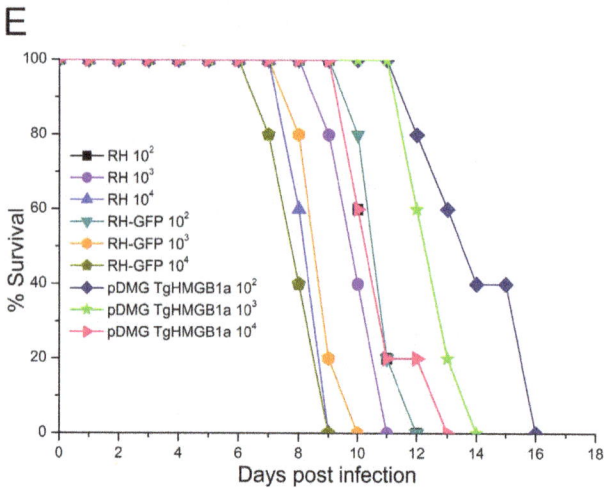

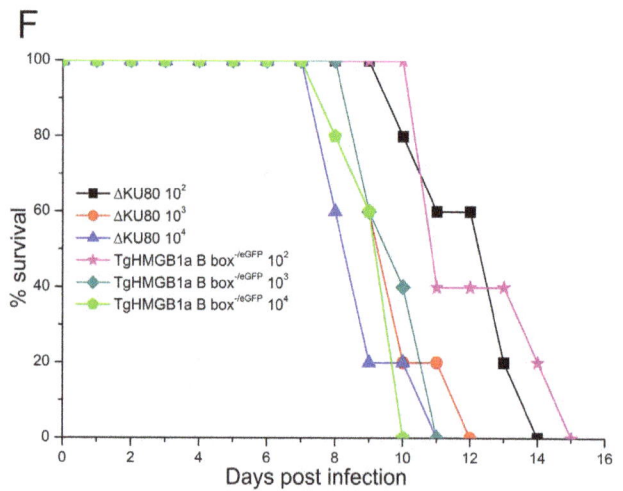

**Figure 7. TgHMGB1a involves to gene transcription regulatory and overexpress but not disrupt TgHMGB1a can delayed the death of mice.** A and B. Quantitative RT-PCR was used to analyze the transcription levels of the indicated genes in TgHMGB1a overexpression (A) and B box-deficient (B) parasites compared with their parental strains. Each bar indicates the relative quantity (RQ) $\pm$ SD. RQs of the transgenic parasites were calibrated using their parental strains (i.e., $\Delta RQ_{transgenic\ parasite} = RQ_{transgenic\ parasite} - RQ_{parental\ strain} (= 1)$). Data presented are representative of three independent experiments, each done in triplicate. Statistical significance was determined using Student's t-test (*$P < 0.1$, **$P < 0.05$, ***$P < 0.005$). C and D. Analysis of TgHMGB1a binds to promoters of indicated genes through chromatin immunoprecipitation (ChIP). C. Regular PCR was performed on the ChIP DNAs from the TgHMGB1a overexpress and RH strains using the promoter-specific primers of indicated genes, normal mice sera, input DNA and TgHMGB1a B box$^{-/eGFP}$ strain were used as ChIP controls, ROP18 coding region was also tested as negative control. PCR products were run on 1.5% agarose gels. D. Quantitative real-time PCR was carried out and the pull-down promoters in RH and TgHMGB1a overexpress strains were normalized against the corresponding input DNA, the promoters level in TgHMGB1a overexpress parasites were represented as log2 functions of relative ratios to RH strain. Data presented are representative of three independent experiments, each done in triplicate. Analysis was carried out using Student's t-test. *$P < 0.1$, **$P < 0.01$, ***$P < 0.001$. E and F. Mouse survival curves for parental and TgHMGB1a transgenic lines. Balb/c mice were separately injected i.p. with the indicated parasites and doses. Mean values shown per group (n = 5), representative of 3 experiments with similar outcomes. Mice infected with the TgHMGB1a overexpression showed a significantly delayed time to death (3 to 5 days) in a low doses ($10^2$ and $10^3$) infection compared to its parental RH or RH-GFP strains ($P < 0.0001$ was considered as statistically significant difference, see the **Data S1**).

to TNF-$\alpha$ secretion is partially but not completely dependent on the bound DNA, however, it's more importantly investigate to reveal how does a nuclear factors contact to host immune system, and actually we haven't detected TgHMGB1a secretion at least in culture supernatant of type I parasites (Data not shown). These results suggest that TgHMGB1s may be related to host inflammatory immune responses in *T.gondii* infection.

In this study, two transgenic parasite strains were generated to examine the roles of TgHMGB1a in *T. gondii*. One strain is the TgHMGB1a B box$^{-/eGFP}$ mutant, and the other is overexpressed TgHMGB1a. We measured the intracellular growth ability of the transgenic parasites using plaque assay, and our results showed slower growth of the TgHMGB1a overexpression strain (**Figure 6**), but no significant changes in the TgHMGB1a B box$^{-/eGFP}$ mutant compared with its parental strains (**Figure S8**). We also demonstrated that overexpressing TgHMGB1a or deleting the TgHMGB1a B box modulated the expression of many virulence factor genes (**Figure 7A and B**). ROP18 and ROP16 were found to be the most important virulence factors in *T. gondii*, parasites that express the Type I ROP18 allele display rapid cell cycle times in cell culture and transfer of the Type I-ROP18 allele into avirulent strains speeds their growth (but not their ability to invade) and dramatically enhances virulence [44], nevertheless, how does ROP18 regulates the parasite growth rate is unknown. ROP18 might influences the parasite cell cycle mechanisms [45] but not be the sole determinant for parasite growth, for which is a so complex process involved a great many genes regulatory and pathways. So, upregulation of ROP18 is not necessarily leads to a faster grow rate, and by contrast, the slower growth rate showed in TgHMGB1a overexpression parasites indicated that TgHMGB1a may be involved to cell cycle regulations. On the other hand, types I and III ROP16, but not type II ROP16, can maintain constitutive activation of STAT3 and STAT6, thereby inhibiting pro-inflammatory responses in macrophages (reviewed by [27]) and conferring a survival advantage to the parasite. In the TgHMGB1a B box$^{-/eGFP}$ parasites, ROP16 was significantly upregulated and may have partially neutralized any negative affect by downregulating other genes in vivo, for instance, the recognition of profilin through the Toll like receptor 11 (TLR-11), that mediates production of IL-12 and triggers a strong Th1-type response to against *T.gondii* infection [46]. The TgHMGB1a overexpression strain caused delayed death in Balb/c mice for that the slower growth rate might attenuate parasitic virulence [30,47], but there were no meaningful changes in the TgHMGB1a B box$^{-/eGFP}$ mutant compared with its parental strains, in agreement with the results obtained in vitro.

Intuitively, all of these results suggested that TgHMGB1a participates in these processes as a negative regulator, which is not consistent with the typical functions of HMGB1 proteins. However, we cannot conclude that TgHMGB1a is a transcriptional repressor in *T. gondii*; at best, we can only show that it is a repressor of parasite proliferation. In fact, in eukaryotes, the association of HMGBs with chromatin is highly dynamic, and the proteins affect the chromatin fiber as architectural factors by transient interactions with nucleosomes [2,48]. It has been proposed that histones and HMGB1s have opposite effects; histone H1 molecules are considered to be general repressors of transcription, whereas HMGBs are often viewed as transcriptional activators [49]. While the linker histones and HMGBs exhibit direct competition in binding to such structures as four-way junction DNA, whether they compete for binding to the nucleosome has not been investigated. The possibility for either opposite or synergistic effects on gene regulation must be considered at this point [49], and future investigations will be necessary to increase our knowledge of the interactions between HMGB1 and histones in *T. gondii*. However, there is no Histone H1 in *T. gondii* but in addition to having a single copy of canonical histones (H2A, H2B, H3 and H4), *T. gondii* encodes five variant histones (centromeric H3 (CenH3), H3.3, H2A. X, H2A. Z, and the parasite-specific H2Bv [50]. Such studies might provide exciting new insights to advance understanding of the transcription-related functions of TgHMGBs.

More direct and very important evidence for the involvement of TgHMGB1a in transcription regulation was generated by quantitative RT-PCR studies, which indicated that the transcription levels of many genes were elevated in the TgHMGB1a overexpression parasites, but were not significantly reduced in the TgHMGB1a B box-deficient strain. Nevertheless, it's more like that TgHMGB1a is negative regulated for ROP16 but promotes transcription of profilin (**Figure 7A and B**), however, it needs to be further investigated. Furthermore, promoter binding assays with ChIP-qPCR analysis showed that TgHMGB1a prefers bind to promoter regions (**Figure 7C and D**) and maybe in non-gene-specific promoter manner (most likely as an architectural factor of the transcription complex). Collectively, we suggest that TgHMGB1a plays an activator role in gene transcription regulation in *T. gondii*, as in mammals, and we supposed that the replication of *T. gondii* was controlled by cell cycle regulatory genes [47,51], which were upregulated in the TgHMGB1a overexpression strain, enhancing the role of inhibition for replication, and then the parasites showed slower growth compare with their parental strain. The mRNA levels of the other HMGB1s in *T. gondii*, TgHMGB1b and TgHMGB1c, were also measured, and only TgHMGB1b showed a slight increase and TgHMGB1c declined. Therefore, it remains unclear whether there is a feedback mechanism involving HMGB1 proteins that affects the synthesis of these genes, with the view of their functions may be partially redundant and compensated for when one is deficient in parasite. However, polymorphisms in HMGB1 genes may support partial redundancy among HMGB1 proteins.

The DNA-related functions of TgHMGB1a are consistent with its nuclear location (**Figure 2**). Interestingly, no canonical NLS was found in TgHMGB1a by bioinformatic predictions, and there were no signal sites identified in the B box, even though B box-deficient TgHMGB1a (**Figure 5 and Fig S9C**) did not concentrate in the nucleus but dispersed throughout the parasite cells (**Figure 6D**). In other species, the acetylation status of HMGB proteins can alter both their DNA-binding properties and their subcellular localizations [52,53]; N-myristoylation and palmitoylation sites have also been shown to contribute to protein localization [54]. Although one N-myristoylation (100–105 aa) and two tyrosine kinase phosphorylation (282–290 and 289–297 aa) sites were identified in TgHMGB1a, further studies will be needed to determine whether PTMs of TgHMGBs affect the localization of these proteins in T. gondii. Investigations of tyrosine kinase phosphorylation sites will be especially interesting, because they have been studied little in this context. Through comparison of TgHMGB1a localization between the intracellular and extracellular, we found an interesting phenomenon that TgHMGB1a showed a translocation after the parasites released to extracellular (**Figure S5 and S6**). Usually, the transcription of most metabolic genes should be reduced when the parasites are outside of the cell, so it may be a protection mechanism of parasites for TgHMGB1s translocated out of the nuclear perhaps because of the non-specific transcription activation of TgHMGB1s.

In conclusion, all of our computational analysis and experimental results were concordant suggested that the TgHMGB1s were substantially similar to multicellular HMGB proteins in both structure and functions. TgHMGB1a was implicated in transcriptional regulation and most likely acts as an activator of many virulence factors. Additional work will be needed to understand the mechanisms of TgHMGB1a and the HMG family proteins in T. gondii, both as nuclear factors and also proinflammatory mediators. To our knowledge, this is the first report of a role for an HMGB1 protein in transcriptional regulation of T. gondii genes that are involved in intracellular parasite replication and maybe affect parasite virulence in mice.

## Supporting Information

**Figure S1 Phylogenetic analysis of HMGBs in various organisms.** A. Phylogenetic analysis based on HMGB proteins from various organisms. HMGB proteins from Apicomplexa and other parasites, humans, mammals, D. rerio, A. thaliana and S. cerevisiae were analyzed based on either the neighbor-joining (NJ) method of distance analysis or the maximum likelihood (ML) method. The bootstrap consensus tree inferred from 1000 replicates is thought to represent the evolutionary history of the taxa analyzed. The tree includes TgHMGB1a (light green circle), TgHMGB1b (light blue circle), TgHMGB1c (light cyan circle) and homologues found in other eukaryotes. Bootstrap values (>50%) from 1000 resamplings are indicated prior to the branch points of the tree. Only bootstrap values >95 were considered to be significant and allowed to form clustered sequences (colored branches). Protein numbers were given according to the EuPathDB and NCBI websites. Also, see Table S1 for the

multiple sequence alignment used to compute the phylogenetic trees. B. Phylogenetic analysis based on the CDS of HMGB1s from various organisms. The coding sequences of HMGB proteins from Apicomplexa and other parasites, human, mammals, D. rerio, A. thaliana and S. cerevisiae were analyzed based on neighbor-joining (NJ) using MEGA 5.22 (1000 bootstrap replicates were performed). The results are consistent with phylogenetic analysis based on the amino acid sequences. TgHMGB1a, b and c are branched three clusters and all they are the closest relatives to HMGB1 of Homo sapiens. Abbreviations are as follows: Hs, Homo sapiens; Xl, Xenopus laevis; Dr, Danio rerio; At, Arabidopsis thaliana; Eh, Entamoeba histolytica; Sc, Saccharomyces cerevisiae; Pf, Plasmodium falciparum; Nc, Neospora caninum; Tg, Toxoplasma gondii; Et, Eimeria tenella; Lm, Leishmania major; Tc, Trypanosoma cruzi; Tb, Trypanosoma brucei; and Sm, Schistosoma mansoni.

**Figure S2 HMG box contained proteins in mice and T. gondii.** Mouse HMGB1 has 2 HMG boxes, whereas all TgHMGB1s but not TGGT_203950have only one HMG box. TGGT1_053220 (ToxoDB ver.9,0) was renamed as TGGT1_210412 in ToxoDB ver.11, and the last exon of TGGT1_053220 has been named TGGT1_210408 in ToxoDB ver.11, however, our results suggested that there might be a combined transcript of this two fragments (TGGT1_210408 and TGGT1_219828), at least, that is TgHMGB1a named in present study. Same to TGGT1_053220, TGGT1_030840 was also renamed, and N-terminal region has been predicted as cyclin-dependent protein serine/threonine kinase regulator subunit protein (TGGT1_219832). The others, such as TGGT_203950 (three HMG box, showed only two box for the long sequence structure) and TGGT_217500 (N-terminal HMG box) should be not classified into HMGB proteins for the low structure similarity and long genetic distance (Figure S1).

**Figure S3 Sequence-based bioinformatic predictions of TgHMGB1a.** A. TgHMGB1a is predicted to lack a signal peptide sequence. B. The four hydrophobic transmembrane domains at the N-terminus are shown. C. Prediction of post-translational modifications of TgHMGB1a. Sites for protein kinase C phosphorylation (14–16 aa), N-glycosylation (83–86 aa), N-myristoylation (100–105 aa), amidation (141–144 aa), casein kinase II phosphorylation (185–188 aa) and tyrosine kinase phosphorylation (282–290 and 289–297 aa) sites were where indicated. Also, proprotein convertase sites were found at the C-terminal.

**Figure S4 TgHMGB1a constantly express in the intracellular parasites but dramatically decreased in extracellular parasites.** The intracellular parasites and freshly released parasites were collected to lysed using RIPA (strong) buffer added cocktail proteinase inhibitor for 30 min on ice, and then analyzed by western blot with anti-TgHMGB1a 4E sera. Meanwhile, freshly released tachyzoites (extracellular 0 h) were incubated additional 0.5, 2 and 6 h in 10% FBS DMEM at 37°C, then centrifuged (800 g, 8 min) and collected the pellets to analyze together with the intracellular parasites. TgActin abs was used to normalize the quantity of per samples.

**Figure S5 Localization of TgHMGB1a in the intracellular stage.** Top panels: TgHMGB1a always concentrated in the parasites nuclear during the course of endodyogeny. Anti-MIC3 antibodies were used as the control (indicated cytoplasm region). Scale bar, 3 μm. Bottom panels: TgHMGB1a is concentrated in the nucleus and less dispersed in the cytoplasm. Anti-SAG1 antibodies were used to label the cytoplasm membrane. Scale bar, 3 μm.

**Figure S6  Localization of TgHMGB1a in the extracellular stage.** Top panels: TgHMGB1a translocates into cytoplasm when the parasites released to extracellular. Anti-MIC3 antibodies were used as the control. Scale bar, 5 μm. Bottom panels: TgHMGB1a translocates into cytoplasm but not enriched in nucleus when the parasites released to extracellular. Anti-SAG1 antibodies were used to label the cytoplasma membrane. Top panel scale bar is 3 μm and bottom panel scale bar is 3 μm.

**Figure S7  EMSAs with cruciform DNA (4H DNA) and recombinant TgHMGB1a 4E protein or recombinant TgGRA1 protein. Increasing concentrations of TgHMGB1a 4E protein (0.1–6 mg/mL) were incubated with 5 nmol of biotinlabeled 4H DNA for 30 min at 25°C, 2 mg/mL TgGRA1 was used as control.** DNA-protein complexes formation were resolved on 6% non-denaturing polyacrylamide gel and transferred to nylon membrane and then detected by Chemiluminescent EMSA Kit. Under identical conditions, only the TgHMGB1a 4E protein can binds cruciform DNA, but not an electrophoretic mobility shift of the 4H probe with TgGRA1 even though in a high concentration.

**Figure S8  Phenotypic analysis of the TgHMGB1a B box-/eGFP mutant strain.** A. Monolayers of HFF were infected with RHΔKU80 or TgHMGB1a B box-/eGFP parasites. Plaques were allowed to form undisturbed and, at 7 days post inoculation, the monolayer was fixed and stained (see Materials and methods). B. An intracellular growth rate assay was carried out after 24 hr of parasite growth. IFA was performed with TgSAG1 abs and the numbers of parasites per vacuole were counted. Each bar indicates the mean ± SD from three independent experiments; n≥150 vacuoles.

**Figure S9  TgHMGB1a overexpression in the RH strain.** A. Schematic of the plasmid transfected into the RH strain for overexpression of TgHMGB1a. TgHMGB1a was fused to GFP, which was under the control of the GRA1 promoter. B. TgHMGB1a overexpression was confirmed by western blotting. As expected, transfected clones showed two TgHMGB1a bands (i.e., one is endogenous TgHMGB1a and the other is TgHMGB1a-GFP). When anti-GFP antibodies were used, only the transfected parasite antigens showed specific binding. As an internal control, polyclonal antibodies against T. gondii immune mapped protein 1 (TgIMP1) were used. C. TgHMGB1a-B box-

GFP was not concentrated in the nucleus. The plasmid of pDMG TgHMGB1a-B box were transfected into RH strain, and the monoclone were obtained by FCM sorting. B box deficient TgHMGB1a has not localized in nuclear of parasites as showed by the fused GFP, and it likely dispersed in cytoplasm. D. To analyze whether the phenotypes changes were specific to the HMG domain of TgHMGB1a in the overexpress strains, pDMG TgHMGB1a-B box, pDMG TgHMGB1a transfected clones and their parental strains (RH and RH-GFP) were examined by plaque assays. The results showed that only plaques of pDMG TgHMGB1a transfected clones were smaller than their parental strains, demonstrating the HMG domain is involved to the phenotype changes. These pictures were representative of 3 experiments with similar outcomes. Scale bar, 0.2 mm.

**Table S1  Sequences were used to phylogenetic analysis.**

**Table S2  Primers used for clone and prokaryotic expression.**

**Table S3  Primers used for TgHMGB1a overexpress and B box mutation.**

**Table S4  Primers used for qRT-PCR.**

**Table S5  Primers of target gene promoters used for ChIP-qPCR.**

**Table S6  Repertoire of high mobility group box proteins in three genotypes *T.gondii*.**

**Data S1  Completed ARRIVE checklist for animal infection experiments.**

## Acknowledgments

We are grateful to Professor Silvia Moreno (University of Georgia, USA) for kindly providing the pTCY vector and RHΔKU80 parasites. We thank Professor Xuenan Xuan (Obihiro University of Agriculture and Veterinary Medicine, Japan) for kindly providing the pDMG plasmid and the RH-GFP strain. We thank Dr Xiangmei Zhou (China Agricultural University) for kindly providing the ana-1 cells.

## Author Contributions

Conceived and designed the experiments: HW QL JL. Performed the experiments: HW TL MZL HZN. Analyzed the data: HW JL QL. Contributed reagents/materials/analysis tools: QL JL. Wrote the paper: HW QL.

## References

1. Sessa L, Bianchi ME (2007) The evolution of High Mobility Group Box (HMGB) chromatin proteins in multicellular animals. Gene 387: 133–140.
2. Stros M (2010) HMGB proteins: interactions with DNA and chromatin. Biochim Biophys Acta 1799: 101–113.
3. Wong B, Masse JE, Yen YM, Giannikopoulos P, Feigon J, et al. (2002) Binding to cisplatin-modified DNA by the Saccharomyces cerevisiae HMGB protein Nhp6A. Biochemistry 41: 5404–5414.
4. Zlatanova J, van Holde K (1998) Binding to four-way junction DNA: a common property of architectural proteins? FASEB J 12: 421–431.
5. Stott K, Watson M, Howe FS, Grossmann JG, Thomas JO (2010) Tail-mediated collapse of HMGB is dynamic and occurs via differential binding of the acidic tail to the A and B domains. J Mol Biol 403: 706–722.
6. Osmanov T, Ugrinova I, Pasheva E (2013) The chaperone like function of the nonhistone protein HMGB1. Biochem Biophys Res Commun 432: 231–235.
7. Travers AA (2003) Priming the nucleosome: a role for HMGB proteins? EMBO Rep 4: 131–136.
8. Zhang Q, Wang Y (2010) HMG modifications and nuclear function. Biochim Biophys Acta 1799: 28–36.
9. Scaffidi P, Misteli T, Bianchi ME (2002) Release of chromatin protein HMGB1 by necrotic cells triggers inflammation. Nature 418: 191–195.
10. Andersson U, Erlandsson-Harris H, Yang H, Tracey KJ (2002) HMGB1 as a DNA-binding cytokine. J Leukoc Biol 72: 1084–1091.
11. Ulloa L, Messmer D (2006) High-mobility group box 1 (HMGB1) protein: friend and foe. Cytokine Growth Factor Rev 17: 189–201.

12. Ferrari S, Ronfani L, Calogero S, Bianchi ME (1994) The mouse gene coding for high mobility group 1 protein (HMG1). J Biol Chem 269: 28803–28808.

13. Thirugnanam S, Munirathinam G, Veerapathran A, Dakshinamoorthy G, Reddy MV, et al. (2012) Cloning and characterization of high mobility group box protein 1 (HMGB1) of Wuchereria bancrofti and Brugia malayi. Parasitol Res 111: 619–627.

14. Cribb P, Perozzi M, Villanova GV, Trochine A, Serra E (2011) Characterization of TcHMGB, a high mobility group B family member protein from Trypanosoma cruzi. Int J Parasitol 41: 1149–1156.

15. Morales M, Onate E, Imschenetzky M, Galanti N (1992) HMG-like chromosomal proteins in Trypanosoma cruzi. J Cell Biochem 50: 279–284.

16. Gnanasekar M, Velusamy R, He YX, Ramaswamy K (2006) Cloning and characterization of a high mobility group box 1 (HMGB1) homologue protein from Schistosoma mansoni. Mol Biochem Parasitol 145: 137–146.

17. de Oliveira FM, de Abreu DSI, Rumjanek FD, Dias-Neto E, Guimaraes PE, et al. (2006) Cloning the genes and DNA binding properties of High Mobility Group B1 (HMGB1) proteins from the human blood flukes Schistosoma mansoni and Schistosoma japonicum. Gene 377: 33–45.

18. Kun JF, Anders RF (1995) A Plasmodium falciparum gene encoding a high mobility group protein box. Mol Biochem Parasitol 71: 249–253.

19. Briquet S, Boschet C, Gissot M, Tissandie E, Sevilla E, et al. (2006) High-mobility-group box nuclear factors of Plasmodium falciparum. Eukaryot Cell 5: 672–682.

20. Kumar K, Singal A, Rizvi MM, Chauhan VS (2008) High mobility group box (HMGB) proteins of Plasmodium falciparum: DNA binding proteins with pro-inflammatory activity. Parasitol Int 57: 150–157.

21. Abhyankar MM, Hochreiter AE, Hershey J, Evans C, Zhang Y, et al. (2008) Characterization of an Entamoeba histolytica high-mobility-group box protein induced during intestinal infection. Eukaryot Cell 7: 1565–1572.

22. Dalrymple BP, Peters JM (1992) Characterization of a cDNA clone from the haemoparasite Babesia bovis encoding a protein containing an "HMG-Box". Biochem Biophys Res Commun 184: 31–35.

23. Morisaki JH, Heuser JE, Sibley LD (1995) Invasion of Toxoplasma gondii occurs by active penetration of the host cell. J Cell Sci 108 (Pt 6): 2457–2464.

24. Appleford PJ, Smith JE (1997) Toxoplasma gondii: the growth characteristics of three virulent strains. Acta Trop 65: 97–104.

25. Sibley LD, Ajioka JW (2008) Population structure of Toxoplasma gondii: clonal expansion driven by infrequent recombination and selective sweeps. Annu Rev Microbiol 62: 329–351.

26. Howe DK, Sibley LD (1995) Toxoplasma gondii comprises three clonal lineages: correlation of parasite genotype with human disease. J Infect Dis 172: 1561–1566.

27. Melo MB, Jensen KD, Saeij JP (2011) Toxoplasma gondii effectors are master regulators of the inflammatory response. Trends Parasitol 27: 487–495.

28. Gaji RY, Behnke MS, Lehmann MM, White MW, Carruthers VB (2011) Cell cycle-dependent, intercellular transmission of Toxoplasma gondii is accompanied by marked changes in parasite gene expression. Mol Microbiol 79: 192–204.

29. Dubey JP (1998) Advances in the life cycle of Toxoplasma gondii. Int J Parasitol 28: 1019–1024.

30. Dzierszinski F, Nishi M, Ouko L, Roos DS (2004) Dynamics of Toxoplasma gondii differentiation. Eukaryot Cell 3: 992–1003.

31. Singh U, Brewer JL, Boothroyd JC (2002) Genetic analysis of tachyzoite to bradyzoite differentiation mutants in Toxoplasma gondii reveals a hierarchy of gene induction. Mol Microbiol 44: 721–733.

32. Dixon SE, Stilger KL, Elias EV, Naguleswaran A, Sullivan WJ (2010) A decade of epigenetic research in Toxoplasma gondii. Mol Biochem Parasitol 173: 1–9.

33. Gaskell EA, Smith JE, Pinney JW, Westhead DR, McConkey GA (2009) A unique dual activity amino acid hydroxylase in Toxoplasma gondii. PLoS One 4: e4801.

34. Saitou N, Nei M (1987) The neighbor-joining method: a new method for reconstructing phylogenetic trees. Mol Biol Evol 4: 406–425.

35. Guindon S, Gascuel O (2003) A simple, fast, and accurate algorithm to estimate large phylogenies by maximum likelihood. Syst Biol 52: 696–704.

36. Tamura K, Peterson D, Peterson N, Stecher G, Nei M, et al. (2011) MEGA5: molecular evolutionary genetics analysis using maximum likelihood, evolutionary distance, and maximum parsimony methods. Mol Biol Evol 28: 2731–2739.

37. El HH, Papoin J, Cerede O, Garcia-Reguet N, Soete M, et al. (2008) Molecular signals in the trafficking of Toxoplasma gondii protein MIC3 to the micronemes. Eukaryot Cell 7: 1019–1028.

38. Bianchi ME, Beltrame M, Paonessa G (1989) Specific recognition of cruciform DNA by nuclear protein HMG1. Science 243: 1056–1059.

39. Koller BH, Smithies O (1992) Altering genes in animals by gene targeting. Annu Rev Immunol 10: 705–730.

40. Sibley LD, Messina M, Niesman IR (1994) Stable DNA transformation in the obligate intracellular parasite Toxoplasma gondii by complementation of tryptophan auxotrophy. Proc Natl Acad Sci U S A 91: 5508–5512.

41. Donald RG, Roos DS (1993) Stable molecular transformation of Toxoplasma gondii: a selectable dihydrofolate reductase-thymidylate synthase marker based on drug-resistance mutations in malaria. Proc Natl Acad Sci U S A 90: 11703–11707.

42. Roos DS, Donald RG, Morrissette NS, Moulton AL (1994) Molecular tools for genetic dissection of the protozoan parasite Toxoplasma gondii. Methods Cell Biol 45: 27–63.

43. Knapp S, Muller S, Digilio G, Bonaldi T, Bianchi ME, et al. (2004) The long acidic tail of high mobility group box 1 (HMGB1) protein forms an extended and flexible structure that interacts with specific residues within and between the HMG boxes. Biochemistry 43: 11992–11997.

44. Taylor S, Barragan A, Su C, Fux B, Fentress SJ, et al. (2006) A secreted serine-threonine kinase determines virulence in the eukaryotic pathogen Toxoplasma gondii. Science 314: 1776–1780.

45. Radke JR, Striepen B, Guerini MN, Jerome ME, Roos DS, et al. (2001) Defining the cell cycle for the tachyzoite stage of Toxoplasma gondii. Mol Biochem Parasitol 115: 165–175.

46. Plattner F, Yarovinsky F, Romero S, Didry D, Carlier MF, et al. (2008) Toxoplasma profilin is essential for host cell invasion and TLR11-dependent induction of an interleukin-12 response. Cell Host Microbe 3: 77–87.

47. Gubbels MJ, White M, Szatanek T (2008) The cell cycle and Toxoplasma gondii cell division: tightly knit or loosely stitched? Int J Parasitol 38: 1343–1358.

48. Varga-Weisz P, van Holde K, Zlatanova J (1994) Competition between linker histones and HMG1 for binding to four-way junction DNA: implications for transcription. Biochem Biophys Res Commun 203: 1904–1911.

49. Zlatanova J, van Holde K (1998) Linker histones versus HMG1/2: a struggle for dominance? Bioessays 20: 584–588.

50. Nardelli SC, Che FY, Silmon DMN, Xiao H, Nieves E, et al. (2013) The histone code of Toxoplasma gondii comprises conserved and unique posttranslational modifications. MBio 4: e913–e922.

51. Gubbels MJ, Lehmann M, Muthalagi M, Jerome ME, Brooks CF, et al. (2008) Forward genetic analysis of the apicomplexan cell division cycle in Toxoplasma gondii. PLoS Pathog 4: e36.

52. Assenberg R, Webb M, Connolly E, Stott K, Watson M, et al. (2008) A critical role in structure-specific DNA binding for the acetylatable lysine residues in HMGB1. Biochem J 411: 553–561.

53. Carneiro VC, de Moraes MR, de Abreu DSI, Da CR, Paiva CN, et al. (2009) The extracellular release of Schistosoma mansoni HMGB1 nuclear protein is mediated by acetylation. Biochem Biophys Res Commun 390: 1245–1249.

54. Cui X, Lei T, Yang DY, Hao P, Liu Q (2012) Identification and characterization of a novel Neospora caninum immune mapped protein 1. Parasitology 139: 998–1004.

55. Huynh MH, Carruthers VB (2009) Tagging of endogenous genes in a Toxoplasma gondii strain lacking Ku80. Eukaryot Cell 8: 530–539.

# 13

# Influence of *Trichobilharzia regenti* (Digenea: Schistosomatidae) on the Defence Activity of *Radix lagotis* (Lymnaeidae) Haemocytes

**Vladimír Skála[1]\*, Alena Černíková[2], Zuzana Jindrová[1,3], Martin Kašný[1,4], Martin Vostrý[1,5], Anthony J. Walker[6], Petr Horák[1]**

1 Charles University in Prague, Faculty of Science, Department of Parasitology, Prague, Czech Republic, 2 Charles University in Prague, Faculty of Science, Institute of Applied Mathematics and Information Technologies, Prague, Czech Republic, 3 Charles University in Prague, 1st Faculty of Medicine, Institute of Immunology and Microbiology, Prague, Czech Republic, 4 Masaryk University, Faculty of Science, Department of Botany and Zoology, Brno, Czech Republic, 5 Institute of Haematology and Blood Transfusion, Prague, Czech Republic, 6 Molecular Parasitology Laboratory, School of Life Sciences, Kingston University, Kingston upon Thames, Surrey, United Kingdom

## Abstract

*Radix lagotis* is an intermediate snail host of the nasal bird schistosome *Trichobilharzia regenti*. Changes in defence responses in infected snails that might be related to host-parasite compatibility are not known. This study therefore aimed to characterize *R. lagotis* haemocyte defence mechanisms and determine the extent to which they are modulated by *T. regenti*. Histological observations of *R. lagotis* infected with *T. regenti* revealed that early phases of infection were accompanied by haemocyte accumulation around the developing larvae 2–36 h post exposure (p.e.) to the parasite. At later time points, 44–92 h p.e., no haemocytes were observed around *T. regenti*. Additionally, microtubular aggregates likely corresponding to phagocytosed ciliary plates of *T. regenti* miracidia were observed within haemocytes by use of transmission electron microscopy. When the infection was in the patent phase, haemocyte phagocytic activity and hydrogen peroxide production were significantly reduced in infected *R. lagotis* when compared to uninfected counterparts, whereas haemocyte abundance increased in infected snails. At a molecular level, protein kinase C (PKC) and extracellular-signal regulated kinase (ERK) were found to play an important role in regulating these defence reactions in *R. lagotis*. Moreover, haemocytes from snails with patent infection displayed lower PKC and ERK activity in cell adhesion assays when compared to those from uninfected snails, which may therefore be related to the reduced defence activities of these cells. These data provide the first integrated insight into the immunobiology of *R. lagotis* and demonstrate modulation of haemocyte-mediated responses in patent *T. regenti* infected snails. Given that immunomodulation occurs during patency, interference of snail-host defence by *T. regenti* might be important for the sustained production and/or release of infective cercariae.

**Editor:** Daniel Doucet, Natural Resources Canada, Canada

**Funding:** The study was financially supported by the Charles University in Prague (Grant GAUK no. 435911, research programmes PRVOUK - P41/PrF, UNCE - 204017 and SVV 260074/2014). The funders had no role in study design, data collection and analysis, decision to publish, or preparation of the manuscript.

**Competing Interests:** The authors have declared that no competing interests exist.

\* Email: mich007@email.cz

## Introduction

Aquatic snails serve as intermediate hosts of many trematodes, including those important in veterinary and human medicine. Compatibility between such parasites and the host snail is partially governed by innate immunological processes that comprise cellular and humoral components. Mobile phagocytic cells called haemocytes play the major role in mediating the cellular defence response whereas lectins are considered as the most essential recognition molecules of humoral response [1], [2]. Haemocyte-mediated defence responses that are important for eliminating foreign invaders such as parasites include phagocytosis, encapsulation, and production of reactive oxygen species (ROS) [1], [3], [4].

Phagocytosis is used to eliminate small non-self particles, primarily bacteria; however, pieces of trematode tegument are also known to be actively engulfed by haemocytes after encapsulation [3]. The phagocytic response also triggers generation of ROS [5], [6]. Among the ROS, hydrogen peroxide ($H_2O_2$) is an important metabolite known for killing sporocysts of the human parasite *Schistosoma mansoni* [4]. At the molecular level, snail haemocyte defence responses are regulated by complex networks of intracellular signalling pathways, including the evolutionarily conserved protein kinase C (PKC) and mitogen-activated protein kinase (MAPK) pathways [7–10]. Activation of PKC, p38 MAPK and/or extracellular signal-regulated kinase (ERK) is required for efficient phagocytosis and $H_2O_2$ production by snail haemocytes; other kinases such as phosphatidylinositol 3-kinase also play a crucial role in these processes [7], [9–12].

During infection, compatible trematodes alter snail host defence responses presumably to help ensure survival and replication of the parasite. Phagocytic activity of haemocytes is decreased e.g. in the gastropods *Biomphalaria glabrata* and *Lymnaea stagnalis* infected with *Echinostoma paraensei* [13] and *Trichobilharzia szidati* [14], respectively. In the prosobranch snail, *Littorina littorea*, infection with *Himasthla elongata* reduces haemocyte ROS production, which correlates with increased haemocyte number in the snail circulation [15]. Such alterations of host defence mechanisms might be caused by trematode-derived components interfering with signalling pathways of snail haemocytes [16]. This hypothesis is supported by results showing that *S. mansoni* excretory-secretory products (ESPs) generated during development of miracidia to mother sporocysts impair $H_2O_2$ production in *B. glabrata* haemocytes [10] and disrupt ERK signalling in these cells [17].

*Radix lagotis* is an important intermediate host of the nasal bird schistosome *Trichobilharzia regenti* [18], [19], a causative agent of cercarial dermatitis in humans [20]. Following penetration into the snail, *T. regenti* miracidia develop to mother sporocysts, which in turn produce daughter sporocysts [21]. This latter stage gives rise to cercariae that are released into the water during the patent phase of infection. As far as immunological aspects of infection are concerned, snail defence responses related to the initiation of *T. regenti* infection, and changes in *R. lagotis* haemocyte activities in the patent phase of infection are unknown.

The present paper combines histological observations of juvenile *R. lagotis* snails infected with *T. regenti* miracidia, with comparisons of haemocyte abundance and haemocyte phagocytic activity and $H_2O_2$ production between uninfected and infected snails in the patent phase of *T. regenti* infection. At the molecular level, basal PKC and ERK phosphorylation in haemocytes from both snail groups was compared and their possible roles in regulation of haemocyte phagocytic activity and $H_2O_2$ production explored. Such complementary approaches provide the first and integrated insight into the immunobiology of *R. lagotis* snails demonstrating modulation of defence responses during infection of snails with the compatible trematode parasite.

## Methods

### Uninfected and *T. regenti*-infected *R. lagotis*

Uninfected *R. lagotis* were maintained in the laboratory at ambient room temperature (19–22°C; RT) in aquaria filled with aerated tap water and were fed fresh lettuce *ad libitum*. Juvenile and adult snails (together with the eggs laid) were reared together.

Juvenile snails with shell heights 5–8 mm were infected with *T. regenti* miracidia obtained as described by Horák *et al.* (1998) [18]. The snails were placed individually into wells of a 24-well culture plate (Nunc) containing tap water and each exposed to 3–8 miracidia for 5 h, with 15 miracidia used to infect each snail for histological analysis. After exposure, the snails were placed in a separate aquarium for 5 weeks, and they were then checked under a direct light source for shedding of *T. regenti* cercariae. Snails releasing cercariae (infected snails) were then maintained in a further separate aquarium.

### Light microscopy

Two juvenile *R. lagotis* were dissected for each infection time point studied, 1, 2, 3, 5, 12, 16, 20, 36, 44, 60 and 92 h post–exposure (p.e.) of snails to *T. regenti* miracidia. The soft body of each snail was carefully removed from its shell and fixed in Bouin-Hollande fixative at RT for 24 h. The specimens were then embedded in JB-4 resin (Polysciences), sections cut to 2 μm with a

microtome (Finesse ME, Shandon Scientific) and stained with Wright-Giemsa (Polysciences). Finally, sections were individually embedded in DPX medium (Sigma), examined under an Olympus BX 51 light microscope and digital images captured using a DP70 digital camera system.

### Transmission electron microscopy

For transmission electron microscopy (TEM), juvenile *R. lagotis* were dissected 5 and 15 h p.e. and fixed in 2.5% glutaraldehyde (Sigma) in complete sterile snail saline (SSS+: 3 mM Hepes, 3.7 mM NaOH, 36 mM NaCl, 2 mM KCl, 2 mM $MgCl_2$, 4 mM $CaCl_2$, pH 7.8, 100 mOsm; [5]) at 4°C for 24 h. The specimens were then post–fixed in 1% $OsO_4$ (Polysciences) in SSS+ for 2 h, washed three times in SSS+, dehydrated in ethanol (50%, 80%, 96%, twice each for 15 min, and 100% three times each for 5 min) and acetone (100%, three times each for 5 min). Subsequently, the tissue was incubated in 100% acetone:Spurr mixture at increasing Spurr concentrations: 2:1 for 2 h, 1:1 for 5 h, 1:2 for 12 h, followed by pure Spurr resin three times each for 12 h. Then, the material in fresh Spurr resin was transferred to plastic capsules and incubated at 60°C for 48 h. The embedded samples were first sectioned at 2 μm thick sections with a Finesse ME microtome, stained with 1% toluidine blue (Polysciences) and observed under a light microscope (Olympus BX 51). When larvae of *T. regenti* were detected, 60–70 nm thick sections were prepared using ultramicrotome Ultracut E (Reichert-Jung). These sections were stained with uranyl acetate and lead citrate [22] and evaluated under TEM JEOL 1011 microscope. Digital images were captured using associated software.

### Haemolymph extraction and enumeration of haemocytes in uninfected and infected *R. lagotis*

Uninfected and infected *R. lagotis* with shell heights 1.0–1.6 cm were selected for haemolymph extraction with infected snails extracted no later than 2 months post-patency. The snails were washed with distilled water, dried, and haemolymph was extracted by head-foot retraction [23].

Haemocyte numbers were quantified for individual uninfected and infected snails. Haemolymph from each snail was pooled on parafilm (Sigma) and diluted 1:1, 2:1, or 3:1 (one part = 10 μl) in incomplete sterile snail saline (SSS-) where 2 mM $MgCl_2$ and 4 mM $CaCl_2$ were omitted, and 2% ethylenediaminetetra-acetic acid (EDTA; Sigma) added (SSS-/EDTA) to reduce haemocyte aggregation/adhesion; SSS-/EDTA buffer was exclusively used for counting haemocytes. Enumeration was carried out with Bürker haemocytometers and haemocyte numbers were expressed as haemocytes/ml of haemolymph. The data were analysed for normality (Shapiro-Wilk normality test) using R 2.13.0 statistical software (www.r-project.org). Spearman's correlation test was used to assess the relationship between shell heights and haemocyte numbers of individual snails. Haemocyte numbers between the snail groups were compared using Wilcoxon signed-rank test (non-parametric two-sample test; Wilcoxon test).

### Preparation of haemocyte monolayers

Haemolymph from uninfected and infected snails (shell heights 1.3–1.6 cm) was extracted in alternating order to ensure similar conditions for both haemolymph types while the monolayers were prepared. Aliquots of haemolymph drawn from the snails were pipetted directly into the wells of a 96-well culture plate (Nunc) containing 50 μl SSS+ to achieve a final volume of 250 μl/well (final ratio: 4 parts haemolymph: 1 part SSS+). Ten to forty snails were required to obtain sufficient haemolymph for each monolayer. Haemocytes were left to settle and adhere to the bottom of the wells for 30 min at RT. Monolayers were then washed with SSS+ (see below) and their quality checked under a microscope (Olympus IX 71). Any wells containing haemocyte clumps or discontinuous monolayers were not used. When haemocyte numbers per well were enumerated, aliquots of haemolymph were also collected on parafilm and diluted with equal amount of SSS-/EDTA; haemocytes were then enumerated as described above.

## Phagocytosis assays

Haemocyte monolayers were washed three times with 250 µl SSS+ and equilibrated in 190 µl SSS+ for 30 min at RT. 10 µl of *Escherichia coli* bioparticles (pHrodo red; Molecular Probes) prepared following manufacturer's instructions were then added to each well and plates incubated at RT in the dark for 2 h. These bioparticles are non-fluorescent outside cells, but become fluorescent in phagosomes. Therefore, no washing was necessary after incubation and intracellular fluorescence was immediately quantified using Tecan Infinite M200 microplate reader at 545 nm excitation and 600 nm emission. The signal of *E. coli* bioparticles alone in wells was also measured in each assay and the value subtracted from all values obtained from wells containing haemocytes and *E. coli* bioparticles.

Phagocytic activity of haemocytes from uninfected and infected snails was then expressed per volume of haemolymph (200 µl) and per 50,000 haemocytes, in case infection altered haemocyte number. Uninfected snails were also used to study the effects of inhibition of PKC and ERK signalling on phagocytic activity. Haemocyte monolayers were pre-incubated for 30 min at RT with 1 µM or 10 µM inhibitor of PKC (GF109203X; Sigma), MEK (U0126; Cell Signalling Technology - CST), which is the immediate upstream activator of ERK, or in DMSO vehicle alone (0.05%; Sigma) prior to adding bioparticles. Effects of inhibition assays were evaluated in terms of haemolymph volume (200 µl).

Using R 2.13.0 statistical software, raw fluorescence intensity data for each measurement were analysed for normality (Shapiro-Wilk normality test). Wilcoxon test was then used to compare the phagocytic activity between uninfected and infected snails, whereas paired t-test was applied to data when assessing the effect of GF109203X and U0126 on phagocytosis by *R. lagotis* haemocytes. For graphic representation, the data for uninfected snails were assigned a value of 100%.

## Hydrogen peroxide assays

Haemocyte monolayers were prepared and haemocyte numbers/well enumerated as described above except that 50 µl haemolymph and 12.5 µl SSS+ were used per well. After washing monolayers twice with 250 µl SSS+, haemocytes were left to equilibrate for 30 min at RT in 100 µl SSS+. $H_2O_2$ output by haemocytes was monitored using the Amplex red hydrogen peroxide/peroxidase assay kit (Molecular Probes) in which Amplex red reacts with $H_2O_2$ to produce the red-fluorescent product, resorufin. Working solutions of the assay mixture that were prepared in SSS+ contained: 0.1 U ml$^{-1}$ horseradish peroxidase (HRP), 50 µM Amplex red reagent, and either 0.1% DMSO or 10 µM PMA (phorbol 12-myristate 13-acetate; Sigma) in DMSO. PMA was used because in other molluscs this phorbol ester increases ROS production by haemocytes [10], [24], [25]. 100 µl of the respective working solution was added to each individual haemocyte monolayer and the plate was incubated in the dark for 30 min at RT. For inhibition assays using uninfected snails, haemocytes were exposed to 5 µM GF109203X, U0126 or

DMSO (vehicle) alone (0.025%) for 30 min at RT prior to adding the working solution containing PMA. The final concentration of DMSO after adding the working solutions was 0.1% in all cases.

Fluorescence was monitored at 520 nm and 615 nm excitation and emission, respectively, in a microplate reader (Tecan Infinite M200) for 60 min. $H_2O_2$ output by uninfected and infected snail haemocytes was evaluated per volume of haemolymph (50 µl) and haemocyte number with adjustment to 50,000 cells. Inhibition assays were evaluated per volume of haemolymph (200 µl).

The data sets were tested for normality (Shapiro-Wilk normality test) and for equality of variances (Two-variances F-test). Two-sample t-test or Wilcoxon test was used to compare basal and PMA-modulated $H_2O_2$ production between uninfected and infected snails. Experiments investigating the effects of PKC and ERK inhibition on $H_2O_2$ production were analysed using either parametric or nonparametric paired tests. Since the tests at different time points are dependent, a Fisher's combination test using inverse normal method [26] was used for further processing of p-values. The resulting test statistic was compared to Pocock's critical value 2.49. If the test statistic was higher than this critical value, a significant difference between data sets was confirmed.

## SDS-PAGE and western blot analysis

Haemocyte monolayers prepared as detailed above were washed three times with 250 µl SSS+, and left to equilibrate in 250 µl SSS+ at RT for 30 min. The SSS+ was then removed and haemocytes lysed by adding 25 µl of hot (95°C) SDS-PAGE sample buffer. Proteins were separated by gel electrophoresis (10% Mini-Protean TGX precast gel; Bio-Rad) and transffered to Immun-Blot PVDF membrane (Bio-Rad) using Trans-Blot turbo blotting system (Bio-Rad). Membranes were blocked with 5% non-fat dried milk (Bio-Rad) in 0.1% Tween/Tris-buffered saline (TTBS) at RT for 45 min, and incubated overnight at 4°C in either anti-phospho-PKC (pan) (βII Ser660) rabbit polyclonal antibodies or anti-phospho-p44/42 MAPK (Erk1/2) (Thr202/Tyr204) (197G2) rabbit monoclonal antibodies (CST) (1:1000 in TTBS). These antibodies were previously validated for detection of exclusively phosphorylated (activated) forms of PKC and ERK in *L. stagnalis* haemocytes [7], [27], and were also used in other studies of molluscs [17]. Following further incubation at RT for 2 h, membranes were washed 3×5 min in TTBS and incubated for 2 h at RT in anti-rabbit IgG HRP-conjugated secondary antibodies (1:4000 in TTBS) (CST). Immunoreactive bands were then visualised using SuperSignal West Dura extended duration substrate (Thermo Scientific) and a LAS 4000 Luminescent image analyser. Blots were stripped in Restore Western blot stripping buffer (Thermo Scientific) for 2 h at RT, and re-probed overnight in p44/p42 MAPK (Erk1/2) antibody (CST) (1:1000 in TTBS), which recognizes ERK regardless of its phosphorylation state. Finally, the blots were stripped and re-probed with anti-actin antibodies (Sigma) (1:4000 in TTBS) for 1 h at RT to confirm equal loading of proteins between lanes.

The intensities of immunoreactive bands were analysed using Multi Gauge 3.2. software. The values for PKC and ERK phosphorylation and for total ERK in haemocytes of uninfected snails were standardised as 100% and differences in PKC and ERK phosphorylation and in total ERK from infected snails calculated. The data were evaluated for normality (Shapiro-Wilk normality test) and for equality of variances (Two-variances F-test). Two-sample t-test was then applied using R 2.13.0 statistical software.

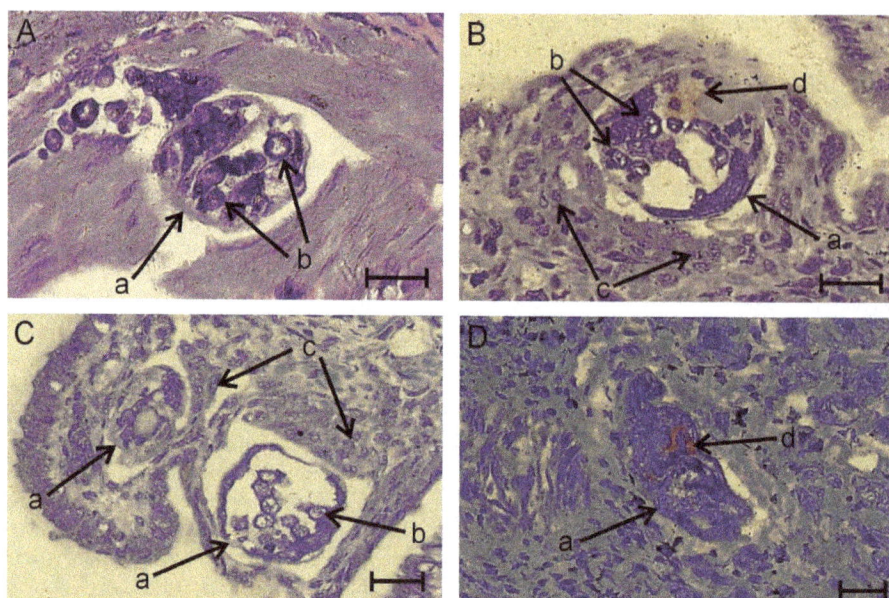

**Figure 1.** *Trichobilharzia regenti* **larvae within the tissue of** *Radix lagotis* **revealed by light microscopy between 1–92 h p.e.; Wright-Giemsa stained sections.** (A) Miracidium of *T. regenti* (a) containing germ cells (b) occurs within the snail tissue without haemocyte infiltration 1 h p.e. (B) and (C) Haemocytes (c) are present in the vicinity of developing *T. regenti* mother sporocyst (a) 2 and 16 h p.e., respectively; germ cells (b) and gland structure (d) of the parasite are visible. (D) The area around *T. regenti* mother sporocyst (a) contains no haemocytes 92 h p.e. Gland structure (d) is located in the body of the parasite. Scale bar = 20 μm. The images shown are representative of the situation seen in all sections observed during these experiments.

## Results

### Histological observations of *R. lagotis* experimentally infected with *T. regenti*

Histological observations of *R. lagotis* experimentally infected with *T. regenti* provided insights into the encapsulation responses within the snail tissue between 1 and 92 h p.e. Haemocytes were not evident in close proximity to the parasite at 1 h p.e. (Figure 1A). However, considerable accumulation of haemocytes was observed close to the developing *T. regenti* between 2 and 16 h p.e. (Figure 1B–C). Haemocytes appeared to surround the developing mother sporocysts irregularly in several layers; however, it was not clear whether the cells were directly attached to the parasite surface. Thereafter, at 20 and 36 h p.e. the haemocytic response against the parasite appeared to decline (data not shown) and while the haemocytes occurred individually in the vicinity of mother sporocysts, they did not accumulate in layers. At the latter time points, 44, 60 and 92 h p.e. no haemocytes were observed close to *T. regenti* (Figure 1D).

Transmission electron microscopy of *T. regenti* mother sporocysts within the snail tissue at 5 and 15 h p.e. (Figure 2; 15 h p.e. shown) showed that the larvae remained apparently undamaged despite numerous haemocytes being adjacent to the parasite (Figure 2A). Furthermore, some haemocytes were in a tight contact with sporocyst surface microvilli, and microtubular aggregates were observed within their phagosomes (Figure 2A–B).

### Haemocyte number in uninfected and *T. regenti*-infected *R. lagotis*

Evaluation of haemocyte number/ml haemolymph in 23 individuals of uninfected and infected *R. lagotis* demonstrated that the concentration of circulating haemocytes did not correlate with the shell height of the snails (Figure 3). Considerable variation in haemocyte number was observed within the extracted

haemolymph for snails of similar size. In uninfected snails, the lowest haemocyte concentration was $4.2 \times 10^4$ cells/ml (shell height 1.40 cm) whereas the highest was $74.9 \times 10^4$ cells/ml (shell height 1.57 cm) (Figure 3). In infected snails, the lowest haemocyte concentration was $4.7 \times 10^4$ cells/ml (shell height 1.04 cm) whereas the highest was $180.4 \times 10^4$ cells/ml (shell height 1.26 cm) (Figure 3). Statistical analysis revealed that mean haemocyte number/ml haemolymph of infected snails was 79% greater than that of uninfected snails ($45.9 \times 10^4$ cells/ml vs. $25.6 \times 10^4$ cells/ml; $p < 0.05$).

### Defence responses of haemocytes from uninfected and *T. regenti*-infected *R. lagotis*

To explore the effects of *T. regenti* infection on haemocyte defence, we measured phagocytic activity and $H_2O_2$ production by haemocytes derived from uninfected and *T. regenti*-infected *R. lagotis*. Haemocyte phagocytic activity was determined by the ability of these cells to internalise *E. coli* bioparticles (Figure 4A). Comparisons made in a physiological context, which consider activity per volume of haemolymph (200 μl), revealed that phagocytosis by haemocytes from infected snails was not significantly different from that of uninfected snails (Figure 4B). However, when the phagocytic activity was compared taking into account the different numbers of haemocytes in the extracted haemolymph, with more haemocytes present as a result of parasite infection, phagocytosis by infected snail haemocytes was reduced significantly to approximately 50% of that of uninfected snails ($p < 0.05$; Figure 4B).

For $H_2O_2$ production we studied basal and PMA-stimulated output by haemocytes from uninfected and infected snails (Figures 5–6). Evaluation per volume of haemolymph (50 μl) revealed that the basal output of $H_2O_2$ by haemocytes from infected snails was similar to that of uninfected snails, despite the infected snails possessing greater numbers of haemocytes/ml

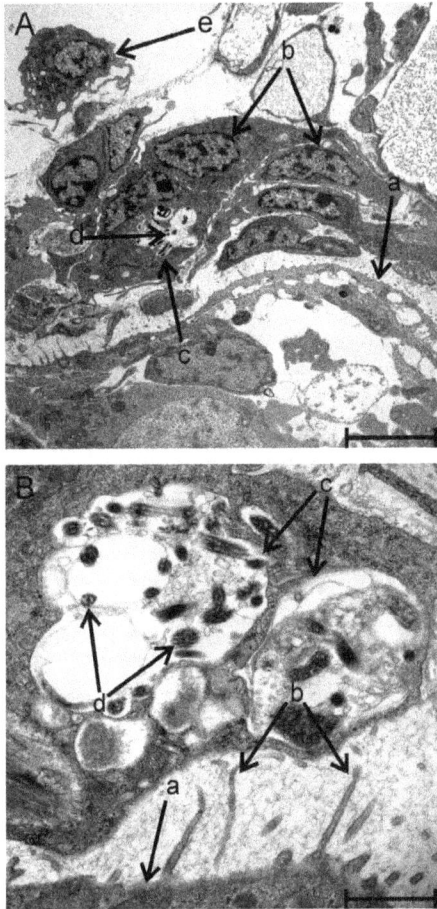

**Figure 2.** *Trichobilharzia regenti* **mother sporocysts within the tissue of** *Radix lagotis* **15 h p.e.; TEM images.** (A) Mother sporocyst of *T. regenti* (a) is surrounded by haemocytes with remarkable nuclei (b). Phagosome (c) of one haemocyte with internalised microtubular aggregates (d) is visible (B in detail). Another haemocyte (e) is located near the parasite. Scale bar = 5 µm. (B) Microvilli (b) are present on the surface of *T. regenti* mother sporocyst (a). Haemocyte adjacent to the sporocyst contains phagosomes (c) with microtubular aggregates (d). Scale bar = 1 µm.

(Figure 5). In contrast, when the data were adjusted for haemocyte number (50,000), the cells from uninfected snails produced significantly more $H_2O_2$ than those from infected snails at each time point after 20 min (p<0.05; Figure 5).

In the presence of 5 µM PMA (an activator of PKC) haemocyte $H_2O_2$ production increased 270% and 240% when considering haemolymph volume (50 µl) in uninfected and infected snails after 60 min, respectively (Figure 6); the difference between snail groups was not statistically significant. In contrast, when considering haemocyte number (50,000) $H_2O_2$ production by haemocytes from uninfected snails in the presence of PMA was approximately 2-fold that of haemocytes from infected snails at all time points studied after 20 min (p<0.01; Figure 6).

## PKC and ERK activation in haemocytes from uninfected and *T. regenti*-infected *R. lagotis*

Because signalling pathways are known to regulate haemocyte defence responses such as phagocytosis and $H_2O_2$ output [7], [9–12], and because these defence responses were supressed in *R. lagotis* haemocytes as a result of *T. regenti* infection, we aimed to determine PKC and ERK activation in haemocyte monolayers derived from uninfected and infected *R. lagotis*. Western blotting of haemocyte proteins with anti-phosphospecific PKC and ERK antibodies, which detect only the active forms of these kinases in snails [7], [8], [27], followed by densitometric analysis of immunoreactive bands from several independent blots revealed that PKC and ERK phosphorylation were reduced by 57% and 55%, respectively, in haemocytes from infected snails when compared to those from uninfected snails (p<0.01; Figure 7A–B). We reasoned, therefore that ERK expression might also be suppressed. However, western blots performed to determine the quantity of ERK in haemocytes using antibodies that detect ERK irrespective of its phosphorylation state (Figure 7C) demonstrated that mean levels of ERK were 24% higher in infected snails when compared to uninfected ones, although this difference was not statistically significant. Unfortunately, lack of a suitable anti-PKC antibody for snails prevented evaluation of total PKC protein levels.

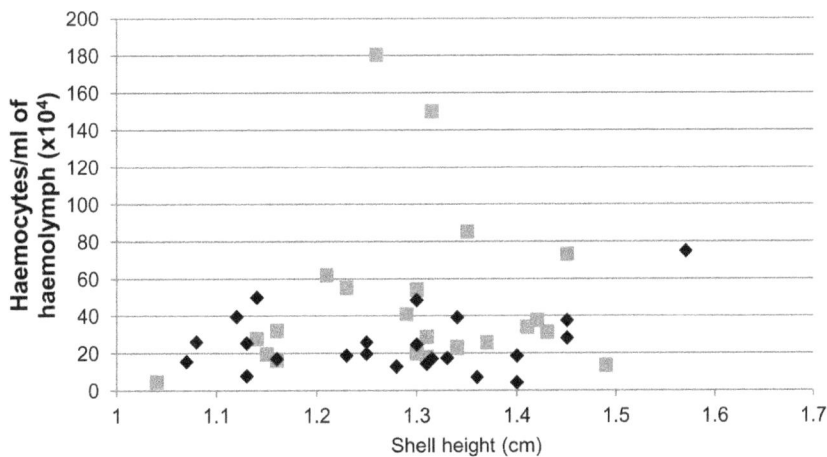

**Figure 3. Number of haemocytes/ml of haemolymph of individual uninfected (black diamond) and** *Trichobilharzia regenti* **infected (grey box)** *Radix lagotis.* The numbers of haemocytes/ml from individual snails with different shell heights were enumerated using a Bürker haemocytometer.

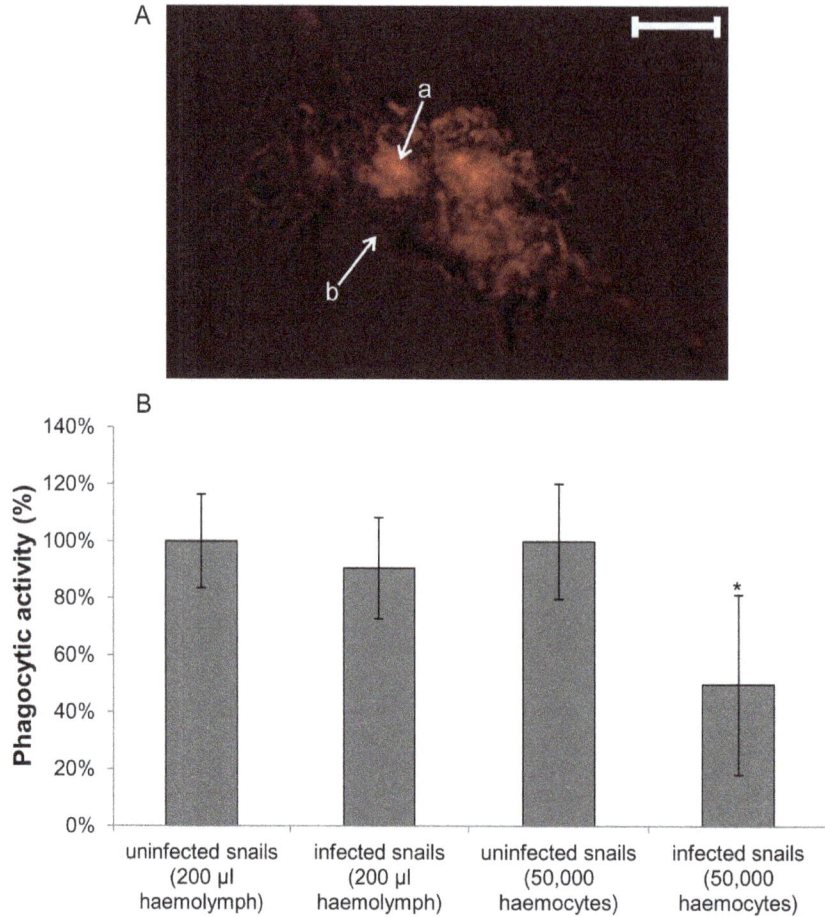

**Figure 4. Phagocytosis of *E. coli* bioparticles by haemocytes from uninfected and *Trichobilharzia regenti* infected *Radix lagotis*.** Phagocytic activities were assessed by incubating *E. coli* bioparticles with haemocyte monolayers and assessing the relative fluorescence of internalised particles after 2 h using a microplate reader. (A) The combined (phase-contrast and fluorescence) image of *E. coli* bioparticles (a) within a haemocyte (b); scale bar = 10 μm. (B) Data were evaluated per volume of haemolymph (200 μl) and per number of haemocytes (50,000) (shown as mean values ± SEM; n = 7) with uninfected snails considered as having 100% activity. *p<0.05 when compared to uninfected snails (50,000 haemocytes); Wilcoxon test.

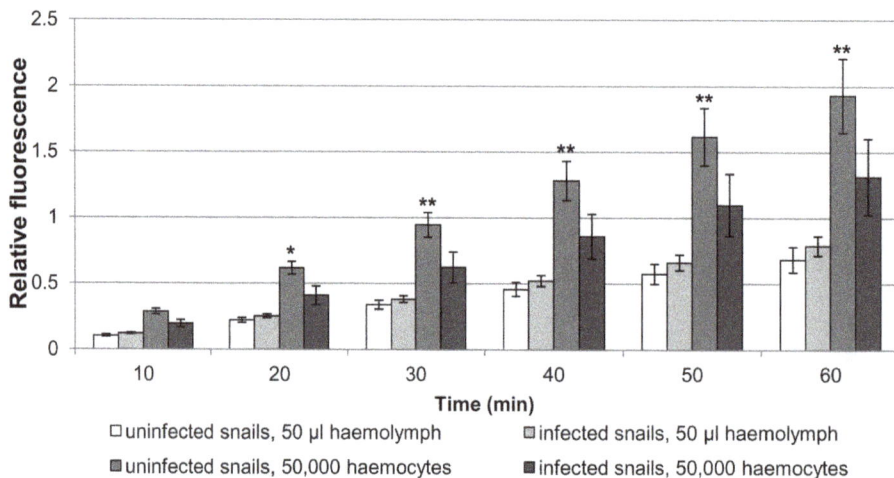

**Figure 5. Basal $H_2O_2$ production in haemocytes from uninfected and *Trichobilharzia regenti* infected *Radix lagotis*.** $H_2O_2$ output by haemocyte monolayers was detected by Amplex red and the intensity of fluorescence measured by microplate reader over 60 min. The mean relative fluorescence values are shown (± SEM; n = 7) and represent the increase in $H_2O_2$ production over time. Data were evaluated per volume of haemolymph (50 μl) and per number of haemocytes (50,000). *p<0.05, **p<0.01, when compared to infected snails (50,000 haemocytes); two-sample t-test or Wilcoxon test combined with Fishers's combination test.

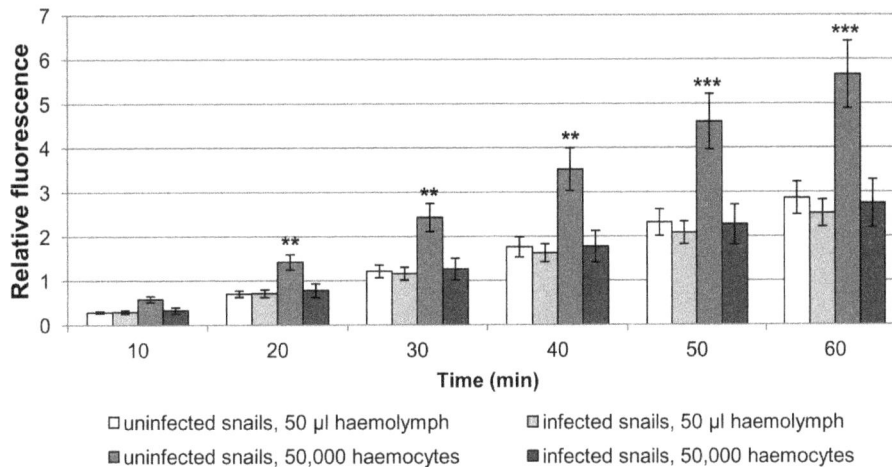

**Figure 6. PMA-stimulated $H_2O_2$ production in haemocytes from uninfected and** *Trichobilharzia regenti* **infected** *Radix lagotis.* $H_2O_2$ output by haemocyte monolayers treated with 5 μM PMA was detected by Amplex red and the intensity of fluorescence was measured by microplate reader over 60 min. The mean relative fluorescence values are shown (± SEM; n = 7) and represent the increase in $H_2O_2$ production over time. Data were evaluated per volume of haemolymph (50 μl) and per number of haemocytes (50,000). \*\*p<0.01, \*\*\*p<0.001, when compared to infected snails (50,000 haemocytes); two-sample t-test or Wilcoxon test combined with Fishers's combination test.

## Effect of PKC and MEK inhibitors on phagocytosis and $H_2O_2$ production

To investigate the possible role of PKC and ERK in the regulation of phagocytosis by *R. lagotis* haemocytes, cells from uninfected snails were incubated with the PKC or MEK inhibitors GF109203X or U0126, respectively, compounds that have been shown to decrease PKC or ERK phosphorylation (activation) in *L. stagnalis* haemocytes [7], [27]. Phagocytosis was blocked in a dose-dependent manner, with 1 μM and 10 μM GF109203X significantly suppressing uptake of bioparticles by approximately 35% and 70%, respectively (p<0.01, p<0.001; Figure 8A). U0126 at 1 μM and 10 μM concentration significantly reduced phagocytic activity of haemocytes by approximately 33% and 67%, respectively (p<0.01, p<0.001; Figure 8B).

Treatment of haemocytes from uninfected *R. lagotis* with 5 μM PMA resulted in a 212% increase in $H_2O_2$ production after 60 min in contrast to an 80% increase in the absence of PMA; thus at this time point PMA stimulated $H_2O_2$ output approximately 2.6-fold when compared to controls (p<0.001; Figure 9). Next, the ability of PKC (GF109203X; 5 μM) and MEK (U0126; 5 μM) inhibitors to affect haemocyte $H_2O_2$ production was tested. GF109203X substantially attenuated $H_2O_2$ release by PMA-stimulated haemocytes when compared to haemocytes treated with DMSO (vehicle) and PMA at all time points (p<0.01, p< 0.001; Figure 9), reducing $H_2O_2$ output to levels similar to those seen under basal conditions. In addition, DMSO did not significantly affect PMA-stimulated $H_2O_2$ production when compared to that of cells treated with PMA only. U0126 also significantly reduced PMA-stimulated $H_2O_2$ production by *R. lagotis* haemocytes (Figure 10). After 60 min, the increase in $H_2O_2$ production as a result of PMA exposure was reduced by 37% (p< 0.001; Figure 10).

## Discussion

### Histological observation of *T. regenti* in *R. lagotis*

We evaluated by histology haemocyte migratory/encapsulation responses triggered in *R. lagotis* by the bird schistosome, *T. regenti*. The timing of this haemocyte response should be interpreted with a tolerance of +5 h, because the snails were exposed to *T. regenti* miracidia for 5 h, after which the specimens were fixed at different time points between 1 and 92 h p.e. Taking this into account, between 1 and 6 h p.e. the response varied where some larvae were encapsulated by haemocytes while others appeared without haemocytic infiltration. In the case of *Biomphalaria alexandrina* infected with *S. mansoni*, haemocytes were also not observed around some miracidia while others underwent encapsulation 6 h p.e. [28]. Haemocytes then surrounded developing *T. regenti* sporocysts in our study between 7 and 21 h p.e., and their occurrence started to fluctuate at the latter time points. This encapsulation, however, did not lead to killing of the parasites that appeared to be morphologically intact. Similarly, compatible *S. mansoni* larvae have also been seen encapsulated but not destroyed by *B. glabrata* haemocytes in in vitro experiments or by *Biomphalaria tenagophila* fibrous cells observed in vivo [29–31]. Haemocytes of *R. lagotis* might be attracted towards *T. regenti* by ciliary plates shed during miracidium-mother sporocyst transformation [21]. A role of *T. szidati* ciliary plates in activating *L. stagnalis* haemocytes has been previously suggested [32], [33]. Furthermore, within haemocytes we observed microtubular aggregates that likely corresponded to the remnants of phagocytosed ciliary plates. Ciliary plates of *S. mansoni* miracidia are also phagocytosed by *B. glabrata* haemocytes [34].

Then, up until 41 h p.e., the haemocytic response against the developing *T. regenti* appeared to decline and no haemocytes were observed in the proximity of larvae between 44 and 97 h p.e. Haemocyte motility might be affected by parasite-derived components such as ESPs, which in the case of *E. paraensei* repel *B. glabrata* haemocytes [35]. Based on our observations, we suggest that the developing sporocysts of *T. regenti* escaped the cellular defence response of *R. lagotis* enabling successful parasite development. However, it is also possible that not all larvae that penetrated the snails were observed and some of these might have been destroyed after encapsulation. Both normally developing and encapsulated sporocysts of *S. mansoni* within *B. glabrata* have previously been observed [36].

Interestingly, in our laboratory-reared *T. regenti*, approximately 90% of *R. lagotis* snails become infected with the parasite while

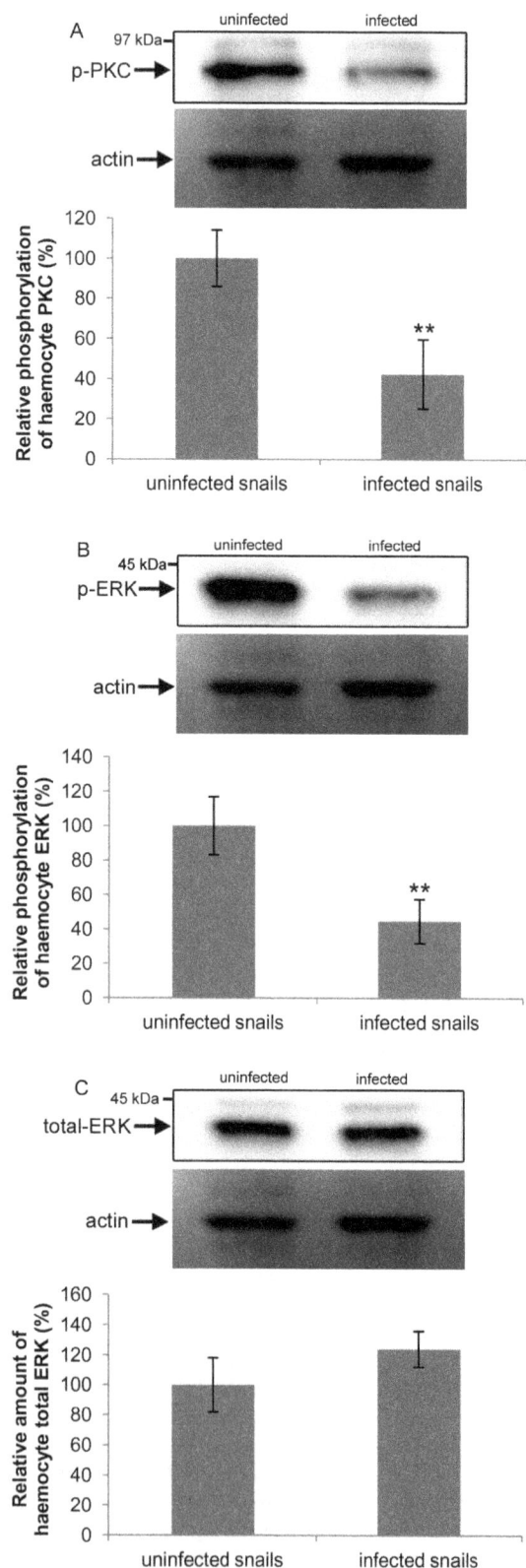

**Figure 7. PKC and ERK phosphorylation and total ERK levels in haemocytes from uninfected and _Trichobilharzia regenti_ infected _Radix lagotis._** Representative blots showing (A) PKC and (B) ERK phosphorylation in adherent haemocytes from uninfected and infected snails. (C) Levels of total ERK in uninfected and infected snails. Band intensities were measured and the mean ($\pm$ SEM) haemocyte PKC and

ERK phosphorylation (n = 10) and total ERK levels (n = 5) calculated (shown in the graphs) with uninfected values considered as 100%. **p< 0.01 when compared to haemocyte PKC and ERK phosphorylation levels in uninfected snails; two-sample t-test.

the remainder appear resistant (data not shown). This phenomenon, reduced compatibility, may be a consequence of long-term passage of the parasite in laboratory conditions [37]. However, reduced compatibility may already arise earlier as shown for the first generation of offspring of _B. alexandrina_ snails susceptible to _S. mansoni_ [38].

## Quantities of circulating haemocytes from uninfected and _T. regenti_-infected _R. lagotis_

Haemocyte numbers/ml haemolymph, phagocytic activity and $H_2O_2$ production were compared between uninfected and infected _R. lagotis_ snails, with infected snails studied during the patent period of infection by _T. regenti_. Correlation analysis revealed that the levels of circulating haemocytes were not influenced by age (shell height) of individual _R. lagotis_ from both groups. In contrast, older specimens of _Lymnaea acuminata_ f. _rufescens_, _Indoplanorbis exustus_ and _Ruditapes decussatus_ have been shown to possess significantly higher haemocyte counts per volume of haemolymph than the younger individuals [39], [40].

Despite variation in haemocyte concentration in uninfected and infected _R. lagotis_, the infected snails had significantly more (1.8-

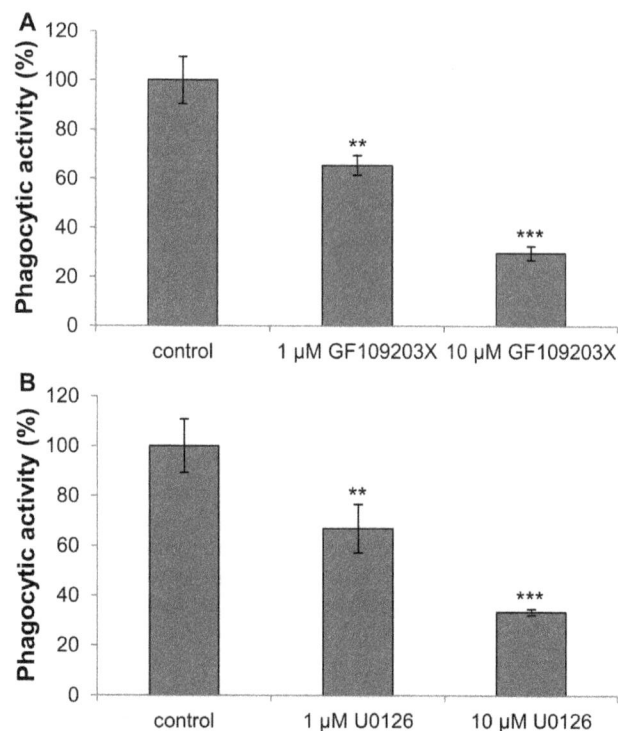

**Figure 8. Effect of PKC (GF109203X) and MEK (U0126) inhibitors on phagocytosis by haemocytes from uninfected _Radix lagotis._** Haemocyte monolayers were pre-incubated with (A) GF109203X, (B) U0126, or vehicle (DMSO; shown as controls) prior to challenge with _E. coli_ bioparticles. The intracellular fluorescence resulting from phagocytosis was measured using a microplate reader and mean ($\pm$ SEM; n = 7) levels of phagocytosis expressed in relation to control (100%) values. **p<0.01, ***p<0.001, when compared to control values; paired t-test.

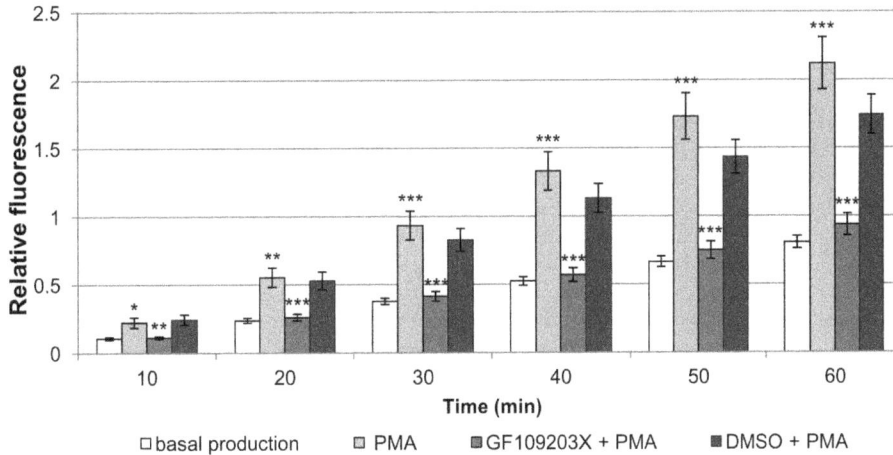

**Figure 9. PMA-stimulated H$_2$O$_2$ production in haemocytes from uninfected *Radix lagotis*, and the effect of PKC inhibition on H$_2$O$_2$ production.** H$_2$O$_2$ output by haemocyte monolayers in the presence of PMA (5 µM), GF109203X (5 µM) and PMA, DMSO (vehicle) and PMA, or SSS+ alone was detected by Amplex red and the intensity of fluorescence was measured by microplate reader over 60 min. The mean (± SEM; n = 7) relative fluorescence values shown represent the increase in H$_2$O$_2$ production over time in the various treatments. *p<0.05, **p<0.01, ***p<0.001, for PMA values compared to basal production, and **p<0.01, ***p<0.001 for GF109203X+PMA compared to DMSO+PMA; paired t-test or paired-samples Wilcoxon test combined with Fishers's combination test.

fold) circulating haemocytes/ml haemolymph, when compared to their uninfected counterparts. Similar differences in haemocyte number were previously found between uninfected and *H. elongata*-infected *L. littorea* [15]. Infection of *B. glabrata* with *E. liei* or *E. paraensei* also results in increased numbers of haemocytes in the circulation [41], [42]. On the other hand, haemocyte concentrations appeared constant in *L. stagnalis* snails infected with *Diplostomum spathaceum* [43], suggesting that increased haemocyte number is not a general response of snails to trematode infection.

## Comparison of defence activities of haemocytes from uninfected and *T. regenti*-infected *R. lagotis* and the influence of PKC and ERK activities

In-vitro experiments with haemocytes from either uninfected or *T. regenti*-infected *R. lagotis* were particularly challenging because preparation of cell monolayers from haemolymph pools maintained on ice as done for *L. stagnalis* and *B. glabrata* [7], [17] was intractable for *R. lagotis*. The majority of haemocytes clumped during such manipulation and, therefore, aliquots of haemolymph expelled during head-foot retraction were transferred directly from the snails to the wells. Furthermore, variation in the numbers of

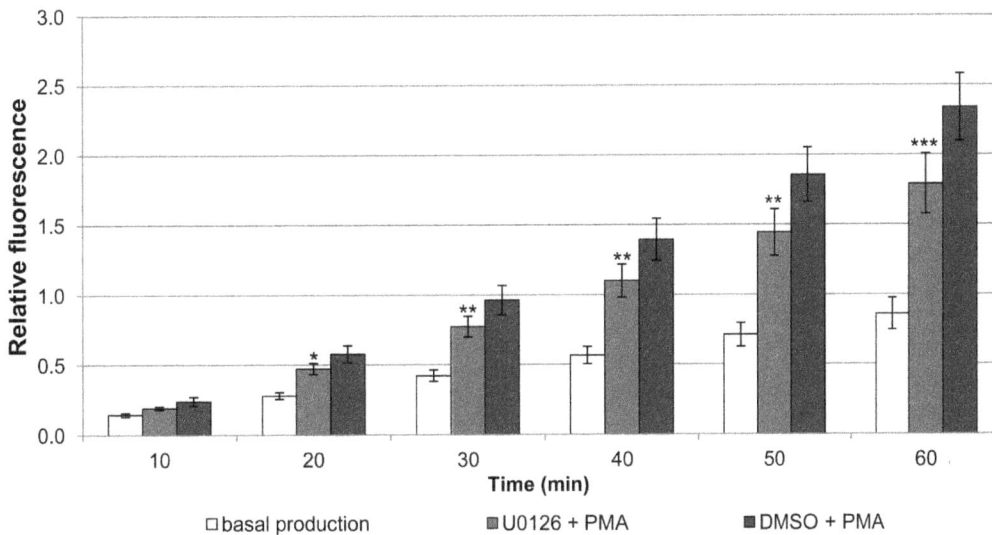

**Figure 10. The effect of MEK inhibition on PMA-stimulated H$_2$O$_2$ production in haemocytes from uninfected *Radix lagotis*.** H$_2$O$_2$ output by haemocyte monolayers in SSS+ alone, U0126 (5 µM) and PMA, or DMSO (vehicle) and PMA was detected by Amplex red and the intensity of fluorescence was measured by microplate reader over 60 min. The mean (± SEM; n = 3 for SSS+ otherwise n = 7) relative fluorescence values shown represent the increase in H$_2$O$_2$ production over time in the various treatments. *p<0.05, **p<0.01 and ***p<0.001 for U0126+PMA compared to DMSO+PMA; paired t-test or paired-samples Wilcoxon test combined with Fishers's combination test.

circulating haemocytes in *R. lagotis* necessitated haemocyte counting in each experiment. Wells with haemolymph from infected snails usually contained almost twice the number of cells in haemolymph from uninfected snails.

Phagocytosis of *E. coli* bioparticles, evaluated using equal numbers of haemocytes (50,000), was approximately 50% lower in infected snails when compared to that of uninfected snails, although when considering haemolymph volume (200 μl) phagocytic activities were similar. Because bioparticles were used in excess and were found free in the incubation medium after exposure to haemocytes, we conclude that the phagocytic activity of haemocytes was not limited by *E. coli* bioparticles availability, but was supressed as a result of *T. regenti* infection. However, it remains to be determined whether individual haemocytes exhibited lower phagocytic activity generally or whether some populations were more affected than others. The increased concentration of haemocytes in infected snails which was 1.8-fold higher in comparison to uninfected snails likely compensated for the overall decreased phagocytic capacity.

Haemocytes obtained from *B. glabrata* or *L. stagnalis* infected with *E. paraensei* or *T. szidati*, respectively, also possess reduced phagocytic activity [13], [14], [44]. Such suppression was observed several days or weeks after exposure to parasites. Furthermore, phagocytic activity of haemocytes was reduced in haemocytes exposed to parasite-derived ESPs [45], [46]. Although specific bioactive molecules of *T. regenti* were not investigated in our study, the phagocytic capacity of *R. lagotis* haemocytes might be affected by products of daughter sporocysts or cercariae as these stages persist in snails in the patent phase of infection.

The PKC and ERK pathways have been found to be essential for efficient phagocytosis by haemocytes of *L. stagnalis*, *B. glabrata* or *Mytilus galloprovincialis* [7], [47], [48]. We therefore explored the possible regulatory role of PKC and ERK in phagocytosis by *R. lagotis* haemocytes. Inhibitors of PKC (GF109203X) and MEK (U0126) significantly blocked haemocyte phagocytic activity in a dose-dependent manner. At 1 μM and 10 μM, GF109203X decreased phagocytosis by 35% and 70% whereas U0126 by 33% and 67%, respectively. This supports the involvement of PKC and ERK in phagocytosis of *E. coli* bioparticles by *R. lagotis* haemocytes. Furthermore, levels of PKC and ERK phosphorylation (activation) were 57% and 55% lower, respectively, in haemocytes from infected snails compared to uninfected snails following adhesion. Thus, the reduced phagocytic activity of haemocytes from infected snails might be caused (at least partly) by supressed PKC and ERK activation in these cells. Because the level of total (phosphorylated and non-phosphorylated) ERK was not reduced in these cells it is possible that the expression of upstream signalling elements might be suppressed; these could include integrin which is known to activate ERK and to be important in cell adhesion [49], [50]. The expression of PKC protein was not studied in the current work since available antibodies are generally ineffective at recognizing PKCs in snail haemocytes (unpublished results).

Infected and uninfected *R. lagotis* haemocytes were further compared in their capacity to generate $H_2O_2$. Amplex red utilized in our study was previously used for monitoring $H_2O_2$ production by snail haemocytes [9], [24]. Basal and PMA-stimulated $H_2O_2$ production did not differ significantly between uninfected and infected snails when considering only the volume of haemolymph (50 μl). On the other hand, basal $H_2O_2$ production calculated per number of haemocytes (50,000) was significantly different, with haemocytes from uninfected snails producing more $H_2O_2$ as early as 20 min. Similarly, PMA-stimulated $H_2O_2$ production by haemocytes from uninfected snails increased significantly with

time from 20 min, being approximately 2-fold higher after 60 min when compared to that of haemocytes from infected snails. The reduced capacity of haemocytes from infected snails to generate $H_2O_2$ might be important for *T. regenti* survival, as $H_2O_2$ was previously shown to be an important ROS involved in in-vitro killing of *S. mansoni* sporocysts [4]. In *L. littorea*, haemocytes from snails infected with *H. elongata* produce 2-fold less superoxide [15], a precursor of $H_2O_2$ [4], [51]. As with phagocytosis, it is possible that *R. lagotis* compensate for decreased $H_2O_2$ generation by haemocytes by increasing their number in the circulation. Nevertheless, whether all haemocytes or their proportion were inhibited remains unknown as well as components of *T. regenti* responsible for such alteration. In *B. glabrata*, PMA-stimulated production of $H_2O_2$ was significantly reduced when haemocytes were simultaneously exposed to PMA and ESPs of *S. mansoni* [10].

As PMA is an activator of PKC, a role of this kinase in the regulation of $H_2O_2$ production by haemocytes from uninfected *R. lagotis* snails was further investigated; participation of ERK signalling in this process using the MEK inhibitor (U0126) was also explored. Haemocytes exposed to GF109203X displayed substantially reduced PMA-stimulated $H_2O_2$ production that was similar to levels comparable with basal (unstimulated) $H_2O_2$ output. U0126 also significantly affected PMA-stimulated $H_2O_2$ output by snail haemocytes, although at less extent than GF109203X. Thus, PKC and ERK appear to play a role in regulating $H_2O_2$ production by *R. lagotis* haemocytes. PKC and ERK signalling were previously found to be crucial in regulation of $H_2O_2$ production by haemocytes of *B. glabrata* [10], [24] and *L. stagnalis* [9]. As already mentioned for haemocytes of infected snails, basal levels of PKC and ERK phosphorylation (activation) were significantly lower than in haemocytes of uninfected snails; lower $H_2O_2$ production by haemocytes from infected snails could therefore be the result of lower PKC and ERK activities in response to the parasite. Our study and a previous report suggesting that ESPs may attenuate PKC and ERK phosphorylation in snail haemocytes [17] support the notion that parasites modulate haemocyte defence pathways at the level of cell signalling [16] and possibly at multiple phases during development.

The present paper provides the first insights into the immunobiology of the snail *R. lagotis*, an important intermediate host of the nasal bird schistosome *T. regenti*. Histological study of the *R. lagotis* response against *T. regenti* showed that haemocytes are able to accumulate near the invading larvae, but they do not destroy the parasite. This enables further development of trematode larvae, leading to patent phase of *T. regenti* infection in snails. The phagocytic activity and capacity for $H_2O_2$ generation were suppressed in haemocytes of infected snails. Importantly, PKC and ERK that appear to regulate such responses in *R. lagotis* were also shown to be less active in haemocytes from infected snails. It is hypothesized that attenuation of both responses in haemocytes is partially compensated by increased concentration of haemocytes in the circulation of infected snails, enabling the snail to fend off other pathogens such as bacteria. Further research is needed to understand how this impacts survival and continued cercarial production of *T. regenti* in *R. lagotis*, and to determine the parasite-derived molecules responsible for alterations in *R. lagotis* haemocyte responses.

## Acknowledgments

We would like to thank Veronika Siegelova, M.Sc., (Charles University in Prague) for laboratory maintenance of snails and parasites.

## Author Contributions

Conceived and designed the experiments: VS ZJ MK MV AJW PH. Performed the experiments: VS ZJ MV. Analyzed the data: VS AC ZJ MV. Contributed reagents/materials/analysis tools: VS AJW PH. Wrote the paper: VS AC ZJ MK MV AJW PH.

## References

1. Van der Knaap WPW, Adema CM, Sminia T (1993) Invertebrate blood cells: morphological and functional aspects of the haemocytes in the pond snail *Lymnaea stagnalis*. Comp Haematol Int 3: 20–26.
2. Adema CM, Hertel LA, Miller RD, Loker ES (1997) A family of fibrinogen-related proteins that precipitates parasite-derived molecules is produced by an invertebrate after infection. Proc Natl Acad Sci U S A 94: 8691–8696.
3. Loker ES, Bayne CJ, Buckley PM, Kruse KT (1982) Ultrastructure of encapsulation of *Schistosoma mansoni* mother sporocysts by haemocytes of juveniles of the 10-R2 strain of *Biomphalaria glabrata*. J Parasitol 68: 84–94.
4. Hahn UK, Bender RC, Bayne CJ (2001) Killing of *Schistosoma mansoni* sporocysts by haemocytes from resistant *Biomphalaria glabrata*: role of reactive oxygen species. J Parasitol 87: 292–299.
5. Adema CM, van Deutekom-Mulder EC, Van der Knaap WPW, Meuleman EA, Sminia T (1991) Generation of oxygen radicals in haemocytes of the snail *Lymnaea stagnalis* in relation to the rate of phagocytosis. Dev Comp Immunol 15: 17–26.
6. Ito T, Matsutani T, Mori K, Nomura T (1992) Phagocytosis and hydrogen peroxide production by phagocytes of the sea urchin *Strongylocentrotus nudus*. Dev Comp Immunol 16: 287–294.
7. Plows LD, Cook RT, Davies AJ, Walker AJ (2004) Activation of extracellular-signal regulated kinase is required for phagocytosis by *Lymnaea stagnalis* haemocytes. Biochim Biophys Acta 1692: 25–33.
8. Plows LD, Cook RT, Davies AJ, Walker AJ (2005) Carbohydrates that mimic schistosome surface coat components affect ERK and PKC signalling in *Lymnaea stagnalis* haemocytes. Int J Parasitol 35: 293–302.
9. Lacchini AH, Davies AJ, Mackintosh D, Walker AJ (2006) Beta-1, 3-glucan modulates PKC signalling in *Lymnaea stagnalis* defence cells: a role for PKC in $H_2O_2$ production and downstream ERK activation. J Exp Biol 209: 4829–4840.
10. Humphries JE, Yoshino TP (2008) Regulation of hydrogen peroxide release in circulating hemocytes of the planorbid snail *Biomphalaria glabrata*. Dev Comp Immunol 32: 554–562.
11. Plows LD, Cook RT, Davies AJ, Walker AJ (2006) Phagocytosis by *Lymnaea stagnalis* haemocytes: A potential role for phosphatidylinositol 3-kinase but not protein kinase A. J Invertebr Pathol 91: 74–77.
12. Zelck UE, Gege BE, Schmid S (2007) Specific inhibitors of mitogen activated protein kinase and PI3-K pathways impair immune responses by haemocytes of trematode intermediate host snails. Dev Comp Immunol 31: 321–331.
13. Noda S, Loker ES (1989) Phagocytic activity of hemocytes of M-line *Biomphalaria glabrata* snails: effect of exposure to the trematode *Echinostoma paraensei*. J Parasitol 75: 261–269.
14. Dikkeboom R, Van der Knaap WPW, Van den Bovenkamp W, Tijnagel JM, Bayne CJ (1988) The production of toxic oxygen metabolites by hemocytes of different snail species. Dev Comp Immunol 12: 509–520.
15. Gorbushin AM, Iakovleva NV (2008) The enigma of the haemogram left-shift in periwinkles infected with trematodes. Fish Shellfish Immunol 24: 745–751.
16. Walker AJ (2006) Do trematode parasites disrupt defence-cell signalling in their snail host? Trends Parasitol 22: 154–159.
17. Zahoor Z, Davies AJ, Kirk RS, Rollinson D, Walker AJ (2008) Disruption of ERK signalling in *Biomphalaria glabrata* defence cells by *Schistosoma mansoni*: implications for parasite survival in the snail host. Dev Comp Immunol 32: 1561–1571.
18. Horák P, Kolářová L, Dvořák J (1998) *Trichobilharzia regenti* n. sp. (Schistosomatidae, Bilharziellinae), a new nasal schistosome from Europe. Parasite 5: 349–357.
19. Huňová K, Kašný M, Hampl V, Leontovyč R, Kuběna A, et al. (2012) *Radix* spp.: Identification of trematode intermediate hosts in the Czech Republic. Acta Parasitol 57: 273–284.
20. Kolářová L, Horák P, Skírnisson K, Marečková H, Doenhoff M (2013) Cercarial dermatitis, a neglected allergic disease. Clin Rev Allergy Immunol 45: 63–74.
21. Horák P, Adema CM, Kolářová L (2002) Biology of the schistosome genus *Trichobilharzia*. Adv Parasitol 52: 155–233.
22. Reynolds ES (1963) The use of lead citrate at high pH as an electron-opaque stain in electron microscopy. J Cell Biol 17: 208–212.
23. Sminia T (1972) Structure and function of blood and connective tissue cells of the fresh water pulmonate *Lymnaea stagnalis* studied by electron microscopy and enzyme histochemistry. Z Zellforsch Mikrosk Anat 130: 497–526.
24. Bender RC, Broderick EJ, Goodall CP, Bayne CJ (2005) Respiratory burst of *Biomphalaria glabrata* hemocytes: *Schistosoma mansoni*-resistant snails produce more extracellular $H_2O_2$ than susceptible snails. J Parasitol 91: 275–279.
25. Arumugam M, Romestand B, Torreilles J, Roch P (2000) In vitro production of superoxide and nitric oxide (as nitrite and nitrate) by *Mytilus galloprovincialis* haemocytes upon incubation with PMA or laminarin or during yeast phagocytosis. Eur J Cell Biol 79: 513–519.
26. Lehmacher W, Wassmer G (1999) Adaptive sample size calculations in group sequential trials. Biometrics 55: 1286–1290.
27. Walker AJ, Plows LD (2003) Bacterial lipopolysaccharide modulates Protein Kinase C signalling in *Lymnaea stagnalis* hemocytes. Biol Cell 95: 527–533.
28. Mohamed AH, Sharaf El-Din AT, Mohamed AM, Habib MR (2011) Tissue response exhibited by *Biomphalaria alexandrina* snails from different Egyptian localities following *Schistosoma mansoni* exposure. Exp Parasitol 127: 789–794.
29. Bayne CJ, Buckley PM, DeWan PC (1980) Macrophagelike hemocytes of resistant *Biomphalaria glabrata* are cytotoxic for sporocysts of *Schistosoma mansoni in vitro*. J Parasitol 66: 413–419.
30. Bayne CJ, Buckley PM, DeWan PC (1980) *Schistosoma mansoni*: cytotoxicity of hemocytes from susceptible snail hosts for sporocysts in plasma from resistant *Biomphalaria glabrata*. Exp Parasitol 50: 409–416.
31. Nacif-Pimenta R, de Mattos ACA, Orfano AS, Barbosa L, Pimenta PFP, et al. (2012) *Schistosoma mansoni* in susceptible and resistant snail strains *Biomphalaria tenagophila*: *in vivo* tissue response and *in vitro* hemocyte interactions. PloS One 7: 1–12.
32. Amen RI, Baggen JMC, Bezemer PD, de Jong-Brink MD (1992) Modulation of the activity of the internal defence system of the pond snail *Lymnaea stagnalis* by the avian schistosome *Trichobilharzia ocellata*. Parasitol 104: 33–40.
33. de Jong-Brink MD, Bergamin-Sassen M, Soto M (2001) Multiple strategies of schistosomes to meet their requirements in the intermediate snail host. Parasitol 123: 129–141.
34. Lie KJ, Jeong KH, Heyneman D (1981) Selective interference with granulocyte function induced by *Echinostoma paraensei* (Trematoda) larvae in *Biomphalaria glabrata* (Mollusca). J Parasitol 67: 790–796.
35. Adema CM, Arguello DF, Stricker SA, Loker ES (1994) A time-lapse study of interactions between *Echinostoma paraensei* intramolluscan larval stages and adherent hemocytes from *Biomphalaria glabrata* and *Helix aspersa*. J Parasitol 80: 719–727.
36. Théron A, Pages JR, Rognon A (1997) *Schistosoma mansoni*: distribution patterns of miracidia among *Biomphalaria glabrata* snail as related to host susceptibility and sporocyst regulatory processes. Exp Parasitol 85: 1–9.
37. Mitta G, Adema CM, Gourbal B, Loker ES, Théron A (2012) Compatibility polymorphism in snail/schistosome interactions: from field to theory to molecular mechanisms. Dev Comp Immunol 37: 1–8.
38. Abou-El-Naga IF, Radwan EH (2012) Defense response of susceptible and resistant *Biomphalaria alexandrina* snails against *Schistosoma mansoni* infection. Rev Biol Trop 60: 1195–1204.
39. Suresh PG, Reju MK, Mohandas A (1994) Factors influencing total haemocyte counts in freshwater gastropods. Comp Haematol Int 4: 17–24.
40. Flye-Sainte-Marie J, Soudant P, Lambert C, Le Goic N, Goncalvez M, et al. (2009) Variability of the hemocyte parameters of *Ruditapes philippinarum* in the field during an annual cycle. J Exp Mar Biol Ecol 377: 1–11.
41. Mounkassa JB, Jourdane J (1990) Dynamics of the leukocytic response of *Biomphalaria glabrata* during the larval development of *Schistosoma mansoni* and *Echinostoma liei*. J Invertebr Pathol 55: 306–311.
42. Loker ES, Cimino DF, Stryker GA, Hertel LA (1987) The effect of size of M line *Biomphalaria glabrata* on the course of development of *Echinostoma paraensei*. J Parasitol 73: 1090–1098.
43. Riley EM, Chappell LH (1992) Effect of infection with *Diplostomum spathaceum* on the internal defense system of *Lymnaea stagnalis*. J Invertebr Pathol 59: 190–196.
44. Nunez PE, Adema CM, de Jong-Brink MD (1994) Modulation of the bacterial clearance activity of haemocytes from the freshwater mollusc, *Lymnaea stagnalis*, by the avian schistosome, *Trichobilharzia ocellata*. Parasitol 109: 299–310.
45. Connors VA, Yoshino TP (1990) In vitro effect of larval *Schistosoma mansoni* excretory-secretory products on phagocytosis-stimulated superoxide production in hemocytes from *Biomphalaria glabrata*. J Parasitol 76: 895–902.
46. Iakovleva NV, Shaposhnikova TG, Gorbushin AM (2006) Rediae of echinostomatid and heterophyid trematodes suppress phagocytosis of haemocytes in *Littorina littorea* (Gastropoda: Prosobranchia). Exp Parasitol 113: 24–29.
47. Humphries JE, Yoshino TP (2003) Cellular receptors and signal transduction in molluscan hemocytes: connections with the innate immune system of vertebrates. Integr Comp Biol 43: 305–312.
48. García-García E, Prado-Álvarez M, Novoa B, Figueras A, Rosales C (2008) Immune responses of mussel hemocyte subpopulations are differentially regulated by enzymes of the PI 3-K, PKC, and ERK kinase families. Dev Comp Immunol 32: 637–653.
49. Schwartz MA (2001) Integrin signaling revisited Trends Cell Biol 11: 466–470.
50. Chen Q, Lin TH, Der CJ, Juliano RL (1996) Integrin-mediated activation of MEK and mitogen-activated protein kinase is independent of Ras. J Biol Chem 271: 18122–18127.
51. Hampton MB, Kettle AJ, Winterbourn CC (1998) Inside the neutrophil phagosome: oxidants, myeloperoxidase, and bacterial killing. Blood 92: 3007–3017.

# 14

# Distinct Strains of *Toxoplasma gondii* Feature Divergent Transcriptomes Regardless of Developmental Stage

**Matthew McKnight Croken[1], Yanfen Ma[2], Lye Meng Markillie[3], Ronald C. Taylor[4], Galya Orr[3], Louis M. Weiss[2,5]\*, Kami Kim[1,2,5]\***

1 Department of Microbiology and Immunology, Albert Einstein College of Medicine, Bronx, New York, United States of America, 2 Department of Pathology, Albert Einstein College of Medicine, Bronx, New York, United States of America, 3 Environmental Molecular Sciences Laboratory, Pacific Northwest National Laboratory, Richland, Washington, United States of America, 4 Computational Biology and Bioinformatics Group, Biological Sciences Division, Pacific Northwest National Laboratory, Richland, Washington, United States of America, 5 Department of Medicine, Albert Einstein College of Medicine, Bronx, New York, United States of America

## Abstract

Using high through-put RNA sequencing, we assayed the transcriptomes of three different strains of *Toxoplasma gondii* representing three common genotypes under both *in vitro* tachyzoite and *in vitro* bradyzoite-inducing alkaline stress culture conditions. Strikingly, the differences in transcriptional profiles between the strains, RH, PLK, and CTG, is much greater than differences between tachyzoites and alkaline stressed *in vitro* bradyzoites. With an FDR of 10%, we identified 241 genes differentially expressed between CTG tachyzoites and *in vitro* bradyzoites, including 5 putative AP2 transcription factors. We also observed a close association between cell cycle regulated genes and differentiation. By Gene Set Enrichment Analysis (GSEA), there are a number of KEGG pathways associated with the *in vitro* bradyzoite transcriptomes of PLK and CTG, including pyrimidine metabolism and DNA replication. These functions are likely associated with cell-cycle arrest. When comparing mRNA levels between strains, we identified 1,526 genes that were differentially expressed regardless of culture-condition as well as 846 differentially expressed only in bradyzoites and 542 differentially expressed only in tachyzoites between at least two strains. Using GSEA, we identified that ribosomal proteins were expressed at significantly higher levels in the CTG strain than in either the RH or PLK strains. This association holds true regardless of life cycle stage.

**Editor:** Laura J. Knoll, University of Wisconsin Medical School, United States of America

**Funding:** Research was supported by National Institutes of Health (NIH) grants AI095094 (LMW), AI087625 (KK), and by grant 40070 (LMW) from Environmental Molecular Sciences Laboratory (EMSL) Pacific Northwest National Laboratory. MMC was supported by the Training Program in Cellular and Molecular Biology and Genetics, funded by NIH T32 GM007491 awarded to the Albert Einstein College of Medicine. A portion of the research was performed using EMSL, a national scientific user facility sponsored by the Department of Energy's Office of Biological and Environmental Research and located at Pacific Northwest National Lab. This work was also supported in part by the Center for AIDS Research at the Albert Einstein College of Medicine and Montefiore Medical Center funded by the National Institutes of Health (NIH AI-051519). The funders had no role in study design, data collection and analysis, decision to publish, or preparation of the manuscript.

**Competing Interests:** The authors have declared that no competing interests exist.

\* Email: louis.weiss@einstein.yu.edu (LMW); kami.kim@einstein.yu.edu (KK)

## Introduction

*Toxoplasma gondii* is an obligate intracellular parasite belonging to the phylum Apicomplexa. It has a complicated life cycle marked by sexual reproduction in the gastrointestinal tract of a feline host and asexual replication in any warm-blooded animal [1]. The asexual cycle itself is divided into a fast-growing tachyzoite stage and slow-growing bradyzoite stage. The bradyzoite form is thought to persist indefinitely within infected hosts within cysts and can reactivate if a host's immune function wanes. The ability to shift between acute and chronic phases of its life cycle is critical for disease pathogenesis and thus the subject of intense investigation [1].

One mark of the success of *Toxoplasma gondii* is its global distribution. It is estimated to infect around 30% of the human population. The parasite is transmitted through water contaminated with cat feces as well as being clonally propagated from animal to animal via carnivorism. The transmission strategies of *T.*

*gondii* has led to a complex structure of populations within the species. Fourteen different haplogroups have been identified around the world with each lineage containing multiple distinct strains [2]. Research has focused on parasites belonging to groups 1, 2, and 3 (also designated types I, II, and III) isolated in North America and Europe. Although these parasites diverged relatively recently (~10 kya) [33], they are marked by distinct differences in phenotype, most prominently virulence. Type I tachyzoites are less able to convert to bradyzoites, thereby causing acute disease in their hosts [34]. Type III strains readily differentiate causing their hosts to become chronically infected, but these strains are infrequently associated with clinical disease in humans [34]. Type II strains tend to be intermediate to types I and III in terms of differentiation competence and virulence [34]. Despite significant differences in phenotypes, there appears to be very little difference in genome sequence [3]. Forward genetics studies have attributed most of the difference in virulence to the ROP18 and ROP5 loci [4] [5] [6].

Few studies currently exist describing the differences among the transcriptomes of *T. gondii* lineages and how these transcriptomes change following differentiation from tachyzoite to bradyzoite [25]. What studies do exist are often stymied by lack of gene annotation. Roughly half of *T. gondii* genes are described only as "hypothetical proteins". Gene Set Enrichment Analysis (GSEA) is a software tool designed to test whether functionally related sets of genes are collectively up or down regulated between experimental conditions [7].

It has been shown that *T. gondii*, like other Apicomplexa, exerts tight control over gene expression [8], but little is known about the effectors that make this control possible. In 2005, a family of genes containing the AP2 DNA-binding domain were identified in the phylum Apicomplexa [9]. Follow up work found 68 AP2 genes in *T. gondii* [10]. Members of this family regulate important life cycle developments in *Plasmodium* [11] [12] [13] [14] and *Toxoplasma* [15]. As many more of these genes are expected to be key transcription factors, observed differences in AP2 expression levels between strains and across life cycle development are of keen importance.

## Results and Discussion

### Inter-strain differences in gene expression much larger than those between tachyzoites and bradyzoites

We analyzed steady state mRNA levels from three different strains of *T. gondii*, RH (Type I, the most common laboratory strain), PLK (Type II, a clone of ME49 the genome reference strain), and CTG (Type III). We grew each strain under both "normal" tachyzoite tissue culture (pH 7, 5% $CO_2$) conditions as well as bradyzoite-cyst inducing stress conditions (pH 8, low $CO_2$), resulting in six total groups. Each condition was sampled in three biological replicates at a single time point. We had expected that this simple experimental design would yield groups of genes linked to the stress response, but the most striking observation is that the genes' steady state RNA levels vary much more between strains than between different stages of their life cycles. Figure 1B illustrates multi-dimensional scaling (MDS) of samples by condition, based on the top five hundred most divergent genes between each condition. The conditions cluster first by strain and then by life cycle stage. While tachyzoites and bradyzoites from different strains have many functional similarities, these results reinforce the divergence of gene expression of these parasites despite relatively modest differences in genome sequence.

The MDS plot only examines those genes with the most divergent expression levels. We also examined differences between conditions for all genes. First, we computed the differences in gene expression between tachyzoites and bradyzoites for each strain (three comparisons). We then examined differences between strains for both tachyzoites and bradyzoites (six more comparisons). The results are illustrated in Figure 1C. For all nine comparisons, the median and the interquartile ranges change very little and remain relatively close to zero. The mRNA levels of the majority of genes are not widely different between the strains and life cycle stage. To measure the total amount of change between compared groups, we calculate the root mean square difference (RMSD) for each of the nine comparisons. The RMSD values, plotted as blue crosses on Figure 1C, are much greater for the inter-strain comparisons than for the alkaline stressed/$CO_2$ starved parasites of the same strain. This supports the observations shown in Figure 1B and further suggests that variation between the groups is driven by relatively small subsets of genes.

Comparisons between transcriptomes of tachyzoites and in vitro bradyzoites of the same strain are consistent with what is known

about the propensity of each of these strains to differentiate. Type III (CTG) parasites readily switch from tachyzoite to bradyzoite and showed the greatest change in mRNA steady state levels, whereas Type I (RH) parasites fail to differentiate under these stress conditions and likewise showed almost no difference in mRNA between tachyzoite and bradyzoite conditions. Type II (PLK) parasites are intermediate to types I and III both in terms of differentiation competence as well as overall changes in its transcriptome. This is an important proof of principle that this type of global transcriptomic analysis does correlate with actual biological states of the parasite.

### Strain specific expression differences

Globally, we see a much greater difference between the transcriptomes of different strains than we do between stressed and unstressed parasites of the same strain (Figures 1B and 1C). To identify which genes are expressed differently between strains, we used edgeR. We compared each strain against the other two under both tachyzoite (unstressed) and bradyzoite (stressed) conditions (FDR 10%). This created six lists of differentially expressed genes that are differentially expressed between strains in either tachyzoites or bradyzoites as well those genes that were differentially expressed in both stages (Figure 2A). The number in the center of the diagram represents the number of genes differentially expressed between all three strains. A full gene list is in Table S2. Genes in the outer intersections differed in only one strain and those not falling into any intersection are cases where the expression level is different between two strains, but the strain's expression level was not significantly different. It is important to note that the number differentially expressed genes between any two strains is always equal to or larger than number of genes affected by differentiation.

If differential gene expression has indeed played an important in the evolution of *T. gondii* lineages, then regulators of transcription are likely candidates as drivers of evolution. Therefore, we examined whether AP2 genes are differentially expressed between strains. mRNA of seven AP2 differed in tachyzoites and mRNA levels of twenty- four AP2 differed among the bradyzoite transcriptomes. Five AP2 genes are differentially expressed between strains regardless of developmental stage. The AP2 genes that are differentially expressed in at least two comparisons are shown in Figure 2A and a full list is available in the Supplementary tables.

Steady state mRNA levels of AP2III-1 are significantly higher in PLK parasites than in either RH or CTG. It is a defining feature of the PLK strain regardless of developmental stage. Compared to PLK and CTG, expression of AP2VIIa-7 is lower in CTG tachyzoites, while expression of AP2IV-2 is higher. In CTG bradyzoites, expression of AP2X-10 and AP2XI-1 is lower than the other strains. RH tachyzoites express AP2IX-9 mRNA at significantly higher levels than either of the other strains examined. This is consistent with this transcription factors' reported role as a repressor of bradyzoite commitment [15], but differs from the pattern of expression initially reported. This difference may reflect that in our study parasites were predominantly the more stressed extracellular forms, in contrast to the intracellular parasites characterized previously. One can hypothesize that high levels of AP2IX-9 contributes to the parasite's ability to withstand stress and its inability to differentiate. Ten AP2 genes are differentially expressed in RH under stress conditions. This may reflect that RH does not differentiate into bradyzoites.

Expression of a number of AP2 genes in our study differs from previous reports that have compared AP2 expression in different strains. Some of these differences may be due to technical reasons

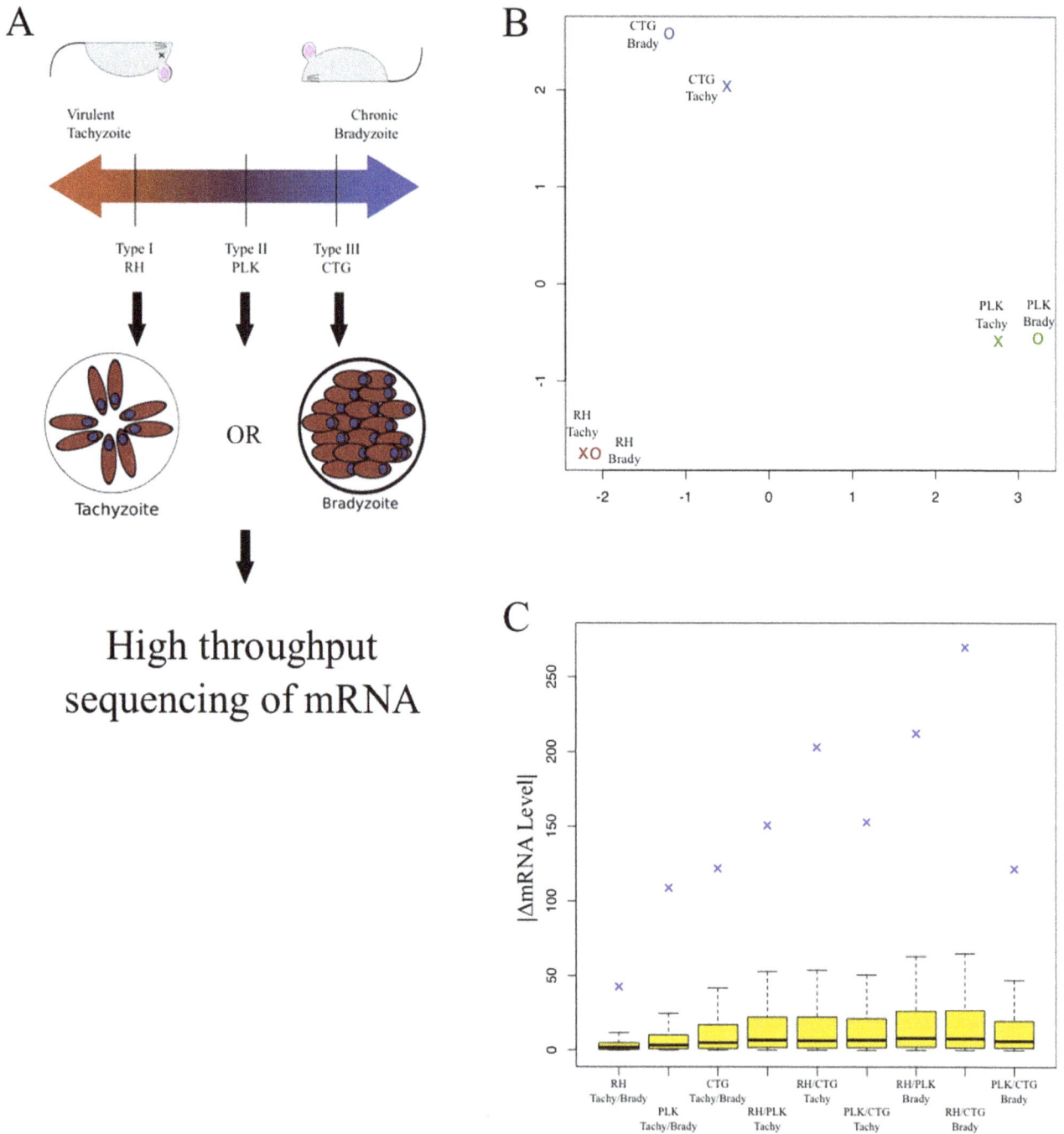

**Figure 1. Patterns of gene expression vary more by strain than by developmental stage.** A) A representative strain from *T. gondii* lineages Types I (RH), II (PLK), and III (CTG) was selected and grown in tissue culture in either pH neutral conditions, conducive to tachyzoite growth or alkaline conditions, inducing bradyzoite differentiation. B) Multi-dimensional scaling (MDS) plot based on pairwise comparisons for each of the six experimental conditions. This is calculated as the root mean square deviation for the 500 most differentially expressed genes between any two conditions. The distance between any two points represents the average difference in expression levels (RPKM) of the most dissimilar genes, relative to differences observed between other conditions. In effect, the MDS plot provides an overview of the total amount of variation between samples. The axes show arbitrary distances. Experimental conditions include parasite strain (red = RH, green = PLK, blue = CTG) and by life cycle stage (X = tachyzoite, O = bradyzoite). Distances calculated using the 'plotMDS' function in the 'limma' Bioconductor package [30]. C) Boxplot represents absolute value of expression level difference between conditions. Interquartile regions are in gold, median differences are plotted as a solid black line. The root mean square deviation for each comparison is represented as a blue cross. From left to right, the first three groups are the intrastrain comparisons, tachyzoite vs. bradyzoite. The next three groups are interstrain comparisons between tachyzoite (unstressed) groups. The final three are are interstrain comparisons between bradyzoite (alkaline-stressed) groups).

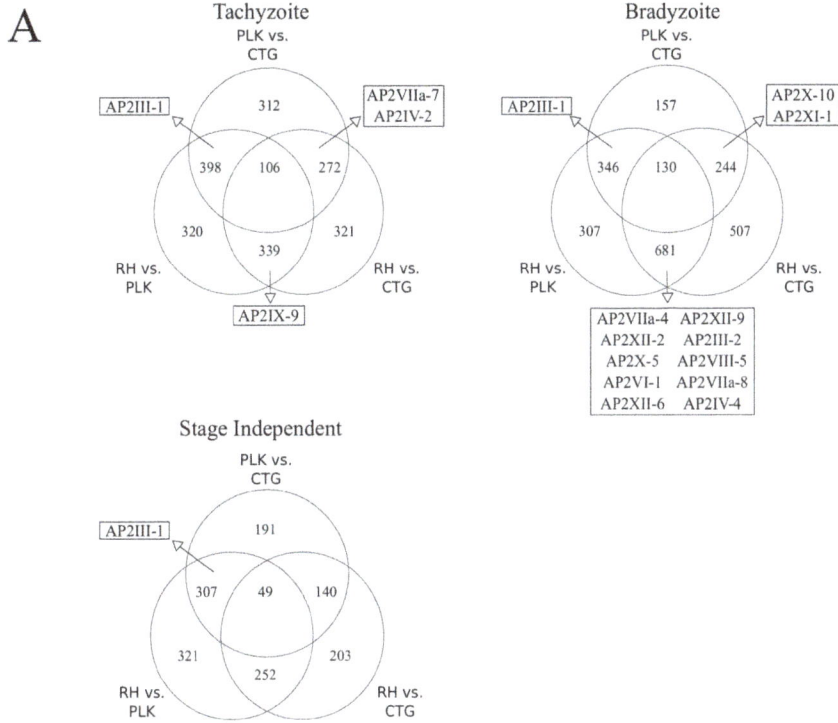

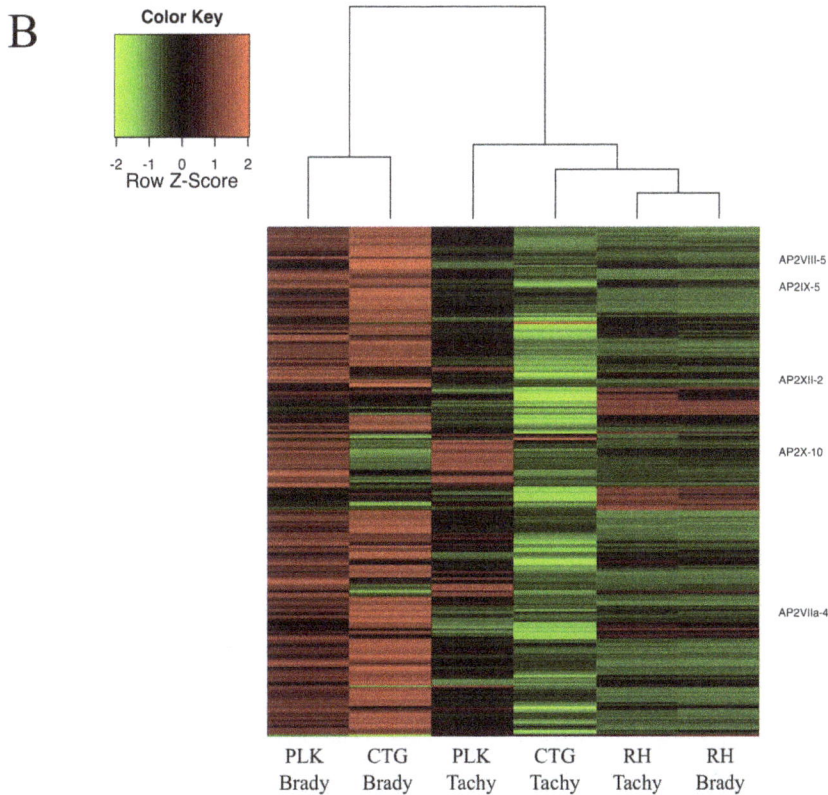

**Figure 2. Hundreds of genes differentially expressed between strains including potential AP2 regulators.** A) Using edgeR, we identify genes that are differentially expressed between strains (FDR <10%), generating three gene lists for tachyzoite condition and three for the bradyzoite condition. We generated Venn diagrams showing the overlap between the lists in the tachyzoite condition and bradyzoite condition. We also compared tachyzoite and bradyzoite gene lists to each other. Venn diagrams shows comparison of genes that appear in both the tachyzoite and the bradyzoite lists and therefore are "stage independent". AP2 containing genes appearing on more than one list (any intersection) are indicated. The

complete set of genes that are differentially expressed are listed in Table S2 and RPKM values for all replicates are listed in Table S1. B) Heat map of genes differentially expressed following CTG bradyzoite differentiation in all six conditions. Red indicates up regulation compared to that gene's expression level under other conditions, whereas green indicates down regulation. Conditions (columns) are clustered based on similarity of expression levels. The five AP2 genes that are differentially expressed are indicated on the right. Heat map was generated using the 'heatmap.2' function in the 'gplots' package for R. Hierarchical clustering of both the rows (genes) and columns (conditions) computed by the 'hclust' function in the R 'stats' package. Based on mean of replicate RPKM values.

such as inaccurate gene models that resulted in incorrect hybridization probes. Other differences are likely due to differences in experimental conditions. As proposed in earlier studies [16] and confirmed by GSEA analysis [22], extracellular tachyzoites represent an intermediate cell cycle arrested state with upregulation of stress-response genes that may be amongst the first that are induced during bradyzoite formation.

If experimental evidence supports the hypothesized role of AP2 genes as bona fide transcription factors, then the genes which they regulate are likely critical in determining phenotypic differences between strains. Differences in expression of AP2 in extracellular vs intracellular parasites may reflect expression differences seen in different biological states.

## Genes and genes pathways associated with *in vitro* stage differentiation

Using the edgeR package from the Bioconductor project, we identified 241 genes that were differentially expressed between CTG tachyzoites and bradyzoites with a false discovery rate (FDR) of 10%. Figure 2B is a heatmap of these genes across all six groups. Both PLK and CTG showed differential expression of these genes, whereas RH parasites are largely insensitive to alkaline stress treatment. Notably, only 33 of the 241 genes (13.7%) are down-regulated in the CTG bradyzoites.

Of the 241 genes related to CTG differentiation identified, there are five genes that are predicted to have an AP2 DNA binding domain [10]. AP2VIIa-4, AP2VIII-5, AP2IX-5, and AP2XII-2 are all up-regulated after 72 hours of *in vitro* bradyzoite conditions, while AP2X-10 is down-regulated. Members of this family have been shown to regulate transcription in malarial parasites, including differentiation to gametocytes [11] [12] [13]. One possibility is that these AP2 transcription factors are important for long-term maintenance of tissue cysts, either promoting the expression of bradyzoite-specific factors or repressing tachyzoite differentiation. Alternatively, these may be only transiently expressed during differentiation, and their temporal expression may reflect a cascade of events that occurs during developmental transitions. Some of the AP2 previously reported to be induced in bradyzoites [22] [23] were down-regulated in stressed CTG, supporting this hypothesis (Table S3). The expression of AP2VIIa-4 and AP2XII-2 are both linked to the cell cycle [8]. Given the associated between cell cycle and differentiation [24], it is not surprising that putative cell cycle transcription factors are differentially expressed following differentiation.

Examining individual genes can be a useful method of analyzing expression data. Understanding how many genes are affected and how much their expression changes provides an important global overview of the transcriptome under different conditions. This type of analysis also provides an unbiased way testing whether particular genes of interest are differentially expressed. In this case, we identified five potential transcription factors belonging to the family AP2. This association is intriguing and suggests a series of genetics experiments that could test if these factors contribute to *in vitro* cyst formation and maintenance.

There are, however, limits to a gene-by-gene analysis. For instance, we were unable to detect differentially expressed genes in either of the RH or PLK strains. RH is a type I parasite and therefore known to be resistant to stress-induced differentiation, but the PLK strain belongs to the type II lineage and is competent to differentiate. Although many of the genes differentially expressed in CTG had a similar pattern of expression PLK, the difference in expression was too low or the variance between replicates too high to achieve statistical significance. The number of required replicates to conclusively discriminate consistent differences is often prohibitively high. In addition, since analyses of this kind involve testing several thousand hypotheses, managing false positive inferences very often makes it impossible to distinguish signal from noise.

Using GSEA, we compared the expression data to a set of genes shown to be differentially expressed after Compound 1 induced differentiation [22]. As Figure 3 shows, we are able to characterize both CTG and PLK, but not RH, as strongly enriched for bradyzoite genes. CTG has quantifiably more enrichment with a normalized enrichment score (NES) of 2.5 (p = 0.000) than PLK with an NES of 1.7 (p = 0.001). This is consistent with a continuum of differentiation competence with the type I lineage very resistant to bradyzoite development, type III differentiating readily, and type II parasites falling somewhere in the middle. Interestingly, the enrichment plot for RH (fig. 3B) is actually shaped like those of PLK, CTG (figs. 3C & 3D), even though the enrichment in RH is not statistically significant. This is consistent with data from other groups that indicates that RH is able to induce many of the stress-associated genes linked with bradyzoite differentiation, but is not able to complete the developmental transition.

## Strain specific metabolic differences

Presently, sequence homology is the primary method of predicting gene function. This poses a difficult problem for divergent eukaryotes like *T. gondii*, where approximately half of protein coding genes remain unannotated. The Kyoto Encyclopedia of Genes and Geneomes (KEGG) curates a collection of molecular pathways [17]. To begin parsing out the biological meaning of many differences in transcriptomes between the strains we used Gene Set Enrichment Analysis (GSEA) software [7] a tool that examines expression data holistically using "functionally related gene sets", rather than testing genes individually. 33 KEGG pathways have been identified in *Toxoplasma* and are an appropriate size for use with GSEA. Tables 1 and 2 summarize the significant results for the interstrain comparisons in tachyzoites and bradyzoites, respectively.

We find very few KEGG pathways enriched in any of the examined conditions. The most prominent result from this analysis is the much higher expression level of ribosomal proteins in CTG. In the absence of other stress, depletion of ribosomal protein RPS13 has been shown to arrest the cell cycle and induce BAG1 expression, but the parasites do not form a mature cyst wall, suggesting a state of partial differentiation [18]. It is not immediately clear why CTG expresses more ribosome components, but translational regulation has been linked to both the stress response and bradyzoite differentiation in *T. gondii* [19].

**A**

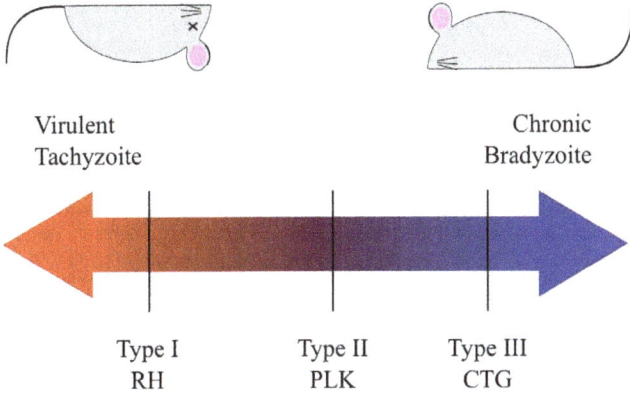

**B**

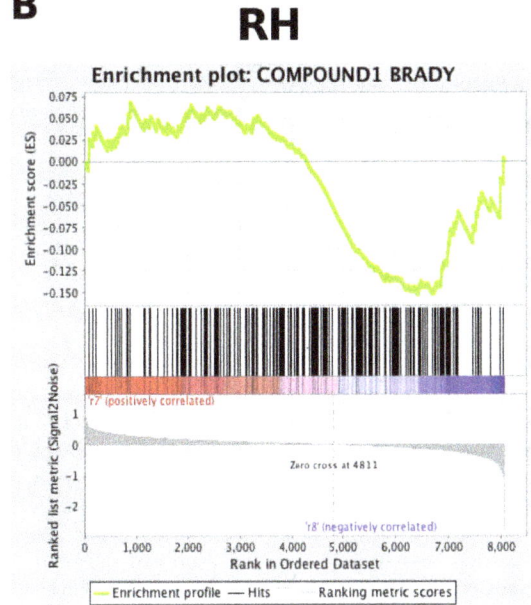

Normalized Enrichment Score:  0.55
P value: 0.996

**C**

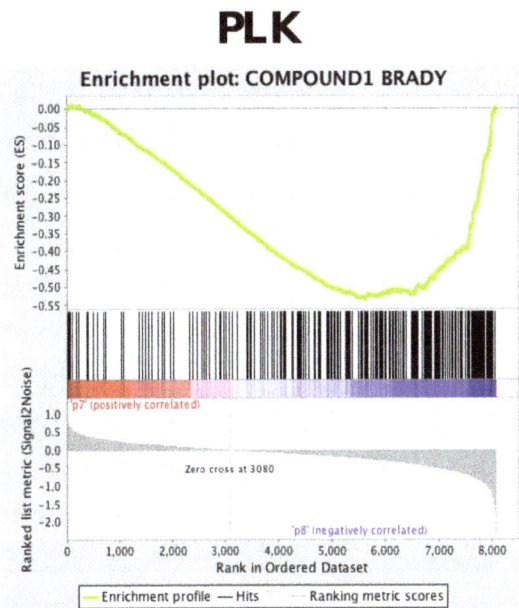

Normalized Enrichment Score:  1.70
P value: 0.001

**D**

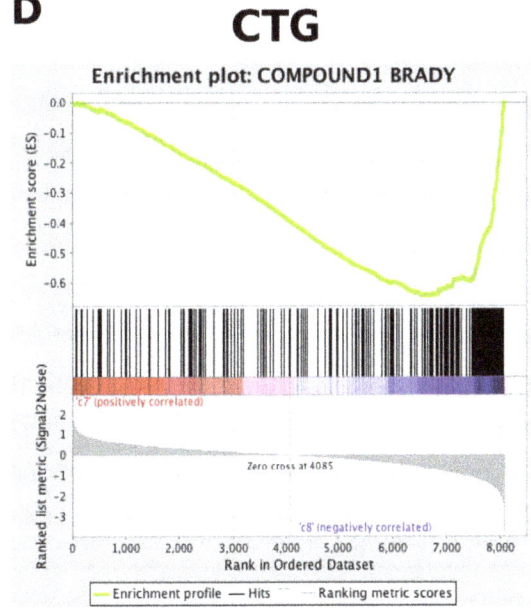

Normalized Enrichment Score:  2.52
P value: 0.000

**Figure 3. GSEA detects bradyzoite-induced genes in PLK and CTG, but not RH parasites.** (A) A schematic of strain virulence of the strains used as a function of ability to differentiate into bradyzoites. (B) GSEA enrichment plot for RH parasites under differentiation conditions compared to compound1 induced genes. Position of black bars indicate ranking of compound 1 genes relative to all other genes. Green line represents strength of enrichment under bradyzoite conditions (right) or tachyzoite conditions (left). (C) Enrichment plot for PLK parasites under differentiation conditions compared to compound 1 induced genes. (D) Enrichment plot for CTG parasites under differentiation conditions compared to compound 1 induced genes.

**Table 1.** Differential expression of KEGG pathways between strains in the tachyzoite stage.

| RH vs. PLK | | RH vs. CTG | | PLK vs. CTG | |
|---|---|---|---|---|---|
| RH | PLK | RH | CTG | PLK | CTG |
| | | | TGO03010: Ribosome | | TGO03010: Ribosome |
| | | | TGO03008: Ribosome Biogenesis in Eukaryote | | TGO03013: RNA Transport |
| | | | TGO03040: Spliceosome | | TGO03008: Ribosome Biogenesis in Eukaryote |
| | | | TGO03013: RNA Transport | | TGO03020: RNA Polymerase |

Thorough annotation of the *Toxoplasma* gene pathways will likely yield more insights into the biological processes underlying the divergence of the strains. The Liverpool Library of Apicomplexan Metabolic Pathways (LAMP) is one such thorough annotation and a promising starting point to address strain-specific differences [20]. However, by GSEA, we were unable to detect enrichment of any LAMP pathway (data not shown). The pathways we tested may not be expressed differently between strains, or the small size of the pathways may contribute to false negative results.

It is possible that despite the large differences in expression we observed between strains, there are not significant changes in the overall biological processes of the parasite between tachyzoites and bradyzoites. Recently Behnke showed that by principal components analysis, the transcriptomes of tachyzoites and bradyzoites group closely when compared to the transcriptomes of merozoite sexual stages harvested from cat intestines or the transcriptome of oocysts [21]. These findings speak to the importance of pathway analysis and global analysis over more traditional gene-by-gene testing.

## Differentiation influences the cell cycle

By examining DNA content, it has been shown that cell cycle arrest accompanies bradyzoite differentiation [24]. More recent work has identified two large sets of cell cycle regulated genes, with one set corresponding to the $G_1$ phase and the other related to genes involved in S-phase and mitosis (S/M) [8]. To test how these cell cycle genes are affected by differentiation stress, we plotted the difference of each gene's expression level between tachyzoite conditions and bradyzoite-inducing stress conditions for each of the three parasite strains.

In keeping with the existing model, Figure 4A shows that S/M genes are more highly expressed in bradyzoite populations while $G_1$ genes are more closely associated with tachyzoites. In bradyzoite differentiated parasites, there is an up regulation of

S/M associated genes, while tachyzoites have higher steady state expression levels of $G_1$-linked genes. The mean difference of each group (red cross) also illustrates these relationships. Further, there is again a clear difference in how each strain is affected by stress (significant difference of means by ANOVA). The cell cycle genes of RH are relatively unaffected by differentiation stress, while PLK experiences significant changes in expression of these genes and CTG more so. These data underscore the fundamental link between cell cycle regulation and life cycle advancement.

In the present model, the "switch" between tachyzoite and bradyzoite occurs in the late S phase [25]. This fits with the observed changes in gene expression, since not every cell cycle regulated gene is influenced the same way following differentiation stress. In Figure 4A, we show that the mean expression values change proportionate to each strain's ability to differentiate, but so does the variance. Not only are many of these gene unaffected by stress, but many of their expression values change in the "wrong" direction, with S/M annotated genes being up regulated in tachyzoites and $G_1$ genes up regulated in bradyzoites.

To understand more precisely how bradyzoite development affects expression of cell cycle genes, we again employed GSEA. This time, we assigned cell-cycle genes into gene sets based on their peak time of expression [26]. In figure 4B, PLK and CTG bradyzoites show clear enrichment of multiple S phase gene sets, while tachyzoites are enriched for $G_1$ gene sets with peak expression between hours four and six. Interestingly, many later $G_1$ gene sets are actually associated with bradyzoites. In this analysis, we also find that the RH cell cycle is affected by the stress conditions, but to a much lesser extent than PLK or CTG.

The cell cycle transcriptome is temporally linked with differentiation stress. Using GSEA, we are able to identify specific cell cycle genes, grouped by time of peak expression, that are linked with bradyzoites or tachyzoites. Once again, the continuum of virulence (Figure 3) is observable. Both the number of the gene sets enriched and the magnitude of the enrichment is great in

**Table 2.** Differential expression of KEGG pathways between strains in treated with alkaline stress bradyzoite induction conditions.

| RH vs. PLK | | RH vs. CTG | | PLK vs. CTG | |
|---|---|---|---|---|---|
| RH | PLK | RH | CTG | PLK | CTG |
| | TGO03030: DNA Replication | | TGO03010: Ribosome | | TGO03010: Ribosome |
| | | | TGO03008: Ribosome Biogenesis in Eukaryotes | | TGO03008: Ribosome Biogenesis in Eukaryotes |
| | | | TGO03030: DNA Replication | | TGO03013: RNA Transport |
| | | | TGO00240: Pyrimidine Metabolism | | TGO00240: Pyrimidine Metabolism |
| | | | | | TGO03020: RNA Polymerase |
| | | | | | TGO03050: Proteasome |

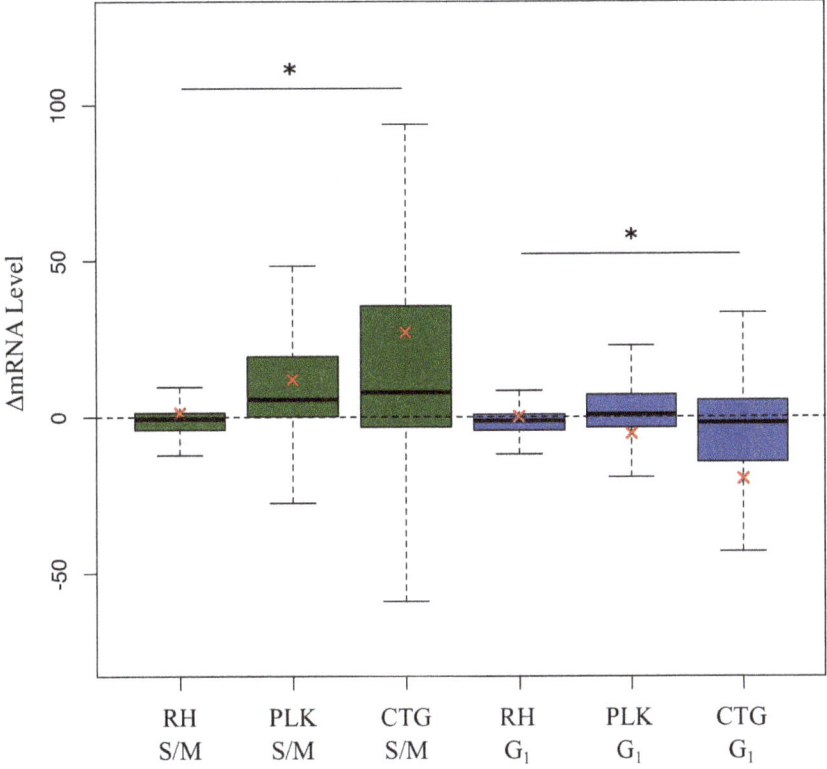

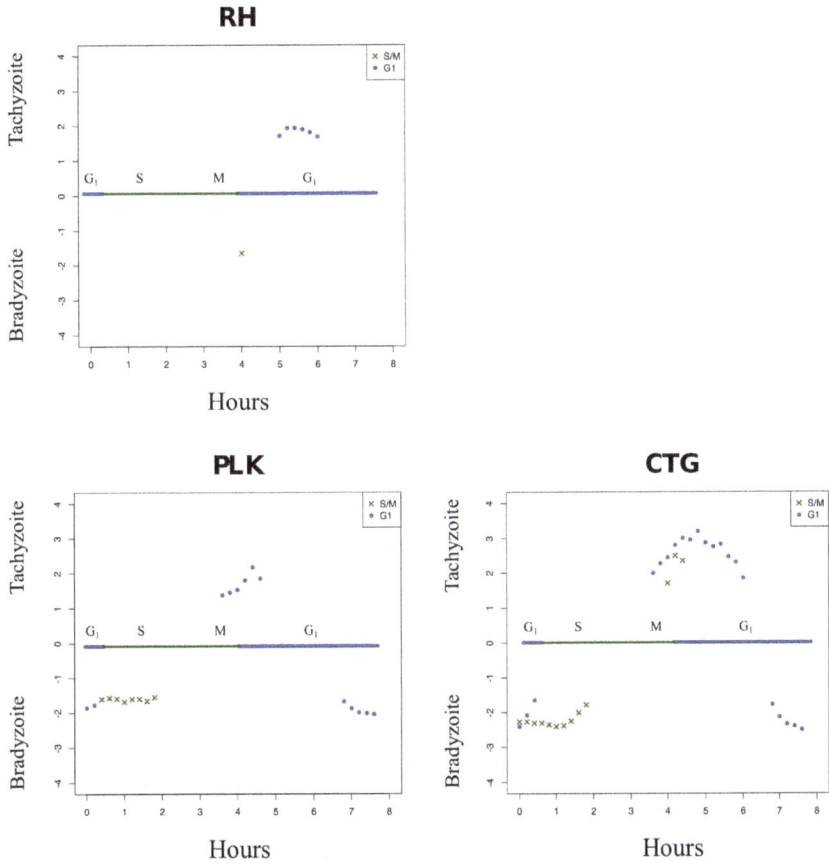

**Figure 4. Expression of cell-cycle genes altered during differentiation.** A) Boxplot represents differences in expression levels of cell cycle-dependent genes following differentiation conditions. Green boxes represent changes in S/M gene expression, blue boxes represent changes in $G_1$ gene expression. A positive difference indicates up regulation of the gene in tachyzoite conditions, while a negative difference in expression values indicates greater expression in the alkaline stress induced bradyzoites. The black bar indicates median value; the red cross indicates mean value. Significance tested by one-way ANOVA. A star (*) indicates: $p < 0.001$. B) Cell cycle genes are annotated as either $G_1$ or S/M [8]. We then sorted genes into groups based on time of peak expression. Each of these gene sets was tested by GSEA [26]. Gene sets with significant (FWER-p value $<0.05$) normalized enrichment scores (NES) are plotted. Positive scores indicate association with the unstressed (tachyzoite) condition, negative scores indicate association with the stress (bradyzoite) condition. Blue and green bar across middle of plots represent an eight hour RH tachyzoite cell cycle. Counter-clockwise from the top, the plots show cell cycle gene sets influenced following RH differentiation, cell cycle gene sets influenced following PLK differentiation, cell cycle gene sets influenced following CTG differentiation. Note that time of expression is based on the RH cell cycle as defined, which is shorter than that of either PLK or CTG.

CTG than PLK. What is most interesting is that the cell cycle of RH parasites does appear to affected by differentiation stress, albeit to a much lesser extent than either of the other two strains. This was the only indication that RH parasites reacted to the alkaline stress conditions and probably speaks to the fact that disruption of the cell cycle precedes the bradyzoite developmental switch.

## Conclusions

In summary our integrative analysis of RNA-seq points to strain specific differences in gene expression that are more prominent than changes in gene expression associated with exposure to alkaline stress. The comparison of strains with different virulence phenotypes and capacity to differentiation using various tools has enabled us to identify pathways and genes that may be conserved between strains as well as those that regulate strain-specific traits.

## Materials and Methods

This work did not involve human subject or animal research. All procedures were approved by the appropriate biosafety committees at the Albert Einstein College of Medicine.

## Parasite culture

Human foreskin fibroblasts (HFF) cells grown in eight 150 mm tissue culture plates were infected with RH (type I), PLK (type II), and CTG (type III) strain in regular medium (pH 7 DMEM with 10% fetal bovine serum, incubated in 5% $CO_2$). The uninvaded RH free parasites were removed 2 hours after inoculation while the CTG and PLK strain free parasites were removed 4 hours later by washing with PBS. Inoculation medium was and then replaced with regular medium (pH 7, 5% $CO_2$) or differentiation medium (pH 8.1 DMEM with 5% fetal bovine serum, 10 mM HEPES, incubated in 0.5% $CO_2$). Thethe RH infection duration was two days while CTG and PLK strain were three days. Tachyzoite preparations were a mix of freshly lysed extracellular parasites and mature vacuoles on the verge of lysis as were alkaline-shocked RH strain. Afterwards, cells were harvested, passed through 27G needle twice to lyse HFF cells and filtered through 3 μm pore polycarbonate membrane to remove HFF cells. Purified parasites were pelleted at 1000 xg for 20 minutes 4°C. Pellets were stored in TRIzol Reagent (Invitrogen) in −80°C.

## Library preparation and sequencing

RNA was extracted using Invitrogen TRIzol Reagent (cat#15596018), followed by genomic DNA removal and cleaning using Qiagen RNase-Free DNase Set kit (cat#79254) and Qiagen Mini RNeasy kit (cat#74104). Agilent 2100 Bioanalyzer was used to assess the integrity of the RNA samples. Only RNA samples having RNA Integrity Number between 9–10 were used. Ambion MicroPoly(A)Purist Kit (cat#AM1919) was used for enrichment of transcripts. The SOLiD Total RNA-Seq Kit (cat#4445374) was used to construct template cDNA for RNA-Seq following the protocol recommended by Applied Biosystems. Briefly, mRNA

was fragmented using chemical hydrolysis followed by ligation with strand specific adapters and reverse transcript to generate cDNA. The cDNA fragments, 150 to 250 bp in size, were subsequently isolated by electrophoresis in 6% Urea-TBE acrylamide gel. The isolated cDNA was amplified through 15 amplification cycles to produce the required number of templates for the SOLiD EZ Bead system, which was used to generate template bead library for the ligation base sequencing by the SOLiD4 instrument.

We aligned sequenced RNA fragments against release 6.1, ME49 strain of *Toxoplasma gondii* [27] using TopHat-1.2 [28]. We set minimum intron size at 30 bp and maximum at 1500 bp, encompassing $>98\%$ of predicted introns [27]. All other parameters were left at their default values. We assigned aligned reads to predicted gene models using BEDTools [29] and generated a table describing how many reads are aligned to each gene.

Using this "counts" table, we generated the unsupervised clustering or multi-dimensional scaling plot in figure 1A with the "plotMDS" function from the Limma package [30] and "predFC" from edgeR package [31]. Both packages are available through the Bioconductor project. Presented data is the mean of three RPKM normalized [32] replicates for each condition. All data generated in this study are accessible at the GEO database under accession number GSE60305. A summary of RPKM for each version 6.1 gene is listed in Table S1. Data have also been provided to the community database www.toxodb.org.

## Detection of differential gene expression

We used the edgeR software package to infer differential expression [31]. Briefly, each gene across all eighteen samples was fitted with a log-linear model, then the regression coefficients for each group (both strain and growth conditions) are compared. Inequality of regression coefficients strongly suggests differential expression. Differentially expressed genes are listed in Table S2.

## Gene Set Enrichment Analysis

GSEA is a pathway analysis tool supported through available through the Broad Institute (http://www.broadinstitute.org/gsea/index. jsp). Subramanian and colleagues describe the algorithm in detail [7]. We permuted by gene sets, not phenotypes and used the "Signal2Noise" ranking metric. Gene sets were based on KEGG pathway annotation [17], while gene sets related to cell cycle and differentiation were developed for implementation of GSEA for *T. gondii* [26].

## Supporting Information

**Table S1   RPKM normalized expression values for all genes, across all samples (TgME49_6.1).** RNA-seq transcriptome data expressed as RPKM for each of 3 biological replicates grown and harvested at pH 7 or pH 8. r = RH; p = PLK; c = CTG.

**Table S2 Membership table for genes differentially expressed in different pairwise comparisons.** The genes are as listed in Table S1, using the version 6.1 gene ID's obtained from www.toxodb.org used to initially map the reads. All genes that were found to be differentially expressed in the comparisons discussed in the text are marked with an 'x'. Cell cycle regulated genes (G1 and S/M) and AP2 genes as defined by Behnke [8] are also indicated. Definition of abbreviations: RH → RH – RH – Tachy vs Brady; PLK → PLK – Tachy vs Brady; CTG → CTG – Tachy vs Brady; rpt → RH vs PLK (Tachyzoite); rct → RH vs CTG (Tachyzoite); pct → PLK vs CTG (Tachyzoite); rpb → RH vs PLK (Bradyzoite); rcb → RH vs CTG (Bradyzoite); pcb → PLK vs CTG (Bradyzoite); g1 → Annotated as G1 cell cycle gene; sm → Annotated as S/M cell cycle gene; ap2 → Predicted or confirmed AP2 DNA-binding domain.

**Table S3 RPKM expression values for annotated AP2 genes.** Product names are those conferred by a community annotation group as available on www.toxodb.org. Gene ID's are

version 6.1. Developmental assignments are per Behnke [8] and Walker Mol Micro [23]. Mean RPKM for each strain and condition are shown: r7 (RH strain pH 7); r8 (RH strain pH 8); p7 (PLK strain pH 7); p8 (PLK strain pH 8); c7 (CTG strain pH 7); c7 (CTG strain pH 8);

## Acknowledgments

Some of this work was published in a thesis submitted in partial fulfillment of the requirements for a Doctor of Philosophy conferred by the Graduate Program in Biomedical Sciences of the Albert Einstein College of Medicine (MMC). We thank Michael White and Bill Sullivan for ongoing discussions throughout the course of this work.

## Author Contributions

Conceived and designed the experiments: MMC GO LMW KK. Performed the experiments: MMC YM LMM. Analyzed the data: MMC LMM RCT GO LMW KK. Contributed reagents/materials/analysis tools: RCT. Wrote the paper: MMC LMW KK.

## References

1. Kim K, Weiss LM (2008) Toxoplasma: the next 100years. Microbes Infect Inst Pasteur 10: 978–984. doi:10.1016/j.micinf.2008.07.015.
2. Khan A, Miller N, Roos DS, Dubey JP, Ajzenberg D, et al. (2011) A monomorphic haplotype of chromosome Ia is associated with widespread success in clonal and nonclonal populations of Toxoplasma gondii. mBio 2: e00228–00211. doi:10.1128/mBio.00228-11.
3. Khan A, Taylor S, Su C, Mackey AJ, Boyle J, et al. (2005) Composite genome map and recombination parameters derived from three archetypal lineages of Toxoplasma gondii. Nucleic Acids Res 33: 2980–2992. doi:10.1093/nar/gki604.
4. Sibley LD, Khan A, Ajioka JW, Rosenthal BM (2009) Genetic diversity of Toxoplasma gondii in animals and humans. Philos Trans R Soc Lond B Biol Sci 364: 2749–2761. doi:10.1098/rstb.2009.0087.
5. Behnke MS, Khan A, Wootton JC, Dubey JP, Tang K, et al. (2011) Virulence differences in Toxoplasma mediated by amplification of a family of polymorphic pseudokinases. Proc Natl Acad Sci USA 108: 9631–9636.
6. Reese ML, Zeiner GM, Saeij JPJ, Boothroyd JC, Boyle JP (2011) Polymorphic family of injected pseudokinases is paramount in Toxoplasma virulence. Proc Natl Acad Sci U S A 108: 9625–9630. doi:10.1073/pnas.1015980108.
7. Subramanian A, Tamayo P, Mootha VK, Mukherjee S, Ebert BL, et al. (2005) Gene set enrichment analysis: a knowledge-based approach for interpreting genome-wide expression profiles. Proc Natl Acad Sci USA 102: 15545–15550.
8. Behnke MS, Wootton JC, Lehmann MM, Radke JB, Lucas O, et al. (2010) Coordinated progression through two subtranscriptomes underlies the tachyzoite cycle of Toxoplasma gondii. PLoS ONE 5: e12354.
9. Balaji S, Babu MM, Iyer LM, Aravind L (2005) Discovery of the principal specific transcription factors of Apicomplexa and their implication for the evolution of the AP2-integrase DNA binding domains. Nucleic Acids Res 33: 3994–4006.
10. Altschul SF, Wootton JC, Zaslavsky E, Yu YK (2010) The construction and use of log-odds substitution scores for multiple sequence alignment. PLoS Comput Biol 6: e1000852.
11. Iwanaga S, Kaneko I, Kato T, Yuda M (2012) Identification of an AP2-family protein that is critical for malaria liver stage development. PLoS ONE 7: e47557.
12. Yuda M, Iwanaga S, Shigenobu S, Kato T, Kaneko I (2010) Transcription factor AP2-Sp and its target genes in malarial sporozoites. Mol Microbiol 75: 854–863.
13. Yuda M, Iwanaga S, Shigenobu S, Mair GR, Janse CJ, et al. (2009) Identification of a transcription factor in the mosquito-invasive stage of malaria parasites. Mol Microbiol 71: 1402–1414.
14. Sinha A, Hughes KR, Modrzynska KK, Otto TD, Pfander C, et al. (2014) A cascade of DNA-binding proteins for sexual commitment and development in Plasmodium. Nature 507: 253–257. doi:10.1038/nature12970.
15. Radke JB, Lucas O, De Silva EK, Ma Y, Sullivan WJ, et al. (2013) ApiAP2 transcription factor restricts development of the Toxoplasma tissue cyst. Proc Natl Acad Sci USA 110: 6871–6876.
16. Lescault PJ, Thompson AB, Patil V, Lirussi D, Burton A, et al. (2010) Genomic data reveal Toxoplasma gondii differentiation mutants are also impaired with respect to switching into a novel extracellular tachyzoite state. PLoS One 5: e14463. doi:10.1371/journal.pone.0014463.
17. Kanehisa M, Goto S, Sato Y, Furumichi M, Tanabe M (2012) KEGG for integration and interpretation of large-scale molecular data sets. Nucleic Acids Res 40: D109–114.
18. Hutson SL, Mui E, Kinsley K, Witola WH, Behnke MS, et al. (2010) T. gondii RP promoters & knockdown reveal molecular pathways associated with proliferation and cell-cycle arrest. PLoS One 5: e14057. doi:10.1371/journal.pone.0014057.
19. Konrad C, Wek RC, Sullivan WJ Jr (2014) GCN2-like eIF2α kinase manages the amino acid starvation response in Toxoplasma gondii. Int J Parasitol 44: 139–146. doi:10.1016/j.ijpara.2013.08.005.
20. Shanmugasundram A, Gonzalez-Galarza FF, Wastling JM, Vasieva O, Jones AR (2013) Library of Apicomplexan Metabolic Pathways: a manually curated database for metabolic pathways of apicomplexan parasites. Nucleic Acids Res 41: D706–713. doi:10.1093/nar/gks1139.
21. Behnke MS, Zhang TP, Dubey JP, Sibley LD (2014) Toxoplasma gondii merozoite gene expression analysis with comparison to the life cycle discloses a unique expression state during enteric development. BMC Genomics 15: 350. doi:10.1186/1471-2164-15-350.
22. Behnke MS, Radke JB, Smith AT, Sullivan WJ Jr, White MW (2008) The transcription of bradyzoite genes in Toxoplasma gondii is controlled by autonomous promoter elements. Mol Microbiol 68: 1502–1518. doi:10.1111/j.1365-2958.2008.06249.x.
23. Walker R, Gissot M, Croken MM, Huot L, Hot D, et al. (2013) The Toxoplasma nuclear factor TgAP2XI-4 controls bradyzoite gene expression and cyst formation. Mol Microbiol 87: 641–655. doi:10.1111/mmi.12121.
24. Radke JR, Guerini MN, Jerome M, White MW (2003) A change in the premitotic period of the cell cycle is associated with bradyzoite differentiation in Toxoplasma gondii. Mol Biochem Parasitol 131: 119–127.
25. White MW, Radke JR, Radke JB (2014) Toxoplasma development - turn the switch on or off? Cell Microbiol 16: 466–472. doi:10.1111/cmi.12267.
26. Croken MM, Qiu W, White MW, Kim K (2014) Gene Set Enrichment Analysis (GSEA) of Toxoplasma gondii expression datasets links cell cycle progression and the bradyzoite developmental program. BMC Genomics 15: 515. doi:10.1186/1471-2164-15-515.
27. Gajria B, Bahl A, Brestelli J, Dommer J, Fischer S, et al. (2008) ToxoDB: an integrated Toxoplasma gondii database resource. Nucleic Acids Res 36: D553–556.
28. Trapnell C, Pachter L, Salzberg SL (2009) TopHat: discovering splice junctions with RNA-Seq. Bioinformatics 25: 1105–1111.
29. Quinlan AR, Hall IM (2010) BEDTools: a flexible suite of utilities for comparing genomic features. Bioinformatics 26: 841–842.
30. Smyth GK (2005) Limma: linear models for microarray data. In: Gentleman R, Carey V, Dudoit S, Irizarry R, Huber W, editors. Bioinformatics and Computational Biology Solutions Using R and Bioconductor. New York: Springer. pp. 397–420.
31. Robinson MD, McCarthy DJ, Smyth GK (2010) edgeR: a Bioconductor package for differential expression analysis of digital gene expression data. Bioinformatics 26: 139–140.
32. Mortazavi A, Williams BA, McCue K, Schaeffer L, Wold B (2008) Mapping and quantifying mammalian transcriptomes by RNA-Seq. Nat Methods 5: 621–628.
33. Su C, Evans D, Cole RH, Kissinger JC, Ajioka JW, et al. (2003) Recent expansion of Toxoplasma through enhanced oral transmission. Science 299: 414–416.
34. Knoll LJ, Tomita T, Weiss LM (2014) Bradyzoite Development. In: Weiss LM, Kim K, editors. *Toxoplasma gondii The Model Apicomplexan: Perspectives and Methods.* 2nd edition. New York: Academic Press. pp. 521–551.

# 15

# Coincidental Loss of Bacterial Virulence in Multi-Enemy Microbial Communities

Ji Zhang[1,2]*, Tarmo Ketola[1], Anni-Maria Örmälä-Odegrip[2], Johanna Mappes[1], Jouni Laakso[1,2]

1 Centre of Excellence in Biological Interactions, Department of Biological and Environmental Science, University of Jyväskylä, Jyväskylä, Finland, 2 Department of Biological and Environmental Science, University of Helsinki, Helsinki, Finland

## Abstract

The coincidental virulence evolution hypothesis suggests that outside-host selection, such as predation, parasitism and resource competition can indirectly affect the virulence of environmentally-growing bacterial pathogens. While there are some examples of coincidental environmental selection for virulence, it is also possible that the resource acquisition and enemy defence is selecting against it. To test these ideas we conducted an evolutionary experiment by exposing the opportunistic pathogen bacterium *Serratia marcescens* to the particle-feeding ciliate *Tetrahymena thermophila*, the surface-feeding amoeba *Acanthamoeba castellanii*, and the lytic bacteriophage Semad11, in all possible combinations in a simulated pond water environment. After 8 weeks the virulence of the 384 evolved clones were quantified with fruit fly *Drosophila melanogaster* oral infection model, and several other life-history traits were measured. We found that in comparison to ancestor bacteria, evolutionary treatments reduced the virulence in most of the treatments, but this reduction was not clearly related to any changes in other life-history traits. This suggests that virulence traits do not evolve in close relation with these life-history traits, or that different traits might link to virulence in different selective environments, for example via resource allocation trade-offs.

**Editor:** Boris Alexander Vinatzer, Virginia Tech, United States of America

**Funding:** This work was supported by the Finnish Academy to JL, 1130724 and 1255572 (URL: www.aka.fi), which had a role in study design, data collection and analysis, decision to publish and preparation of the manuscript. This work was also supported by the CoE in Biological Interactions to JM, 252411 (URL: https://www.jyu.fi/bioenv/en/divisions/coe-interactions), which had a role in study design, data collection and analysis, decision to publish and preparation of the manuscript. This work was also supported by the Finnish Cultural Foundation to JZ (URL: www.skr.fi), which had a role in data collection and analysis, decision to publish and preparation of the manuscript. This work was also supported by the Ellen and Artturi Nyyssö nen Foundation to JZ (URL: www.eans.fi), which had a role in study design, data collection and analysis.

**Competing Interests:** The authors have declared that no competing interests exist.

* Email: Ji.Zhang@Helsinki.Fi

## Introduction

Compared to the vast knowledge on the prevention and treatment of bacterial infectious disease, relatively little is known about how the virulence of bacteria has evolved. Virulence evolution is often exemplified as a tug of war between the multicellular host and the pathogen, where the virulence (the degree of host damage or mortality caused by the pathogen) [1] evolves solely through host-pathogen interaction [2–4]. Contrary to this idea, the "coincidental evolution of virulence hypothesis" suggests that virulence evolves indirectly due to selection forces that are not related to the host-pathogen interaction *per se*, but because of selection that occurs outside host environments [2,3,5,6]. This is a plausible expectation when considering opportunistic, environmentally growing bacterial pathogens because they typically live in a complex web of interactions with biotic and abiotic selection pressures that might not be directly connected to their potential hosts [7].

In the natural environment, top-down regulation by bacteriophages and protozoans are two major biotic causes of bacterial mortality [8,9]. In order to survive, bacteria have evolved wide arrays of defence mechanisms against their natural enemies [10,11]. These adaptations have also been suggested to alter the virulence of the bacteria [11–13]. For example, a biofilm-forming ability can effectively lower predation pressure by ciliate predators that prey in the open water. However, the biofilm-forming ability of many bacteria can also be directly linked to the virulence of bacteria as it can prevent macrophage phagocytosis inside the multicellular host [11,14–16].

In addition to the means that prevent predator ingestion in the first place, bacteria have evolved ways to survive the ingestion process and even benefit from it [11]. Survival and reproduction inside protozoan predators, especially in amoebae, may have even contributed to the evolution of several bacterial pathogens [11]. Therefore, virulence could have evolved via adaptations to survive inside protozoan food vacuoles, which could then promote survival within phagocytes in the immune system [17,18]. Perhaps the most typical example of this type of evolution is *Legionella pneumophila* causing Legionnaires' disease. This species is sometimes found as a parasite of free-living amoeba [19–21]. However, infection of the human body is an evolutionary dead end for *L. pneumophila* because human-to-human transmission is unlikely [22,23]. This suggests that the virulence traits of *L. pneumophila* are not evolved from human-bacteria interaction, but rather "coincidentally" via amoeba-bacteria interaction

[2,24]. In fact this linkage is assumed strong enough that the virulence of bacterial clones are frequently assayed indirectly via amoebae resistance tests [25–31].

Bacteriophages can also have a profound impact on the evolution of bacterial virulence. Bacteriophages are known to carry important virulence genes [32–34]. For example, they have been found to contain genes encoding exotoxins and other virulence factors that can be horizontally transferred into the bacterial genome [35,36]. Moreover, bacteria can alter their cell surface antigens to evade phage adsorption [10], whilst host immune systems rely on bacterial surface antigens to identify bacterial invaders [37,38]. Thus bacteriophage selected bacterial surface antigens could indirectly affect host entry, either positively or negatively.

Although protozoan predators and bacteriophages could potentially contribute to elevated bacterial virulence, outside-host defensive adaptations can also be costly and traded off with virulence related traits [39,40]. For example, when bacteria experience protozoan predation, the motility of bacteria that is sometimes positively linked to virulence [41,42] can trade off with anti-predator traits resulting in lowered virulence [39]. It has also been shown that elevated outside-host temperature can select for higher virulence in *Serratia marcescens*, while coevolution with phage can counteract this effect [43]. Moreover high virulence in *Salmonella typhimurium* can be costly in terms of reduced growth in the outside host environment because of the expression of virulence factor (type III secretion system) in a non-host environment [44]. The nutritional conditions of the bacterial growth environment can also significantly affect bacterial metabolism and the expression of virulence factors [45–47]. For example, it has been found that the virulence of the pathogenic fungi was negatively correlated to the carbon-to-nitrogen (C:N) ratio of the culturing medium [48–50]. Therefore, if a similar correlation occurs for bacteria, then the costly virulence traits might be selected against during a prolonged period in a non-host environment. In conclusion, the environmental lifestyle can attenuate or strengthen the virulence depending on the selection forces in the system [7].

Although predators are supposed to play an important role in the evolution of virulence, experiments testing this theory are rare and the studies that do exist only consider a single predator system [40,51,52]. However, in a natural environment it is more conceivable that several predators are present simultaneously, potentially complicating the picture considerably. To test how virulence and other life-history traits evolve in complex enemy communities, we cultured the facultative pathogen *S. marcescens* either alone or with three types of common bacterial predators (amoeba, ciliate and bacteriophage in all seven possible combinations) in a simulated pond water environment for 8 weeks. *S. marcescens* is a gram-negative opportunistic pathogen infecting a broad spectrum of hosts, including plants, corals, nematodes, insects, fish and mammals [57,58]. They can also be found free-living in soil, freshwater, and marine ecosystems [59,60] making it likely that *S. marcescens* frequently encounters parasitic and predatory organisms. Notably, *S. marcescens* is also capable of re-entering the environment after decomposing the host. This creates the possibility that the pathogen virulence is selected in nature by both environmental and host-pathogen interactions. During the experiment we followed the population dynamics of the prey bacterium. Due to the presumed importance of predators on the evolution of the bacterial virulence [28,53,54], the amoeba densities were also followed throughout the experiment. After the evolution experiment, a library containing 384 differentially evolved clones was built to detect changes in virulence, growth ability, biofilm-forming ability and amoeba resistance. The virulence of the ancestor and the evolved bacteria, *S. marcescens* Db11 was quantified in the fruit fly (*Drosophila melanogaster*) hosts via an oral infection model [55]. Since phagocytes play a vital role in the clearance of the Db11 from the hemolymph in this animal model [55], we believed that choosing the bacterial strain and infection model was relevant to our study. We hypothesized that if *S. marcescens* Db11 gained amoeba-resistance in the presence of amoeba predation, this resistance could be used to fight against phagocytes in the hemolymph, and thus gain higher virulence. With data from a multi-predator experiment we can test if the bacterial virulence is selected for, a result that is expected in the presence of bacterial enemies (phage, ciliate and amoebae), especially amoebae. However, it is also plausible that selection pressures by bacterial enemies could select against virulence [39,40].

## Methods

### Study species

*Serratia marcescens* Db11 [56,57] was initially isolated from a dead fruit fly and was kindly provided by Prof. Hinrich Schulenburg. The predatory particle feeding ciliate, *Tetrahymena thermophila* (strain ATCC 30008) has a short generation time of ca. 2 h [61] and was obtained from American Type Culture Collection. It is routinely maintained in PPY (Proteose Peptone Yeast Medium) at 25°C [51,62]. The free-living amoeba, *Acanthamoeba castellanii* (strain CCAP 1501/10) has a generation time ca. 7 h [62] and was obtained from Culture Collection of Algae and Protozoa (Freshwater Biological Association, The Ferry House, Ambleside, United Kingdom) and routinely maintained in PPG (Proteose Peptone Glucose Medium) [63] at 25°C. Obligatory lytic bacteriophage Semad11, capable of infecting *S. marcescens* Db11, was isolated from a sewage treatment plant in Jyväskylä, Finland in 2009. No specific permission was required for collection or location of the bacteriophage. Semad11 is a T7-like bacteriophage belonging to Podoviridae (A.-M. Örmälä-Odegrip, unpublished data).

The evolution experiment was performed in New Cereal Leaf - Page's Amoeba Saline Solution (NAS) medium which was prepared as follows: 1 g of cereal grass powder (Aldon Corp., Avon, NY) was boiled in 1 liter of $dH_2O$ for 5 minutes, and then filtered through a glass fiber filter (GF/C, Whatman). After cooling, 5 ml of both PAS stock solutions I and II were added before being made up to a final volume of 1 litre with deionized water [64,65].

Before the experiment started, the organisms were cultured separately and prepared as follows: bacterial culture, a single colony of *S. marcescens* was seeded to 80 ml of NAS medium in a polycarbonate Erlenmeyer flask capped with a membrane filter (Corning). The flask was incubated at 25°C on a rotating shaker (120 rpm) for 48 hours. The amoeba and ciliate cells were harvested and washed twice in 40 ml of PAS (Page's Amoeba Saline) with centrifugation at $1200 \times g$ for 15 min to pellet the cells. After the centrifugation, cells were suspended in PAS and adjusted to a final concentration of ca. 10 cells $\mu l^{-1}$. To prepare the bacteriophage stock, LB-Soft agar (0.7%) from semi-confluent plates was collected and mixed with LB (4 ml per plate), and incubated for 3.5 h at 37°C. Debris was removed by centrifugation for 20 min at $9682 \times g$ at 5°C. Stock was filtered with 0.2 μm Acrodisc Syringe Filters (Pall). The bacteriophage stock was diluted 1:100,000 in NAS medium, giving approximately $10^6$ plaque-forming unit (PFU) $ml^{-1}$.

## Evolution experiment

The bacterium *S. marcescens* was either cultured alone or in a co-culture with the ciliates, amoebae and bacteriophages enemies 8 combinations (B, BA, BC, BP, BAC, BAP, BCP and BACP; B: bacteria; A: amoebae; C: ciliate; P: phage; Figure 1) for 8 weeks. Each treatment was replicated in 8 flasks. The experiment was initiated in 25 cm$^2$ polystyrene flasks with 0.2 µm hydrophobic filter membrane caps (Sarstedt). Each flask was inoculated with 1 ml of the appropriate microorganism suspension and then the total volume was adjusted to 15 ml with NAS medium. The static liquid cultures were incubated at 25°C and 50% of the medium were replaced weekly with fresh NAS medium, making the system a pulsed resource type [66,67]. Static liquid culture would create a spatial structuration that was similar to the pond water environment. All samples were taken just before the weekly medium renewal (Figure 1).

## Measurements during the evolution experiment

**Bacterial biomass dynamics.** Bacterial biomass in the free water phase was measured from 5 separate 200 µl samples from each flask on 100-well Honeycomb plates (Oy Growth Curves Ab Ltd). The amount of biomass was measured as optical density (OD) at 460–580 nm wavelength using Bioscreen C spectrophotometer (Oy Growth Curves Ab Ltd). The measurements were repeated 10 times at 5 min intervals. The mean of the measurements was used in the data analysis. To measure the amount of *S. marcescens* biofilm attached to the flask walls after 8 weeks had expired, 15 ml of 1% crystal violet solution (Sigma-Aldrich) was injected to the flasks. After 10 minutes, the flasks were rinsed 3 times with distilled water, and then 15 ml of 96% ethanol was added to flasks to dissolve crystal violet from the walls for 24 hours [68]. The amount of biofilm was quantified with the OD of the crystal violet-ethanol solution at 460–580 nm with Bioscreen C spectrophotometer [66].

**Amoeba population dynamics.** To follow the population dynamics of the amoeba, we measured the density of amoeba cells attached on the flask well. This measurement largely reflects the amoeba population dynamics in the flasks since the proportion of floating cells and cysts would be minimal after 7 days culture in the static cultures. In brief the flasks were carefully flipped and images (total area 5.23 mm$^2$) of the flask wall were digitized with an Olympus SZX microscope (32× magnification). The amoeba cells

attached to the flask wall were counted with a script developed in our lab for the Image Pro Plus software (v. 7.0) (Material S1). To determine the ciliate density by the end of the experiment, 250 µl of open water sample was mixed with 10 µl Lugol solution and injected into a glass cuvette rack (depth 2.34 mm). For each sample, 8 randomly placed images (total area 41.84 mm$^2$) were digitized with an Olympus SZX microscope (32× magnification). The cell numbers in each image were counted with an Image Pro Plus script [69].

**Detecting phage presence.** To detect if the bacteriophages were present in the microcosms throughout the experiment and to detect possible contamination, we took 3 independent 500 µl samples from all flasks at the end of evolution experiment. The samples were treated with chloroform and centrifuged to remove bacteria, amoebas and ciliates. 10 µl supernatant drops were then added to 1.5% agar plates. The upper layer of the each plate was covered with 0.7% LB-agar that was mixed with 200 µl of overnight grown *S. marcescens* Db11 ancestor cells. The plates were incubated overnight in 25°C and the presence of phage plaques were checked.

## Measurements after the evolution experiment

**Amoeba plaque test.** After the evolution experiment was finished, half of the flasks from each treatment were randomly sampled to test for any resistance of the bacteria to amoeba predation. The test was adapted from a Wildschutte *et al.* [70] briefly, the flasks from the evolution experiments were shaken vigorously before 1 ml of the culture was transferred to a new tube containing 7 ml of dH$_2$O. The tubes were mixed thoroughly and then centrifuged at 250 g for 10 min to bring down the suspended protozoan cells. 1 ml of the supernatant was spread evenly on to LN agar plates (PAS with 0.2% peptone, 0.2% glucose and 1% agar). A total of 10$^5$ predatory amoeba cells (washed twice in PAS) suspended in 15 µl PAS solution were added to a sterile paper disk, and then placed in the middle of the plate. All the plates were incubated at 25°C for 8 days and then photographed. The images of the plates were used to measure plaque sizes with Image Pro Plus software (v. 7.0). A large plaque size indicates a small amoeba predation resistance.

**Growth and biofilm forming ability of the individual clones.** After the evolution experiment was complete, liquid samples from each replicate population of all treatments were

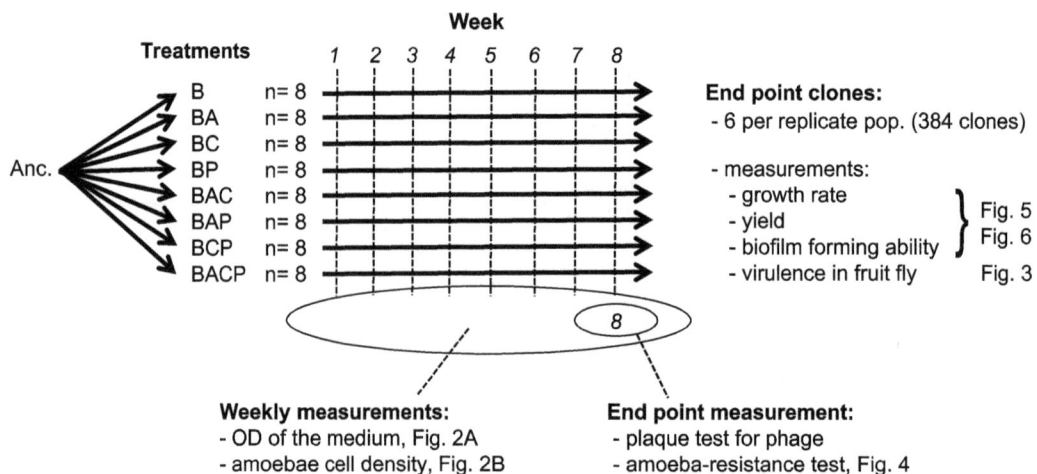

**Figure 1. Schematic overview of our experimental evolution study, number of replicate populations, legends for treatments (Anc.: ancestral bacterial strain DB 11; B: bacteria; A: amoebae; C: ciliate; P: phage), and descriptions of different measurements.**

streaked on three Luria–Bertani (LB) agar plates. They were incubated for 48 hours at 25°C, two bacterial colonies were randomly picked from each plate and inoculated to 5 ml LB liquid medium. The clones were grown at 25°C overnight on a shaker (120 rpm). To make stock cultures of the clone library, 200 µl of the liquid culture of each strain was mixed with 200 µl of 80% glycerol on 100-well Honeycomb plates in a randomized order and stored at −80°C. Prior to clonal growth measurements stock cultures from the clone library were inoculated to 100-well Honeycomb plates, directly from the freezer, with a plate replicator (EnzyScreen). Each well of the 100-well Honeycomb plates contained 400 µl fresh LB liquid medium. The OD of each well was measured continuously without shaking for 30 hours in 5 min intervals to estimate the maximum growth rate and maximum population size of the clones. After 30 hours, 100 µl of 1% crystal violet solution (Sigma-Aldrich) was added to each well to quantify amount of biofilm that was produced. After 10 minutes incubation in 25°C, the plates were rinsed with distilled water 3 times and then 400 µl of 96% ethanol was added to each well and left for 24 hours to dissolve the crystal violet from the walls [68]. The amount of biofilm was quantified by measuring the OD of the crystal violet-ethanol solution at 460–580 nm with Bioscreen C spectrophotometer [66].

Identical measurements were also recorded with NAS medium either with or without the protozoan predators (amoeba or ciliate). The abovementioned clones were first grown in 400 µl LB liquid medium in 100-well Honeycomb plates. After incubation at 25°C for 24 h in static cultures, 10 µl of the bacterial culture was transferred to 100-well Honeycomb plates. Each well contained 390 µl NAS mixed with or without NAS washed amoeba (5–10 cells/µl) or ciliate cells (0.5–1 cells/µl). Subsequent measurements for growth and biofilm assay were performed as described above.

Estimation of growth rate and yield was based on the Matlab script written by TK that fits linear regression to 25 time-points along a sliding data window with background correction, using ln-transformed OD data. The maximal growth rate is determined by finding the largest slope of linear regression within all fitted regressions for the particular clone. The yield was determined as the highest average OD over the 25 data point window.

**Virulence of the evolved clones.** Stock cultures of the clone library were inoculated to 100-well Honeycomb plates, filled with 400 µl fresh LB liquid medium with plate replicator (EnzyScreen). For a positive control the ancestor *S. marcescens* Db11 was added to two separate wells in each plate and used in the subsequent infection experiment. After 24 h incubation at 25°C without shaking, 800 µl of the bacterial culture was mixed with the 800 µl of 100 mM sucrose solution. The mixture was absorbed to cotton dental roll (Top Dent, Lifco Dental, Enköping, Sweden) folded on the bottom of a standard 75×23 mm fly vial (Sarstedt, Nümbrecht, Germany). 1600 µl of 100 mM sucrose solution was used as a negative control. Ten *D. melanogaster* adults (2–3 days old) from a large laboratory colony (Oregon R, kindly provided by Christina Nokkala from the University of Turku) were transferred to each vial and plugged with cotton. This was done for all the bacterial clones. Deaths of flies were monitored over next 4 days at 3–6 h intervals.

## Statistical analysis

Changes in bacterial density and amoeba density were compared using repeated measurements ANOVA. The effects of the evolutionary treatments on bacterial virulence were quantified with Cox regression by fitting evolutionary treatment and identity of population as categorical covariates. The amoeba plaque test and the amount of biofilm at the end of the evolution experiment

were compared using ANOVA. All the analyses were done with SPSS v. 19 (IBM).

Life history and defensive traits of evolved clones were tested with ANOVA including treatment (all possible combinations of predators) as a fixed factor and population identity as a random factor. From the data we tested effects of treatments on growth rate, yield and biofilm, as well as growth rate and yield under the influence of ciliate or amoebae presence. Coevolution of traits was studied with MANOVA and subsequent eigenanalysis to reveal if changes in certain traits would lead to corresponding changes in other traits across the treatments. Thus, this analysis allows pinpointing strongly interconnected traits [71]. MANOVA and eigenanalysis was performed with MATLAB function manova1 (R2012a, Mathworks; Statistics toolbox) for population averaged trait values. Values used for MANOVA for virulence were hazard function coefficients averaged over the populations within each treatment.

Two replicated populations (one in the treatment BC and one in BAC) were found contaminated by Semad11 phages. Moreover, in two replicates of the treatment BACP phages were not detected by plaque assay. All the other samples from phage containing treatments formed phage plaques on the ancestor Db11 bacterial lawn. This confirms that phages did not go extinct during the experiment. The aforementioned 4 flasks were excluded from the data analysis.

## Results

### The population dynamics of bacterial prey and amoeba predators during the experiment

The presence of predators generally reduced the bacterial biomass in free water phase (OD of the medium: $F_{7, 52} = 674.620$, p<0.001; Figure 2A). The ciliates reduced the biomass most dramatically: on average by 27% during the weeks 1–8 when compared to the control (B). Biomass reduction by ciliate and phage (BCP), and amoeba and ciliate (BAC) communities was 25%. Amoeba (BA) and amoeba and phage (BAP) communities reduced bacteria biomasses by 20%. The bacteriophage (BP) reduced the biomass only by 2%. The pairwise-comparisons of the rest of the treatments were significant after Bonferroni correction, except BCP vs. BC, and BACP vs. BAC.

The amount of biofilm produced in each treatment was different at week 8 when measured directly from the microcosm walls (ANOVA, $F_{7, 52} = 39.101$, p<0.001). The highest amount of biofilm was found in the presence of ciliates (BC) and the lowest amount of biofilm was found in the treatments BAC and BACP. Detailed pairwise comparison can be found in Table S1.

The amoeba population sizes declined in all treatments after the initial increase during the first week ($F_{3, 28} = 280.257$, p<0.001; Figure 2B). The amoeba population sizes were higher in amoeba (BA) and amoeba and phage (BAP) treatments (on average 30 cells µl$^{-1}$) throughout the 8-week evolution experiment. Adding phage to the amoeba treatment did not change the population dynamics of the amoeba (Fisher's LSD: BA vs. BAP, p = 0.485). However, ciliates reduced the amoeba population sizes: on average only 5 cells µl$^{-1}$ were found throughout the experiment in treatment BAC, and on average 8 cells/µl in treatment BACP.

### Virulence

In order to explore if past selection with predators had influenced virulence we utilized the *Drosophila* oral infection assay. The treatment group that had evolved with ciliates and phages (BCP) had clearly lower virulence than the rest of the evolved treatment groups (p<0.01 in all pairwise comparisons;

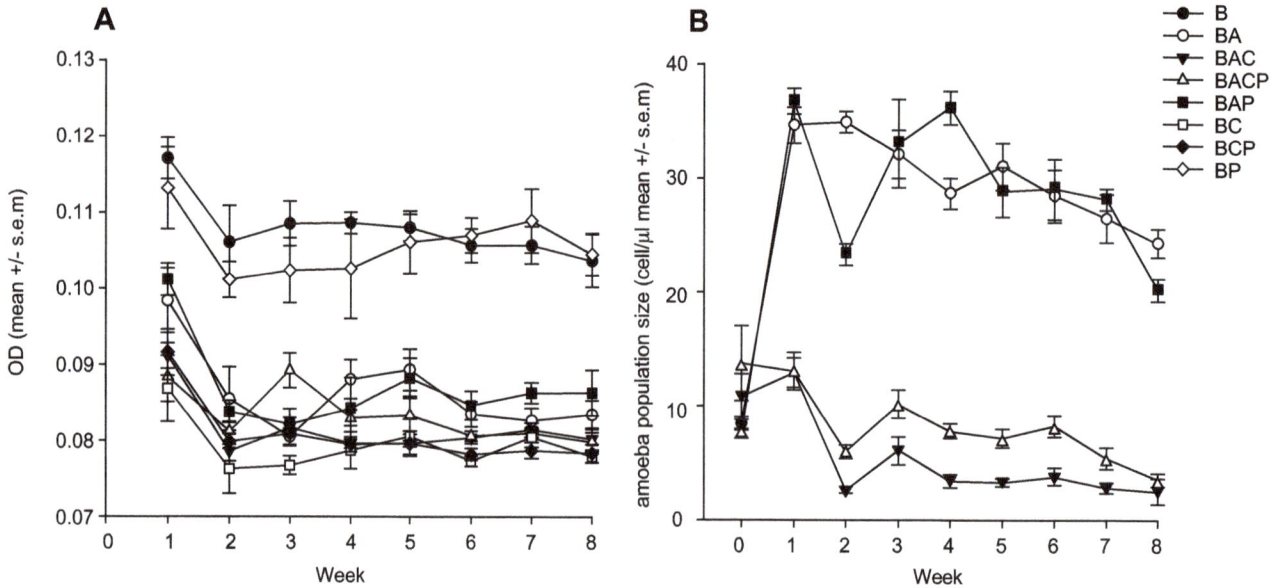

**Figure 2. Bacterial biomass dynamics (A) and amoebae population dynamics (B) during the eight-week evolution experiment.** The bacteria were reared alone or in several combinations of bacterial enemies (Anc.: ancestral bacterial strain DB 11; B: bacteria; A: amoebae; C: ciliate; P: phage). See Table S1 for pairwise comparisons.

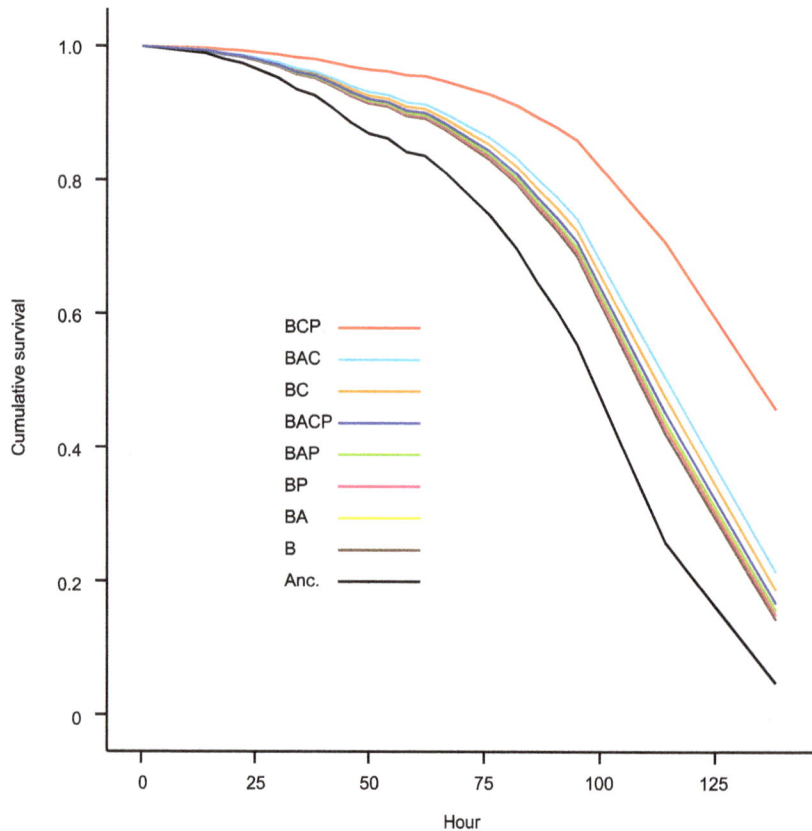

**Figure 3. Cumulative survival curves of the fruit flies that were infected with evolved and ancestral bacterial clones.** Anc.: ancestral bacterial strain DB 11; B: bacteria; A: amoebae; C: ciliate; P: phage. The survival curves represented the pooled survival data of the 480 fly individuals for each treatment (10 flies per vial, 6 clones per population and 8 replicates per treatment). The treatment codes are in the order of the increasing virulence. See Table S2 for pairwise comparisons.

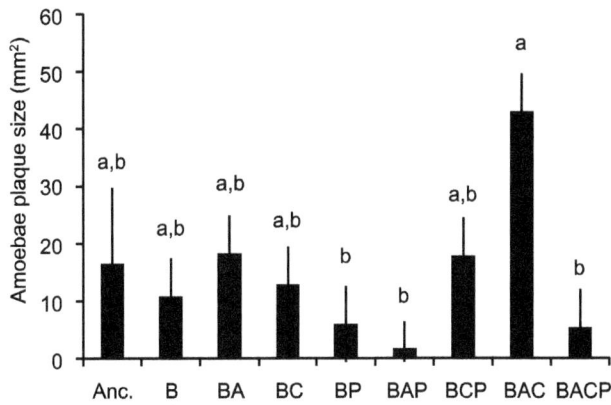

**Figure 4. Sensitivity of evolved bacteria on amoebae predation measured using amoeba plaque test.** Anc.: ancestral bacterial strain DB 11; B: bacteria; A: amoebae; C: ciliate; P: phage. Sensitivity is measured as a plaque size (mm$^2$) in bacterial lawn caused by the introduced amoeba in semi-solid agar plate. Letters indicate if treatment means are statistically similar (p>0.05), after Bonferroni correction for multiple comparisons. Tests are based on the post hoc comparisons of estimated marginal means for treatments ANOVA. All bars correspond to 4 randomly picked samples from 8 replicate populations.

Figure 3; Table S2). In this group the between population variation was also very high and statistically significant. In the majority of the other groups, the between population variation was clearly non-significant (p<0.135, but in BCP, p<0.001; BAC p = 0.003). Moreover, all treatments had lower virulence than the ancestor (p<0.01 in all pairwise comparisons; Figure 3; Table S2). There was no statistical support for the difference in virulence between any other treatment groups.

## Sensitivity to amoebae predation

The amoeba plaque test revealed that the sensitivity to amoeba predation (measured with the area of visible plaque formed on bacterial lawn) was highest in bacteria co-cultured with amoeba and ciliates (BAC: p<0.05 in all pairwise comparisons; Figure 4;

Table S1). The lowest amoeba sensitivity was found in the treatment where the amoeba were reared with phages (BAP; Figure 4) or with ciliates and phages (BACP; BAP vs. BACP: p = 0.66, Figure 4). The detailed result of pairwise comparisons can be found in the Table S1. Although the treatment group that had evolved with ciliates and phages (BCP) had the lower virulence than the other evolved treatment groups, its sensitivity to amoeba predation did not differ from the others in pairwise comparisons (Table S1), which suggests no clear link between amoebae predation and virulence.

## Life-history traits

Treatments did not influence the maximal growth rate strongly (Table 1), however low growth rates were found in clones that had evolved with ciliates (BC; Figure 5A; Table S2). The ancestor clones had the largest yield. Evolutionary treatments did not differ greatly from each other but the clones that had evolved with phages had the lowest yield (BP; Figure 5B; Table S2). Evolutionary changes occurred most dramatically in the biofilm forming ability. The highest biofilm forming abilities were found from the clones that had evolved alone (B) or with amoeba and ciliate (BAC). Intermediate biofilm production was found in ancestral clones (Anc.) or if clones had evolved with amoebae (BA) or with all enemies (BACP). The lowest biofilm production was found if clones that had evolved with ciliate (BC), phage (BP), amoebae and phage (BAP) or with ciliate and phage (BCP) (Figure 5C; Table S1).

From the defensive traits the strongest changes were observed in growth rate and yield when the bacteria were co-cultured with amoebae. In both of the traits ancestor bacteria deviated from the clones that had undergone evolutionary treatments; ancestors had higher growth rate co-cultured with amoebae but lower yield than evolved clones (Figure 6; Table S2). Similarly, from the evolved groups the highest growth rate was found from the group that had evolved with phage and amoebae (BAP), whereas its yield with amoebae was lowest (Figure 6; Table S2). Growth rate measurements did not indicate that treatments affected the resistance of clones against ciliate predators. However, yield with ciliates was lowest if bacteria had evolved alone (B) or with amoebae and

**Table 1.** Estimated (ANOVA) evolutionary effects of different combinations of enemies (treatment), and population identity on bacterial virulence against *Drosophila melanogaster*, and on bacterial life-history traits, measured alone or with amoebae or ciliate.

| | Treatment | | | Population | | |
|---|---|---|---|---|---|---|
| | Wald | df | p | Wald | df | p |
| Virulence | 48.6 | 8 | <0.001 | 169.58 | 52 | <0.001 |
| | F | df1,2 | p | F | df | p |
| Growth rate | 2.465 | 8,32.007 | 0.033 | 0.799 | 52,306 | 0.836 |
| Yield | 2.582 | 8,39.627 | 0.023 | 1.349 | 52,306 | 0.066 |
| Biofilm | 12.987 | 8,37.001 | <0.001 | 1.097 | 52,306 | 0.311 |
| | F | df1,2 | p | F | df | p |
| Growth with amoebae | 9.565 | 8,40.485 | <0.001 | 1.456 | 52,306 | 0.029 |
| Yield with amoebae | 3.595 | 8,46.032 | 0.003 | 2.885 | 52,306 | <0.001 |
| Growth with ciliate | 1.296 | 8,39.223 | 0.274 | 1.304 | 52,306 | 0.091 |
| Yield with ciliate | 4.757 | 8,40.432 | <0.001 | 1.449 | 52,306 | 0.031 |

Wald denotes Wald's test statistics, and F corresponds to F-test statistics, df denote degrees of freedom and p indicate statistical significance.

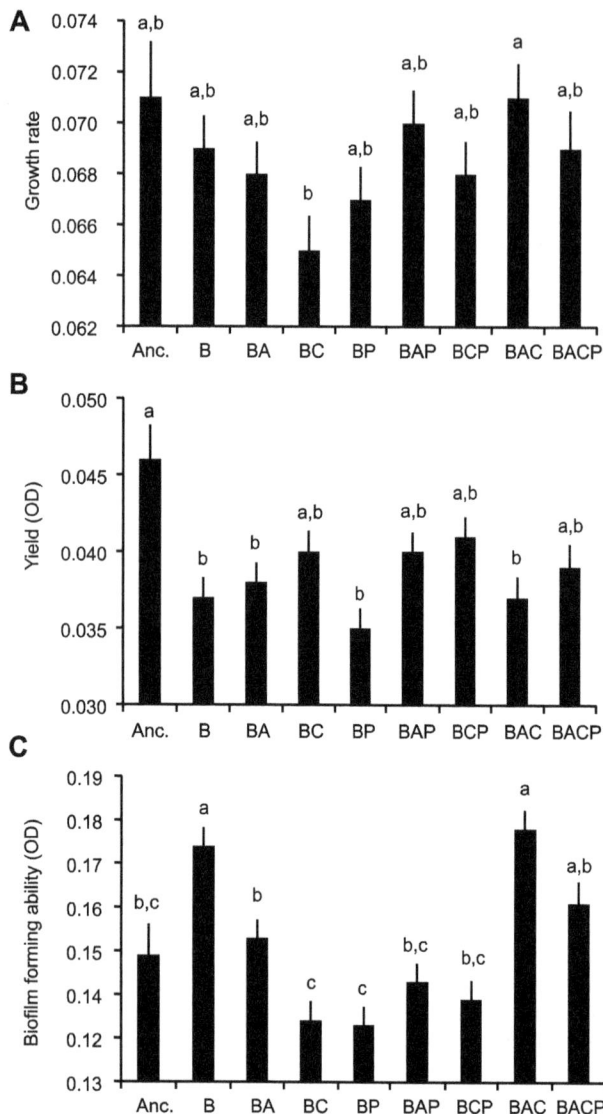

**Figure 5. Growth rate (panel A), yield (panel B) and biofilm forming ability (panel C) differences between ancestral clones and clones that have evolved alone (B) or in different combinations of bacterial enemies.** Anc.: ancestral bacterial strain DB 11; B: bacteria; A: amoebae; C: ciliate; P: phage. Letters indicate if treatment means are considered statistically similar (p>0.05) after Bonferroni correction for multiple comparisons. Tests are based on the post hoc comparisons of estimated marginal means for treatments of ANOVA testing the effects of treatment and population identity on these traits. Bars correspond to measurements of 6 clones from 8 replicate populations, in ancestor n = 16).

ciliate (BAC), and highest if clones had evolved with ciliate (BC) and ciliate and phage (BCP) (Figure 6C; Table 1; Table S2).

## Coevolved traits

Based on the individual traits in different treatment combinations, it is difficult to get an idea of how co-ordinate traits evolved. Therefore we analysed all traits in multivariate ANOVA, followed by eigen-analysis. In this analysis we found that two dimensions dictated multivariate evolution amongst evolutionary treatments (support for two dimensions p = 0.028, for 1 dimension p<0.001). The first dimension was characterized by variation in biofilm and

ciliate defence. Those treatments that exerted strong positive selection on biofilm production had a lower yield with ciliate in free water. The second dimension of trait evolution was formed by those treatments that had increased growth rate without predators and also had a lower growth rate with ciliate. Neither of the two "major" eigen-functions linked virulence to other traits (Table 2).

When similar analysis was performed with information from the amoeba plaque test, a similar result was found, again supporting what was found for two multivariate dimensions (support for two dimensions p = 0.023, for 1 dimension p<0.001). The amoeba plaque test contributed moderately to a second eigenvector. This eigenvector had a slightly different composition than the analysis without the amoeba plaque test as the ltter contained all microcosm replicates. If anything, a higher resistance against amoebae was associated with higher biofilm forming ability and higher growth rate with ciliate. However, the amoeba plaque test clearly did not predict virulence. Since the amoeba plaque test was performed for a subset of the populations, their inclusion in more detailed measurements resulted in a smaller dataset. We base our discussion on the analysis of the larger and thus more reliable eigen-analysis without the amoeba plaque test (Table 2).

## Discussion

Predators such as ciliates and amebae, and parasitic phages are expected to be the main determinants of bacterial mortality in the natural environment [8,9]. In addition to selection exerted by these predators and parasites on defensive traits, selections by bacterial natural enemies have often been suggested to lead to increased bacterial virulence [2,5,72] to such a degree that amoebae-resistance has been used as a direct proxy of strains' virulence [25,27–31,73]. However, we did not find evidence to suggest such a relationship as most of the experimental treatments had attenuated virulence. Moreover, there was no indication of coevolution of other life-history traits with virulence in evolved strains.

Contradictory to the theory that protozoan predators and phage parasites, amoebae in particular, play a strong role in the evolution of high virulence [11–13,53,72,74], we found that virulence attenuated in all of the evolved populations regardless of the presence of the enemies. This is in accordance with the previous study with S. marcescens that suggest that ciliates can select for attenuated virulence [39,40]. However, here we show that this is also the case with amoeba and phages [43], and under selection by multiple enemies at the same time. By far the strongest decrease in virulence was found in clones that had evolved with phages and ciliates. However, the between population variation in this group was very high and statistically significant. This was shown in separate analyses designed to test the amount of between population variance within treatments. In most of the other groups the between population variation was clearly non-significant (p<0.135, but in BCP, p<0.001; BAC p = 0.003). The between population variation is often seen as a signature of drift, mutation accumulation and lack of directional selection on traits [40,75–78]. Thus, we propose that the strong decline of virulence in ciliate-phage treatment was primarily caused by the decay of unused traits through random mutation accumulation [40,76].

To find out if evolutionary changes in virulence could be linked to changes in traits that are important for fitness outside the host, we measured growth parameters and resistance against amoeba and ciliates. However, none of the traits seemed to be determining the level of virulence amongst evolved strains. Thus, it is clear that resistance to protozoan predators or changes in other measured

Coincidental Loss of Bacterial Virulence in Multi-Enemy Microbial Communities

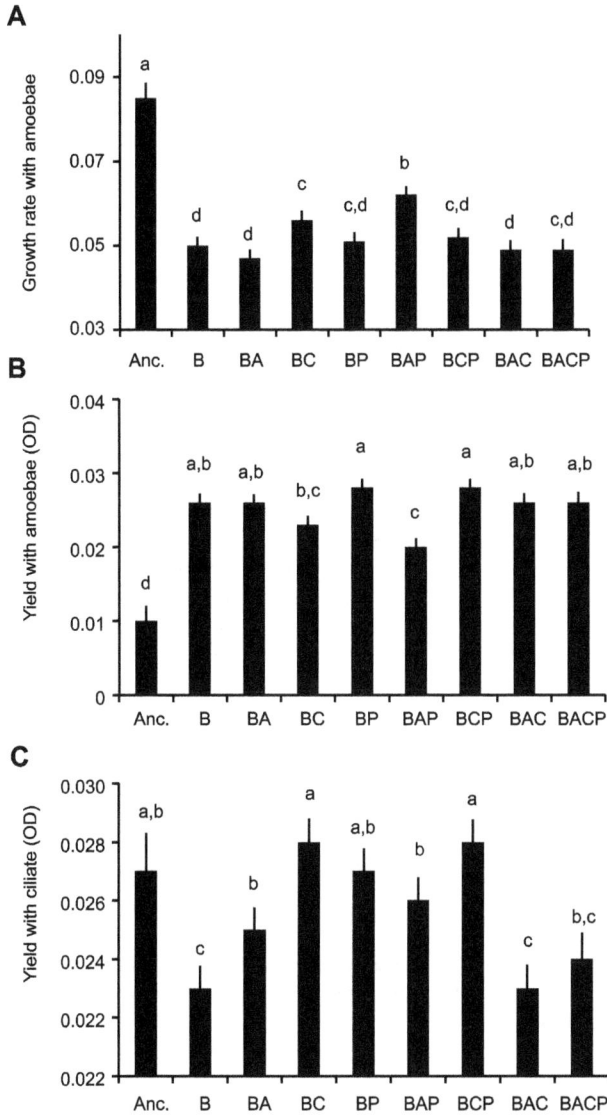

**Figure 6. Growth rate with amoebae (panel A), yield with amoebae (panel B) and yield with ciliate (panel C) differences between ancestral clones and clones that have evolved alone or in different combinations of bacterial enemies.** Anc.: ancestral bacterial strain DB 11; B: bacteria; A: amoebae; C: ciliate; P: phage. Letters indicate if treatment means are considered statistically similar after Bonferroni correction for multiple comparisons. Tests are based on the post hoc comparisons of estimated marginal means for treatments of ANOVA testing the effects of treatment and population identity on these traits. Bars correspond to measurements of 6 clones from 8 replicate populations, in ancestor n = 16).

life-history traits are poor indicators for virulence in *S. marcescens*. However, in other bacterial species, virulence correlated positively with bacterial defences against predators [5,12,17,26,72] (but see [39,40,77]). In addition, several lines of research have emphasized the role of growth rate with virulence [79,80] (but see [44,77]). However, it could be that virulence might be traded off with whatever trait that is under selection in the given environment. This could have led to the clear lack of a connection between virulence and life history traits in an experiment where selection pressures are different between each treatment. This is a plausible outcome if virulence is traded off with life-history traits via finite

**Table 2.** Eigenvectors describing dimensions of multivariate evolution under different kind of enemies (loadings, i.e. correlations, of original variables to new composite variable, based on MANOVA.

| | Eig. 1. | Eig. 2. | Eig. 3. | Eig. 4. | Eig. 5. | Eig. 6. | Eig. 7. | Eig. 8. | Eig. 9. |
|---|---|---|---|---|---|---|---|---|---|
| Biofilm | **0.753** | 0.326 | −0.205 | 0.215 | 0.229 | 0.376 | −0.343 | −0.024 | 0.183 |
| Yield | 0.131 | 0.257 | −0.268 | 0.671 | −0.023 | −0.155 | 0.150 | −0.199 | −0.149 |
| Yield with amoeba | 0.081 | −0.223 | −0.276 | −0.200 | 0.330 | 0.094 | 0.452 | −0.459 | 0.678 |
| Yield with ciliate | **−0.579** | −0.253 | −0.032 | 0.472 | −0.203 | 0.111 | −0.665 | 0.115 | −0.026 |
| Growth rate | 0.018 | **−0.518** | 0.498 | 0.241 | −0.664 | −0.375 | −0.106 | 0.245 | 0.022 |
| Growth with amoeba | −0.150 | 0.228 | 0.489 | −0.143 | 0.321 | 0.501 | 0.335 | −0.357 | 0.511 |
| Growth with ciliate | 0.222 | **0.610** | −0.483 | −0.356 | 0.075 | −0.209 | 0.184 | 0.319 | 0.460 |
| Virulence | 0.037 | −0.149 | 0.080 | 0.171 | 0.176 | 0.236 | 0.202 | 0.666 | 0.063 |
| Eigenvalues | 4.692 | 0.9220 | 0.5689 | 0.1967 | 0.1039 | 0.0394 | 0.0173 | <0.001 | <0.001 |
| % explained | 71.74 | 14.10 | 8.70 | 3.01 | 1.59 | 0.60 | 0.27 | <0.001 | <0.001 |

Below the eigenvectors are the eigenvalues i.e. amount of variation explained by eigenvectors and the percentage of the total variation explained. First eigen function (Eig1) describes contribution of individual traits to the largest difference between the treatments. In eigenvectors that are considered significant (Eig 1 and 2, see results) the largest contributors to the evolutionary differences are highlighted with bold.

resources. Then, any energetically costly trait under strong selection in outside-host environments could lead to virulence attenuation. Although we did not find a strong connection with virulence traits and life-history traits, we found that regardless of the evolutionary treatments, high biofilm-forming ability was closely linked with a low yield in a condition of co-culturing with ciliates in free water, and high growth rate was linked with low growth rate with ciliates. The first eigenvector (biofilm vs. yield with ciliates) could indicate that growth in biofilms does not lead to good protection against predation in free water, or that exposure to predation leads to biofilm formation only when predators are present and thus reduces cells in a free water environment. However, these findings from the eigen analysis effectively mean that there are evolutionary constraints between different life-history traits that remain unchangeable regardless of the evolutionary treatments. Moreover, since the predators (amoeba and ciliates) were effectively reducing the population densities in the long run (Figure 2A) the lack of selection on life history traits is not a plausible explanation for the obtained results.

When we compared ancestor's life history traits to evolved strains, it seemed that ancestors grew better with amoebae and were also the most virulent clone (Figure 3). However, another amoebae-resistance measurement, yield with amoebae, was actually lower for the ancestor strain than evolved strains. Moreover, in the amoebae-plaque test the ancestor strain did not excel in comparison to evolved strains. These contrasting results from measurements that should indicate the ability to resist amoebae suggest that there is a weak indication that amoebae-resistance evolved simultaneously with virulence. Therefore we suggest that the culture conditions could attenuate virulence, without any clear changes in life history traits. Interestingly, several experimental evolution studies have previously found that bacterial virulence decreases due to the exposure to the outside-host environment [39,40,81].

Alternatively, it has been suggested that if traits are not needed under particular conditions then their alleles become harmful and accumulate which leads to unused trait decay [76]. *S. marcescens* strain Db11 was isolated from a dead *Drosophila* fly over thirty years ago and has been routinely grown in highly protein-rich LB medium. Yet, the virulence of the strain has been maintained from lab to lab [55,57,82]. It is possible that the protein-enriched culture condition like LB medium (LB containing 10 g/l tryptone and 5 g/l yeast extract) was somehow needed for *S. marcescens* Db11 to maintain its virulence. However, a more likely scenario could be that our experimental conditions (low concentration of high C:N ratio plant detritus) selected against virulence in the 8 week evolution experiment. In addition, there might be other unknown factors that could of affected the virulence. Although predators in general effectively lowered population sizes of the bacteria (Figure 2A), spatial heterogeneity and potentially other niches created by the static cultures might lower the strength of

selection on defensive traits. For example, the biofilm could act as a protection against protozoan predation, and some bacteria might have not been under selection at all.

To summarize, we found no support for the idea that enemies outside the host could select for higher virulence, as all experimental treatments led towards lower bacterial virulence. Among evolved strains virulence was not linked to other life-history characters, suggesting that selective pressures from protozoan predators (ciliates and amoebae) and parasitic phages did not dictate virulence evolution. In conclusion, our dataset offered a case against coincidental evolution of the virulence hypothesis that expects outside-host selections, especially amoebae predation, would lead to higher bacterial virulence [2,3,5,72].

## Supporting Information

**Table S1 Pairwise comparisons of experimental treatment differences on amoeba population sizes, biomass in the free water phase and attached biofilm at the end of the experiment.** Significant pairwise comparisons after Bonferroni correction are high-lighted with bold (critical $\alpha$: $0.00178 = 0.05 \div 28$, but amoebae population size critical $\alpha$: $0.008$ $(0.05/6)$). (B = bacteria alone, BC = with ciliate; BA with amoebae; BP with phage etc.. Anc. stands for ancestor Db11 strain).

**Table S2 Pairwise comparisons of experimental treatment differences on measured virulence, growth and defensive traits.** Significant pairwise comparisons after Bonferroni correction are highlighted with bold (critical $\alpha$: $0.00138$ $(0.05/36)$) (B = bacteria alone, BC = with ciliate; BA with amoebae; BP with phage etc.. Anc. stands for ancestor Db11 strain).

**Material S1 The macro for the Image-Pro Plus program to automatically count protozoan cells in microscopic images.**

## Acknowledgments

We thank Kalevi Viipale for intensive discussions on the conceptual issues. We also thank Angus Buckling, Anna-Liisa Laine, Waldron Samuel and Lauri Mikonranta for commenting on the manuscript.

## Author Contributions

Conceived and designed the experiments: JZ TK AMÖ JM JL. Performed the experiments: JZ AMÖ. Analyzed the data: JZ TK AMÖ JM JL. Contributed reagents/materials/analysis tools: JZ TK AMÖ JM JL. Wrote the paper: JZ TK AMÖ JM JL.

## References

1. Casadevall A, Pirofski LA (2003) The damage-response framework of microbial pathogenesis. Nat Rev Microbiol 1: 17–24.
2. Levin BR (1996) The evolution and maintenance of virulence in microparasites. Emerg Infect Dis 2: 93–102.
3. Levin BR, Svanborg Eden C (1990) Selection and evolution of virulence in bacteria: an ecumenical excursion and modest suggestion. Parasitology 100 Suppl: S103–115.
4. May RM, Anderson RM (1983) Epidemiology and Genetics in the Coevolution of Parasites and Hosts. P Roy Soc Lond B Bio 219: 281–313.
5. Adiba S, Nizak C, van Baalen M, Denamur E, Depaulis F (2010) From grazing resistance to pathogenesis: the coincidental evolution of virulence factors. PloS one 5: e11882.
6. Coombes BK, Gilmour MW, Goodman CD (2011) The evolution of virulence in non-o157 shiga toxin-producing *Escherichia coli*. Front Microbiol 2: 90.

7. Brown SP, Cornforth DM, Mideo N (2012) Evolution of virulence in opportunistic pathogens: generalism, plasticity, and control. Trends Microbiol 20: 336–342.
8. Jürgens K, Matz C (2002) Predation as a shaping force for the phenotypic and genotypic composition of planktonic bacteria. Antonie Van Leeuwenhoek 81: 413–434.
9. Suttle CA (2005) Viruses in the sea. Nature 437: 356–361.
10. Labrie SJ, Samson JE, Moineau S (2010) Bacteriophage resistance mechanisms. Nat Rev Microbiol 8: 317–327.
11. Matz C, Kjelleberg S (2005) Off the hook - how bacteria survive protozoan grazing. Trends Microbiol 13: 302–307.
12. Brüssow H (2007) Bacteria between protists and phages: from antipredation strategies to the evolution of pathogenicity. Mol Microbiol 65: 583–589.

13. Greub G, Raoult D (2004) Microorganisms resistant to free-living amoebae. Clin Microbiol Rev 17: 413–433.
14. Hall-Stoodley L, Costerton JW, Stoodley P (2004) Bacterial biofilms: from the natural environment to infectious diseases. Nat Rev Microbiol 2: 95–108.
15. Jousset A (2012) Ecological and evolutive implications of bacterial defences against predators. Environ Microbiol 14: 1830–1843.
16. Thurlow LR, Hanke ML, Fritz T, Angle A, Aldrich A, et al. (2011) *Staphylococcus aureus* biofilms prevent macrophage phagocytosis and attenuate inflammation in vivo. J Immunol 186: 6585–6596.
17. Al-Quadan T, Price CT, Abu Kwaik Y (2012) Exploitation of evolutionarily conserved amoeba and mammalian processes by *Legionella*. Trends Microbiol 20: 299–306.
18. Gao LY, Harb OS, AbuKwaik Y (1997) Utilization of similar mechanisms by *Legionella pneumophila* to parasitize two evolutionarily distant host cells, mammalian macrophages and protozoa. Infect Immun 65: 4738–4746.
19. Abu Kwaik Y, Gag LY, Stone BJ, Venkataraman C, Harb OS (1998) Invasion of protozoa by *Legionella pneumophila* and its role in bacterial ecology and pathogenesis. Appl Environ Microb 64: 3127–3133.
20. Ohno A, Kato N, Sakamoto R, Kimura S, Yamaguchi K (2008) Temperature-dependent parasitic relationship between *Legionella pneumophila* and a free-living amoeba (*Acanthamoeba castellanii*). Appl Environ Microb 74: 4585–4588.
21. Rowbotham TJ (1980) Preliminary report on the pathogenicity of *Legionella pneumophila* for freshwater and soil amoebae. J Clin Pathol 33: 1179–1183.
22. Fields BS, Benson RF, Besser RE (2002) *Legionella* and Legionnaires' disease: 25 years of investigation. Clin Microbiol Rev 15: 506–526.
23. Muder RR, Yu VL, Woo AH (1986) Mode of transmission of *Legionella pneumophila*. A critical review. Arch Intern Med 146: 1607–1612.
24. Ensminger AW, Yassin Y, Miron A, Isberg RR (2012) Experimental Evolution of *Legionella pneumophila* in Mouse Macrophages Leads to Strains with Altered Determinants of Environmental Survival. Plos Pathog 8.
25. Bonifait L, Charette SJ, Filion G, Gottschalk M, Grenier D (2011) Amoeba Host Model for Evaluation of *Streptococcus suis* Virulence. Appl Environ Microb 77: 6271–6273.
26. Cosson P, Soldati T (2008) Eat, kill or die: when amoeba meets bacteria. Curr Opin Microbiol 11: 271–276.
27. Froquet R, Lelong E, Marchetti A, Cosson P (2009) *Dictyostelium discoideum*: a model host to measure bacterial virulence. Nat Protoc 4: 25–30.
28. Greub G, La Scola B, Raoult D (2004) Amoebae-resisting bacteria isolated from human nasal swabs by amoebal coculture. Emerg Infect Dis 10: 470–477.
29. Hasselbring BM, Patel MK, Schell MA (2011) *Dictyostelium discoideum* as a Model System for Identification of *Burkholderia pseudomallei* Virulence Factors. Infect Immun 79: 2079–2088.
30. Lelong E, Marchetti A, Simon M, Burns JL, van Delden C, et al. (2011) Evolution of *Pseudomonas aeruginosa* virulence in infected patients revealed in a *Dictyostelium discoideum* host model. Clin Microbiol Infec 17: 1415–1420.
31. Smith MG, Gianoulis TA, Pukatzki S, Mekalanos JJ, Ornston LN, et al. (2007) New insights into *Acinetobacter baumannii* pathogenesis revealed by high-density pyrosequencing and transposon mutagenesis. Genes & development 21: 601–614.
32. Brüssow H, Canchaya C, Hardt WD (2004) Phages and the evolution of bacterial pathogens: from genomic rearrangements to lysogenic conversion. Microbiol Mol Biol Rev 68: 560–602, table of contents.
33. Hacker J, Hentschel U, Dobrindt U (2003) Prokaryotic chromosomes and disease. Science 301: 790–793.
34. Hacker J, Kaper JB (2000) Pathogenicity islands and the evolution of microbes. Annu Rev Microbiol 54: 641–679.
35. Boyd EF (2012) Bacteriophage-encoded bacterial virulence factors and phage-pathogenicity island interactions. Adv Virus Res 82: 91–118.
36. Casas V, Maloy S (2011) Role of bacteriophage-encoded exotoxins in the evolution of bacterial pathogens. Future Microbiol 6: 1461–1473.
37. Bell JK, Mullen GE, Leifer CA, Mazzoni A, Davies DR, et al. (2003) Leucine-rich repeats and pathogen recognition in Toll-like receptors. Trends Immunol 24: 528–533.
38. Sahly H, Keisari Y, Crouch E, Sharon N, Ofek I (2008) Recognition of bacterial surface polysaccharides by lectins of the innate immune system and its contribution to defense against infection: the case of pulmonary pathogens. Infect Immun 76: 1322–1332.
39. Friman VP, Lindstrom C, Hiltunen T, Laakso J, Mappes J (2009) Predation on multiple trophic levels shapes the evolution of pathogen virulence. PloS one 4: e6761.
40. Mikonranta L, Friman V-P, Laakso J (2012) Life History Trade-Offs and Relaxed Selection Can Decrease Bacterial Virulence in Environmental Reservoirs. PloS one 7: e43801.
41. Josenhans C, Suerbaum S (2002) The role of motility as a virulence factor in bacteria. Int J Med Microbiol 291: 605–614.
42. Lertsethtakarn P, Ottemann KM, Hendrixson DR (2011) Motility and Chemotaxis in *Campylobacter* and *Helicobacter*. Nat Rev Microbiol, Vol 65 65: 389–410.
43. Friman VP, Hiltunen T, Jalasvuori M, Lindstedt C, Laanto E, et al. (2011) High temperature and bacteriophages can indirectly select for bacterial pathogenicity in environmental reservoirs. PloS one 6: e17651.
44. Sturm A, Heinemann M, Arnoldini M, Benecke A, Ackermann M, et al. (2011) The cost of virulence: retarded growth of *Salmonella typhimurium* cells expressing type III secretion system 1. Plos Pathog 7: e1002143.
45. Friedman ME, Kautter DA (1962) Effect of nutrition on the respiratory virulence of *Listeria monocytogenes*. J Bacteriol 83: 456–462.
46. Heckly RJ, Blank H (1980) Virulence and viability of *Yersinia pestis* 25 years after lyophilization. Appl Environ Microbiol 39: 541–543.
47. Midelet-Bourdin G, Leleu G, Copin S, Roche SM, Velge P, et al. (2006) Modification of a virulence-associated phenotype after growth of *Listeria monocytogenes* on food. J Appl Microbiol 101: 300–308.
48. Ali S, Huang Z, Ren SX (2009) Media composition influences on growth, enzyme activity, and virulence of the entomopathogen hyphomycete *Isaria fumosoroseus*. Entomol Exp Appl 131: 30–38.
49. Safavi SA, Shah FA, Pakdel AK, Reza Rasoulian G, Bandani AR, et al. (2007) Effect of nutrition on growth and virulence of the entomopathogenic fungus *Beauveria bassiana*. FEMS Microbiol Lett 270: 116–123.
50. Wu JH, Ali S, Huang Z, Ren SX, Cai SJ (2010) Media Composition Influences Growth, Enzyme Activity and Virulence of the Entomopathogen *Metarhizium anisopliae* (Hypocreales: Clavicipitaceae). Pak J Zool 42: 451–459.
51. Friman VP, Hiltunen T, Laakso J, Kaitala V (2008) Availability of prey resources drives evolution of predator-prey interaction. Proc Biol Sci 275: 1625–1633.
52. Hosseinidoust Z, van de Ven TG, Tufenkji N (2013) Evolution of *Pseudomonas aeruginosa* virulence as a result of phage predation. Appl Environ Microbiol 79: 6110–6116.
53. Molmeret M, Horn M, Wagner M, Santic M, Abu Kwaik Y (2005) Amoebae as training grounds for intracellular bacterial pathogens. Appl Environ Microbiol 71: 20–28.
54. Steinert M, Heuner K (2005) *Dictyostelium* as host model for pathogenesis. Cell Microbiol 7: 307–314.
55. Nehme NT, Liegeois S, Kele B, Giammarinaro P, Pradel E, et al. (2007) A model of bacterial intestinal infections in *Drosophila melanogaster*. Plos Pathog 3: e173.
56. Kurz CL, Chauvet S, Andres E, Aurouze M, Vallet I, et al. (2003) Virulence factors of the human opportunistic pathogen *Serratia marcescens* identified by in vivo screening. EMBO J 22: 1451–1460.
57. Flyg C, Kenne K, Boman HG (1980) Insect pathogenic properties of *Serratia marcescens*: phage-resistant mutants with a decreased resistance to *Cecropia* immunity and a decreased virulence to *Drosophila*. J Gen Microbiol 120: 173–181.
58. Grimont PA, Grimont F (1978) The genus *Serratia*. Annu Rev Microbiol 32: 221–248.
59. Mahlen SD (2011) *Serratia* infections: from military experiments to current practice. Clin Microbiol Rev 24: 755–791.
60. Sutherland KP, Porter JW, Turner JW, Thomas BJ, Looney EE, et al. (2010) Human sewage identified as likely source of white pox disease of the threatened Caribbean elkhorn coral, *Acropora palmata*. Environ Microbiol 12: 1122–1131.
61. Kiy T, Tiedtke A (1992) Mass Cultivation of *Tetrahymena thermophila* Yielding High Cell Densities and Short Generation Times. Appl Microbiol Biot 37: 576–579.
62. Kennedy GM, Morisaki JH, Champion PA (2012) Conserved mechanisms of *Mycobacterium marinum* pathogenesis within the environmental amoeba *Acanthamoeba castellanii*. Appl Environ Microbiol 78: 2049–2052.
63. Page FC (1976) An Illustrated Key to Freshwater and Soil Amoebae: With Notes on Cultivation and Ecology. Freshwater Biological Association.
64. La Scola B, Mezi L, Weiller PJ, Raoult D (2001) Isolation of *Legionella* anisa using an amoebic coculture procedure. J Clin Microbiol 39: 365–366.
65. Page FC (1988) A New Key to Freshwater and Soil Gymnamoebae with instructions for culture. Freshwater Biological Association.
66. Friman VP, Laakso J (2011) Pulsed-resource dynamics constrain the evolution of predator-prey interactions. Am Nat 177: 334–345.
67. Friman VP, Laakso J, Koivu-Orava M, Hiltunen T (2011) Pulsed-resource dynamics increase the asymmetry of antagonistic coevolution between a predatory protist and a prey bacterium. J Evol Biol 24: 2563–2573.
68. O'Toole GA, Kolter R (1998) Initiation of biofilm formation in *Pseudomonas fluorescens* WCS365 proceeds via multiple, convergent signalling pathways: a genetic analysis. Mol Microbiol 28: 449–461.
69. Laakso J, Loytynoja K, Kaitala V (2003) Environmental noise and population dynamics of the ciliated protozoa *Tetrahymena thermophila* in aquatic microcosms. Oikos 102: 663–671.
70. Wildschutte H, Wolfe DM, Tamewitz A, Lawrence JG (2004) Protozoan predation, diversifying selection, and the evolution of antigenic diversity in *Salmonella*. Proc Natl Acad Sci U S A 101: 10644–10649.
71. Potvin C (2001) ANOVA Experimental Layout and Analysis. In: Scheiner SM and Gurevitch J, editors. Design and Analysis of Ecological Experiments. New York: Oxford University Press. pp. 69–75.
72. Steinberg KM, Levin BR (2007) Grazing protozoa and the evolution of the *Escherichia coli* O157: H7 Shiga toxin-encoding prophage. Proc Biol Sci 274: 1921–1929.
73. Cosson P, Zulianello L, Join-Lambert O, Faurisson F, Gebbie L, et al. (2002) *Pseudomonas aeruginosa* virulence analyzed in a *Dictyostelium discoideum* host system. J Bacteriol 184: 3027–3033.
74. Casadevall A (2008) Evolution of intracellular pathogens. Nat Rev Microbiol 62: 19–33.
75. Cooper TF, Lenski RE (2010) Experimental evolution with *E. coli* in diverse resource environments. I. Fluctuating environments promote divergence of replicate populations. BMC Evol Biol 10: 11.

76. Hall AR, Colegrave N (2008) Decay of unused characters by selection and drift. J Evol Bio 21: 610–617.

77. Ketola T, Mikonranta L, Zhang J, Saarinen K, Ormala AM, et al. (2013) Fluctuating temperature leads to evolution of thermal generalism and preadaptation to novel environments. Evolution 67: 2936–2944.

78. Travisano M, Mongold JA, Bennett AF, Lenski RE (1995) Experimental tests of the roles of adaptation, chance, and history in evolution. Science 267: 87–90.

79. Chesbro WR, Wamola I, Bartley CH (1969) Correlation of virulence with growth rate in *Staphylococcus aureus*. Can J Microbiol 15: 723–729.

80. West SA, Buckling A (2003) Cooperation, virulence and siderophore production in bacterial parasites. Proc Biol Sci 270: 37–44.

81. Gomez P, Buckling A (2011) Bacteria-phage antagonistic coevolution in soil. Science 332: 106–109.

82. Zhang J, Friman VP, Laakso J, Mappes J (2012) Interactive effects between diet and genotypes of host and pathogen define the severity of infection. Ecology and Evolution 2: 2347–2356.

# 16

# Computational Design of Protein-Based Inhibitors of *Plasmodium vivax* Subtilisin-Like 1 Protease

**Giacomo Bastianelli[1,2], Anthony Bouillon[3,4¤], Christophe Nguyen[5], Dung Le-Nguyen[5], Michael Nilges[1,2*⑨], Jean-Christophe Barale[3,4*⑨¤]**

1 Institut Pasteur, Unité de Bioinformatique Structurale, Département de Biologie Structurale et Chimie, Paris, France, 2 CNRS UMR 3528, Paris, France, 3 Institut Pasteur, Unité d'Immunologie Moléculaires des Parasites, Département de Parasitologie et de Mycologie & CNRS URA 2581, Paris, France, 4 CNRS, URA2581, Paris, France, 5 SYSDIAG, CNRS UMR3145 CNRS-BioRad, Montpellier, France

## Abstract

*Background:* Malaria remains a major global health concern. The development of novel therapeutic strategies is critical to overcome the selection of multiresistant parasites. The subtilisin-like protease (SUB1) involved in the egress of daughter *Plasmodium* parasites from infected erythrocytes and in their subsequent invasion into fresh erythrocytes has emerged as an interesting new drug target.

*Findings:* Using a computational approach based on homology modeling, protein–protein docking and mutation scoring, we designed protein–based inhibitors of *Plasmodium vivax* SUB1 (PvSUB1) and experimentally evaluated their inhibitory activity. The small peptidic trypsin inhibitor EETI-II was used as scaffold. We mutated residues at specific positions (P4 and P1) and calculated the change in free-energy of binding with PvSUB1. In agreement with our predictions, we identified a mutant of EETI-II (EETI-II-P4LP1W) with a *Ki* in the medium micromolar range.

*Conclusions:* Despite the challenges related to the lack of an experimental structure of PvSUB1, the computational protocol we developed in this study led to the design of protein-based inhibitors of PvSUB1. The approach we describe in this paper, together with other examples, demonstrates the capabilities of computational procedures to accelerate and guide the design of novel proteins with interesting therapeutic applications.

**Editor:** Laurent Rénia, Agency for Science, Technology and Research - Singapore Immunology Network, Singapore

**Funding:** This work was partly supported by MEST-CT-05-020311, a Marie Curie Early Stage Research Training Fellowship (EIMID) of the Framework Program 6 by the European Commission. AB is a fellow of the "Direction Generale pour l'Armement" from the French Ministry of Defense. This work was partly supported by the "Fond dédié: combattre les maladies parasitaires" granted by Sanofi-Aventis and the French Ministry of Research and the Institut Carnot-Pasteur Maladies Infectieuses. The funders had no role in study design, data collection and analysis, decision to publish, or preparation of the manuscript. SYSDIAG, CNRS UMR3145 CNRS-BioRad provided support in the form of salaries and operating costs for authors CN and DLN, but did not have any additional role in the study design, data collection and analysis, decision to publish, or preparation of the manuscript. The specific roles of these authors are articulated in the "author contributions" section.

**Competing Interests:** The authors have the following interests. This work was partly supported by the "Fond dédié: combattre les maladies parasitaires" granted by the French Ministry and Sanofi-Aventis. Christophe Nguyen and Dung Le-Nguyen are employed by SYSDIAG, CNRS UMR3145 CNRS-BioRad. There are no patents, products in development or marketed products to declare.

* Email: nilges@pasteur.fr (MN); jcb@pasteur.fr (JCB)

⑨ These authors participated equally in the supervision of this work.

¤ Current address: Institut Pasteur, Unité de Biologie et Génétique du Paludisme, Team Malaria Targets and Drug Development, Département de Parasitologie et de Mycologie, Paris, France

## Introduction

With more than 400 millions infections worldwide, malaria remains a major public health issue, principally in sub-Saharan Africa. An effective vaccine would help reduce disease burden, but the best candidates are still in development or evaluation phase [1,2]. The rapid development of multidrug-resistant *Plasmodium* [3] parasites necessitates accelerating the discovery of novel anti-malarial compounds to meet the needs of the agenda for malaria control and eradication [4].

In humans, *Plasmodium sp.* development comprises different stages, with the asexual intra–erythrocytic forms being responsible for the symptoms of the disease, such as fever, anemia, and cerebral malaria that can lead to death [5]. The erythrocyte invasion by *Plasmodium* merozoites critically depends on protease activities involved in both the daughter parasites egress from erythrocytes, and invasion into another erythrocyte. The parasite subtilisin-like protein 1 (SUB1) plays a critical role during both the hepatic and erythrocytic phases of *Plasmodium* biological cycle and is hence considered an interesting multi-stage target for developing a new class of anti–malarials [6] [7].

Most of the ancient therapies against *Plasmodium* are based on small molecules such as chloroquine, quinolones, antifolate, artemisinin derivatives, or atovaquone. The development of new classes of active molecules such as protein–based drugs or peptidomimetics [8,9] is an active and promising field of research. Among protein–based drugs, dermaseptin S4 (DS4) was shown to irreversibly inhibit the *in vitro* parasite growth through a cytotoxic hemolytic activity. Dermaseptin S3 acts in a similar manner as DS4 but did not present hemolytic activity through a cytotoxic hemolytic activity [10].

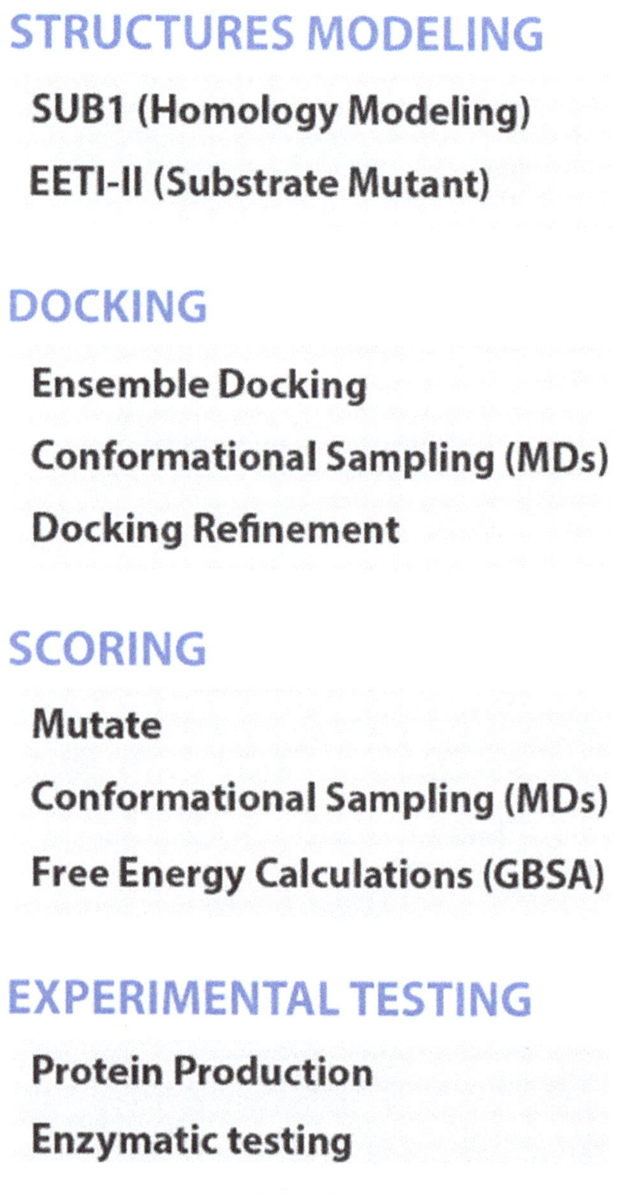

# STRUCTURES MODELING

## SUB1 (Homology Modeling)

## EETI-II (Substrate Mutant)

# DOCKING

## Ensemble Docking

## Conformational Sampling (MDs)

## Docking Refinement

# SCORING

## Mutate

## Conformational Sampling (MDs)

## Free Energy Calculations (GBSA)

# EXPERIMENTAL TESTING

## Protein Production

## Enzymatic testing

**Figure 1. Computational Protein Design Strategy.** Step 1: Prediction of the structure of the enzyme (PvSUB1) by comparative modelling and of the scaffold for mutational analysis (EETI-II-sub) by replacing one of the loops with a substrate sequence. Step 2: docking of EETI-II-sub to the target protein by ensemble docking procedure with several conformations from molecular dynamics simulations for each protein partner, and refinement of the best solutions. Step 3: mutation of the scaffold, conformational sampling and scoring of the mutants. Step 4: experimental testing by an enzymatic inhibitory assay on the recombinant enzyme of PvSUB1.

In the design of protein–based drugs, most approaches use combinatorial libraries based on different screening methods such as phage [11], ribosome [12] or mRNA display [13]. Their use is wide–spread, in particular for selecting high-affinity protein binders, despite their limitations due to the library size and the large quantities of the target protein needed to perform screening. Moreover, when the selection is not based on binding but on inhibiting a crucial enzyme of the biological cycle, a rather complex selection system has to be employed. Computational protein design can be used to reduce the sequence/structure space that needs to be explored and thus accelerate the process of screening and selection of target inhibitors.

Here, we present a strategy for the computational design of protein-based inhibitors targeting the subtilisin–like 1 protease of the human parasite *Plasmodium vivax* (PvSUB1). PvSUB1 can be expressed as a recombinant active enzyme [14] [15], and a specific enzymatic assay allows one to evaluate specific inhibitors. To search for potential inhibitors of PvSUB1, we used a computational design strategy, employing as scaffold the small protein EETI-II (*Ecballium elaterium* trypsin inhibitor II) [16], a trypsin inhibitor extracted from *Ecballium elaterium*. The family of cystein–knot proteins, to which EETI-II belongs, and in particular the cyclotides [17], possesses interesting biochemical properties [18]. EETI-II is composed of 28 amino-acids and its three-dimensional structure is tightly constrained by 3 disulphide bridges that contribute to its rigidity and biological stability [19]. We opted for this scaffold because several studies showed the possibility to engineer this protein to obtain specific mutants [20], *via* the extension of the EETI bioactive loop [21] or by changing its sequence to change its specificity towards the targeted enzyme [22] [23] [24] [25].

Compared to studies using an iterative computational design procedure focused on electrostatic binding contributions and single mutants [26], or on re–designing a scaffold protein to bind to a specified region on a target protein [27], we here faced the additional challenge that the 3D structure of the target itself or a close sequence homologue was not known. Nonetheless, the use of state–of–the–art structure prediction, docking and scoring methods allowed us to successfully identify mutants of the scaffold EETI-II that inhibited the target PvSUB1 enzyme.

## Results and Discussion

The computational protein design approach involved four steps (see Figure 1). The first step was the modeling of the structure of the enzyme (PvSUB1) and the scaffold (EETI-II). Because of the lack of an experimental PvSUB1 structure, we built structures based on sequence homology. We also generated the model of a mutant of EETI-II containing the substrate sequence of PvSUB1, which we called EETI-II- sub. The second step was the docking of EETI-II-sub to the target protein. We employed an ensemble docking procedure with several conformations obtained from molecular dynamics (MD) simulations for each protein partner to implicitly include flexibility in the docking, and refined the best docking solutions by molecular dynamics to obtain high-quality structures of the complex. The third step aimed at identifying mutants of EETI-II-sub that had higher binding affinity towards PvSUB1. In this step, we mutated residues in EETI-II-sub at the protein–protein interface of the complex, ran conformational sampling of the mutant with molecular dynamics, and calculated the free energy of binding via implicit solvent models based on the Generalized Born approximation (GBSA). The last step consisted in the experimental testing of the inhibitor by an enzymatic inhibitory assay specific for the PvSUB1 recombinant enzyme.

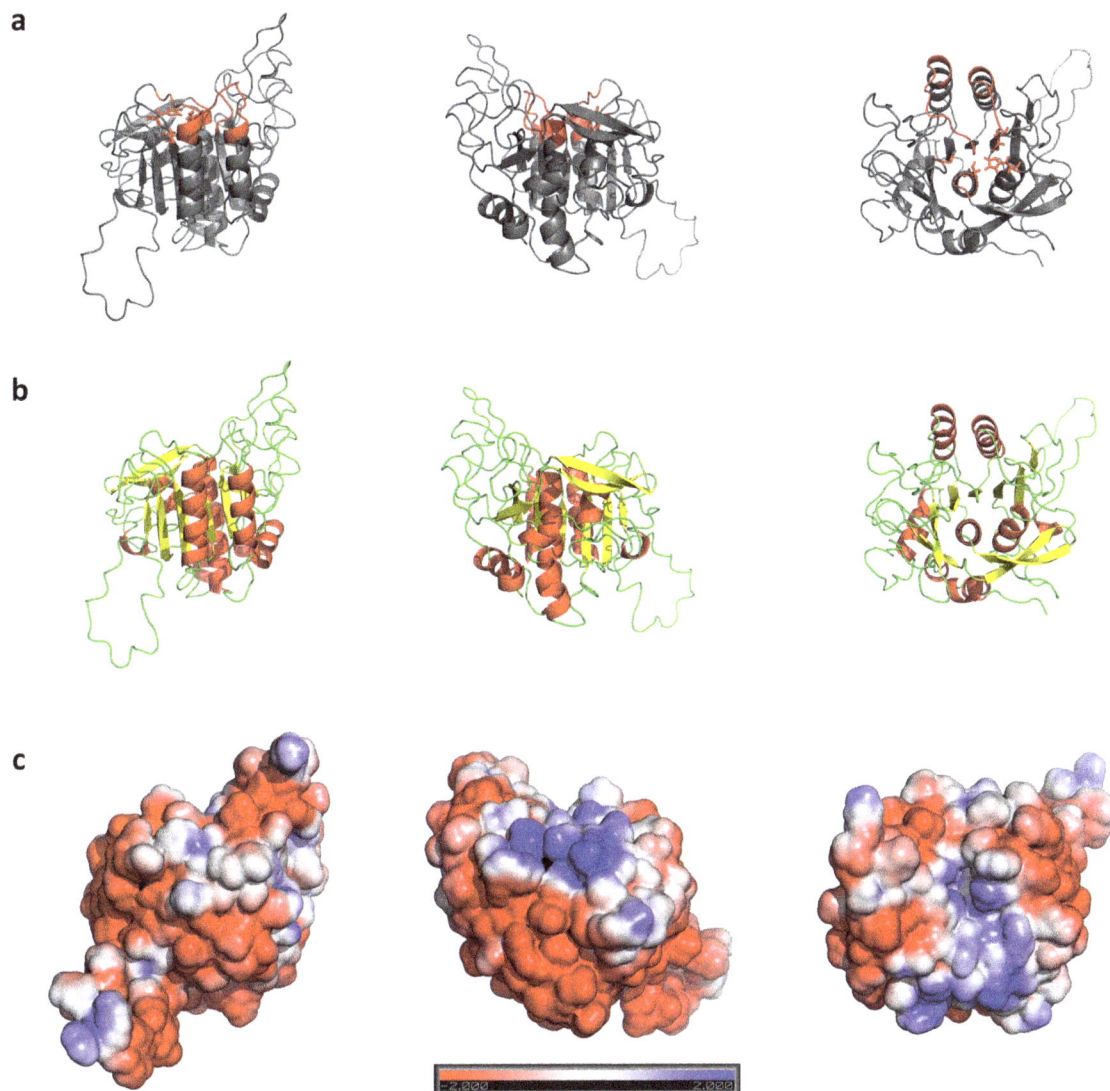

**Figure 2. 3D model of PvSUB1 catalytic domain. A:** Highlighted in red is the region forming the substrate binding pocket and red sticks correspond to the residues that form the catalytic triad; **B:** Cartoon representation of secondary structures; **C:** APBS surface electrostatic representation.

**Table 1.** Catalytic site distances along MD simulations.

| Distance | 1R0R | 1T02 | PvSUB1 |
|---|---|---|---|
| HIS@CA-SER@CA | 8.52±0.34 | 9.38±0.36 | 9.89±0.24 |
| HIS@CB-SER@CB | 6.56±0.43 | 7.55±0.41 | 7.91±0.31 |
| HIS@CA-ASP@CA | 7.78±0.52 | 7.91±0.29 | 8.33±0.18 |
| HIS@CB-ASP@CB | 6.59±0.66 | 6.73±0.41 | 7.46±0.24 |
| SER@CA-ASP@CA | 10.24±0.35 | 9.18±0.32 | 10.35±0.25 |
| SER@CB-ASP@CB | 7.42±0.42 | 8.83±0.49 | 8.72±0.34 |
| SER@OG-HIS@NE2* | 7.54±1.11 | 4.69±1.19 | 5.11±0.71 |
| ASP@CG-HIS@ND1 | 8.52±0.34 | 9.38±0.36 | 6.74±0.54 |
| ASN@CG-SER@CB | 6.70±0.71 | 7.5±0.91 | 6.52±0.41 |

Values are expressed in Å. The distance SER@OG-HIS@NE2 shows the largest fluctuation for both subtilisins with known structure and for our models, consistent with variations of this distance in subtilisin crystal structures (Table 2). Catalytic triad: Asp316, His372, Ser549; Asp137, His168, Ser325 and Asp139, His171, Ser328 for PvSUB1, 1R0R and 1TO2 respectively.

**Table 2.** Subtilisin catalytic site geometries.

| Distance | lower-range | upper-range |
|---|---|---|
| HIS@CA-SER@CA | 8.3 | 8.72 |
| HIS@CB-SER@CB | 6.44 | 6.89 |
| HIS@CA-ASP@CA | 7.22 | 7.46 |
| HIS@CB-ASP@CB | 5.83 | 6.61 |
| SER@CA-ASP@CA | 9.87 | 10.11 |
| SER@CB-ASP@CB | 8.15 | 8.53 |
| SER@OG-HIS@NE2 | 2.57 | 3.36 |
| ASP@CG-HIS@ND1 | 3.15 | 3.35 |
| ASN@CG-SER@CB | 6.34 | 6.74 |
| SER@OH-surface | 1.58 | 1.60 |

Smallest (lower-range) and longest (upper-range) measures from experimentally determined structures used as templates in the modeling of PvSUB1. The asparagine (Asn) is the residue forming the oxyanion hole. Catalytic triads of PvSUB1, bacterial subtilisins Carlsberg (1R0R) and BPN (1TO2): Asp316, His372, Ser549; Asp137, His168, Ser325 and Asp139, His171, Ser328, respectively.

## Modeling and molecular dynamics simulations of PvSUB1

In order to generate a reliable 3D-model of PvSUB1, we used the procedure described in our previous publication where we modeled the structure of PfSUB1, a close homologous of PvSUB1 [14]. A similar homology modeling strategy generated 3D-models of PfSUB1 used to identify small-molecule inhibitors of PfSUB1 with an *in silico* screening approach [15]. The particular challenge of obtaining a high quality 3D model of PvSUB1 was the low sequence identity with the available templates (only ~30%, just above the "twilight zone" for homology modeling). Using state of the art modeling methods, it is possible to generate homology models with a Cα RMSD<1.0 Å when the sequence identity with the template is >50%. With sequence identity below 25%, larger divergences from the target structure can appear, making the model less precise [28]. However, our previous analysis had shown that major divergences between the PvSUB1 sequence and the structural templates were localized in regions distant from the catalytic groove that binds the substrate, and that the sequence identity in the substrate binding area is >30% [14,15]. In addition, to evaluate the stability and the overall quality of the PvSUB1 model (Figure 2), we performed multiple molecular dynamics simulations of our model PvSUB1 and of two of the templates used in the modeling, subtilisins BPN (1TO2) and Carlsberg (1R0R).

Tables 1 and 2 show distance ranges among the residues of the catalytic triad, in the MD simulations of PvSUB1 and in the crystal structures of the templates. All distances fell within the ranges observed in the experimental X–ray crystal structures apart from the distances HIS372@CB-ASP316@CB and ASP316@CG-HIS372@ND1, which are slightly outside (less than 1 Å). This is consistent with the fact that this distance shows the highest variation in subtilisin X–ray crystal structures (Table 2). Stability can be also measured by analyzing the RMSD along the MD trajectories from the starting structure. In Figure 3.A, we plotted the average RMSDs obtained for the 5 MD trajectories of 10 ns each.

The trajectories of the PvSUB1 model diverge more than those of BPN (1TO2) and Carlsberg (1R0R). This is primarily due to the regions in the model for which there is no structural information in the templates. In the model, these regions are unstructured, solvent exposed and distant from the binding pocket. When we removed them from the analysis, the RMSD reduced to values

similar to those observed in the MD trajectories of subtilisin BPN (1TO2). The trend of the RMSD from the average structure shows a stabilization of the model along the MD trajectories (Figure 3.B). The per–residue fluctuation (RMSF) in Figure 3.C shows some very flexible regions, for example the region 50–80, where structural information in the templates is absent.

Most of residues forming the binding region (orange rectangles) were much less flexible than the rest, similar to what we observed for the other two subtilisins (data not shown). The fact that the model of the binding region showed similar stability in MD simulations as the X–ray crystal structures is an indication that there are no major errors in the model of this region. Obviously, despite the care we took in generating and validating our model, there may be structural errors with an effect on the success of our computational design procedure. For this reason, we included an additional step to refine the model of the structure of PvSUB1.

## Refinement of PvSUB1 model

To obtain a refined structure of the PvSUB1 model, we performed MD simulations of the complex of PvSUB1 and its substrate hexapeptide (P4-VGADDV-P2'). The hexapeptide was docked according to the X-ray structure of Subtilisin E with its pro-peptide (1SCJ), and refined with MD, where we restrained a few distances between the protein and the peptide (see [14] for details). Even though subtilisins do not undergo major conformational changes upon binding [29], small rearrangements might take place at the interface at the level of side–chains for example. The refinement allowed us to obtain bound–like conformations of PvSUB1, which facilitated the subsequent docking step (Figure 4). Figure 5 shows the catalytic triad Ser-His-Asp of the PvSUB1 model.

## Docking of EETI-II-sub

The wild type EETI-II did not inhibit PvSUB1 (Table 3). We then replaced the EETI-II residues involved in the binding to the protease catalytic groove with the sequence of the PvSUB1 natural hexapeptide substrate. This EETI-II-sub mutant inhibited PvSUB1 activity with a Ki >0.75 mM (Table 3). We derived snapshots (50 for each protein) for the ensemble docking of EETI-II-sub onto PvSUB1 by a cluster analysis (based on the residues at the interface) of the 5×2 ns molecular dynamics trajectories to obtain the best representative structures from the simulations.

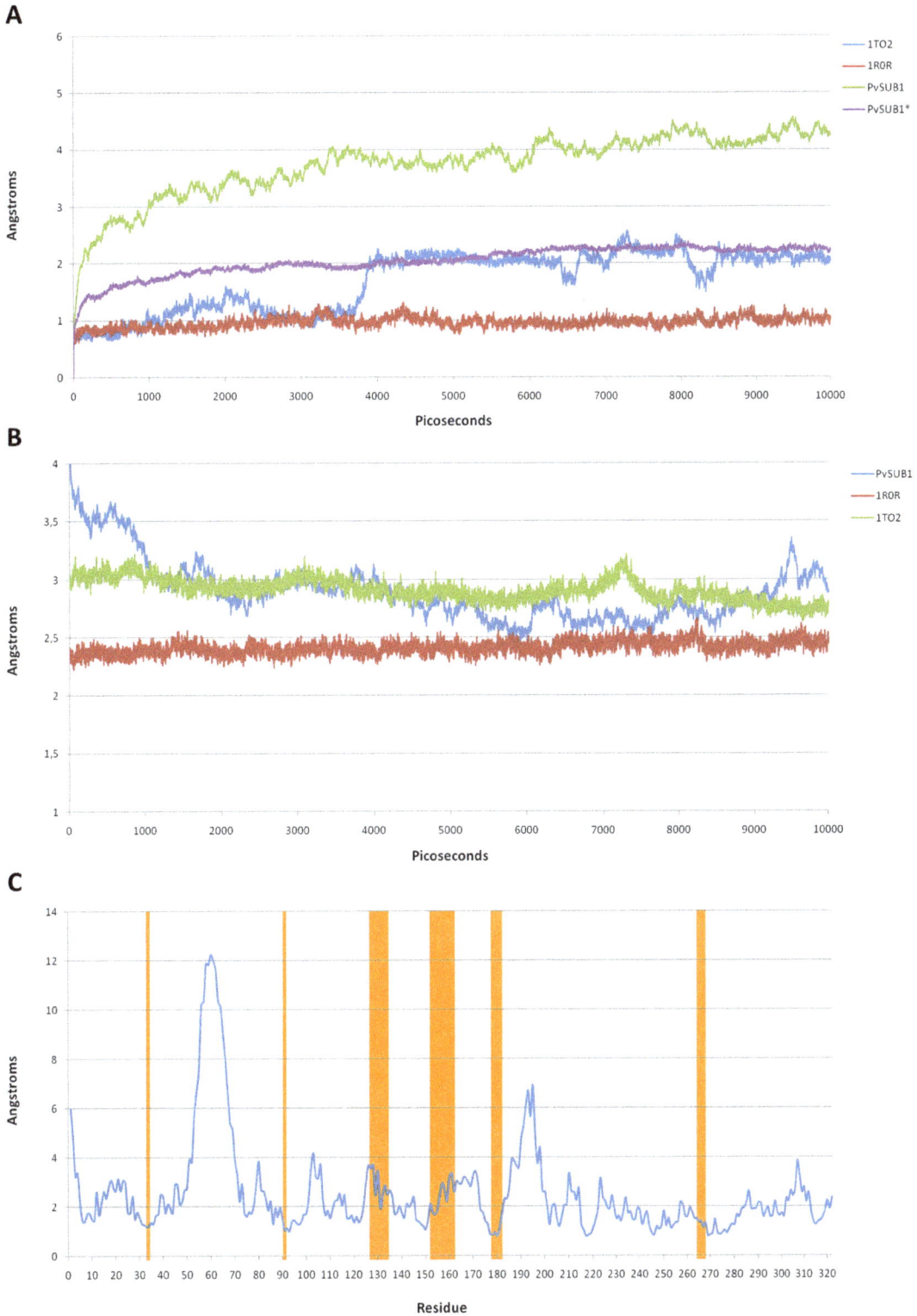

**Figure 3. PvSUB1 molecular dynamics simulations. A:** Average RMSD values for PvSUB1 and the 3D structure of two homologous bacterial subtilisins (1TO2, 1ROR). PvSUB1* shows the RMSD calculated without the regions missing template structural information; **B:** Fluctuation of the RMSD from the average structure. **C:** Root mean square fluctuation (RMSF) on a per-residue basis. In orange are highlighted PvSUB1 residues involved in the substrate-binding region.

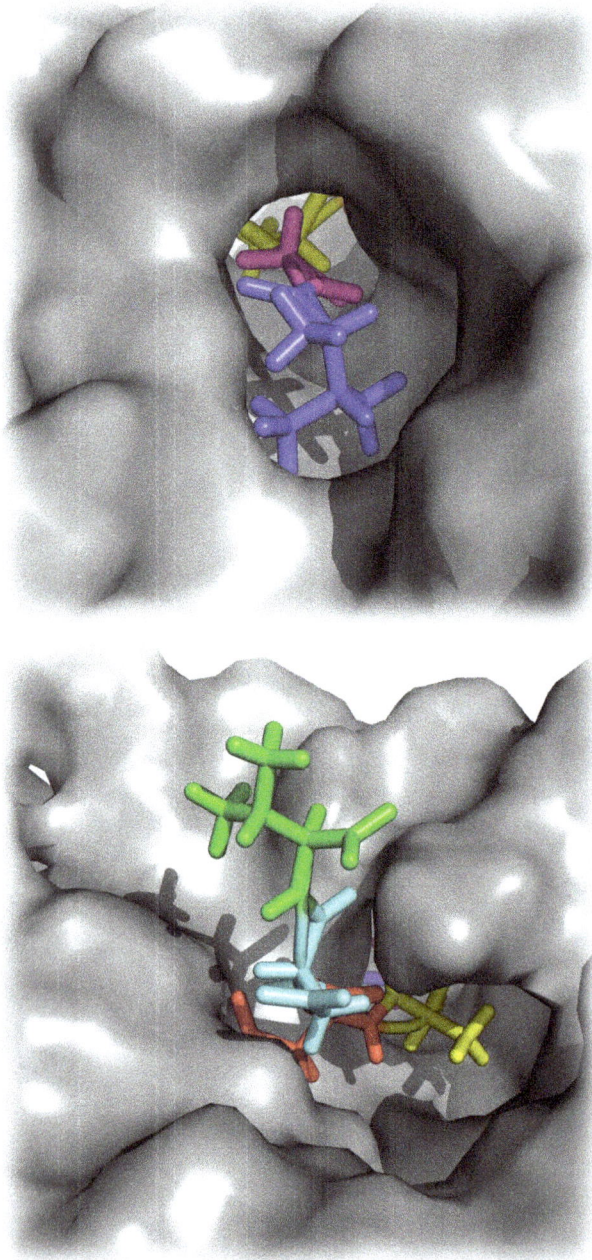

**Figure 4. Docking of PvSUB1 hexapeptide substrate into PvSUB1 catalytic groove.** Blue: P4, Violet: P3, Yellow: P2, Red: P1, Cyan: P1', Green: P2'.

## Scoring mutants

A preliminary free energy calculation was performed with snapshots from multiple MD simulations of EETI-II-sub docked onto PvSUB1 to obtain more consistent MM/GBSA results [30]. A free-energy decomposition (Figure 8) shows the contribution of each single residue to the total free energy of binding.

The biggest contribution to the free energy of binding came from the main-chain contacts of residues P4, P3, P2 and P1. This is in agreement with previous observations of important interactions between a protein-inhibitor and a serine protease active site, where important contacts are made by main-chain atoms [31]. For the case of EETI-II-sub the highest contribution originates from the cysteine in P3 and its main-chain, accounting for −4.34 kcal/mol.

We then tried to identify the most favorable mutations that could improve the binding affinity of EETI- II-sub to PvSUB1. The cysteine in P3 cannot be mutated because its side-chain is involved in a disulfide bridge that has an important function in stabilizing the EETI-II scaffold and maintaining the loop rigid, whereas the alanine in P2 already contributes with −4.17 kcal/mol to the total binding energy. We also looked at the parasite sequences that are natural substrates of PvSUB1 or PfSUB1 to suggest positions to introduce mutations in the EETI-II scaffold. Table 4 lists these sequences (experimentally derived or deducted from sequence alignments) for PfSUB1 and PvSUB1. While the sequences of several PfSUB1 substrates were experimentally determined [32,33], few were identified for PvSUB1 [15]. Considering the evolutionary proximity of *P. vivax* and *P. falciparum*, with active sites displaying >60% sequence identity [15], these predicted sequences can be considered reliable. Comparing the cleavage sites, we observed that only alanine and glycine appeared in P2, suggesting that only small residues are tolerated in this position. Position P1′ has a negative contribution to the energy and therefore is an interesting position to mutate. However, considering the lack of a specific pocket for this residue (Figure 7) we can consider this position almost a secondary contact residue [31] and we decided to keep the wild–type residue. Finally, SUB1 cleavage site sequences have a fairly high similarity at the P1 and P4 positions (Table 4) and we therefore focused on these positions to mutate the EETI-II inhibitory loop. It is worth mentioning that the contribution of these P1 and P4 positions within the substrate-PfSUB1 interaction has recently been experimentally established [34].

We performed 10×100 ps MD simulations and MM/GBSA free energy calculations for all possible residues in position P4 and P1 independently, assuming that the effect of the two mutations would be additive. The free energy calculations for the mutants in P4 showed that hydrophobic and bulky residues were preferred for this position (Figure 9.A). This result fits with the fact that pocket S4 is composed of six hydrophobic residues (L131, M134, F153, I161, F162, P205) and seems to have enough space to accommodate larger hydrophobic side-chains than valine (Figure 10). Position P1 instead presents as favorable mutations aromatic residues with polar groups (Tyr, Trp), glutamate and positively charged residues (Lys, Arg). Surprisingly we found as favorable mutations some positively charged residues, whereas most of sequences recognized by the homologous PfSUB1 present negatively charged (Asp, Glu) or neutral polar (Gln, Asn) side-chains at P1. This might be explained by either the low substrate specificity common to some subtilisins or imprecisions in the structure of the complex.

Figure 6 shows the score distribution among the solutions, with the 5 solutions with the highest geometric score (see Methods) highlighted with red circles. The relatively low geometric score indicated that most docking solutions were sub-optimal, although we had used an elaborate ensemble docking procedure. To optimize and refine the docking results we took the best 5 docking poses (red circles in 6) and refined them by a restrained molecular dynamics procedure similar to what we used for the refinement of the PvSUB1 model. This procedure helped re-establish native contacts when compared to regular unrestrained MD simulations. Subsequently we selected the solution that fulfilled all distances. Figure 7 shows the docked complex.

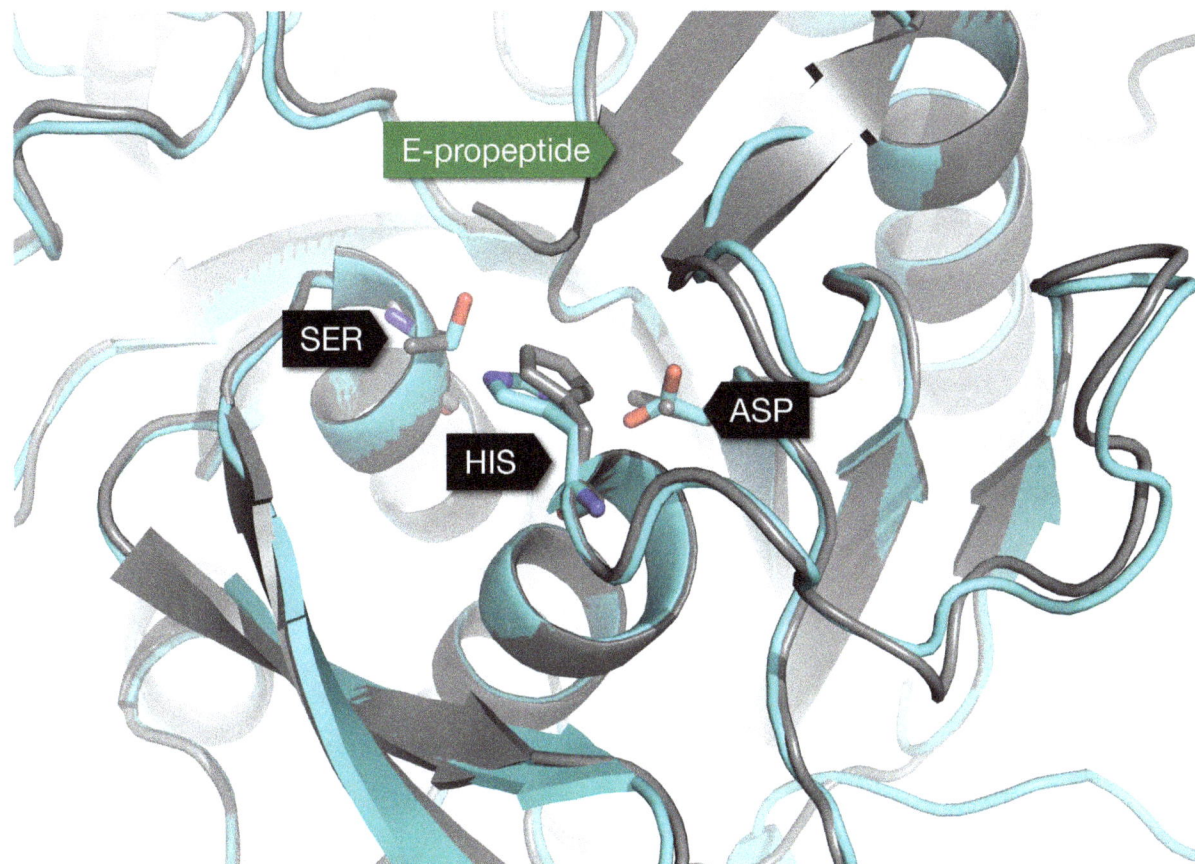

**Figure 5. Structural alignment of the obtained PvSUB1 model (cyan) with the 3D-structure of Subtilisin E (gray, PDB 1SCJ) that was used as a template in the homology modelling.** The catalytic triads in both proteins are highlighted with a stick representation. PvSUB1 catalytic triad: Asp 316, His 372 and Ser 549. Subtilisin E catalytic triad: Asp 32, His 64, Ser 221.

## Evaluation of EETI-II mutants on PvSUB1 enzymatic activity

All mutants of EETI-II were produced by chemical synthesis, folded and purified by reverse–phase HPLC. In Table 3 we present the results of PvSUB1 inhibition of the different synthesized EETI-II mutants. We initially tested mutants in position P4 according to our scoring results and found that the EETI-II with a leucine in P4 (EETI-II-P4L) inhibits PvSUB1 with a Ki of 147 μM, i.e., about one order of magnitude higher that the

**Table 3.** Sequence and inhibitory activity of EETI-II mutants on PvSUB1.

| Name of tested EETI-II | EETI-II active site sequences $P_4$ $P_3$ $P_2$ $P_1$ $P'_1$ $P'_2$ | Ki on PvSUB1 (mM) |
|---|---|---|
| EETI-II-WT | G C P R I L | NI |
| PvS1-WT | V C A D D V | >0.75 |
| PvS1-$P_{4W}$ | W C A D D V | >0.75 |
| PvS1-$P_{4P}$ | P C A D D V | >0.75 |
| PvS1-$P_{4M}$ | M C A D D V | >0.75 |
| PvS1-$P_{4L}$ | L C A D D V | 0.15±0.03 |
| PvS1-$P_{4I}$ | I C A D D V | 0.6±0.01 |
| PvS1-$P_{4L}$ $P_{1E}$ | L C A E D V | 0.34±0.05 |
| PvS1-$P_{4L}$ $P_{1K}$ | L C A K D V | 0.39±0.14 |
| PvS1-$P_{4L}$ $P_{1R}$ | L C A R D V | 0.75±0.035 |
| PvS1-$P_{4L}$ $P_{1Y}$ | L C A Y D V | 0.24±0.02 |
| PvS1-$P_{4L}$ $P_{1W}$ | L C A W D V | 0.08±0.01 |

Active site sequences of the tested EETI-II mutants and their Ki for PvSUB1. NI: No Inhibition.

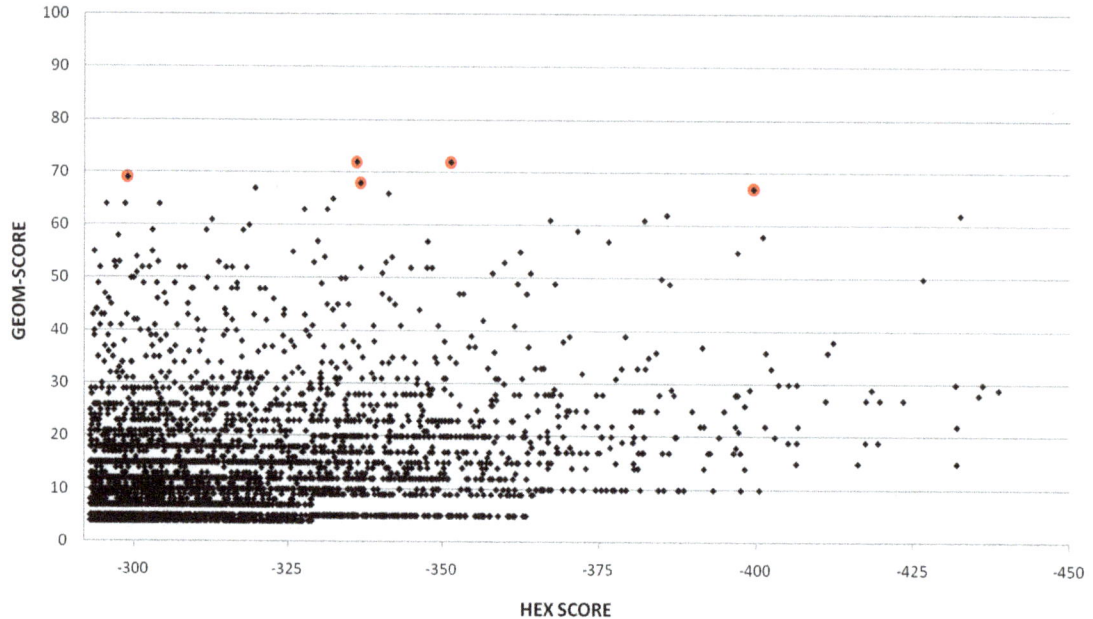

**Figure 6. Docking results.** The red circles indicate the docking poses that have been selected for refinement.

valine (Table 3). This could be explained by the higher flexibility of the leucine and its bulkier side-chain, which could fit more tightly into the hydrophobic S4 pocket (Figure 10). The fact that the isoleucine mutant shows a $Ki$ of 591 µM, which is three times less than that of leucine, could be the effect of its beta–branched side–chain. We decided to keep the leucine mutant in P4 and test mutations in P1. Among all the EETI-II-P4L mutants for the P1 position, only one showed a better $Ki$ of 86 µM, the EETI-II-P4L-P1W, which contained a tryptophane in P1. All other mutants had higher $Ki$ suggesting the importance of keeping the aspartate in this position. We found the comparison of the active mutants with the list of substrate sequences of PvSUB1 in Table 4 particularly intriguing. In position P4 of the substrate there is a preference for valine and threonine, while leucine is only present in 2 out of 18 sequences; the mutant with a leucine in P4 is more than 5x more active on PvSUB1 than the one with a valine (PvS1-WT). This might be explained by the presence of important structural constraints (disulphide bonds) that are present in EETI-II compared to a more flexible conformation of the substrate sequence [29]. According to the cleavage sites predicted by sequence alignment of the substrates, we observe a prevalence of negatively charged residues such as glutamate and aspartate while there was a preference for aromatic residues in the designed inhibitors, with the tryptophan having the best $Ki$. Even in this case, the explanation might lie in the structural constraints that are present in the inhibitors and/or a set of new interactions at position P1 (Figure 10).

## Conclusions

Subtilisin-like proteins of *Plasmodium* are promising biological targets for developing novel therapeutics. One of these proteins, SUB1 plays an essential role in both the hepatic and erythrocytic stages of *Plasmodium*, making this enzyme a particularly interesting drug target. With the aim to develop an inhibitor of *Plasmodium vivax* SUB1 (PvSUB1), we redesigned a protein scaffold, the natural trypsin inhibitor EETI-II. Despite challenges in this project (the 3D structure of the target protein had to be

modeled from homologues with only 30% sequence identity), the computational procedure allowed us to predict mutants that proved to be inhibitors of PvSUB1 in experimental tests.

However, some predicted good inhibitors did not show any improvements in binding to PvSUB1 when experimentally tested. This might be caused by the flexibility in the binding as proteases and their S# pockets are inherently flexible and known to alter their shape to accommodate various substrates. Imprecisions in the homology model of PvSUB1 can also have obvious consequences for the precision of the docking and of the energetic analysis. The free energy analysis based on implicit solvent methods is in itself approximate and neglects important factors influencing binding. It is encouraging that, nonetheless, we have obtained protein–based inhibitors of PvSUB1 and this opens new ways to further improve our best mutants by computational or experimental protein engineering protocols. The protein design approach described in this work demonstrates the capabilities of computational procedures to accelerate and guide the design of novel proteins with potential future therapeutic applications.

## Methods

### Modeling PvSUB1

Models of PvSUB1 were generated and validated with the same protocol as in a previous study [14]. Binding pocket residues of PvSUB1 were defined according to the interaction of canonical inhibitors with subtilisins: 152–162, 127–134, 265–268, 179–182, 90 (His), 34 (Asp), 267 (Ser). Regions were structural information from template structures is lacking are 307–324, 1–9, 51–73, 190–199, 227–231. The models were further evaluated by performing molecular dynamic simulations and comparing their dynamic behavior to that of two subtilisins 3D structures used in the homology modeling (Carlsberg: 1R0R, BPN: 1TO2). The validated model of PvSUB1 was further refined in order to obtain bound-like conformations for the ensemble docking. We docked the hexapeptide of the sequence recognized by PvSUB1 to its structure, using 1SCJ (subtilisin E + pro-peptide) as template. We performed restrained molecular dynamic simulation to refine the

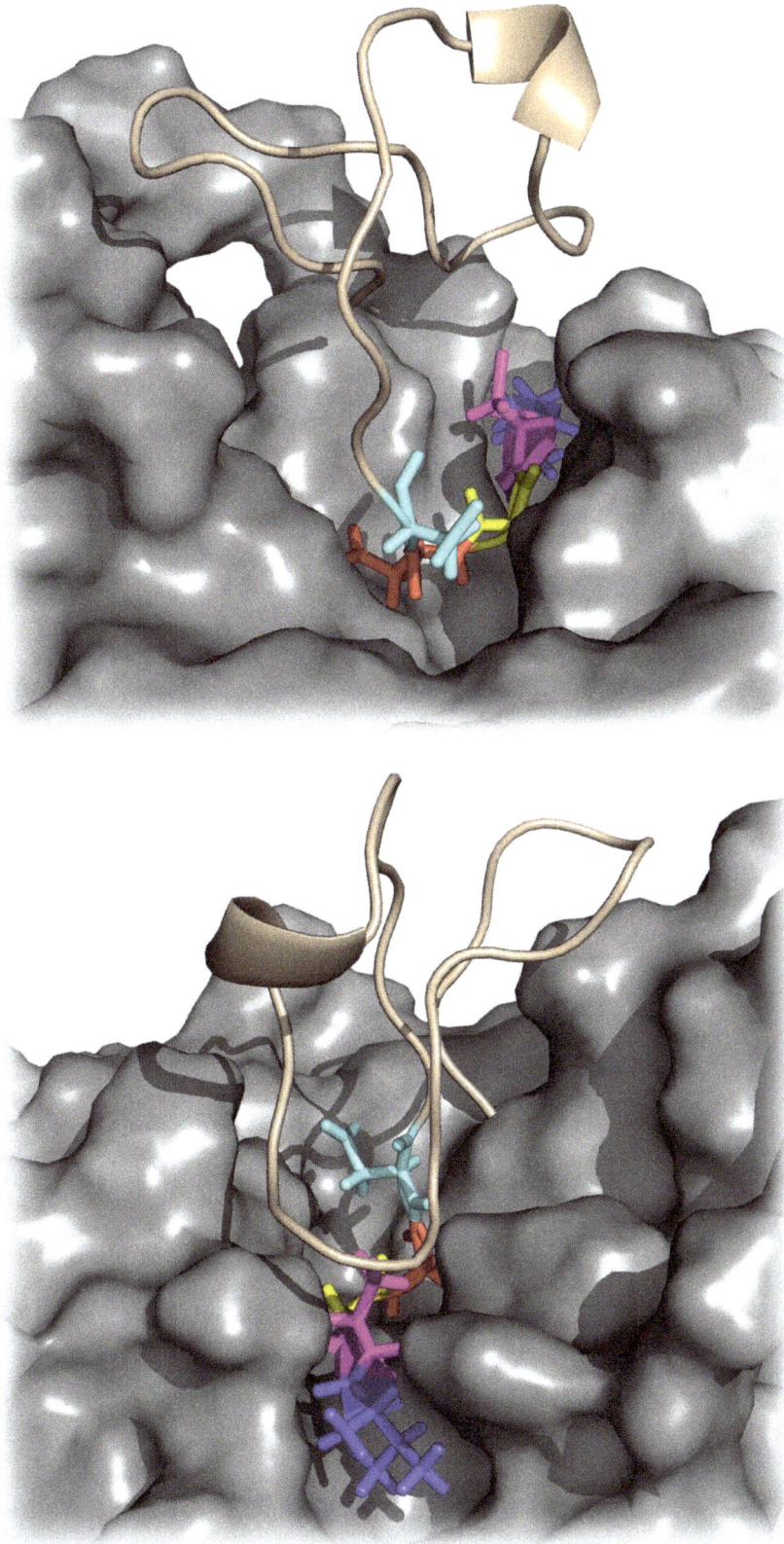

**Figure 7. EETI-II-sub docked to PvSUB1.** Blue: P4, Violet: P3, Yellow: P2, Red: P1, Cyan: P1'.

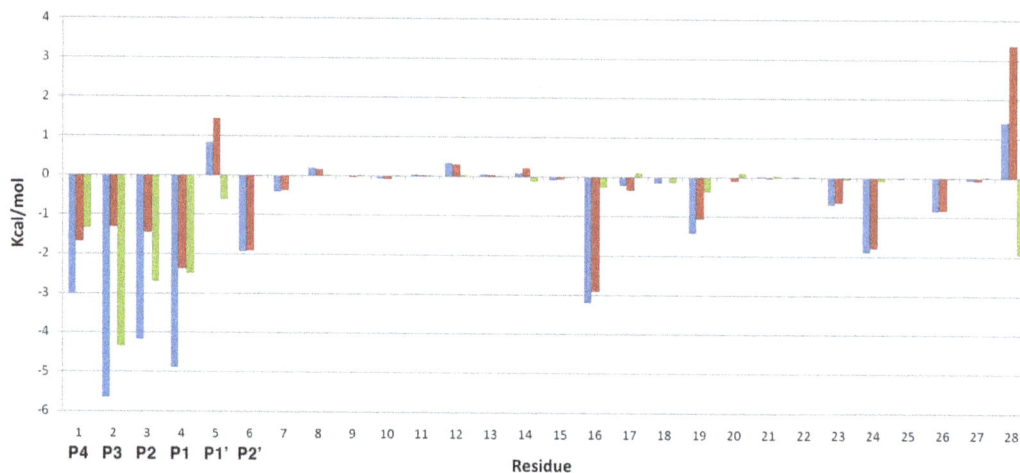

**Figure 8. Free energy decomposition.** Blue: All atoms, Red: Side chain atoms, Green: Backbone atoms. The largest contribution to the free energy of binding comes from the main-chain contacts of residues P4, P3, P2 and P1. The highest contribution comes from the cysteine in P3 and its main-chain, accounting for −4.34 kcal/mol.

complex, using the protocol described in the Section "Docking Refinement". Surface electrostatic distributions on models were calculated with the APBS [35] module implemented in Pymol.

**EETI-II.** The structure of EETI-II (2IT7) was retrieved from the PDB (http://www.rcsb.org) [36] and mutations at position P1 and P4 of the inhibitory loop were generated with the toolkit MMTSB [37]. We built the EETI-II substrate-like mutant (EETI-II-sub) by replacing its inhibitory loop (GCPRIL) with the sequence recognized by PvSUB1 (VGADDV). 3D images of the protein complexes were rendered with the molecular modeling software Pymol.

## Molecular Dynamics (MD) Simulations

All molecular dynamics simulations were performed with the SANDER module from the AMBER9 [38] package and the force–field ff99SB [39]. After minimization *in vacuo* the complexes were hydrated with TIP3P [40] water molecules and neutralized by adding an appropriate number of monovalent counterions. The MD unit cell was a truncated octahedral box with a minimum distance of 10 Å between the solute and the cell boundary. To minimize the water molecules we ran a two stages minimization protocol in which we first applied positional restraints with an energy constant of 10 kcal/(mol Å$^2$) to the solute followed by a stage with 1 kcal/(mol Å$^2$) energy constant. Both minimization stages consisted of 10 steepest-descent and 490 conjugate-gradient steps. During the equilibration/heating and production dynamics all covalent bonds to hydrogen atoms were constrained with the SHAKE [41] algorithm, and we used a time step of 2 fs. We used periodic boundary conditions with a distance cutoff of 8.0 Å for the direct part of the non-bonded interaction and PME [42] (Particle Mesh Ewald) to account for long-range electrostatic interactions. The minimized system was then thermalized and equilibrated by heating from 0 to 300°K over 20 ps under constant-volume conditions followed by 10 ps at constant-pressure. The production MD phase was launched from the final configuration after equilibration with a relaxation time of 2.0 ps for heat bath coupling and a pressure relaxation time of 2.0 ps.

## Docking

The pool of conformers used for the ensemble docking was obtained from multiple (5×2 ns) MD trajectories of PvSUB1 (receptor) and EETI-II (ligand). We extracted snapshots every 10 ps of MD for a total of 1000 snapshots of each protein and clustered these structures with a single-linkage algorithm implemented in GROMACS [43] (g-cluster tool), where the RMSD was calculated only for the binding interface residues. We selected for cross-docking the centroids of each of the 50 clusters identified. The ensemble docking was performed with the rigid-body docking software HEX [44] (version 4.5) where the search was restricted to the binding pocket by positioning EETI-II structures around the interface and limiting the search to 30° for twist range, receptor and ligand range. We used a shape + electrostatic correlation type while the other parameters of HEX were left as default. The best 5000 docking solutions according to the HEX score were selected and re-ranked by a mixed score based on the geometry of the interaction between a canonical loop inhibitor and a subtilisin. This score (geometric score) was composed for 40% of the HEX score (based on surface complementarity) and for 60% of an empirical score defined by the conserved distances between atoms in the inhibitory loop of a canonical inhibitor and a subtilisin. A mixed score permits to take into consideration the shape complementary and the conserved structural feature of the interaction. The best 5 solutions were selected for refinement.

## Docking Refinement

The refinement is based on a protocol that uses restrained molecular dynamics simulations. The refinement procedure consisted of a total of 360.000 steps based on specific distance restraints at the ligand/receptor interface and used the NMR refinement tools in AMBER9. The chosen conserved distance restraints were 267SER@OG-P1@C (lower bounds = 2.5 Å, lower-intermediate = 3.0 Å, intermediate- upper = 4.0 Å, upper bounds = 5.5 Å), 154SER@HN-P3@O, 154SER@O-P3@HN, 129GLY@HN-P4@O, 129GLY@O-P4@HN, 127LYS@O-P2@HN (lower bounds = 1.5 Å, lower-intermediate = 2.0 Å, intermediate- upper = 3.0 Å, upper bounds = 3.5 Å). In all cases, an energy constant of 20 kcal/mol Å$^2$ was employed. The refinement consisted of three phases. In the first phase (120.000 steps), the

**Table 4.** SUB1 natural substrates.

| *Plasmodium* protein containing a SUB1 processing site | P₄P₃P₂P₁ ↓ P′₁P′₂ | References |
|---|---|---|
| **Pro-region first maturation site of PfSUB1** | **VSAD ↓ NI** | [48] |
| **Pro-region second maturation site of PfSUB1** | **VSAD ↓ NI** | [48] |
| PfSERA1 site 1 | IKAE ↓ AE | [6] |
| PfSERA2 site 1 | TKGE ↓ DD | [6] |
| PfSERA3 site 1 | VKAA ↓ SV | [6] |
| PfSERA4 site 1 | ITAQ ↓ DD | [6] |
| **PfSERA5 site 1** | **IKAE ↓ TE** | [6] [49] |
| PfSERA6 site 1 | VKAQ ↓ DD | [6] |
| PfSERA7 site 1 | FKGE ↓ DE | [6] |
| PfSERA9 site 1 | VKGS ↓ TE | [6] |
| PfSERA1 site 2 | IYSQ ↓ ED | [6] |
| PfSERA2 site 2 | IWGQ ↓ ET | [6] |
| PfSERA3 site 2 | LYGQ ↓ EE | [6] |
| PfSERA4 site 2 | VYGQ ↓ DT | [6] |
| **PfSERA5 site 2** | **IFGQ ↓ DT** | [6] [49] |
| PfSERA6 site 2 | VHGQ ↓ SN | [6] |
| PfSERA7 site 2 | ISAQ ↓ DE | [6] |
| PfSERA9 site 2 | VHGQ ↓ SG | [6] |
| PvSERA1 site 1 | TKGE ↓ DE | |
| PvSERA2 site 1 | MKAQ ↓ DE | |
| PvSERA3 site 1 | AKGE ↓ DE | |
| PvSERA4 site 1 | RKAQ ↓ QQ | |
| PvSERA5 site 1 TKGE ↓ DE | | |
| **PfSERA5 site 3** | VRGD ↓ TE | [6] |
| **PfMSP1 clone 3D7 junction MSP1₈₃ -MSP1₃₀** | **LVAA ↓ SE** | [50] |
| **PfMSP1 clone 3D7 junction MSP1₃₀ -MSP1₃₈** | **ITGT ↓ SS** | [50] |
| **PfMSP1 clone 3D7 junction MSP1₃₈ -MSP1₄₂** | **VTGE ↓ AI** | [50] |
| **PfMSP1 clone FCB1 junction MSP1₃₀ -MSP1₃₈** | **VSAN ↓ DD** | [51] |
| **PfMSP1 clone FCB1 junction MSP1₃₈ -MSP1₄₂** | **VTGE ↓ AV** | [51] |
| **PcMSP1 junction MSP1₈₃ -MSP1₃₀** | ATGE ↓ SE | [52] |
| **PcMSP1 junction MSP1₃₀ -MSP1₃₈** | VSAE ↓ SE | [52] |
| **PcMSP1 junction MSP1₃₈ -MSP1₄₂** | ANAQ ↓ ST | [52] |
| PvMSP1 clones Sal1 and (Belem) junction MSP1₈₃ -MSP1₃₀ | LRGA(S) ↓ SA | |
| PvMSP1 clones Sal1 and Belem junction MSP1₃₀ -MSP1₃₈ | VGGN ↓ SE | |
| PvMSP1 clones Sal1 and Belem junction MSP1₃₈ -MSP1₄₂ TTGE ↓ AE | | |
| **PfMSP6 clone 3D7** | **VQAN ↓ SE** | [31] |
| **PfMSP7 clone 3D7 site 1** | **VKAQ ↓ SE** | [31] |
| **PfMSP7 clone 3D7 site 2** | **TQGQ ↓ EV** | [31] |
| **PfRAP-1 clone 3D7** | **IVGA ↓ DE** | [32] |
| **PfMSRP2 clone 3D7** | **LKGE ↓ SE** | [32] |
| **Pro-region first maturation site of PvSUB1** | **VGAD ↓ NI** | [15] |
| **Pro-region second maturation site of PvSUB1** | **SHAA ↓ SS** | [15] |
| **Pro-region third maturation site of PvSUB1** | **HLAG ↓ SK** | [15] |
| **Pro-region first maturation site of PbSUB1** | **VGAD ↓ SI** | [15] |
| Pro-region first maturation site of PySUB1 | VGAD ↓ SI | [15] |

The table shows the sequences of the cleavage sites recognized by SUB1 in in *Plasmodium falciparum* and *Plasmodium vivax*. Cleavage site sequences in bold characters have been experimentally determined, while the ones in normal characters are deducted from sequence alignments. The arrow indicates the site of cleavage between the P1 and the P1'. Cleavage site sequences have a fairly high similarity in particular at the P1, P2 and P4 positions.

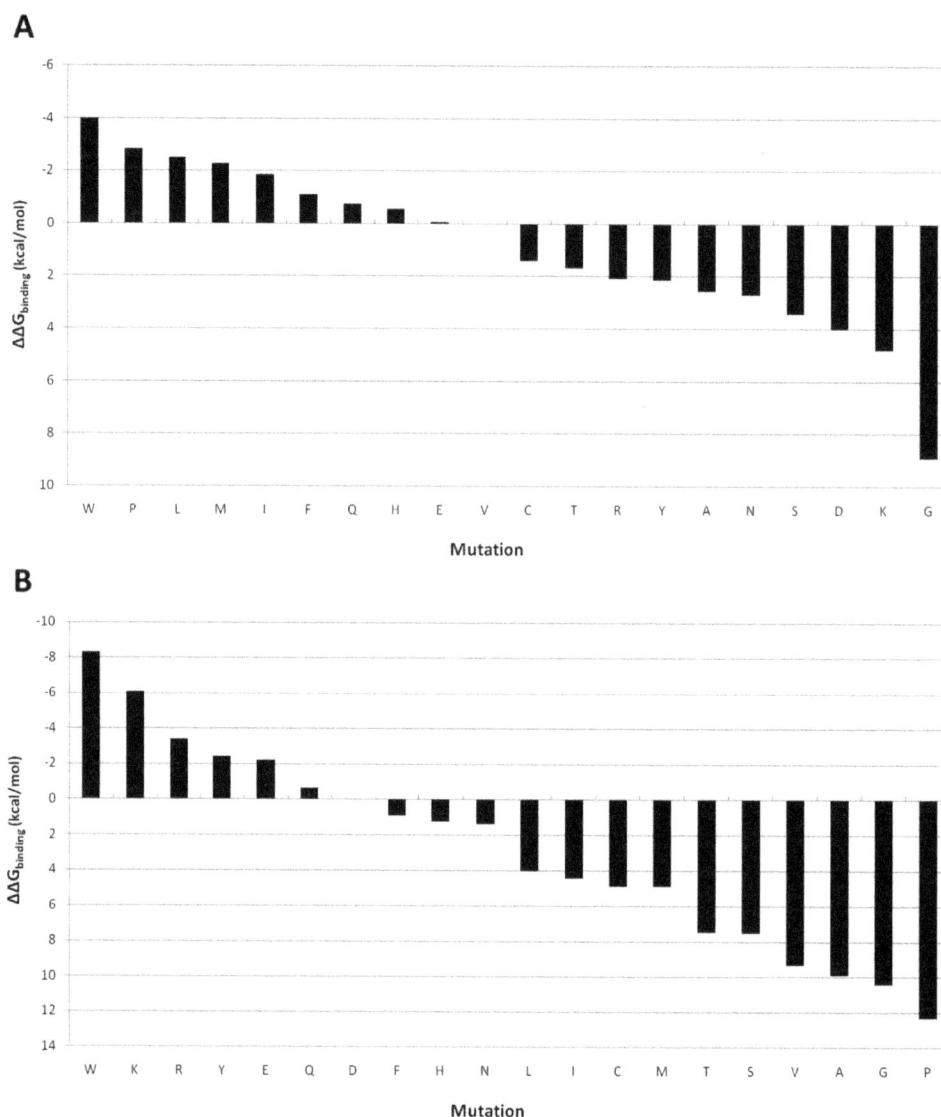

**Figure 9. Scoring mutations on P4 and P1. A:** mutants in position P4. The mutational profile of P4 shows that hydrophobic and bulky residues are preferred for this position. **B:** mutants in position P1. Position P1 instead prefers aromatic residues with polar groups (Tyr, Trp), glutamate and positively charged residues (Lys, Arg).

receptor (PvSUB1) was kept rigid by applying a Cartesian restraint with an energy constant of 10 kcal/(mol $\mathring{A}^2$) and the distance restraints were switched on/off every 15.000 steps. In the second phase a Cartesian restraint with an energy constant of 10 kcal/(mol $\mathring{A}^2$) was applied only to heavy main-chain atoms keeping side-chains fully flexible, and in the third phase the energy constant of this Cartesian restraint was reduced to 0.01 kcal/(mol $\mathring{A}^2$). During these phases, the ligand (EETI-II or hexapeptide) was kept completely flexible. An additional 1 ns of regular MD simulation was performed to allow the system to relax into its final configuration. In the validation, this resulted in a high accuracy complex structure, with no distortions at the interface.

## Scoring

The MM/GBSA protocol [45] in AMBER9 was used to calculate the relative free energy of each mutant. The default GBSA model used in the calculations was that of Tsui and Case [46] with an external dielectric of 80 and internal dielectric of 1.0.

For calculating the nonpolar contribution, the surface tension coefficient was set to 0.0072 and the surface offset to 0.0. The solvent accessible surface area was calculated with the ICOSA method. We calculated the relative free energy of binding from snapshots extracted each 10 ps from $10 \times 100$ ps trajectories.

## Protein Production

**EETI-II and mutants.** All peptides (desalted, 35%–60% pure as assessed by HPLC) were obtained from GenScript Corporation, Piscataway, NJ, USA. In a typical procedure, peptide (50 mg) was dissolved in 75 mL of $KH_2PO_4$ buffer (0.2 M, pH 8.2) and allowed to air-oxidize at room temperature under gentle stirring. Monitoring was achieved with Ellman's test [47] and analytical HPLC (column ACE C18, 5 µm×4.6 mm, eluent A: 0.1% $TFA/H_2O$, eluent B: 60% $CH_3CN/H_2O$/0.1% TFA) with a 30 min linear gradient of 25% to 55% B at 1 mL flow rate (monitoring at 210 nm). When Ellman's tests are negative and the HPLC monitoring shows no more trace of starting materials (after

A

B

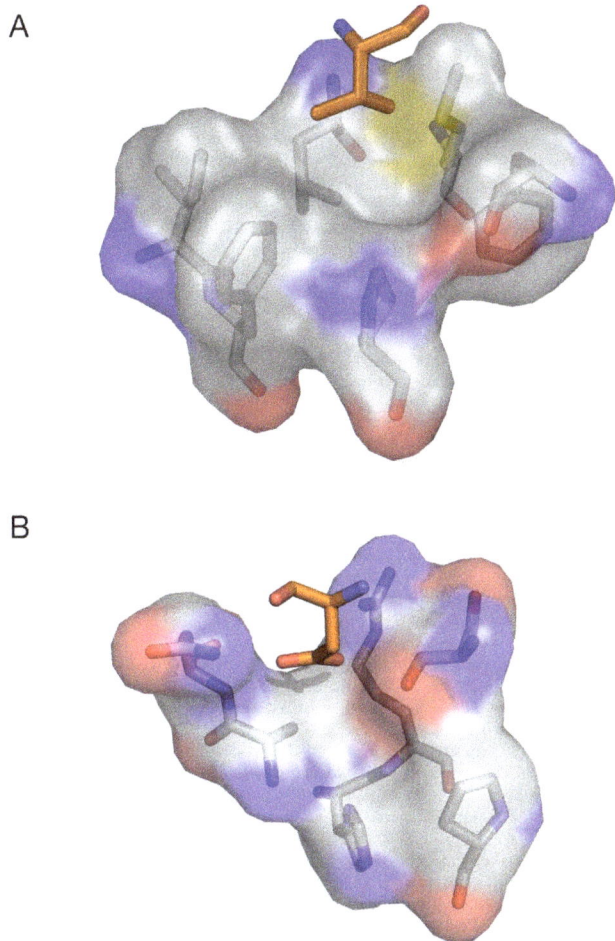

**Figure 10. Residues forming the S1 and S4 pockets.** The residue P4 (A) and P1 (B) of EETI-II are shown with an orange stick representation.

3 to 5 days), the reaction mixture was centrifuged and the supernatant loaded onto a preparative HPLC column (Merck Lichrospher C₁₈, 10 μm, 250×25 mm). Elution was achieved with a 90 min linear gradient of 25% to 55% B at 10 mL flow rate (monitoring at 220 nm). The fractions containing the oxidized peptide were combined and lyophilized to yield 15 to 30% of the desired peptide. Successful oxidation was confirmed by mass spectrometry (MALDI-Tof, Bruker Biflex III).

**Production and purification of the PvSUB1 recombinant enzyme.** The production and purification of the PvSUB1 (Genbank accession number FJ536585) recombinant enzyme was performed essentially as previously described [14,15]. Briefly for large-scale protein production, *Spodoptera frugiperda* Sf9 insect cells (1L at 3×106 cells/mL, Invitrogen) were infected for 72 h with recombinant PvSUB1-recombinant baculovirus at a Multiplicity Of Infection (MOI) of 10 in Insect XPRESS medium (Lonza) supplemented with 50 μg/mL gentamycin and 0.5 μg/

mL tunicamycin (Sigma-Aldrich). Culture supernatant containing the secreted PvSUB1 recombinant protein was harvested, centrifuged 30 min at 2150 g to remove cells and cellular debris and concentrated/diafiltrated against D-PBS 0.5 M NaCl, 5 mM Imidazole (loading buffer). The proteins were purified on an AKTA purifier system (GE Healthcare). The sample was loaded onto a 3 mL TALON Metal affinity resin (Clontech Laboratories) equilibrated in loading buffer. After extensive washes with loading buffer, the bound protein was eluted with a linear gradient of 5–200 mM imidazole in D-PBS, 0.5 M NaCl. Fractions containing PvSUB1 were pooled and concentrated by using Amicon Ultra 15 (10000 MWCO) and size-fractionated onto a HiLoad 16/60 Superdex 75 column equilibrated with 20 mM Tris pH 7.5, 100 mM NaCl. Fractions were monitored by absorbance (280 nm) and analyzed by Coomassie blue staining of SDS-PAGE gels and enzyme activity assay. The fractions containing the recombinant enzyme activity were pooled and the protein concentration was determined using the BCA Protein Assay following manufacturers recommendations (Bio Basic). Purified PvSUB1 recombinant protein was stored at −20°C following the addition of 30% v/v of pure Glycerol.

### Enzymatic Test

For the kinetic assays we used the purified recombinant PvSUB1 enzyme and its specific peptide sub- strate whose sequence is deduced from PvSUB1 auto-maturation site: KLVGADDVSLA, with cleavage occurs between the two aspartates for PvSUB1. The KLVGADDVSLA sequence was coupled to the fluorophore/quencher dyes Dabsyl/Edans (Exc/ Em 360/500 nm) at each edge. The enzymatic assays were performed in 20 mM Tris pH 7.5 and 25 mM CaCl₂ at 37°C as previously described [15]. For the determination of the Ki, the compounds, previously resuspended in ultra-pure distilled water at 10 mM, were tested at ten different concentrations ranging from 1 mM to 2 μM following sequential 1:2 dilutions. The final mixture was distributed in duplicate into a 384-well black microtiter plate (Thermo Scientific) and the fluorescence was monitored every 3 minutes for 90 min at 37°C in a Labsystems Fluoroskan Ascent spectro-fluorometer. The slope of the linear part of the kinetic was determined in an Excel (Microsoft) spreadsheet. Every steps of the enzymatic assay were done on ice to make sure that the protein was not active before the measure of the fluorescence. The Ki and IC50 values were determined (N = 3) using GraphPad Prism software.

### Acknowledgments

We are grateful to Odile Puijalon for constant support and critical reading of the manuscript. This work is dedicated to the memory of our colleague Dung Le-Nguyen.

### Author Contributions

Conceived and designed the experiments: GB DLN MN JCB. Performed the experiments: GB AB CN DLN. Analyzed the data: GB AB CN DLN MN JCB. Contributed reagents/materials/analysis tools: GB AB CN DLN MN JCB. Contributed to the writing of the manuscript: GB DLN MN JCB.

### References

1. Greenwood BM, Targett GA (2011) Malaria vaccines and the new malaria agenda. Clin Microbiol Infect 17: 1600–1607.
2. Agnandji ST, Lell B, Soulanoudjingar SS, Fernandes JF, Abossolo BP, et al. (2011) First results of phase 3 trial of RTS,S/AS01 malaria vaccine in African children. N Engl J Med 365: 1863–1875.
3. Noedl H, Socheat D, Satimai W (2009) Artemisinin-resistant malaria in Asia. N Engl J Med 361: 540–541.
4. Alonso PL, Brown G, Arevalo-Herrera M, Binka F, Chitnis C, et al. (2011) A research agenda to underpin malaria eradication. PLoS Med 8: e1000406.
5. Mishra SK, Newton CR (2009) Diagnosis and management of the neurological complications of *falciparum* malaria. Nat Rev Neurol 5: 189–198.

6. Yeoh S, O'Donnell RA, Koussis K, Dluzewski AR, Ansell KH, et al. (2007) Subcellular discharge of a serine protease mediates release of invasive malaria parasites from host erythrocytes. Cell 131: 1072–1083.

7. Tawk L, Lacroix C, Gueirard P, Kent R, Gorgette O, et al. (2013) A Key Role for *Plasmodium* Subtilisin-like SUB1 Protease in Egress of Malaria Parasites from Host Hepatocytes. J Biol Chem 288: 33336–33346.

8. Keizer DW, Miles LA, Li F, Nair M, Anders RF, et al. (2003) Structures of phage-display peptides that bind to the malarial surface protein, apical membrane antigen 1, and block erythrocyte invasion. Biochemistry 42: 9915–9923.

9. Zhu S, Hudson TH, Kyle DE, Lin AJ (2002) Synthesis and in vitro studies of novel pyrimidinyl peptidomimetics as potential antimalarial therapeutic agents. J Med Chem 45: 3491–3496.

10. Dagan A, Efron L, Gaidukov L, Mor A, Ginsburg H (2002) In vitro antiplasmodium effects of dermaseptin S4 derivatives. Antimicrob Agents Chemother 46: 1059–1066.

11. Smith GP, Petrenko VA (1997) Phage Display. Chem Rev 97: 391–410.

12. He M, Taussig MJ (2002) Ribosome display: cell-free protein display technology. Brief Funct Genomic Proteomic 1: 204–212.

13. Roberts RW, Szostak JW (1997) RNA-peptide fusions for the in vitro selection of peptides and proteins. Proc Natl Acad Sci U S A 94: 12297–12302.

14. Bastianelli G, Bouillon A, Nguyen C, Crublet E, Petres S, et al. (2011) Computational reverse-engineering of a spider-venom derived peptide active against *Plasmodium falciparum* SUB1. PLoS One 6: e21812.

15. Bouillon A, Giganti D, Benedet C, Gorgette O, Petres S, et al. (2013) In Silico Screening on the Three-dimensional Model of the *Plasmodium vivax* SUB1 Protease Leads to the Validation of a Novel Anti-parasite Compound. J Biol Chem 288: 18561–18573.

16. Heitz A, Chiche L, Le-Nguyen D, Castro B (1989) 1 H 2D NMR and distance geometry study of the folding of *Ecballium elaterium* trypsin inhibitor, a member of the squash inhibitors family. Biochemistry 28: 2392–2398.

17. Craik DJ, Daly NL, Waine C (2001) The cystine knot motif in toxins and implications for drug design. Toxicon 39: 43–60.

18. Craik DJ, Clark RJ, Daly NL (2007) Potential therapeutic applications of the cyclotides and related cystine knot mini-proteins. Expert Opin Investig Drugs 16: 595–604.

19. Kolmar H (2009) Biological diversity and therapeutic potential of natural and engineered cystine knot miniproteins. Curr Opin Pharmacol 9: 608–614.

20. Le-Nguyen D, Mattras H, Coletti-Previero MA, Castro B (1989) Design and chemical synthesis of a 32 residues chimeric microprotein inhibiting both trypsin and carboxypeptidase A. Biochem Biophys Res Commun 162: 1425–1430.

21. Christmann A, Walter K, Wentzel A, Kratzner R, Kolmar H (1999) The cystine knot of a squash-type protease inhibitor as a structural scaffold for *Escherichia coli* cell surface display of conformationally constrained peptides. Protein Eng 12: 797–806.

22. Hilpert K, Wessner H, Schneider-Mergener J, Welfle K, Misselwitz R, et al. (2003) Design and characterization of a hybrid miniprotein that specifically inhibits porcine pancreatic elastase. J Biol Chem 278: 24986–24993.

23. Reiss S, Sieber M, Oberle V, Wentzel A, Spangenberg P, et al. (2006) Inhibition of platelet aggregation by grafting RGD and KGD sequences on the structural scaffold of small disulfide-rich proteins. Platelets 17: 153–157.

24. Souriau C, Chiche L, Irving R, Hudson P (2005) New binding specificities derived from Min-23, a small cystine-stabilized peptidic scaffold. Biochemistry 44: 7143–7155.

25. Wentzel A, Christmann A, Kratzner R, Kolmar H (1999) Sequence requirements of the GPNG beta-turn of the *Ecballium elaterium* trypsin inhibitor II explored by combinatorial library screening. J Biol Chem 274: 21037–21043.

26. Lippow SM, Wittrup KD, Tidor B (2007) Computational design of antibody-affinity improvement beyond in vivo maturation. Nat Biotechnol 25: 1171–1176.

27. Jha RK, Leaver-Fay A, Yin S, Wu Y, Butterfoss GL, et al. (2010) Computational design of a PAK1 binding protein. J Mol Biol 400: 257–270.

28. Chung SY, Subbiah S (1996) A structural explanation for the twilight zone of protein sequence homology. Structure 4: 1123–1127.

29. Otlewski J, Jelen F, Zakrzewska M, Oleksy A (2005) The many faces of protease-protein inhibitor interaction. EMBO J 24: 1303–1310.

30. Genheden S, Ryde U (2010) How to obtain statistically converged MM/GBSA results. J Comput Chem 31: 837–846.

31. Komiyama T, VanderLugt B, Fugere M, Day R, Kaufman RJ, et al. (2003) Optimization of protease-inhibitor interactions by randomizing adventitious contacts. Proc Natl Acad Sci U S A 100: 8205–8210.

32. Koussis K, Withers-Martinez C, Yeoh S, Child M, Hackett F, et al. (2009) A multifunctional serine protease primes the malaria parasite for red blood cell invasion. Embo J 28: 725–735.

33. Silmon de Monerri NC, Flynn HR, Campos MG, Hackett F, Koussis K, et al. (2011) Global identification of multiple substrates for *Plasmodium falciparum* SUB1, an essential malarial processing protease. Infect Immun 79: 1086–1097.

34. Fulle S, Withers-Martinez C, Blackman MJ, Morris GM, Finn PW (2013) Molecular Determinants of Binding to the *Plasmodium* Subtilisin-like Protease 1. J Chem Inf Model.

35. Baker NA, Sept D, Joseph S, Holst MJ, McCammon JA (2001) Electrostatics of nanosystems: application to microtubules and the ribosome. Proc Natl Acad Sci U S A 98: 10037–10041.

36. Berman HM, Westbrook J, Feng Z, Gilliland G, Bhat TN, et al. (2000) The Protein Data Bank. Nucleic Acids Res 28: 235–242.

37. Feig M, Karanicolas J, Brooks CL, 3rd (2004) MMTSB Tool Set: enhanced sampling and multiscale modeling methods for applications in structural biology. J Mol Graph Model 22: 377–395.

38. Case DA, Darden TA, Cheatham TEI, Simmerling CL, Wang J, et al. (2006) AMBER9. University of California, San Francisco.

39. Hornak V, Abel R, Okur A, Strockbine B, Roitberg A, et al. (2006) Comparison of multiple Amber force fields and development of improved protein backbone parameters. Proteins 65: 712–725.

40. Jorgensen WL, Chandrasekhar J, Madura JD (1983) Comparison of simple potential functions for simulating liquid water. J Chem Phys 79: 926.

41. Ryckaert J, Ciccotti G, Berendsen H (1977) Numerical integration of the cartesian equations of motion of a system with constraints: molecular dynamics of n-alkanes. J Comput Phys 23: 327–341.

42. Darden T, York D, Pederson L (1993) Particle mesh ewald-an nlog(n) method for ewald sums in large systems. J Chem Phys 98: 10089–10092.

43. Van Der Spoel D, Lindahl E, Hess B, Groenhof G, Mark AE, et al. (2005) GROMACS: fast, flexible, and free. J Comput Chem 26: 1701–1718.

44. Ritchie DW, Kemp GJ (2000) Protein docking using spherical polar Fourier correlations. Proteins 39: 178–194.

45. Gohlke H, Kiel C, Case DA (2003) Insights into protein-protein binding by binding free energy calculation and free energy decomposition for the Ras-Raf and Ras-RalGDS complexes. J Mol Biol 330: 891–913.

46. Tsui V, Case DA (2000) Theory and applications of the generalized Born solvation model in macromolecular simulations. Biopolymers 56: 275–291.

47. Ellman GL (1959) Tissue sulfhydryl groups. Arch Biochem Biophys 82: 70–77.

48. Sajid M, Withers-Martinez C, Blackman MJ (2000) Maturation and specificity of *Plasmodium falciparum* subtilisin-like protease-1, a malaria merozoite subtilisin-like serine protease. J Biol Chem 275: 631–641.

49. Debrabant A, Maes P, Delplace P, Dubremetz JF, Tartar A, et al. (1992) Intramolecular mapping of *Plasmodium falciparum* P126 proteolytic fragments by N-terminal amino acid sequencing. Molecular and Biochemical Parasitology 53: 89–96.

50. Stafford WH, Blackman MJ, Harris A, Shai S, Grainger M, et al. (1994) N-terminal amino acid sequence of the *Plasmodium falciparum* merozoite surface protein-1 polypeptides. Mol Biochem Parasitol 66: 157–160.

51. Heidrich HG, Miettinin-Bauman A, Eckerskorn C, Lottspeich F (1989) The N-terminal amino acid sequences of the *Plasmodium falciparum* (FCBI) merozoite surface antigen of 42 and 36 kilodaltons, both derived from the 185–195 kilodalton precursor Molecular and Biochemical Parasitology 34: 147–154.

52. O'Dea KP, McKean PG, Harris A, Brown KN (1995) Processing of the *Plasmodium chabaudi chabaudi* AS merozoite surface protein 1 in vivo and in vitro. Molecular and Biochemical Parasitology 72: 111–119.

# 17

# Sporozoite Immunization of Human Volunteers under Mefloquine Prophylaxis Is Safe, Immunogenic and Protective: A Double-Blind Randomized Controlled Clinical Trial

Else M. Bijker[1,9], Remko Schats[2,9], Joshua M. Obiero[1], Marije C. Behet[1], Geert-Jan van Gemert[1], Marga van de Vegte-Bolmer[1], Wouter Graumans[1], Lisette van Lieshout[3,4], Guido J. H. Bastiaens[1], Karina Teelen[1], Cornelus C. Hermsen[1], Anja Scholzen[1], Leo G. Visser[2], Robert W. Sauerwein[1]*

1 Radboud university medical center, Department of Medical Microbiology, PO Box 9101, 6500 HB Nijmegen, The Netherlands, 2 Leiden University Medical Center, Department of Infectious Diseases, PO Box 9600, 2300 RC Leiden, The Netherlands, 3 Leiden University Medical Center, Department of Medical Microbiology, PO Box 9600, 2300 RC Leiden, The Netherlands, 4 Leiden University Medical Center, Department of Parasitology, PO Box 9600, 2300 RC Leiden, The Netherlands

## Abstract

Immunization of healthy volunteers with chloroquine ChemoProphylaxis and Sporozoites (CPS-CQ) efficiently and reproducibly induces dose-dependent and long-lasting protection against homologous *Plasmodium falciparum* challenge. Here, we studied whether chloroquine can be replaced by mefloquine, which is the only other licensed anti-malarial chemoprophylactic drug that does not affect pre-erythrocytic stages, exposure to which is considered essential for induction of protection by CPS immunization. In a double blind randomized controlled clinical trial, volunteers under either chloroquine prophylaxis (CPS-CQ, n = 5) or mefloquine prophylaxis (CPS-MQ, n = 10) received three sub-optimal CPS immunizations by bites from eight *P. falciparum* infected mosquitoes each, at monthly intervals. Four control volunteers received mefloquine prophylaxis and bites from uninfected mosquitoes. CPS-MQ immunization is safe and equally potent compared to CPS-CQ inducing protection in 7/10 (70%) versus 3/5 (60%) volunteers, respectively. Furthermore, specific antibody levels and cellular immune memory responses were comparable between both groups. We therefore conclude that mefloquine and chloroquine are equally effective in CPS-induced immune responses and protection.

*Trial Registration:* ClinicalTrials.gov NCT01422954

Editor: Steffen Borrmann, Kenya Medical Research Institute - Wellcome Trust Research Programme, Kenya

Funding: This trial was funded by The Netherlands Organisation for Health Research and Development (ZonMw, project 95110086) and the Dioraphte foundation (project 12010100). AS received an EMBO long-term fellowship. The funders had no role in study design, data collection and analysis, decision to publish, or preparation of the manuscript.

Competing Interests: The authors have declared that no competing interests exist.

* Email: Robert.Sauerwein@radboudumc.nl

9 These authors contributed equally to this work.

## Introduction

Malaria remains one of the most important infectious diseases worldwide and still causes approximately 207 million cases and 627,000 deaths every year [1]. Anti-disease immunity against malaria is not easily induced: in endemic areas this takes many years of repeated exposure to develop [2], and sterile protection against infection does not seem to be induced at all [3]. Also candidate vaccines have shown only limited protective efficacy so far [4,5]. Novel vaccines and drugs can be tested for efficacy at an early stage of clinical development in Controlled Human Malaria Infection (CHMI) studies, exposing a small number of healthy volunteers to *Plasmodium falciparum* by bites from infected *Anopheles* mosquitoes. Immunization of healthy volunteers under chloroquine ChemoProphylaxis with Sporozoites (CPS-CQ im-

munization) efficiently, reproducibly and dose-dependently induces protection against homologous CHMI [6,7], shown in a subset of volunteers to last for more than 2 years [8]. CPS-CQ immunization requires exposure to bites from only a total of 30–45 *P. falciparum* infected mosquitoes to induce 89–95% protection [6,7,9]. In contrast, protection by immunization with radiation-attenuated sporozoites (RAS) requires a minimum of 1000 infected mosquito bites [10], or intravenous injection of five times 135,000 cryopreserved sporozoites [11].

The unprecedented efficiency of the CPS immunization regime may relate to its design: in contrast to RAS, CPS immunization allows full liver stage development and exposure to early blood-stages. Moreover, chloroquine is known for its immunomodulatory capacities [12–14] that may play a role in induction of protection, which is mediated by pre-erythrocytic immunity [9] including antibodies directed against sporozoites [15–17], and likely T cells targeting liver-stages [7]. Next to chloroquine, mefloquine (MQ) is the only licensed drug for chemoprophylaxis that does not affect pre-erythrocytic stage development [18]. We therefore aimed to assess whether chloroquine could be replaced by mefloquine for CPS immunization. In a double blind randomized controlled clinical trial we assessed safety, immunogenicity and protection against challenge for CPS-MQ compared to CPS-CQ.

## Methods

### Study subjects

Healthy subjects between 18 and 35 years old with no history of malaria were screened for eligibility based on medical and family history, physical examination and standard hematological and biochemical measurements. Urine toxicology screening was negative in all included subjects; none of the subjects were pregnant or lactating. Serological analysis for HIV, hepatitis B, hepatitis C and *P. falciparum* asexual blood-stages was negative in all subjects. All subjects had an estimated 10-year risk smaller than 5% of developing a cardiac event as estimated by the Systematic Coronary Evaluation System adjusted for the Dutch population [19]. None of the subjects had travelled to a malaria-endemic area during or within 6 months prior to the start of the study. All subjects provided written informed consent before screening. The Central Committee for Research Involving Human Subjects of The Netherlands approved the study (NL 37563.058.11). Investigators complied with the Declaration of Helsinki and Good Clinical Practice including monitoring of data. This trial is registered at ClinicalTrials.gov, identifier NCT01422954. The protocol for this trial and supporting CONSORT checklist are available as supporting information (**Checklist S1** and **Protocol S1**).

### Study design and procedures

This single center, double blind randomized controlled trial was conducted at Leiden University Medical Center (Leiden, the Netherlands) from April 2012 until April 2013 (**Figure 1**). Twenty subjects were randomly divided into three groups by an independent investigator using a computer-generated random-number table. Subjects, investigators and primary outcome assessors were blinded to the allocation. Subjects in the CPS-CQ group (n = 5) received a standard prophylactic regimen of chloroquine consisting of a loading dose of 300 mg on the first and fourth day and subsequently 300 mg once a week for 12 weeks. Subjects in the CPS-MQ group (n = 10) and the control group (n = 5) received mefloquine prophylaxis starting with a loading split dose regimen to limit potential side-effects: 125 mg twice per week for a duration of 3 weeks and subsequently 250 mg once a week for 12 weeks. Chloroquine and mefloquine were administered as capsules, indistinguishable from each other. During this period all subjects were exposed to the bites of 8 *Anopheles* mosquitoes three times at monthly intervals, starting 22 days after start of mefloquine prophylaxis and 8 days after start of chloroquine prophylaxis. Volunteers in the CPS-CQ and CPS-MQ groups received bites from mosquitoes infected with the *P. falciparum* NF54 strain, control subjects received bites from uninfected mosquitoes. The immunization dose was based on our previous dose-de-escalation trial [7] and aimed to establish partial protection in the CPS-CQ group in order to enable detection of either improved or reduced protection in the CPS-MQ group. Sample sizes were calculated based on the expected difference of 4 days in prepatent period between the CPS-CQ and CPS-MQ groups, a standard deviation of 1.6 and 2.3 days respectively, an α

of 5% and a power of 0.90. This calculation resulted in a CPS-CQ group of 4 and a CPS-MQ group of 8 subjects. To account for possible dropouts based on (perceived) side effects we included one and two extra volunteers in the CPS-CQ and CPS-MQ groups respectively. The control group was included as infectivity control for the challenge infection.

On days 6 to 10 after each immunization by mosquito exposure, all subjects were followed on an outpatient basis and peripheral blood was drawn for blood smears, standard hematological measurements, cardiovascular markers and retrospective qPCR.

Twenty weeks after the last immunization, sixteen weeks after discontinuation of prophylaxis, all subjects were challenged by the bites of five mosquitoes infected with the homologous NF54 *P. falciparum* strain, according to previous protocols [20]. After this challenge-infection, all subjects were checked twice daily on an outpatient basis from day 5 up until day 15 and once daily from day 16 up until day 21 for symptoms and signs of malaria. Thick blood smears for parasite detection were made during each of these visits after challenge, hematological and cardiovascular markers were assessed daily. As soon as parasites were detected by thick smear, subjects were treated with a standard curative regimen of 1000 mg atovaquone and 400 mg proguanil once daily for three days according to Dutch national malaria treatment guidelines. If subjects remained thick smear negative, they were presumptively treated with the same curative regimen on day 21 after challenge infection. All subjects were followed closely for 3 days after initiation of treatment and complete cure was confirmed by two negative blood smears after the last treatment dose. Chloroquine and mefloquine levels were measured retrospectively in citrate-plasma from the day before challenge by liquid chromatography (detection limit for both chloroquine and mefloquine: 5 µg/L) [21].

*Anopheles stephensi* mosquitoes for immunizations and challenge-infection were reared according to standard procedures at the insectary of the Radboud university medical center. Infected mosquitoes were obtained by feeding on NF54 gametocytes, a chloroquine- and mefloquine-sensitive *P. falciparum* strain, as described previously [22]. After exposure of volunteers, all blood-engorged mosquitoes were dissected to confirm the presence of sporozoites. If necessary, feeding sessions were repeated until the predefined number of infected or uninfected mosquitoes had fed.

### Endpoints

The primary endpoint was prepatent period, defined as the time between challenge and first positive thick blood smear. Secondary endpoints were parasitemia and kinetics of parasitemia as measured by qPCR, adverse events and immune responses.

### Detection of parasites by thick smear

Blood was sampled twice daily from day 5 until day 15 and once daily from day 16 up until day 21 after challenge and thick smears were prepared and read as described previously [9]. In short, approximately 0.5 µl of blood were assessed by microscopy and the smear was considered positive if two unambiguous parasites were seen.

### Quantification of parasitemia by qPCR

Retrospectively, parasitemia was quantified by real-time quantitative PCR (qPCR) on samples from day 6 until day 10 after each immunization and from day 5 until day 21 after challenge as described previously [23], with some modifications. Briefly, 5 µl Zap-Oglobin II Lytic Reagent (Beckman Coulter) was added to 0.5 ml of EDTA blood, after which the samples were mixed and stored at −80°C. After thawing, samples were spiked with the

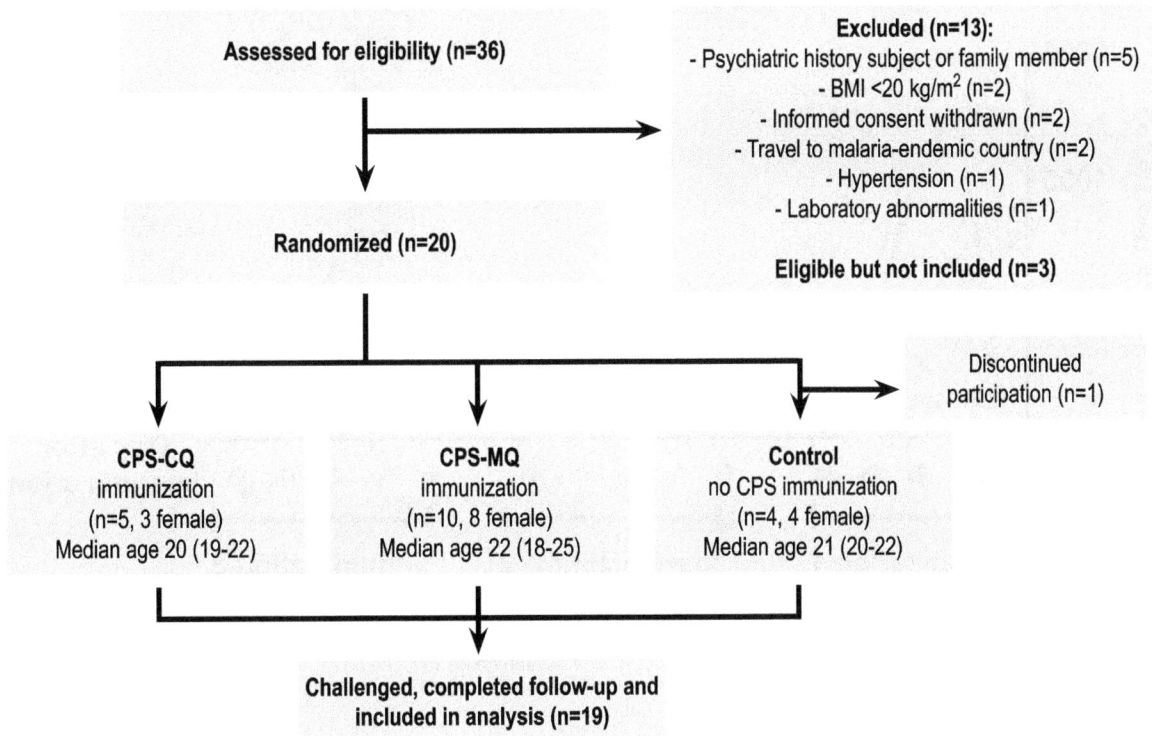

**Figure 1. Study flow diagram.** Thirty-six subjects were screened for eligibility, of whom twenty were included in the trial and randomized over three groups. One control subject was excluded after initiation of chemoprophylaxis but before the first immunization because of an unexpected visit to a malaria-endemic area during the study period. In a double-blind fashion, fifteen subjects received either CPS-CQ or CPS-MQ immunization and four control subjects received bites from uninfected mosquitoes and mefloquine prophylaxis. Subjects received a challenge infection by bites of five infected mosquitoes sixteen weeks after discontinuation of prophylaxis.

extraction control Phocine Herpes Virus (PhHV) and DNA was extracted with a MagnaPure LC isolation instrument. Isolated DNA was resuspended in 50 µl $H_2O$, and 5 µl was used as template. For the detection of *P. falciparum*, the primers as described earlier [23] and the TaqMan MGB probe AAC AAT TGG AGG GCA AG-FAM were used. For quantification of PhHV the primers GGGCGAATCACAGATTGAATC, GCG-GTTCCAAACGTACCAA and the probe Cy5-TTTTTATGTGTCCGCCACCATCTGGATC were used. The sensitivity of qPCR was 35 parasites/ml of whole blood.

## Adverse events and safety lab

Adverse events (AEs) were recorded as following: mild events (easily tolerated), moderate events (interfering with normal activity), or severe events (preventing normal activity). Fever was recorded as grade 1 ($>37\cdot5°C-38\cdot0°C$), grade 2 ($>38\cdot0°C-39\cdot0°C$) or grade 3 ($>39\cdot0°C$). Platelet and lymphocyte counts were determined in EDTA-anti-coagulated blood with the Sysmex XE-2100 (Sysmex Europe GmbH, Norderstedt, Germany). D-dimer concentrations were assessed in citrate plasma by STA-R Evolution (Roche Diagnostics, Almere, The Netherlands).

## Immunological analyses

In order to assess cellular immune memory responses, peripheral blood mononuclear cell (PBMC) re-stimulation assays were performed as described previously [7]. PBMCs were collected, frozen in fetal calf serum containing 10% dimethylsulfoxide, and stored in vapor phase nitrogen before initiation of prophylaxis (baseline; B) and one day before the challenge infection (C-1).

After thawing, PBMCs were re-exposed *in vitro* to *P. falciparum*-infected red blood cells (*Pf*RBC) and incubated for 24 hours at 37°C in the presence of a fluorochrome-labeled antibody against CD107a. Uninfected red blood cells (uRBCs) were used as a negative control. During the last 4 hours of incubation, 10 µg/ml Brefeldin A and 2 µM Monensin were added, allowing cytokines to accumulate within the cells. As a positive control, 50 ng/ml PMA and 1 mg/ml ionomycin were added for the last four hours of incubation. After 24 h stimulation, cells were further stained with a viability marker and fluorochrome-labeled antibodies against CD3, CD4, CD8, CD56, γδ-T cell receptor, IFNγ and granzyme B (**Table S1** [7]). For each volunteer, cells from all time points were tested in a single experiment: thawed and stimulated on the same day and stained the following day. Samples were acquired on a 9-color Cyan ADP (Beckman Coulter) and data analysis was performed using FlowJo software (version 9.6.4; Tree Star). A representative example showing the full gating strategy is shown in **Figure S1.** Gating of cytokine-positive cells was performed in a standardized way by multiplying a fixed factor with the 75 percentile of the geometric Mean Fluorescent Intensity (MFI) of cytokine negative PBMCs for each volunteer, time point and stimulus. Responses to uRBC were subtracted from the response to *Pf*RBC for each volunteer on every time point.

Plasma for the assessment of malaria-specific antibodies was collected and stored at baseline (B), 27 days after the first immunization (I1; one day before the second immunization), 27 days after the second immunization (I2; one day before the third immunization), and one day before the challenge infection (C-1). Antibody titers were assessed as described previously [17]. In

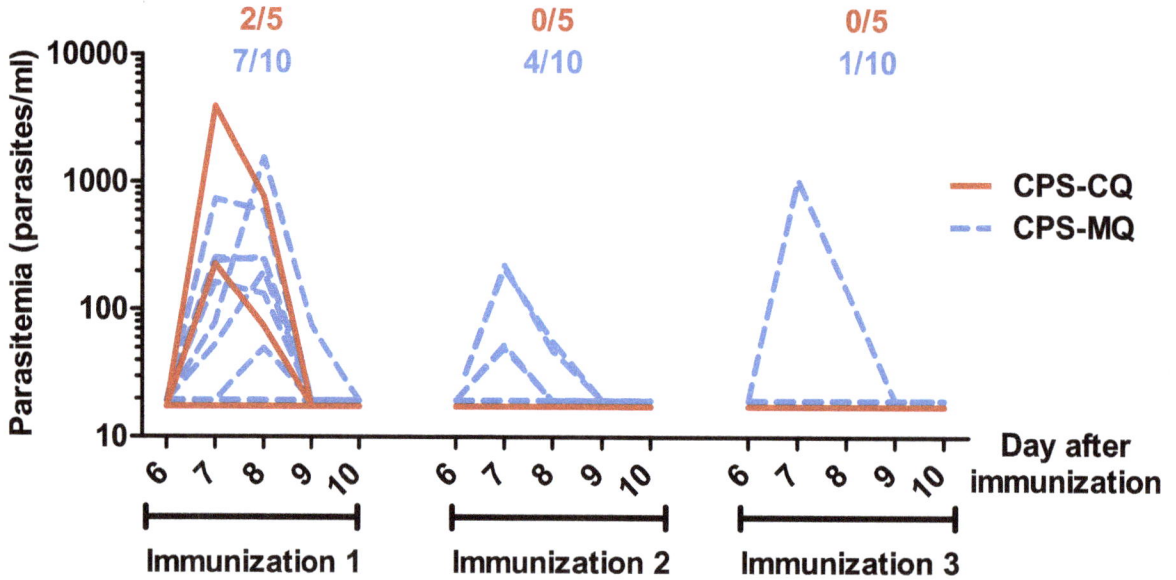

**Figure 2. Parasitemia during CPS immunization.** Parasitemia was determined retrospectively, once daily from day 6 until day 10 after each immunization, by real-time quantitative PCR (qPCR). Each line represents an individual subject from the CPS-MQ (dashed blue lines) or CPS-CQ group (red lines). The number of subjects with a positive qPCR/total number of volunteers in the CPS-MQ (blue) and CPS-CQ (red) groups after each immunization are shown above the graph. Values shown as 17.5 on the log-scale were negative (i.e. half the detection limit of the qPCR: 35 parasites/ml).

summary, serially diluted citrate plasma was used to perform standardized enzyme-linked immunosorbent assay (ELISA) in NUNC Maxisorp plates (Thermo Scientific) coated with 1 μg/ml circumsporozoite protein (CSP), liver-stage antigen-1 (LSA-1) or merozoite surface protein-1 (MSP-1) antigen, diluted in PBS. Bound IgG was detected using horseradish peroxidase (HRP)

conjugated anti-human IgG) (Thermo Scientific, 1/60000) and Tetramethylbenzidine (all Mabtech). Spectrophotometrical absorbance was measured at 450 nm. OD values were converted into AUs by four-parameter logistic curve fit using Auditable Data Analysis and Management System for ELISA (ADAMSEL-v1.1, http://www.malariaresearch.eu/content/software; accessed 27 October 2014). Levels of antibodies were calculated in relation to a pool of 100 sera from adults living in a highly endemic area in Tanzania (HIT serum [24]), which was defined to contain 100 arbitrary units (AU) of IgG directed against each antigen.

### Statistical analyses

The proportion of protected subjects in the CPS-CQ versus CPS-MQ group was tested with the Fisher's exact test using Graphpad Quickcalcs online and the 95% confidence interval (CI) of protection for each group was calculated by modified Wald Method [25]. Further statistical analyses were performed with GraphPad Prism 5. Differences in prepatent period and time from qPCR positivity until thick smear positivity were tested by Mann Whitney test. Antibody levels are shown as individual titers with medians and differences between time points were analyzed by Friedman test with Dunn's multiple comparison post-hoc test. Induction of cellular immune responses was tested for CPS-CQ and CPS-MQ groups separately by Wilcoxon matched-pairs signed rank test (B versus C-1). A p-value of <0.05 was considered statistically significant. Analyses of parasitemia were performed on log transformed data, the geometric mean peak parasitemia after each immunization was calculated using the maximum parasitemia for each subject.

### Results

#### Safety of CPS-CQ and CPS-MQ immunization

Twenty out of 36 screened subjects (median age 21 years; range 18–25) were included in the study (**Figure 1**). One control subject was excluded between start of prophylaxis and the first immuni-

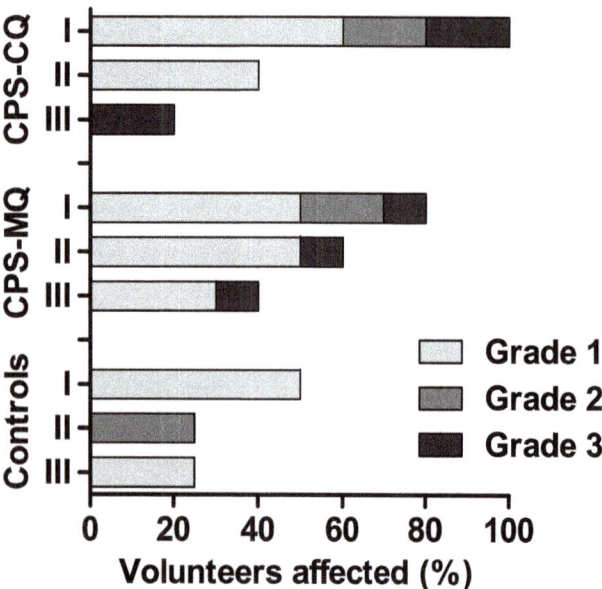

**Figure 3. Adverse events during CPS immunization.** Percentage of volunteers in each group experiencing possibly or probably related AE after the first (I), second (II) and third (III) immunization. AEs were evaluated at each visit and graded for severity as described in the methods paragraph: mild (light grey), moderate (dark grey) and severe (black). Only the highest intensity per subject is listed. No Serious Adverse Events occurred.

**Table 1.** Protection against challenge infection after CPS-CQ and CPS-MQ immunization.

| Group | Protection | | | Unprotected volunteers | | | | | |
| | | | | Day of positivity after challenge[c] | | | | ΔTS-qPCR+[c] | p |
| | Number[a] | Percentage[b] | | Thick smear | p | qPCR | p | | |
| CPS-CQ | 3/5 | 60 (23–88) | | 14.0 (14.0–14.0) | | 11.3 (10.5–12.0) | | 2.8 (2.0–3.5) | |
| CPS-MQ | 7/10 | 70 (39–90) | | 12.0 (11.0–12.0) | 1.0[d] | 10.0 (9.0–10.0) | 0.10[f] | 2.0 (2.0–2.0) | 0.40[f] |
| Control | 0/4 | 0% (0–55) | | 8.5 (7.0–12.0) | 0.03[e] | 6.3 (5.0–9.5) | 0.048[g] | 2.5 (1.5–2.5) | 0.70[g] |

[a]Presented as protected/total number of subjects.
[b]Presented as % protected (95% CI by modified Wald Method).
[c]Presented as median (range) days.
[d,e]p-value calculated by Fisher's exact test comparing [d]CPS-MQ versus CPS-CQ or [e]control versus all CPS-immunized subjects.
[f,g]p-value calculated by Mann Whitney test comparing [f]CPS-MQ versus CPS-CQ or [g]control versus all CPS-immunized subjects (both excluding protected subjects).

zation because of an unexpected intermittent visit to a malaria-endemic area. Thick blood smears performed from day 6 up until day 10 after each immunization remained negative in all volunteers. As determined retrospectively by qPCR, 2/5 subjects in the CPS-CQ group and 7/10 subjects in the CPS-MQ group showed sub-microscopic parasitemia after the first immunization (geometric mean peak parasitemia for positive subjects: 948 parasites/ml [range 228–3938] and 256 parasites/ml [range 48–1559] respectively, **Figure 2**). After the second immunization, four CPS-MQ subjects showed sub-microscopic parasitemia (geometric mean peak parasitemia for positive subjects 104 parasites/ml [range 48–223]), while none of the CPS-CQ subjects showed parasitemia. After the third immunization, only one CPS-MQ subject showed parasitemia by qPCR (peak parasitemia 1059 Pf/ml).

After the first immunization, all subjects (5/5) in the CPS-CQ group and almost all CPS-MQ subjects (8/9) experienced possibly or probably related AEs. One subject in each group had a grade 3 AE (headache and vomiting, respectively). Two control volunteers reported mild AEs (**Figure 3** and **Table S2**). After the second immunization, two CPS-CQ volunteers and six volunteers in the CPS-MQ group had mild AEs. Two control subjects experienced moderate and severe headache, respectively. After the third immunization, one volunteer in the CPS-CQ group and four CPS-MQ volunteers had AEs; one control subject experienced mild AEs (**Figure 3** and **Table S2**). One CPS-CQ subject reported moderate sleeping problems while taking chloroquine prophylaxis. One control subject had moderate problems with initiation of sleep and another control subject experienced vivid dreams under mefloquine prophylaxis. Other than mild to moderate dizziness and sleep related AEs, which all resolved after chemoprophylaxis was stopped, no neuropsychiatric AEs occurred. No serious adverse events occurred.

During immunization, one subject each in the CPS-CQ, CPS-MQ and control groups showed platelet counts below the lower limit of normal ($150 \times 10^9$/L); lowest values $105 \times 10^9$/L, $116 \times 10^9$/L and $131 \times 10^9$/L, respectively. Three, five and two subjects from the CPS-CQ, CPS-MQ and control groups respectively, showed leukocyte counts below the lower limit of normal ($4 \times 10^9$/L); mean lowest value during immunization period: $3.8 \times 10^9$/L [SD 1.2], $4.0 \times 10^9$/L [SD 1.1] and $4.2 \times 10^9$/L [SD 0.7] respectively. No subject developed leukocyte counts lower than $2.0 \times 10^9$/L. One volunteer in each group showed leukocyte counts above the upper limit of normal ($10 \times 10^9$/L; highest values $10.8 \times 10^9$/L, $13.8 \times 10^9$/L and $10.1 \times 10^9$/L respectively). After the first immunization, 3/5 CPS-CQ subjects, 7/10 in the CPS-MQ group and none in the control group developed elevated d-dimer levels (>500 ng/ml). After the second immunization, six CPS-MQ subjects but none in the CPS-CQ or control groups showed elevated d-dimer levels. After the third immunization, three CPS-MQ subjects showed elevated d-dimer levels, while none of the subjects in the other groups did.

## Protection against challenge infection

In the CPS-CQ group 3/5 subjects and in the CPS-MQ group 7/10 volunteers were protected against challenge infection (Fisher's exact test p = 1.0). All control subjects became thick smear positive (median day 8.5, range 7–12, p = 0.03 versus CPS-immunized subjects; **Table 1**). None of the protected subjects showed parasitemia by qPCR at any time point during follow-up (**Figure 4**). The median prepatent period was not significantly different between the CPS-CQ and CPS-MQ groups, neither when protected subjects were arbitrarily set at a prepatent period of 21 days (p = 1.00), nor when comparing unprotected subjects

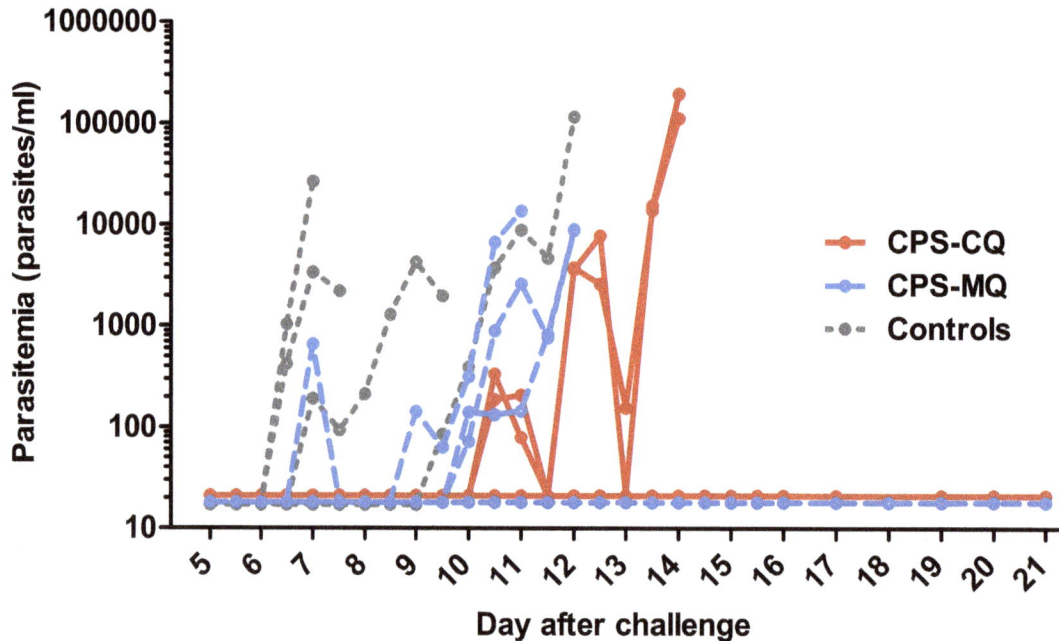

**Figure 4. Parasitemia after challenge infection.** Parasitemia was assessed retrospectively by real-time quantitative PCR (qPCR) twice daily from day 5 until day 15 and once daily up until day 21 after challenge. Each line represents an individual subject. Red lines represent CPS-CQ immunized volunteers (n = 5), dashed blue lines CPS-MQ immunized subjects (n = 10) and dotted grey lines malaria-naive control subjects (n = 4). Values shown as 17.5 on the log-scale were negative (i.e. half the detection limit of the qPCR: 35 parasites/ml).

only (p = 0.1). The median chloroquine plasma concentration on the day before challenge infection was 9 µg/L (range 7–10) in the CPS-CQ group, and the median mefloquine concentration was 24 µg/L (range 5–116) in the mefloquine groups.

### Immunogenicity of CPS-CQ and CPS-MQ

Antibodies against the pre-erythrocytic antigens CSP and LSA-1 and the cross-stage antigen MSP-1 were assessed by ELISA. Antibodies against CSP were induced in both CPS-CQ and CPS-MQ immunized volunteers (p<0.05 and p<0.01 respectively, on C-1; **Figure 5A and 5B**), but not significantly higher in protected compared to unprotected subjects (p = 0.88 and p = 0.48 respectively). Antibodies against LSA-1 were only significantly induced in CPS-MQ immunized volunteers on I2 (p<0.001; **Figure 5C and 5D**), although not higher in protected subjects (p = 0.39). Anti-MSP-1 antibodies by CPS immunization were not statistically significant increased in either group (**Figure 5E and 5F**).

IFNγ production by both adaptive and innate cell subsets in response to *in vitro P. falciparum* re-stimulation was induced by both CPS-CQ and CPS-MQ (**Figure S2**), without a clear quantitative or qualitative difference between the study groups. Next, CD107a expression by CD4 T cells and granzyme B production by CD8 T cells, both associated with protection in a previous CPS-CQ trial [7], were assessed by flow cytometry. Four out of 5 CPS-CQ and 8/10 CPS-MQ immunized subjects showed induction of CD107a expression by CD4 T cells upon *in vitro* re-stimulation after immunization (**Figure 6A and 6B**). Although volunteer numbers were too low to reach statistical significance, the magnitude of this response appeared to be associated with protection for CPS-CQ (**Figure 6A**), while for CPS-MQ it was not (**Figure 6B**). Granzyme B production by CD8 T cells was not

significantly induced in either CPS-CQ or CPS-MQ group, nor was it associated with protection (**Figure 6C and 6D**).

After challenge, MSP-1 specific antibodies were boosted in all unprotected volunteers (fold change median 20.4 (range 7.1–33.6), 76.0 (5.7–106.3) and 7.7 (2.9–15.3) for CPS-CQ, CPS-MQ and control groups respectively). None of the protected subjects showed an increase in MSP-1 antibody levels on C+35 compared to C-1 (median fold change 1.0 (range 1.0–1.3) and 1.0 (0.6–2.4) for CPS-CQ and CPS-MQ groups, respectively).

### Discussion

Immunization of healthy volunteers with *P. falciparum* sporozoites while taking mefloquine prophylaxis is safe, induces both humoral and cellular immune responses and protects against homologous malaria challenge.

Although most volunteers experienced AEs after the first immunization, their frequency declined after subsequent immunizations in line with a reducing number of volunteers developing parasitemia. The majority of AEs was mild, with only 10–20% of subjects experiencing a grade 3 AEs after each immunization. In general, the reported neurologic and psychiatric side effects of mefloquine are a major concern limiting its acceptability and clinical application. In this study, mild to moderate dizziness and sleep-related complaints occurred in a small number of subjects in both chloroquine and mefloquine groups. Although this study was not powered to detect differences in AEs, frequency of neuropsychiatric AEs did not appear to differ between both drugs. This is in line with most reports in literature comparing AEs of mefloquine or chloroquine (with or without proguanil) for chemo-prophylactic use [26–29] although one study found more neuropsychiatric AEs in subjects taking mefloquine by retrospective questionnaire [30]. Taking the small sample size into consideration, both CPS-CQ

## CPS-CQ

## CPS-MQ

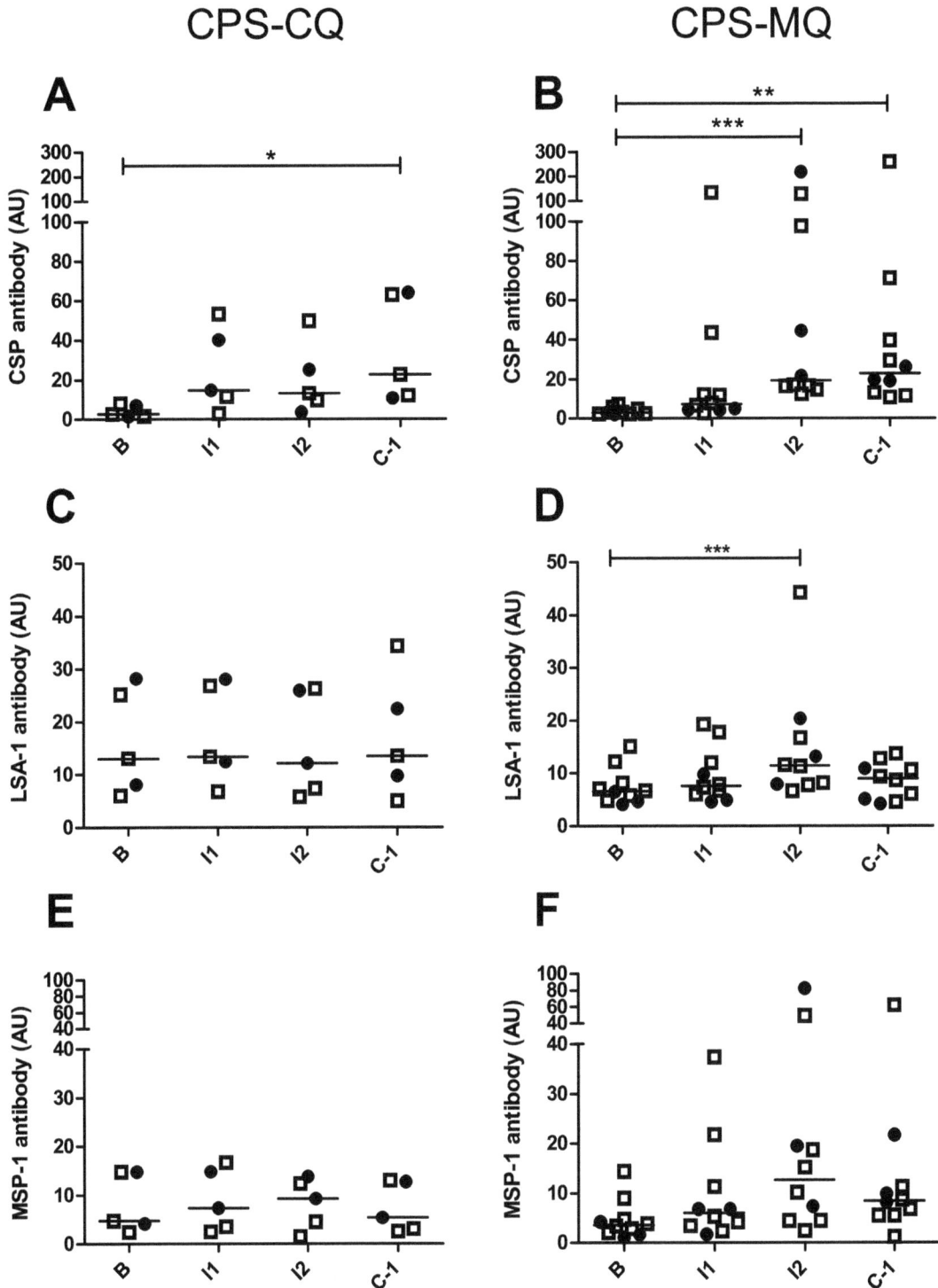

**Figure 5. Antibody responses induced by CPS-CQ and CPS-MQ immunization.** Antibodies against CSP (A and B; in AU), LSA-1 (C and D), and MSP-1 (E and F) were analyzed at baseline (B), 28 days after the first (I1) and second (I2) immunization and one day before challenge (C-1; 20 weeks after the last immunization) for all CPS-CQ (A, C and E, n = 5) and CPS-MQ (B, D and F, n = 10) immunized volunteers. Data are shown as individual titers with medians. Open squares indicate protected subjects, filled circles indicate unprotected subjects. Differences between the time points were analyzed by Friedman test with Dunn's multiple comparison post-hoc test. Significant differences are indicated by asterices with * (p< 0.05), ** (p<0.01), *** (p<0.001).

and CPS-MQ immunization regimens appear to be reasonably well tolerated and safe. In 2013, however, after completion of this study, the U.S. Food and Drug Administration (FDA) issued a boxed warning for mefloquine, stating that neurologic side effects might be permanent. This might lead to adjustment of prophylaxis guidelines and limitation of mefloquine use where

# CPS-CQ                              CPS-MQ

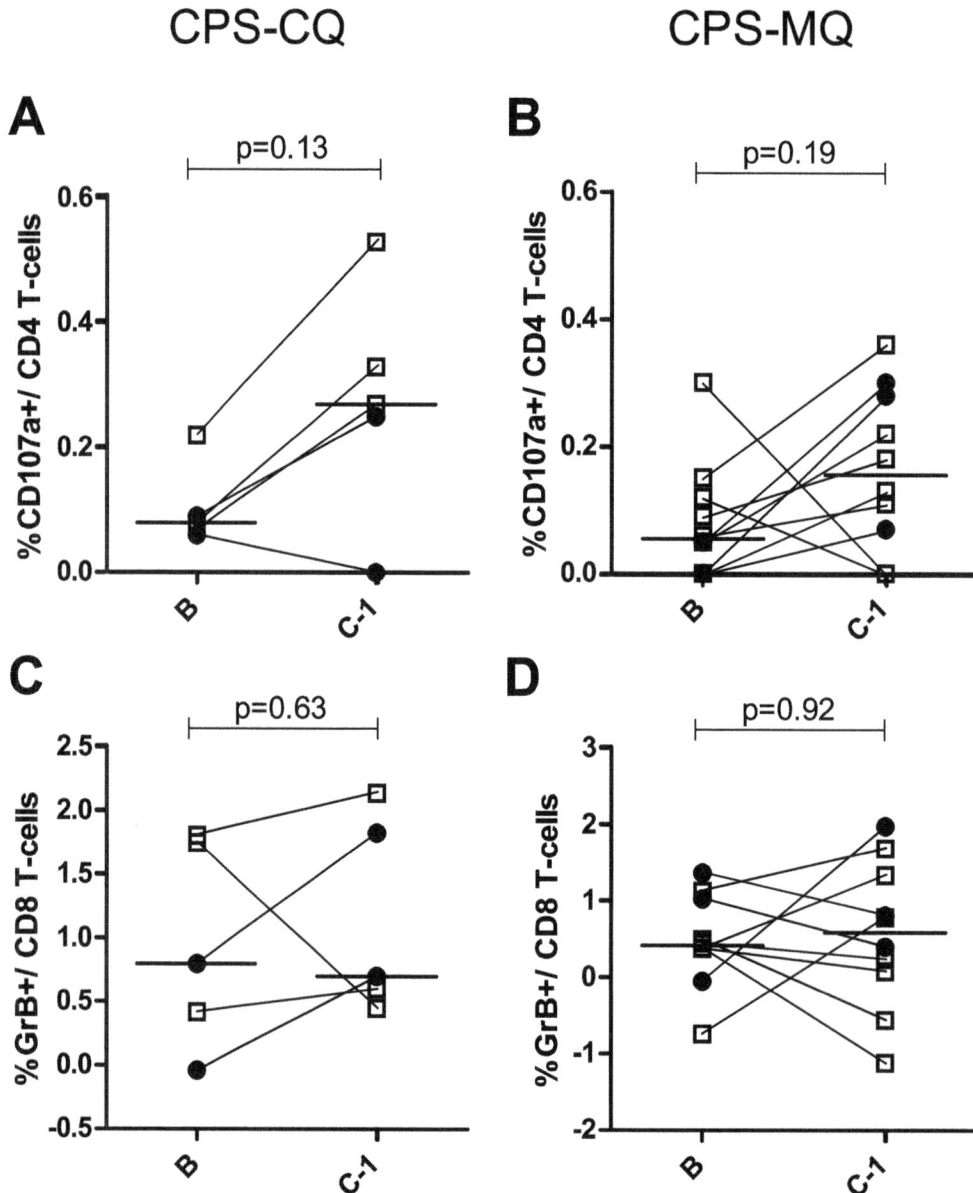

**Figure 6. Cellular immune responses: CD107a expression by CD4 T cells and granzyme B production by CD8 T cells.** CD107a expression by CD4 T cells after *Pf*RBC re-stimulation, corrected for uRBC background in CPS-CQ (A) and CPS-MQ (B) groups; granzyme B production by CD8 T cells after *Pf*RBC re-stimulation, corrected for uRBC background in CPS-CQ (C) and CPS-MQ (D) groups. Symbols and lines represent individual subjects before immunization (B) and one day before challenge (C-1). Open squares indicate protected subjects, filled circles indicate unprotected subjects. Differences between B and C-1 for all subjects were tested by Wilcoxon matched-pairs signed rank test.

alternatives are available, as for now it remains a recommended antimalarial prophylactic for several target groups [31].

In previous studies we showed that 19/20 subjects (95%) were protected after bites from 45 infected mosquitoes, 8/9 (89%) after bites from 30 and 5/10 (50%) after bites from 15 infected mosquitoes during chloroquine prophylaxis [6,7,9]. The 60–70% protection observed in the current CPS-CQ and CPQ-MQ groups, immunized with bites from 24 mosquitoes, demonstrates the reproducibility of CPS immunization and indicates a linear relationship between immunization dose and protection. This confirms the consistency of the CPS approach and is remarkable, given the assumed variation in the number of sporozoites injected by mosquitoes [32]. This study further establishes CPS immunization as a worthwhile immunization protocol to relatively easily

induce protection and create differentially protected cohorts to study target antigens and correlates of protection, both of which would be highly valuable tools in the search for *P. falciparum* vaccines and biomarkers of protection [33].

Although the study was not powered to detect these differences, there are hints suggestive of more efficient induction of protection by CPS-CQ compared to CPS-MQ: i) the two unprotected CPS-CQ volunteers showed a longer prepatent period than the CPS-MQ subjects (14 versus 12 days, Mann-Whitney test p = 0.13); ii) induction of immunity required less immunizations in the CPS-CQ group i.e. none of these subjects showed blood-stage parasites after the second immunization while subjects in the CPS-MQ group still developed parasitemia after the second and third immunization. If there is a difference between CPS-CQ and CPS-

MQ in protective efficacy, it is small, but possibly detectable in larger cohorts or when the immunization dose is further reduced.

Induction of anti-circumsporozoite antibodies by CPS-CQ is consistent with previous work, but neither anti-LSA-1, nor MSP-1 antibodies were induced by CPS-CQ in the current study [17]. Antibodies against the latter antigens are dose-dependently induced [17], and the current immunization regime using bites from 3×8 P. falciparum-infected mosquitoes might have been insufficient [7]. The induction of cellular P. falciparum-specific memory responses, as reflected by IFNγ production, is in line with previous CPS-CQ studies, even though limited sample size hampered statistical significance for some cell types. Interestingly, CD107a expression by CD4 T cells upon in vitro re-stimulation, associated with protection in a previous CPS-CQ study [7], appeared again to be associated with protection in the CPS-CQ group, but not the CPS-MQ group. Granzyme B production by CD8 T cells upon in vitro re-stimulation did not appear to be a reproducible marker of protection in this second CPS study [7]. Whether this might be related to immunization dose remains to be investigated in future CPS trials.

The striking efficiency of CPS immunization might at least be partly due to the established immune modulating properties of the 4-amino-quinoline chloroquine [12], possibly reflected by the more efficient induction of degranulating CD4 T cells. Chloroquine has been shown to increase cross-presentation in hepatitis B vaccination and influenza [13,14], and thus may enhance cellular immune responses considered essential for protection against liver-stages [12]. For mefloquine, a 4-methanolquinoline, this immune-modulating property has, to our knowledge, not been reported. A possible strategy to assess whether chloroquine and/or mefloquine indeed have immune enhancing effects on whole sporozoite immunization would be to compare immunization with RAS in the presence or absence of these drugs.

Mefloquine or chloroquine plasma concentrations were still detectable in all volunteers one day before the challenge infection. Possible contributing effects of these remaining drug levels to the protective efficacy outcome were considered in several ways; i) The interval between first qPCR and thick smear positivity, as proxy for parasite multiplication, was 2.8 in the CPS-CQ group, 2.0 in the CPS-MQ group and 2.5 in the control group. This interval is similar to previous CHMI studies with the NF54 P. falciparum strain in the absence of prophylactic drug levels [7,34]; ii) the two volunteers with the highest mefloquine levels (116 and 77 µg/L) were control subjects who became thick smear positive with only a minimal delay in patency within the time-frame of historical controls [35]; iii) plasma chloroquine and mefloquine levels at C-1 were in all volunteers well below the minimum therapeutic concentration (CQ: 30 µg/L [36]) or the concentration at which breakthrough infections are observed in non-immune people (MQ<406–603 µg/L [37]). iv) We cannot rule out that protected subjects experienced transient parasitemia after challenge, which was cleared in the first blood-stage cycle by remaining drug levels. But because parasitemia was not detected by qPCR in any of the protected subjects at any time point after challenge potential parasitemia must have been below the qPCR detection limit of 35 parasites/ml, indicating a reduction of at least 92% in liver load, given a geometric mean height of the first peak or parasitemia in non-immune historical controls of 456 parasites/ml [35]; v) None of the protected subjects showed a boost in anti-MSP-1 antibodies after challenge while all unprotected subjects did, suggesting that protected subjects did not experience blood-stage parasitemia after challenge. [9]. From these combined data we believe that remaining drug concentrations are unlikely to have contributed to the observed protection, although this cannot be formally excluded.

A review of rodent studies using different attenuation methods for whole sporozoite immunization shows that increased development of the parasite in the liver, but absence of blood-stage parasitemia during immunization is associated with the highest protective efficacy [38]. It would therefore be interesting to investigate CPS immunization with alternative antimalarials with varying targets in the parasite life cycle. CPS immunization with causal prophylactic drugs affecting liver-stages, e.g. primaquine, will likely results in a reduction of AEs because of reduced or absent blood-stage exposure. Whether antigen-exposure is sufficient to induce protection when the liver-stage is abrogated, remains to be answered.

In conclusion, we show that immunization of healthy volunteers under mefloquine prophylaxis with P. falciparum sporozoites is safe, immunogenic and protective. These findings could have important implications for malaria vaccine development and further development of CPS approaches.

## Supporting Information

**Figure S1  Gating strategy. (A)** Representative flow cytometry plots for a uRBC stimulated sample from one volunteer at baseline (before immunization). Singlet viable PBMCs were subdivided into (i) CD56hi NK cells, (ii) CD56dim NK cells, (iii) NKT cells, (iv) γδT cells, (v) CD8 T cells, (vi) CD4 T cells. **(B)** Gating of IFNγ, CD107a and granzyme B positive cells for uRBC, PfRBC and PMA/ionomycin re-stimulated cells at baseline. For uRBC and PfRBC stimulation CD4 T cells are shown, for PMA/ionomycin total viable PBMCs. Within each sample, gating of cytokine-positive cells was performed in a standardized way by multiplying a fixed factor with the 75 percentile of the geometric Mean Fluorescent Intensity (MFI) of cytokine negative PBMCs.

**Figure S2  Cellular immune responses: IFNγ production.** IFNγ production by different cell subtypes in response to in vitro re-stimulation with PfRBC (corrected for uRBC background), before immunization (B) and one day before challenge (C-1). Differences between B and C-1 were tested by Wilcoxon matched-pairs signed rank test.

**Table S1  Antibodies used for flow cytometry.**

**Table S2  Possibly and probably related adverse events during CPS-CQ and CPS-MQ immunization.**

**Protocol S1  Trial protocol.**

**Checklist S1  CONSORT checklist.**

## Acknowledgments

We thank the study participants for their commitment and contribution to malaria research. We thank K. Suijk-Benschop, J. Fehrmann-Naumann, M. Kortekaas, G. Hardeman, C. Prins, E. Jonker, and S. ten Velden-Schipper for blood collection and support; the LUMC department of Medical Microbiology for facilitating parasitological diagnosis and M. Erkens, J. van der Slot, H. Gerritsma, F. van de Sande, J. van Schie, E. Brienen, J. Schelfaut, J. Verweij, J. Kromhout, E. van Oorschot and M. Beljon for reading many thick smears. We appreciate the expert reviews and guidance of the Safety Monitoring Committee: J.A. Romijn, M. de Boer and M. Laurens. We thank R. Siebelink-Stoter for culturing parasites;

J. Klaassen, L. Pelser-Posthumus, J. Kuhnen, and A. Pouwelsen for breeding and infecting mosquitoes and for assistance with CHMIs; M. Willems and A. Teirlinck for PBMC collection; T. Arens for assistance with qPCR analysis; C. Janse for his unlimited hospitality; M. van Meer for help with the ELISAs; P. Houzé for measurement of mefloquine and chloroquine levels; J. Wiersma for assistance with CHMIs; M. Bootsma for providing cardiac expertise; A. Jansens for administrative support; CromSource for monitoring; the staff from the LUMC Central Clinical Biochemistry and Haematology Laboratories and the LUMC Pharmacy; M. Vos and B. van Schaijk for molecular analysis of the parasite batches and TropIQ for testing parasite batches.

## Author Contributions

Conceived and designed the experiments: EMB RS LGV AS RWS. Performed the experiments: EMB RS JMO MCB GJvG MvdVB WG LvL GJHB KT CCH AS. Analyzed the data: EMB RS JMO CCH. Contributed to the writing of the manuscript: EMB RS AS LGV RWS.

## References

1. World Health Organization (2013) World malaria report 2013.
2. Doolan DL, Dobano C, Baird JK (2009) Acquired immunity to malaria. Clin Microbiol Rev 22: 13–36.
3. Tran TM, Li S, Doumbo S, Doumtabe D, Huang CY, et al. (2013) An Intensive Longitudinal Cohort Study of Malian Children and Adults Reveals No Evidence of Acquired Immunity to Plasmodium falciparum Infection. Clin Infect Dis 57: 40–47.
4. RTS'S Clinical Trials Partnership, Agnandji ST, Lell B, Fernandes JF, Abossolo BP, et al. (2012) A phase 3 trial of RTS, S/AS01 malaria vaccine in African infants. N Engl J Med 367: 2284–2295.
5. Crompton PD, Pierce SK, Miller LH (2010) Advances and challenges in malaria vaccine development. J Clin Invest 120: 4168–4178.
6. Roestenberg M, McCall M, Hopman J, Wiersma J, Luty AJ, et al. (2009) Protection against a malaria challenge by sporozoite inoculation. N Engl J Med 361: 468–477.
7. Bijker EM, Teirlinck AC, Schats R, van Gemert G-J, van de Vegte-Bolmer M, et al. (2014) Cytotoxic Markers Associate with Protection against Malaria in Human Volunteers Immunized with Plasmodium falciparum Sporozoites. Journal of Infectious Diseases: jiu293.
8. Roestenberg M, Teirlinck AC, McCall MBB, Teelen K, Makamdop KN, et al. (2011) Long-term protection against malaria after experimental sporozoite inoculation: an open-label follow-up study. Lancet 377: 1770–1776.
9. Bijker EM, Bastiaens GJ, Teirlinck AC, van Gemert GJ, Graumans W, et al. (2013) Protection against malaria after immunization by chloroquine prophylaxis and sporozoites is mediated by preerythrocytic immunity. Proc Natl Acad Sci U S A 110: 7862–7867.
10. Hoffman SL, Goh LM, Luke TC, Schneider I, Le TP, et al. (2002) Protection of humans against malaria by immunization with radiation-attenuated Plasmodium falciparum sporozoites. J Infect Dis 185: 1155–1164.
11. Seder RA, Chang LJ, Enama ME, Zephir KL, Sarwar UN, et al. (2013) Protection Against Malaria by Intravenous Immunization with a Nonreplicating Sporozoite Vaccine. Science.
12. Sauerwein RW, Bijker EM, Richie TL (2010) Empowering malaria vaccination by drug administration. Curr Opin Immunol 22: 367–373.
13. Accapezzato D, Visco V, Francavilla V, Molette C, Donato T, et al. (2005) Chloroquine enhances human CD8+ T cell responses against soluble antigens in vivo. J Exp Med 202: 817–828.
14. Garulli B, Di Mario G, Sciaraffia E, Accapezzato D, Barnaba V, et al. (2013) Enhancement of T cell-mediated immune responses to whole inactivated influenza virus by chloroquine treatment in vivo. Vaccine 31: 1717–1724.
15. Behet MC, Foquet L, van Gemert GJ, Bijker EM, Meuleman P, et al. (2014) Sporozoite immunization of human volunteers under chemoprophylaxis induces functional antibodies against pre-erythrocytic stages of Plasmodium falciparum. Malar J 13: 136.
16. Felgner PL, Roestenberg M, Liang L, Hung C, Jain A, et al. (2013) Pre-erythrocytic antibody profiles induced by controlled human malaria infections in healthy volunteers under chloroquine prophylaxis. Sci Rep 3: 3549.
17. Nahrendorf W, Scholzen A, Bijker EM, Teirlinck AC, Bastiaens GJ, et al. (2014) Memory B-cell and antibody responses induced by Plasmodium falciparum sporozoite immunization. Journal of Infectious Diseases: jiu354.
18. CDC (2014) Accessed 27-OCT-2014.
19. Nederlandsche Internisten Vereeniging, Hartstichting N, voor Cardiologie NV, voor Heelkunde NV, voor Neurologie NV, et al. (2006) Multidisciplinaire Richtlijn Cardiovasculair risicomanagement.
20. Verhage DF, Telgt DS, Bousema JT, Hermsen CC, van Gemert GJ, et al. (2005) Clinical outcome of experimental human malaria induced by Plasmodium falciparum-infected mosquitoes. Neth J Med 63: 52–58.
21. Lejeune D, Souletie I, Houze S, Le bricon T, Le bras J, et al. (2007) Simultaneous determination of monodesethylchloroquine, chloroquine, cyclo-guanil and proguanil on dried blood spots by reverse-phase liquid chromatography. J Pharm Biomed Anal 43: 1106–1115.
22. Ponnudurai T, Lensen AH, Van Gemert GJ, Bensink MP, Bolmer M, et al. (1989) Infectivity of cultured Plasmodium falciparum gametocytes to mosquitoes. Parasitology 98 Pt 2: 165–173.
23. Hermsen CC, Telgt DS, Linders EH, van de Locht LA, Eling WM, et al. (2001) Detection of Plasmodium falciparum malaria parasites in vivo by real-time quantitative PCR. Mol Biochem Parasitol 118: 247–251.
24. Roestenberg M, McCall M, Hopman J, Wiersma J, Luty AJ, et al. (2009) Protection against a malaria challenge by sporozoite inoculation. The New England journal of medicine 361: 468–477.
25. GraphPad (Last accessed: 4 Dec 2013) QuickCalcs. Accessed.
26. Lobel HO, Miani M, Eng T, Bernard KW, Hightower AW, et al. (1993) Long-term malaria prophylaxis with weekly mefloquine. Lancet 341: 848–851.
27. Steffen R, Fuchs E, Schildknecht J, Naef U, Funk M, et al. (1993) Mefloquine compared with other malaria chemoprophylactic regimens in tourists visiting east Africa. Lancet 341: 1299–1303.
28. Croft AM, Clayton TC, World MJ (1997) Side effects of mefloquine prophylaxis for malaria: an independent randomized controlled trial. Trans R Soc Trop Med Hyg 91: 199–203.
29. Schlagenhauf P, Tschopp A, Johnson R, Nothdurft HD, Beck B, et al. (2003) Tolerability of malaria chemoprophylaxis in non-immune travellers to sub-Saharan Africa: multicentre, randomised, double blind, four arm study. BMJ 327: 1078.
30. Barrett PJ, Emmins PD, Clarke PD, Bradley DJ (1996) Comparison of adverse events associated with use of mefloquine and combination of chloroquine and proguanil as antimalarial prophylaxis: postal and telephone survey of travellers. BMJ 313: 525–528.
31. Centers for Disease Control and Prevention http://wwwnc.cdc.gov/travel/yellowbook/2014/chapter-3-infectious-diseases-related-to-travel/malaria#3939. Accessed 27 October 2014.
32. Ponnudurai T, Lensen AH, van Gemert GJ, Bolmer MG, Meuwissen JH (1991) Feeding behaviour and sporozoite ejection by infected Anopheles stephensi. Trans R Soc Trop Med Hyg 85: 175–180.
33. Duffy PE, Sahu T, Akue A, Milman N, Anderson C (2012) Pre-erythrocytic malaria vaccines: identifying the targets. Expert review of vaccines 11: 1261–1280.
34. Roestenberg M, de Vlas SJ, Nieman A-E, Sauerwein RW, Hermsen CC (2012) Efficacy of preerythrocytic and blood-stage malaria vaccines can be assessed in small sporozoite challenge trials in human volunteers. J Infect Dis 206: 319–323.
35. Roestenberg M, O'Hara GA, Duncan CJA, Epstein JE, Edwards NJ, et al. (2012) Comparison of clinical and parasitological data from controlled human malaria infection trials. PLoS One 7: e38434.
36. Rombo L, Bergqvist Y, Hellgren U (1987) Chloroquine and desethylchloroquine concentrations during regular long-term malaria prophylaxis. Bull World Health Organ 65: 879–883.
37. Palmer KJ, Holliday SM, Brogden RN (1993) Mefloquine. Drugs 45: 430–475.
38. Nganou-Makamdop K, Sauerwein RW (2013) Liver or blood-stage arrest during malaria sporozoite immunization: the later the better? Trends Parasitol 29: 304–310.

# 18

# Caspase-1/ASC Inflammasome-Mediated Activation of IL-1β–ROS–NF-κB Pathway for Control of *Trypanosoma cruzi* Replication and Survival Is Dispensable in NLRP3$^{-/-}$ Macrophages

**Nilay Dey[1]\*, Mala Sinha[2], Shivali Gupta[1], Mariela Natacha Gonzalez[5], Rong Fang[3], Janice J. Endsley[1], Bruce A. Luxon[2], Nisha Jain Garg[1,3,4]\***

1 Department of Microbiology and Immunology, University of Texas Medical Branch (UTMB), Galveston, Texas, United States of America, 2 Department of BioChemistry & Molecular Biology, UTMB, Galveston, Texas, United States of America, 3 Department of Pathology, UTMB, Galveston, Texas, United States of America, 4 Faculty of the Institute for Human Infections and Immunity and the Center for Tropical Diseases, UTMB, Galveston, Texas, United States of America, 5 Instituto Nacional de Parasitología "Dr. Mario Fatala Chaben", Ciudad Autónoma de Buenos Aires, Argentina

## Abstract

In this study, we have utilized wild-type (WT), ASC$^{-/-}$, and NLRP3$^{-/-}$ macrophages and inhibition approaches to investigate the mechanisms of inflammasome activation and their role in *Trypanosoma cruzi* infection. We also probed human macrophages and analyzed published microarray datasets from human fibroblasts, and endothelial and smooth muscle cells for *T. cruzi*-induced changes in the expression genes included in the RT Profiler Human Inflammasome arrays. *T. cruzi* infection elicited a subdued and delayed activation of inflammasome-related gene expression and IL-1β production in mφs in comparison to LPS-treated controls. When WT and ASC$^{-/-}$ macrophages were treated with inhibitors of caspase-1, IL-1β, or NADPH oxidase, we found that IL-1β production by caspase-1/ASC inflammasome required reactive oxygen species (ROS) as a secondary signal. Moreover, IL-1β regulated NF-κB signaling of inflammatory cytokine gene expression and, subsequently, intracellular parasite replication in macrophages. NLRP3$^{-/-}$ macrophages, despite an inability to elicit IL-1β activation and inflammatory cytokine gene expression, exhibited a 4-fold decline in intracellular parasites in comparison to that noted in matched WT controls. NLRP3$^{-/-}$ macrophages were not refractory to *T. cruzi*, and instead exhibited a very high basal level of ROS (>100-fold higher than WT controls) that was maintained after infection in an IL-1β-independent manner and contributed to efficient parasite killing. We conclude that caspase-1/ASC inflammasomes play a significant role in the activation of IL-1β/ROS and NF-κB signaling of cytokine gene expression for *T. cruzi* control in human and mouse macrophages. However, NLRP3-mediated IL-1β/NFκB activation is dispensable and compensated for by ROS-mediated control of *T. cruzi* replication and survival in macrophages.

**Editor:** Herbert B. Tanowitz, Albert Einstein College of Medicine, United States of America

**Funding:** This work was supported in part by grants from the National Institute of Allergy and Infectious Diseases (NIAID R01AI054578) and National Heart Lung and Blood Institute (R01HL094802) of the National Institutes of Health (NIH) to NJG. ND was a recipient of a grant (R03AI208810) from NIAID/NIH. The funders had no role in study design, data collection and analysis, decision to publish, or preparation of the manuscript.

**Competing Interests:** The authors have declared that no competing interests exist.

\* Email: nidey@utmb.edu (ND); nigarg@utmb.edu (NJG)

## Introduction

Chagas disease affects 11–18 million people world-wide [1]. Upon exposure to *Trypanosoma cruzi* (*T. cruzi* or *Tc*), infected individuals exhibit an acute phase of Chagas disease that lasts for a couple of months and is characterized by symptoms such as fever, fatigue, body aches, diarrhea, and vomiting. After the control of parasitemia, a majority of infected patients enter an indeterminate chronic phase that is marked by a lack of clinical symptoms of the acute phase. Ten to thirty years after initial infection, 30–40% of indeterminate phase patients progress to develop chagasic cardiomyopathy [2].

The studies in experimental models have shown that macrophages (mφs), as well as dendritic and natural killer cells, play an important role in control of *T. cruzi* infection [3–5]. The interaction of *T. cruzi* with mφs and other cell types involved in the innate immune response are mediated by pattern recognition receptors (PRRs) such as toll-like receptors (TLRs). Upon recognition of pathogen-assoCiated molecular patterns (PAMPs), TLRs transmit the signal via cytoplasmic domains for the recruitment of cytosolic adaptor molecules, including myeloid differentiation primary-response protein 88 (MyD88), and subsequently induce nuclear factor κB (NFκB) activation, leading to the production of inflammatory cytokines and linking an innate response to an adaptive immune response (reviewed in [4]). *T. cruzi*-derived glycosylphosphatidylinositols and mucins have been shown to serve as PAMPs in engaging TLR signaling of a cytokine response. Others have demonstrated that TLR4 and TLR9 are engaged by parasite-derived glycosylinositol phospholipids and DNA, respectively, during the activation of host innate immune response leading to regulation of infection [6,7]. *T. cruzi* also expresses cruzipain, a kinin-releasing cysteine protease, which induces dendritic cells maturation via activation of bradykinin (BK) $B_2$ receptors ($B_2$R) [8,9].

A newly discovered family of PRRs is named Nucleotide-binding oligomerization domain (NOD) like receptors (NLRs) [10,11]. NLRs have a tripartite domain structure and are characterized by the presence of a central nucleotide-binding oligomerization domain (NOD), also called NACHT domain, present in neuronal apoptosis inhibitor proteins (NAIP) and a C-terminal leucine-rich repeats (LRRs) domain of variable length (20–29 amino acids). The N-terminal effector binding region consists of a protein-to-protein interaction domain, i.e., Pyrin domain (PYD), a caspase recruitment domain (CARD), or baculovirus inhibitor of an apoptosis protein repeat (BIR) domain. Based upon the presence of PYD, CARD and BIR effector domains, NLRs are classified as NLRP, NLRC, and NAIP, respectively [11,12]. Currently known members of the NLR family in humans include seven NLRCs (NLRC1-NLRC5, NLRX, and CIITA or NLRA), fourteen NLRPs (NLRP1-NLRP14), and seven NAIPs (NAIP1-NAIP7). The multi-meric protein macromolecules formed by NLRs are named inflammasomes. The most studied NLRP1 and NLRP3 inflammasomes recruit ASC (apoptosis-assoCiated, speck-like protein containing a CARD domain) and caspase-1 proteins. The ASC-dependent cleavage and activation of caspase-1 results in the formation of an active complex responsible for converting to active forms of pro-IL-1β (31 kDa to 17 kDa) and pro-IL-18 (24 kDa to 18 kDa) [13] and the activation of the inflammatory cytokine response.

In the context of pathogens invading the heart, it is recognized that besides innate immune cells, both endothelial and vascular smooth muscle cells (VSMCs) can also sense and respond to pathogens (or PAMPs) [14–16]. CardiomyoCytes, the main type of cells in the heart, and heart resident fibroblasts also express TLRs and/or NLRs [17,18].

In this study, we have utilized wild type (WT), ASC$^{-/-}$ and NLRP3$^{-/-}$ mφs and inhibitory approaches to investigate the mechanisms of inflammasome activation and their role in the context of *T. cruzi* infection. We also probed the RT Profiler PCR Array System to identify the inflammasome-related changes induced by *T. cruzi* infection of human mφs and analyzed the published microarray datasets from *T. cruzi*-infected fibroblasts, and smooth muscle and endothelial cells for the change in expression of the 84 genes included in the inflammasome arrays. Our data demonstrate that *T. cruzi* infection, in comparison to treatment with LPS, elicits a subdued activation of inflammatory gene expression and IL-1β production in mφs. Yet, caspase-1/ASC inflammasome-dependent activation of the IL-1β – reactive oxygen species (ROS) – NF-κB pathway played an important role in control of *T. cruzi* replication in mφs. Further, NLRP3 controlled the ROS levels in mφs, and NLRP3 deficiency resulted in a potent increase in ROS-mediated parasite killing in infected mφs. To the best of our knowledge, this is the first study demonstrating a double-edged role of NLRP3 in determining mφ activation of ROS and cytokine response, both of which are required for clearance of *T. cruzi* infection.

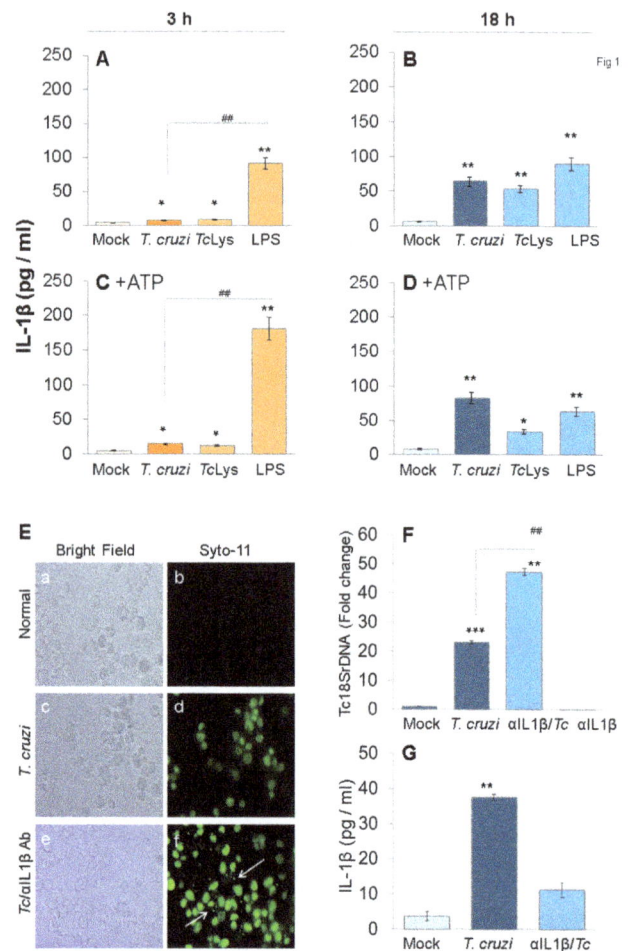

Figure 1. IL-1β production in macrophages infected by *T. cruzi*. (A–D) PMA-differentiated THP-1 mφs were incubated with *T. cruzi* trypomastigotes (cell: parasite ratio, 1:3), *Tc* lysate (10 µg protein/10^6 cells) or LPS (100 ng/ml) for 3 h (A&C) and 18 h (B&D). In some experiments, ATP was added during last 30 min of incubation (C&D). IL-1β release in supernatants was determined by ELISA. (E–G) IL-1β contributes to parasite control in mφs. THP-1 mφs were incubated with SYTO®11-labeled *T. cruzi* in the presence or absence of anti-IL-1β antibody for 18 h. (E) SYTO®11 fluorescence as an indicator of parasite uptake (shown by arrows) was determined by using an Olympus BX-15 microscope equipped with a digital camera (magnification 40X). (F) Quantitative PCR analysis of parasite burden in infected mφs by using *Tc18SrDNA*-specific oligonucleotides (normalized to human *GAPDH*). (G) Addition of anti-IL-1β antibody depletes secreted IL-1β levels in *T. cruzi*-infected mφs. In all figures, data are representative of three independent experiments and presented as mean ± SD. Significance is shown by *normal versus infected and #treated/infected versus infected (*,#p<0.05, **,##p<0.01, and ***,###p<0.001).

## Materials and Methods

### Ethics statement

All animal experiments were performed according to the National Institutes of Health Guide for Care and Use of Experimental Animals and approved by the UTMB's Animal Care and Use Committee (protoCol # 08-05-029).

### Mice, parasites, and cells

C57BL/6 female mice (6–8-weeks old) were purchased from Harlan Labs (Indianapolis, IN). NLRP3$^{-/-}$ and ASC$^{-/-}$ mice

**Table 1.** Inflammasome-related differential gene expression in THP-1 macrophages in response to *T. cruzi* infection (± ATP) in comparison to normal controls.

| *Tc* vs control at 3h | | | *Tc* + ATP vs control at 3 h | | | *Tc* + ATP vs *Tc* at 3 h | | |
|---|---|---|---|---|---|---|---|---|
| Gene name | ddCt log ratio | p value | Gene name | ddCt log ratio | p value | Gene name | ddCt log ratio | p value |
| CXCL1 | −4.10 | 0.000 | CXCL1 | −4.48 | 0.000 | ACTB | −2.41 | 0.002 |
| TNF | −2.86 | 0.000 | TNF | −3.68 | 0.000 | CXCL2 | −2.21 | 0.001 |
| NFKBIA | −2.46 | 0.000 | CXCL2 | −5.36 | 0.000 | PTGS2 | −4.73 | 0.020 |
| CXCL2 | −3.14 | 0.000 | NFKBIA | −3.24 | 0.000 | TNF | −0.82 | 0.028 |
| CIITA | 3.07 | 0.001 | TXNIP | 2.46 | 0.000 | CIITA | −1.68 | 0.037 |
| BCL2 | 0.95 | 0.005 | ACTB | −2.79 | 0.001 | NLRP3 | −0.65 | 0.051 |
| NFKB1 | −1.03 | 0.007 | RIPK2 | −1.99 | 0.001 | | | |
| RIPK2 | −1.43 | 0.008 | NFKB1 | −1.25 | 0.002 | | | |
| TXNIP | 1.56 | 0.007 | MAPK3 | 1.42 | 0.006 | | | |
| MAPK1 | 0.79 | 0.010 | PSTPIP1 | 1.11 | 0.006 | | | |
| PSTPIP1 | 1.03 | 0.010 | NLRP3 | −0.96 | 0.007 | | | |
| B2M | 1.44 | 0.023 | BCL2 | 0.86 | 0.009 | | | |
| SUGT1 | 0.86 | 0.026 | SUGT1 | 1.03 | 0.010 | | | |
| TAB2 | 0.66 | 0.032 | PYCARD | 1.10 | 0.015 | | | |
| CCL5 | −1.12 | 0.042 | IL1B | −1.75 | 0.018 | | | |
| PEA15 | 0.57 | 0.043 | RPL13A | 0.78 | 0.025 | | | |
| PYCARD | 0.86 | 0.049 | IRF1 | 0.78 | 0.034 | | | |
| CCL2 | −2.92 | 0.053 | CCL2 | −2.92 | 0.053 | | | |
| RPL13A | 0.65 | 0.056 | CCL5 | −1.06 | 0.053 | | | |

| *Tc* vs control at 18h | | | *Tc* + ATP vs control at 18 h | | | *Tc* + ATP vs *Tc* at 18 h | | |
|---|---|---|---|---|---|---|---|---|
| Gene name | ddCt log ratio | p value | Gene name | ddCt log ratio | p value | Gene name | ddCt log ratio | p value |
| CXCL1 | −3.67 | 0.000 | TNF | −3.97 | 0.000 | NAIP | −4.79 | 0.001 |
| TNF | −3.60 | 0.000 | CXCL1 | −3.35 | 0.000 | MAPK12 | 1.40 | 0.011 |
| HSP90B1 | 2.13 | 0.000 | HSP90B1 | 2.41 | 0.000 | CHUK | 1.14 | 0.049 |
| CXCL2 | −4.07 | 0.000 | CXCL2 | −4.34 | 0.000 | TIRAP | 2.57 | 0.076 |
| GAPDH | 1.61 | 0.000 | PYCARD | 2.10 | 0.000 | | | |
| MAPK1 | 1.40 | 0.000 | MAPK1 | 1.21 | 0.000 | | | |
| CTSB | 1.86 | 0.000 | GAPDH | 1.39 | 0.001 | | | |
| CCL5 | −2.26 | 0.000 | CCL5 | −2.17 | 0.001 | | | |
| NAIP | 5.05 | 0.001 | TAB2 | 1.20 | 0.001 | | | |
| PYCARD | 1.64 | 0.001 | CTSB | 1.56 | 0.001 | | | |
| SUGT1 | 1.21 | 0.003 | MAP3K7 | 1.15 | 0.002 | | | |
| PSTPIP1 | 1.16 | 0.004 | MAPK12 | 1.75 | 0.003 | | | |
| TRAF6 | 1.66 | 0.005 | PSTPIP1 | 1.14 | 0.005 | | | |
| MAPK3 | 1.26 | 0.012 | CHUK | 1.70 | 0.006 | | | |
| TAB2 | 0.81 | 0.011 | NFKBIA | −1.26 | 0.008 | | | |
| CFLAR | −1.77 | 0.016 | B2M | 1.60 | 0.013 | | | |
| IL1B | −1.79 | 0.016 | P2RX7 | −2.18 | 0.013 | | | |
| MAP3K7 | 0.86 | 0.015 | CASP4 | 0.79 | 0.018 | | | |
| TNFSF14 | 1.60 | 0.016 | NLRC5 | 3.65 | 0.016 | | | |
| ACTB | 1.60 | 0.025 | PEA15 | 0.70 | 0.017 | | | |
| CASP4 | 0.70 | 0.032 | TNFSF14 | 1.57 | 0.018 | | | |
| CCL2 | −3.34 | 0.030 | RPL13A | 0.81 | 0.022 | | | |
| HSP90AA1 | 1.12 | 0.027 | ACTB | 1.60 | 0.025 | | | |
| IRAK1 | −3.86 | 0.029 | HSP90AA1 | 1.12 | 0.027 | | | |
| P2RX7 | −1.85 | 0.030 | IRF1 | −0.83 | 0.026 | | | |
| RPL13A | 0.75 | 0.032 | PTGS2 | −4.56 | 0.024 | | | |

**Table 1.** Cont.

| Tc vs control at 18h | | | Tc + ATP vs control at 18 h | | | Tc + ATP vs Tc at 18 h | | |
|---|---|---|---|---|---|---|---|---|
| Gene name | ddCt log ratio | p value | Gene name | ddCt log ratio | p value | Gene name | ddCt log ratio | p value |
| PANX1 | 1.00 | 0.034 | TIRAP | 3.29 | 0.028 | | | |
| PEA15 | 0.56 | 0.047 | CARD6 | 2.02 | 0.031 | | | |
| CARD6 | 1.79 | 0.052 | CCL2 | −3.32 | 0.031 | | | |
| | | | IRAK1 | −3.85 | 0.030 | | | |
| | | | PANX1 | 0.98 | 0.035 | | | |
| | | | SUGT1 | 0.81 | 0.035 | | | |
| | | | TRAF6 | 1.12 | 0.040 | | | |
| | | | MAPK3 | 0.91 | 0.055 | | | |

The 96-well RT Profiler Human Inflammasome PCR Arrays (SA Biosciences/Qiagen) were probed in triplicate with cDNA from THP-1 macrophages infected with *T. cruzi* (*Tc*) for 3 h or 18 h (with or without ATP) as described in Materials and Methods. The Ct values from qPCR data were analyzed by using open source HTqPCR v.1.7 software package (v.2.13). All array data were normalized by Quantile method and filtered to exclude genes that exhibited Ct values>35. The relative expression level of each target gene in treated cells was calculated using the formula, fold change = 2-ΔΔCt, where ΔCt represents the Ct (sample) - Ct (control). LimmaCt in HT-qPCR package was employed for contrast analysis of all the groups included in experiment and identification of genes that were overall differentially expressed (p<0.05).

(C57BL/6 background) were a gift from Dr V. Dixit (Genentech, San Francisco, CA) and bred at the UTMB animal facility. *T. cruzi* (SylvioX10/4 strain) trypomastigotes were maintained and propagated by the continuous *in vitro* passage of parasites in monolayers of C2C12 cells (an immortalized mouse myoblast cell line). *T. cruzi* isolate and C2C12 cells were purchased from American Type Culture Collection (ATCC, Manassas VA).

Single-cell suspensions of bone marrow (BM) - derived monoCytes from WT, ASC$^{-/-}$ and NLRP3$^{-/-}$ mice (C57BL/6 background) were added to petri dishes ($10^6$ cells/ml) in complete RPMI media containing 20 ng/ml murine macrophage–colony stimulating factor (M-CSF, eBioscience, San Diego, CA) and incubated at 37°C in 5% $CO_2$ for 10 days to support differentiation to mφs. The differentiated BM mφs were maintained in the presence of 5 ng/ml M-CSF during experimental use. THP-1 human monoCytes were differentiated into mφs by overnight incubation with 50 ng/ml phorbol-12-myristate-13-acetate (PMA), and then rested at 37°C/5% $CO_2$ for 48 h in RPMI complete media containing 10% FBS. RAW 264.7 murine mφs were routinely cultured in DMEM with 10% FBS.

In general, primary or cultured mφs ($0.5–1\times10^6$ cells/ml) were seeded in Nunc Lab-Tek II chamber slides or 24-well or 6-well plates, infected with *Tc* trypomastigotes (cell: parasite ratio, 1:3) for 2 h, washed to remove free parasites, and then incubated for 3, 6, 12, and 18 h. In some experiments, 5 mM ATP was added during the last 30 min of incubation. When monitoring the role of inflammasomes or NADPH oxidase (NOX2)-mediated ROS in parasite control, infected mφs were incubated in the presence of 1 μg/ml anti-IL-1β antibody (Santa Cruz, Dallas TX); 20 μM Ac-YVAD-CHO (caspase-1 inhibitor, Enzo Life Sc., Farmingdale, NY); 100 ng/ml cycloheximide (inhibits protein biosynthesis); 50 μM glibenclamide (bloCks the maturation of caspase-1 and pro-IL-1β by inhibiting $K^+$ efflux and also inhibits NLRP3 inflammasome activation (Imgenex, San Diego, CA); 10 μM diphenylene iodonium (DPI) or 30 μM apoCynin (inhibitors of NOX2/ROS); and 1 mM N-acetylcysteine (NAC, ROS scavenger). Macrophages incubated with media alone or LPS (100 ng/ml) were used as controls. Cells and culture supernatants were stored at −80°C.

## Probing the RT Profiler Human Inflammasome PCR Arrays

THP-1 cells, with and without *T. cruzi* infection and treatments, were harvested with 500 μl Bio-Rad cell lysis/RNA extraction buffer. Total RNA was extracted by using an Aurum DNA-free RNA isolation kit (Bio-Rad, Hercules, CA) and measured at 260 and 280 nm for determination of purity and concentration. The cDNA probes were generated by reverse transcription of 5 μg total RNA by using the Bio-Rad iScript cDNA synthesis kit.

The 96-well RT Profiler Human Inflammasome PCR Arrays (SA Biosciences/Qiagen, Valencia, CA) containing primer pairs for 84 key genes involved in the function of inflammasomes and NLR signaling were probed with 2 μl of cDNA template in the presence of 6.5 μl of dNTPs, $MgCl_2$, and stabilizers (iQ SYBR Supermix, Bio-Rad), and PCR was carried out on a LightCycler 480 Multiple Plate System. A total of 42 arrays were probed with 14 research samples in triplicate, and datasets were analyzed by Web-based PCR Array Data Analysis software (SA Biosciences) for threshold cycle (Ct) value determination.

The Ct values from qPCR data were analyzed by using the open source HTqPCR v.1.12 software package [19]. Briefly, all array data were imported into HTqPCR, normalized by the quantile method and then filtered to exclude genes that exhibited Ct values >35. The relative expression level of each target gene in infected cells was calculated by using the formula, fold change = $2^{-\Delta\Delta C_t}$, where $\Delta C_t$ represents the $C_t$ (*Tc*-infected or LPS-treated sample) - $C_t$ (control). The LimmaCt routine in the HT-qPCR package was employed for contrast analysis of all groups included in the experiment and for the identification of genes that were overall differentially expressed. LimmaCt utilizes the Linear Model for Microarray data (limma) R package to fit linear models for analyzing designed experiments and for the assessment of differential expressions in microarray data to perform contrast analysis between the different experimental groups. LimmaCt uses Empirical Bayesian methods from the eBayes function in limma to provide stable results even when the number of arrays is small. The basic statistics used for significance analysis consists of moderated t-statistics with the same interpretation as ordinary t-statistics computed for each gene and contrasts, except that the standard errors are shrunk towards a common value by using a Bayesian model. The eBayes function computes moderated F-

**A**  *Tc* vs. cont @ 3h    *Tc* vs. cont @18h    Fig.2

7 | 11 | 18

Green: Up regulated
Red: Down regulated

NFKB1A
CIITA
BCL2
NFKB1
RIPK2
TXNIP
B2M

CXCL1 | SUGT1
TNF | TAB2
CXCL2 | CCL5
MAPK1 | PEA15
PSTPIP1 | PYCARD
| CCL2

HSP90B1 | TNFSF14
GAPDH | ACTB
CTSB | CASP4
NAIP | HSP90AA1
TRAF6 | IRAK1
MAPK3 | P2RX7
CFLAR | RBL13A
IL1B | PANX1
MAP3K7 | CARD6

**B**  *Tc* vs. cont @ 3h    *Tc*+ATP vs. cont @ 3h

4 | 15 | 7

MAPK1
B2M
TAB2
PEA15

MAPK3
NLRP3
IL1B
IRF1
RPL13A
PTGS2

CXCL1 | SUGT1
NFKBIA | CCL5
BCL2 ↑ | PYCARD
NFKB1 | CCL2
RIPK2 | TNF
TXNIP | CXCL2
PSTPIP1↑ | CIITA↑
| ACTB

**C**  *Tc* vs. cont @ 18h    *Tc*+ATP vs. cont @ 18h

3 | 27 | 8

MAPK3
CFLAR
IL1B

CHUK
NFKB1A
B2M
NLRC5
IRF1
PTGS2
TIRAP

CARD6 | HSP90AA1 | PANX1 | TRAF6
CASP4 | HSP90B1 | PEA15 | RPL13A
CCL2 | IRAK1 | PSTPIP1 | GAPDH
CCL5 | MAP3K7 | PYCARD | ACTB
CTSB | TAB2 | SUGT1 | NAIP ↑
CXCL1 | MAPK1 | TNF | MAPK12
CXCL2 | P2RX7 | TNFSF14 |

**D**  LPS vs. cont @ 3h    LPS+ATP vs. control @ 3h

1 | 30 | 4

TRAF6

TAB1
PEA15
CIITA
PANX1

CXCL1 | MAPK1 | CTSB | BIRC2
CXCL2 | BIRC3 | B2M | CCL2
TNF | RPL13A | IRF2 | ACTB
RIPK2 | GAPDH | HSP90AA1 | NLRC5↑
NFKB1 | PSTPIP1 | SUGT1 | FADD
NFKBIA | CCL5 | PTGS2 | TXNIP
HSP90B1 | NLRP3 | NAIP |
PYCARD | NFKBIB | TNFSF14 |

**E**  LPS vs. cont @ 18h    LPS+ATP vs. cont @ 18h

4 | 23 | 5

MAP12
PYCARD
RTC
PSTPIP1

PANX1
ACTB
IKBKG
B2M
TNFSF14

CXCL1 | IL1B | MAPK3
CCL5 | CTSB | TRAF6
MAP3K7 | CCL2 | CFLAR
TAB2 | PTGS2 | RELA
HSP90B1 | RPL13A | CHUK
CXCL2 | BIRC3 | PEA15
TNF | HSP90AA1 | NFKB1A
CASP1 | MAPK1 |

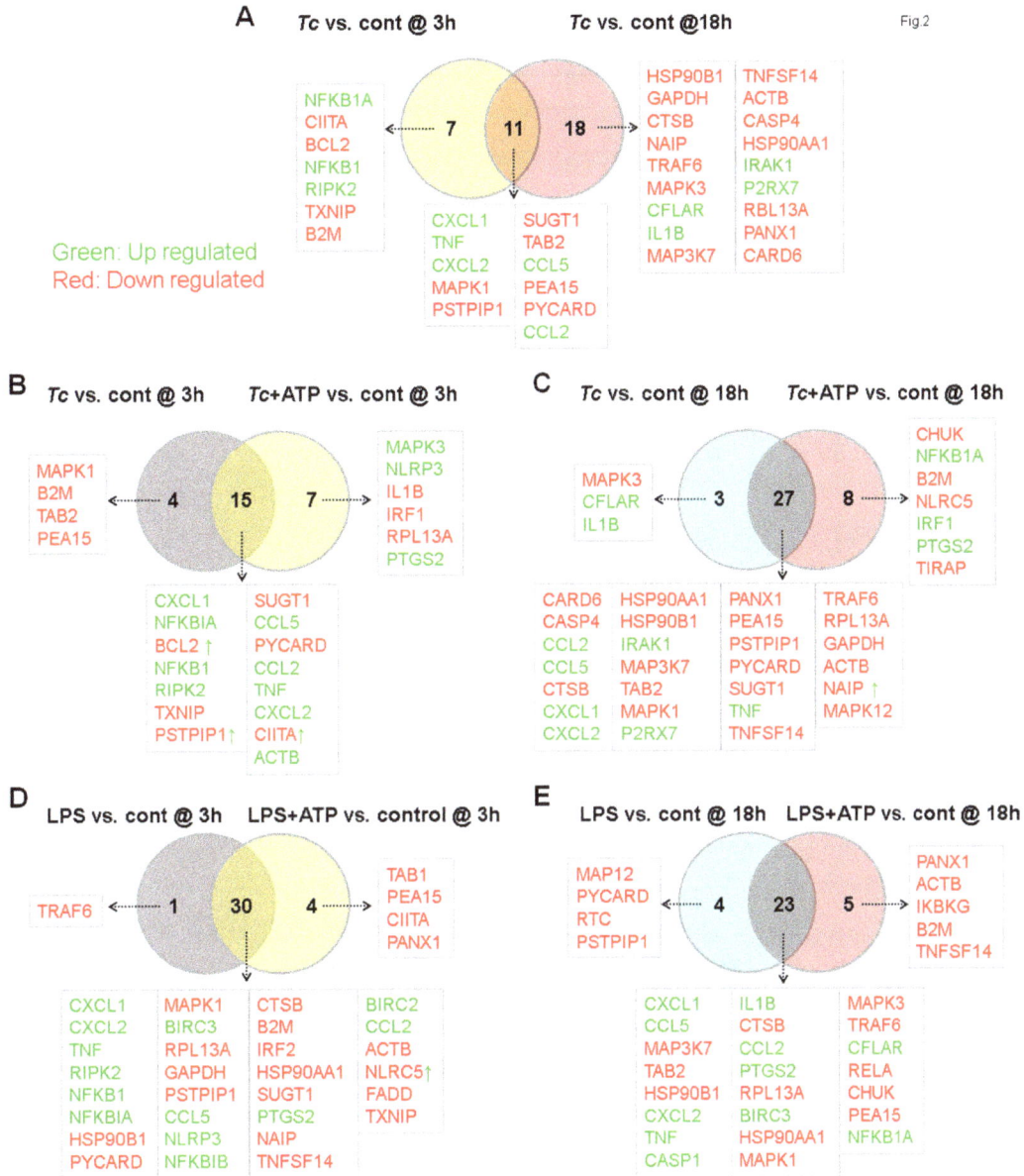

**Figure 2. Venn diagram of inflammasome-related differential gene expression in mφs infected with *T. cruzi* (±ATP).** THP-1 mφs were incubated with *T. cruzi* or LPS (± ATP treatment) as in Fig. 1. Total RNA was isolated, and cDNA was used as a template to probe the expression of 95 genes (including house-keeping genes) in the RT$^2$ Profiler Inflammasome PCR Arrays. The differential mRNA level was captured by quantitative RT-PCR, normalized to housekeeping genes, and HTqPCR was employed to attain the statistically significant differential expression in treated- versus-control samples (Table 1 and Table S2). Shown are Venn diagrams of comparative analysis of gene expression in *T. cruzi*-infected mφs at 3 h versus 18 h (A), effect of ATP stimulus on gene expression at 3 h (B) and 18 h (C) pi, and comparative effect of ATP stimulus on gene expression in LPS-treated mφs at 3 h (D) and 18 h (E). Differential up-regulation (green) and down-regulation (red) of genes with respect to controls is presented. Genes presenting as red with green arrow in B–F showed decreased expression without ATP but were up-regulated by ATP treatment (and vice versa).

statistics which combines all of the t-statistics from all the contrasts to calculate the significance of a gene(s). This F-statistic determines if gene(s) are differentially expressed across any contrast. The p-value is calculated based on the moderated t-statistics and F-statistics followed by an FDR adjustment. LimmaCt with the top table module was used to gain an appreciation of the change in differential expression of a particular gene over time (or treatment), and significance in changes in gene expression was accepted at p<0.05.

In some experiments, quantitative RT-PCR was performed for IL-1β, IL-18, CXCL1, and TNF-α mRNA levels and house-keeping genes (GAPDH and β-actin) with gene-specific primer pairs (Table S1) and iQ SYBR Supermix on a C1000 Touch Thermal Cycler (Bio-Rad). The relative expression level of each target gene was calculated by using the formula, fold change = $2^{-\Delta\Delta C_t}$, where $\Delta C_t$ represents the $C_t$ (infected sample) - $C_t$ (control).

## Functional analysis

Datasets for differential expression of the genes included in RT Profiler Human Inflammasome PCR Arrays in human foreskin fibroblasts (HFF), human microvascular endothelial cells

**Table 2.** Ingenuity iReport analysis of inflammasome-related datasets in THP-1 macrophages infected by *T. cruzi* for 3 h or 18 h (± ATP).

| Pathways | Gene name | FoCus mols | P value |
|---|---|---|---|
| **_T. cruzi_ infection of THP-1 mφs vs control (3h)** | | | |
| *Top biological and molecular functions* | | | |
| 1) Inflammatory responses/infection | ↓ B2M, ↓ BCL2, ↑ CCL2, ↑ CCL5, ↓ CIITA, ↑ CXCL1, ↑ CXCL2, ↑ NFKB1, ↑ NFKB1A, ↓ PYCARD, ↑ RIPK2, ↑ TNF, ↓ TXNIP | 13 | 5.91E-12–2.49E-04 |
| 1a. Recruitment of neutrophils/phagoCytes | ↑ CCL2, ↑ CCL5, ↑ CXCL1, ↑ CXCL2, ↓ PYCARD, ↑ RIPK2, ↑ TNF | 7 | 2.77E-10 |
| 1b. Activation of leukoCyte/lymphoCytes | ↓ BCL2, ↑ CCL2, ↑ CCL5, ↑ CXCL1, ↑ CXCL2, ↑ NFKB1, ↑ NFKB1A, ↓ PYCARD, ↑ TNF | 9 | 1.89E-07 |
| 2) Cell death of immune cells (decreased) | ↓ B2M, ↓ BCL2, ↑ CCL2, ↑ CCL5, ↓ CIITA, ↑ CXCL1, ↑ CXCL2, ↓ MAPK1, ↑ NFKB1, ↑ NFKB1A, ↓ PEA15, ↓ PYCARD, ↑ RIPK2, ↓ TAB2, ↑ TNF, ↓ TXNIP | 16 | 4.50E-10 |
| 2a. Macrophage cell death (decreased) | ↓ BCL2, ↑ CCL2, ↑ CCL5, ↑ CXCL1, ↑ CXCL2, ↑ NFKB1, ↑ NFKB1A, ↓ PYCARD, ↑ TNF | 9 | 3.78E-13 |
| *Toxicity analysis* | | | |
| Decreases MPT, mitoChondrial swelling | ↓ B2M, ↓ BCL2, ↑ NFKB1, ↓ PYCARD, ↑ TNF | 5 | 7.27E-08 |
| **_T. cruzi_ infection of THP-1 mφs vs control (3 h + ATP)** | | | |
| *Top biological and molecular functions* | | | |
| 1) Inflammatory responses/infection | ↓ BCL2, ↑ CCL2, ↑ CCL5, ↑ CXCL1, ↑ CXCL2, ↓ IL1B, ↓ IRF1, ↑ NFKB1, ↑ NFKB1A, ↓ NLRP3, ↑ PTGS2, ↓ PYCARD, ↑ RIPK2, ↑ TNF, ↓ TXNIP | 16 | 7.54E-14–1.79E-05 |
| 1a. Recruitment/migration of phagoCytes/neutrophils | ↑ CCL2, ↑ CCL5, ↑ CXCL1, ↑ CXCL2, ↓ IL1B, ↑ MAPK3, ↑ NLRP3, ↓ PYCARD, ↑ RIPK2, ↑ TNF | 9 | 7.69E-09 |
| 1b. Activation of lymphoCyte/leukoCytes | ↓ BCL2, ↑ CCL2, ↑ CCL5, ↑ CXCL1, ↓ IL1B, ↓ IRF1, ↑ NFKB1, ↑ PTGS2, ↓ PYCARD, ↑ RIPK2 ↑ TNF, ↓ TXNIP | 12 | 3.44E-12 |
| 2) Cell death of immune cells/necrosis (decreased) | ↓ BCL2, ↑ CCL2, ↑ CCL5, ↑ CXCL1, ↑ CXCL2, ↓ IL1B, ↓ IRF1, ↑ MAPK3, ↑ NFKB1, ↑ NFKB1A, ↑ NLRP3, ↑ PTGS2, ↓ PYCARD, ↑ RIPK2, ↑ TNF, ↓ TXNIP | 16 | 9.06E-17–1.99E-05 |
| 2a. Cell death of phagoCytes, myeloid cells (decreased) | ↓ BCL2, ↑ CCL2, ↑ CCL5, ↑ CXCL1, ↑ CXCL2, ↓ IL1B, ↓ IRF1, ↑ NFKB1, ↑ NFKB1A, ↓ PYCARD, ↑ TNF | 11 | 9.06E-17 |
| *Toxicity analysis* | | | |
| Decreased MPT, mitoChondrial swelling | ↓ B2M, ↓ BCL2, ↑ NFKB1, ↓ PYCARD, ↑ TNF | 5 | 7.27E-08 |
| **_T. cruzi_ infection of THP-1 mφs vs control (18 h)** | | | |
| *Top biological and molecular functions* | | | |
| 1) Cell death of immune cells (decreased) | ↓ ACTB, ↓ CASP4, ↑ CCL2, ↑ CCL5, ↑ CFLAR, ↓ CTSB, ↑ CXCL1, ↑ CXCL2, ↓ GAPDH, ↓ HSP90AA1, ↓ HSP90B1, ↑ IL1B, ↑ IRAK1, ↓ MAP3K7, ↓ MAPK1, ↓ MAPK3, ↑ NAIP, ↑ P2RX7, ↓ PEA15, ↓ PYCARD, ↓ TAB2, ↑ TNF, ↓ TNFS14, ↓ TRAF6 | 24 | 1.48E-12–2.60E-04 |
| 1a.Cell death of myeloid/phagoCytes | ↓ CASP4, ↑ CCL2, ↑ CCL5, ↑ CFLAR, ↑ CXCL1, ↑ CXCL2, ↑ IL1B, ↑ P2RX7, ↓ PYCARD, ↑ TNF | 10 | 1.72E-12 |
| 2) Inflammation/infectious disease | ↓ CASP4, ↑ CCL2, ↑ CCL5, ↑ CXCL1, ↑ CXCL2, ↓ HSP90AA1, ↑ IL1B, ↓ MAPK1, ↓ PANX1, ↑ P2RX7, ↓ PYCARD, ↑ TNF, ↓ TNFS14, ↓ TRAF6 | 14 | 1.08E-06 |
| 2a. Activation of leukoCyte/lymphoCytes | ↑ CCL2, ↑ CCL5, ↑ CXCL1, ↓ HSP90B1, ↑ IL1B, ↓ MAP3K7, ↓ PYCARD, ↑ TNF, ↓ TNFS14 | 9 | 2.95E-07 |
| 2b. Migration of phagoCytes/neutrophils | ↑ CCL2, ↑ CCL5, ↓ CTSB, ↑ CXCL1, ↑ CXCL2, ↑ IL1B, ↑ P2RX7, ↑ TNF | 8 | 2. 07E-07 |
| *Toxicity analysis* | | | |
| Gene regulation by PPARα | ↓ HSP90AA1, ↑ IL1B, ↓ MAP3K7, ↓ MAPK1, ↓ MAPK3, ↑ TNF, ↓ TRAF | 7 | 4.16E-10 |
| **_T. cruzi_ infection of THP-1 mφs vs control (18 h + ATP)** | | | |
| *Top biological and molecular functions* | | | |
| 1) Cell death (decreased) | ↓ ACTB, ↓ B2M, ↓ CASP4, ↑ CCL2, ↑ CCL5, ↓ CHUK, ↓ CTSB, ↑ CXCL1, ↑ CXCL2, ↓ GAPDH, ↓ HSP90AA1, ↓ HS90B1, ↑ IRAK1, ↑ IRF1, ↓ MAP3K7, ↓ MAPK1 ↓ MAPK12, ↑ NFKB1A, ↑ P2RX7, ↓ PEA15, ↑ PTGS2, ↓ PYCARD, ↓ TAB2, ↑ TNF, ↓ TNFS14, ↓ TRAF6 | 26 | 6.53E-12–4.09E-04 |
| 1a.Cell death of myeloid/phagoCytes | ↓ CASP4, ↑ CCL2, ↑ CCL5, ↑ CFLAR, ↑ CXCL1, ↑ CXCL2, ↑ IL1B, ↑ P2RX7, ↓ PYCARD, ↑ TNF | 10 | 1.48E-12 |
| 2) Inflammation/Infectious disease | ↓ ACTB, ↓ B2M, ↓ CASP4, ↑ CCL2, ↑ CCL5, ↓ CHUK, ↓ CTSB, ↑ CXCL1, ↑ CXCL2, ↓ HSP90B1, ↑ IRAK1 ↓ MAP3K7, ↑ NFKB1A, ↑ PTGS2, ↑ P2RX7, ↓ PYCARD, ↓ TIRAP, ↑ TNF, ↓ TNFS14, ↓ TRAF6 | 21 | 3.33E-10–4.43E-04 |

**Table 2.** Cont.

| Pathways | Gene name | FoCus mols | P value |
|---|---|---|---|
| 2a. Activation of leukoCyte/lymphoCytes | ↓ B2M, ↑ CCL2, ↑ CCL5, ↑ CXCL1, ↑ CXCL2, ↓ MAP3K7, ↑ TNF | 7 | 9.31E-06 |
| 2b. Migration of phagoCytes/neutrophils | ↑ CCL2, ↑ CCL5, ↓ CHUK, ↓ CTSB, ↑ CXCL1, ↑ TNF ↑ CXCL2, ↑ NFKB1A, ↑ P2RX7, ↑ PTGS2, ↓ TIRAP | 11 | 2.43E-04 |
| *Toxicity analysis* | | | |
| Gene regulation by PPARα | ↓ CHUK, ↓ HSP90AA1, ↓ MAP3K7, ↓ MAPK1, ↓ MAP3K7, ↑ NFKB1A, ↑ PTGS2, ↑ TNF, TRAF6 | 9 | 2.11E-11 |

All differentially expressed proteins identified in THP-1 macrophages infected with *T. cruzi* for 3 or 18 h (±ATP) (listed in Table 1) were uploaded into Ingenuity Pathway Analysis (IPA) to interpret datasets in the context of biological proCesses and function, and pathway and molecular networks. Presented are the top networks with a p value <0.01 to which maximal number of the differentially expressed proteins identified in chagasic plasma (bolded letters) were assoCiated with. FoCus molecules are the number of differentially expressed plasma proteins assoCiated with an individual network.

(HMVEC), and human vascular smooth muscle cells (HVSMC), infected with *T. cruzi* for 24 h, were obtained from the HG_U133 plus 2.0 Affymetrix array analysis data posted at Gene Expression Omnibus [20]. The selected differentially expressed gene datasets from *Tc*-infected HFF, HMVEC, and VSMC cells were submitted to Ingenuity iReport Analysis (Ingenuity Systems, Redwood city, CA). The iReport retrieves a set of biological information such as gene name, sub-cellular loCation, tissue specificity, function, assoCiation with disease, and integrates into networks and signaling pathways with biological meaning and significance. An e-value was calculated by estimating the probability of a random set of genes having a frequency of annotation for that term greater than the frequency obtained in the real set, and a threshold of e value $<10^{-3}$ was set to retrieve significant molecular functions and biological processes.

### Parasite infectivity and replication in mφs

*T. cruzi* trypomastigotes were labeled with 5 μM SYTO®11 (binds DNA, Molecular Probes-Invitrogen, Eugene, OR) or 5 μM carboxyfluorescein succinimidyl ester (CFSE, binds amines, Invitrogen) for 20 min at 37oC. THP-1- or BM- derived mφs were infected and incubated with labeled *T. cruzi* trypomastigotes,

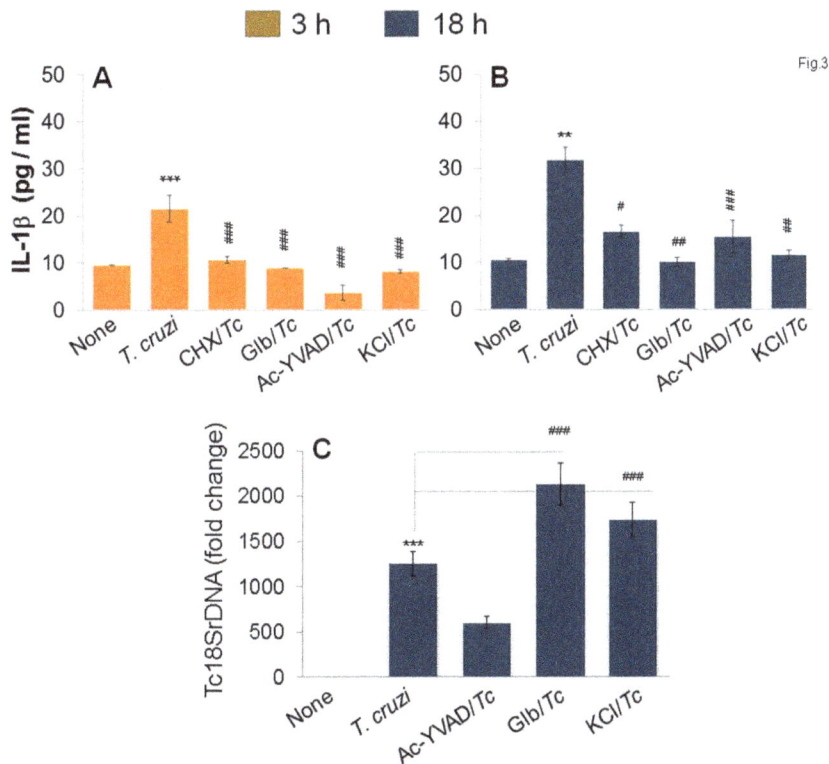

**Figure 3. NLRP3/caspase-1 inflammasome is the major source of IL-1β for parasite control in mφs.** THP-1 mφs were incubated with *T. cruzi* in the presence or absence of cycloheximide (CHX,), glibenclamide (Glb), Ac-YVAD-CHO, and KCl for 3 h *(A)* and 18 h *(B&C)*. Macrophages incubated with media alone were used as controls. *(A&B)* IL-1β release in supernatants was determined by an ELISA. *(C)* Quantitative PCR analysis of parasite burden in infected macrophages using *Tc18SrDNA*-specific oligonucleotides.

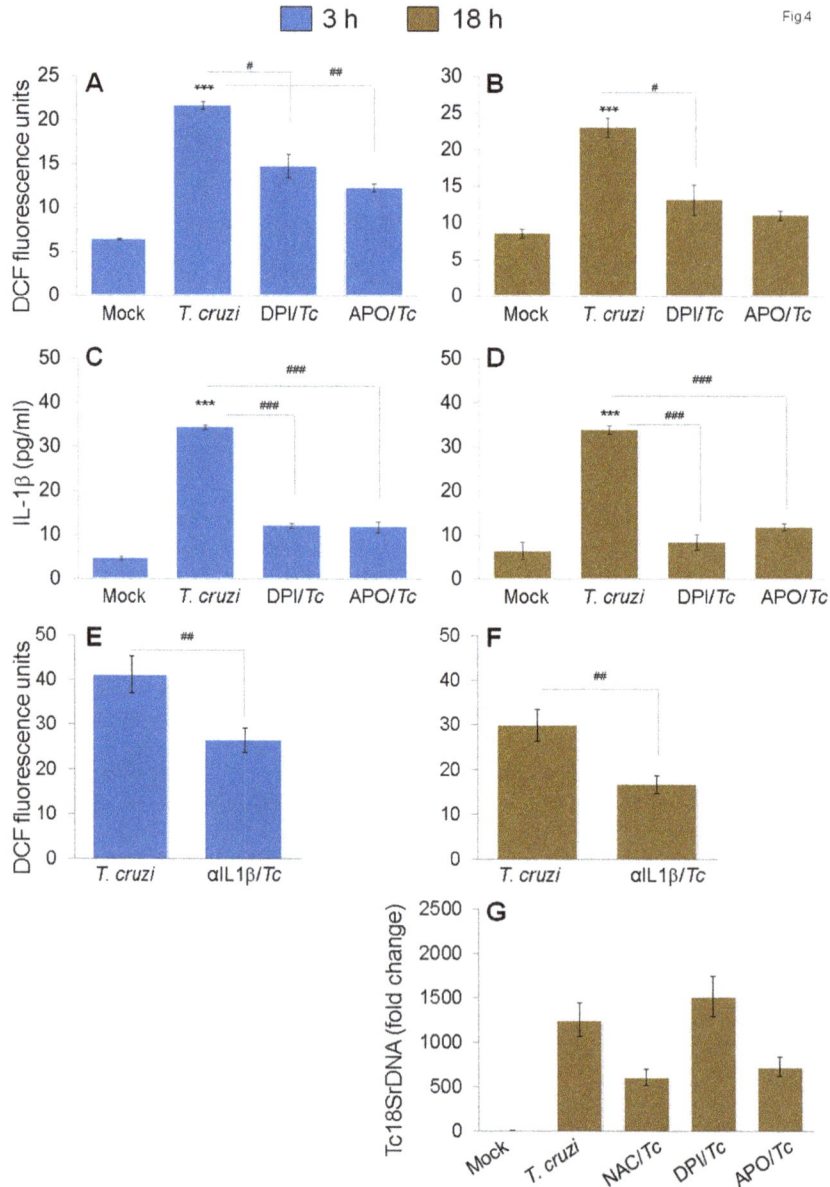

**Figure 4. Feedback cycle of NOX2/ROS and IL-1β activation in mφs infected by *T. cruzi*.** THP-1 mφs were infected with *T. cruzi* as in Fig. 1, and incubated for 3 h *(A,C,E)* or 18 h *(B,D,F,G)* in presence of NOX2 inhibitors (diphenylene iodinium (DPI) or apoCynin), ROS scavenger (N-acetylcysteine (NAC)) or IL-1β antibody. *(A–D)* NOX2 inhibitors decreased ROS and IL-1β levels in infected mφs. Shown are *(A&B)* H₂DCFDA oxidation by intracellular ROS, resulting in formation of fluorescent DCF by fluorimetry and *(C&D)* IL-1β release in supernatants determined by an ELISA. *(E&F)* Treatment with anti-IL-1β antibody decreased the ROS levels in infected mφs. *(G)* Effect of ROS inhibitors on intracellular parasite burden, as determined by qPCR, in infected mφs.

as above. Cells were washed, and SYTO®11 or CFSE fluorescence as an indicator of parasite uptake was determined by using an Olympus BX-15 microscope equipped with a digital camera (magnification 40X). Cells infected with CFSE-labeled parasites were also fixed with 2% paraformaldehyde and visualized on a FACSCalibur flow cytometer (BD Biosciences, San Jose, CA) acquiring 20,000 events. Further analysis was performed by using FlowJo software (ver. 7.6.5, Tree-Star, San Carlo, CA). Mean Fluorescence intensity (MFI) of CSFE positive cells was used as a relative marker of parasites per cell.

Total DNA from normal and infected cells was isolated by using TRIzol reagent (Life Technologies, Grand Island, NY). Total DNA (100 ng) was used as a template in a quantitative PCR (qPCR) on an iCycler thermal cycler with SYBR Green Supermix (Bio-Rad) and oligonucleotide pairs specific for $Tc$18S ribosomal DNA (Table S1). Data were normalized to murine or human GAPDH, and fold change calculated as $2^{-\Delta\Delta C_t}$, where $\Delta\Delta C_t$ represents the $C_t$ (sample) - $C_t$ (control).

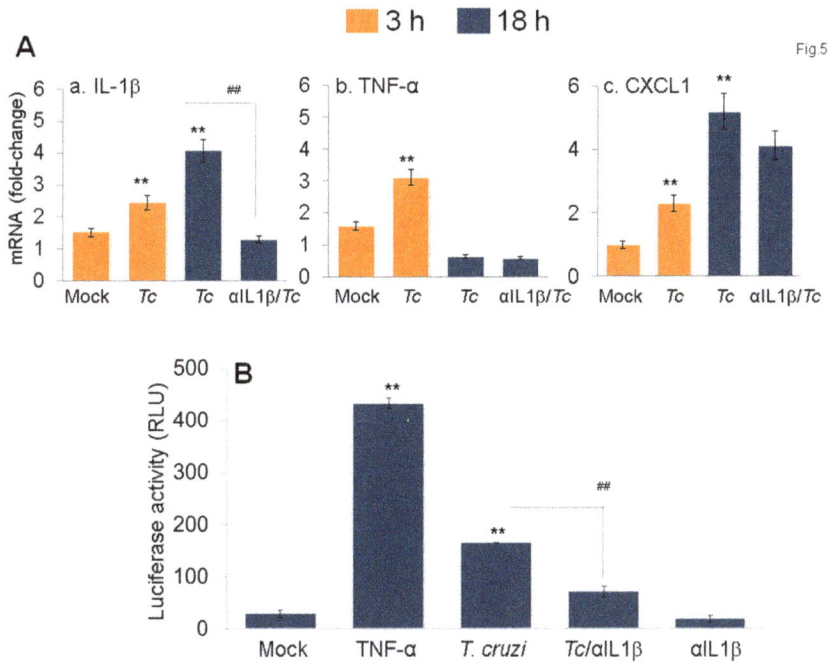

**Figure 5. IL-1β signals NF-κB activation and inflammatory cytokine gene expression in infected mφs.** *(A)* The mRNA levels for IL-1β (panel a), TNFα (panel b) and CXCL1 (panel c) cytokines were measured in *T. cruzi*-infected THP-1 mφs at 3h and 18h pi by quantitative RT-PCR. *(B)* RAW 264.7 macrophages were transiently transfected with pGL4.NF-κB-Luc reporter plasmid and pREP7-Rluc plasmid (transfection efficiency control) as described in Materials and Methods. Transfected cells were infected with *T. cruzi* and incubated in the presence or absence of anti-IL-1β antibody. Mφs incubated with 10 ng/ml recombinant TNF-α for 6 h were used as positive controls. The relative NF-κB transcriptional activity was measured by firefly luciferase activity and normalized to *Renilla* luciferase activity. The transcriptional activity of NF-κB in normal cells was considered as baseline and valued at 1.

## Activation of mφs by *T. cruzi*: ROS, nitric oxide (˙NO) and cytokines levels

Mφs were infected with *T. cruzi* for 1 h, washed to remove free parasites, and then incubated for up to 18 h, as above. Cells were incubated with 5 μM Carboxymethyl-2′,7′-dichlorodihydrofluorescein diacetate (CM-H₂DCF-DA, from Life Technologies, $Ex_{495}/Em_{527}$, fluoresces green upon oxidation by intracellular ROS) for 30 min, and the fluorescence was recorded using a SpectraMax M5 microplate reader. In some experiments, cells were incubated for 30 min with 5 μM dihydroethidium. (DHE, $Ex_{518nm}/Em_{605nm}$, fluoresces red upon oxidation and binds DNA). Micrographs of DCF or DHE fluorescence were visualized on an Olympus BX-15 microscope, and images were captured by using a mounted digital camera (magnification 40X).

ROS release in supernatants was determined by an Amplex red assay. Briefly, 50 μl of supernatants from infected mφs were added in triplicate to 96-well, flat-bottomed plates, and mixed with a similar volume of 100 μM 10-acetyl 3, 7-dihydroxyphenoxazine (Amplex Red, Life Technologies) and 0.3 U/ml horseradish peroxidase. The $H_2O_2$-dependent oxidation of Amplex red to red fluorescent resorufin ($Ex_{563nm}/Em_{587nm}$) was recorded as above (standard curve: 50 nM - 5 μM $H_2O_2$) [21].

The ˙NO level (indicator of iNOS activity) was monitored by the Greiss reagent assay and by using the Nitrate/Nitrite Colorimetric Assay Kit (Cayman, Ann Arbor, MI). Briefly, culture supernatants (50 μl) were reduced with 0.01 unit/100 μl of nitrite reductase, and incubated for 10 min with 100 μl of 1% sulfanilamide prepared in 5% phosphoric acid/0.1% N-(1-napthyl) ethylenediamine dihydroChloride (1:1, v/v). After incubation for 10 min,

formation of diazonium salt was monitored at 545 nm (standard curve: 2–50 μM sodium nitrite).

Culture supernatants from mφs incubated with *T. cruzi* (± inhibitors) were also utilized for measuring IL-1β and IL-18 release by using OptEIA ELISA kits (eBioscience), according to the manufacturer's instructions.

## Transient transfection and luciferase assay

RAW 264.7 cells were plated in 6-well plates, and, when at > 70% confluency, transfected with pGL4.NF-κB-Luc reporter plasmid (3 μg/well, Promega, San Diego, CA) by using JetPEI transfection reagent (Polyplus transfection, New York, NY), according to instructions provided by the manufacturer. The pREP7-Rluc plasmid (500 ng) expressing renilla luciferase was co-transfected into RAW macrophages and used as an internal control reporter. After 30 h of transfection, cells were washed, replenished with complete medium, and infected with *T. cruzi* (± inflammasome inhibitors) for 18 h (positive control: 10 ng/ml recombinant TNF-α for 6 h). The relative NF-κB transcriptional activity was detected by using a Steady-Glo luciferase assay system (Promega) and recorded on a luminometer (Turner Biosystems, Sunnyvale, CA).

## Statistical Analysis

All experiments were conducted at least twice with triplicate observations per sample per experiment. All data were analyzed by using Graph Pad InStat ver.3 software and expressed as mean ± SD. Data were analyzed by the Student's *t* test (comparison of 2 groups) and 1-way analysis of variance (ANOVA) with Tukey's post-hoC test (comparison of multiple groups). Significance is

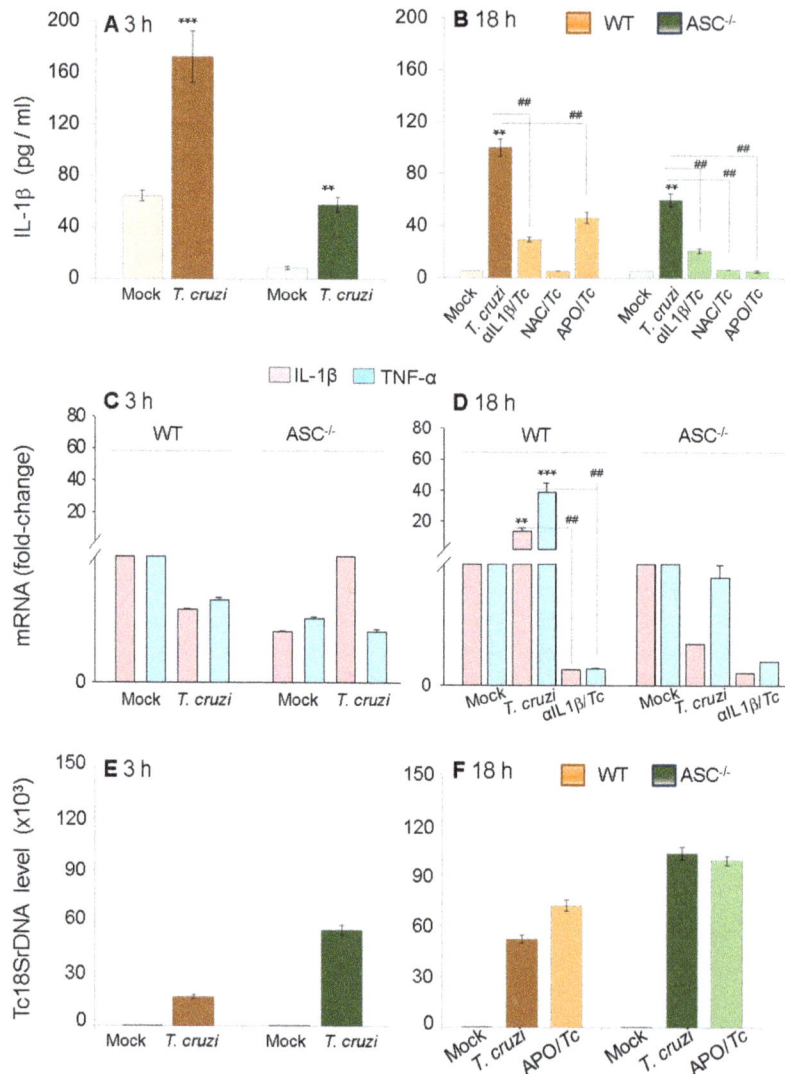

**Figure 6. ASC$^{-/-}$ mφs are compromised in the IL-1β–ROS–NF-κB pathway for control of *T. cruzi*.** Bone-marrow-derived macrophages were isolated from matched WT and ASC$^{-/-}$ mice. Primary mφs were infected with *T. cruzi* and incubated for 3 h or 18 h in the presence or absence of anti-IL-1β Ab or ROS scavengers (as in Figs.1&4). Shown are IL-1β release measured by an ELISA *(A&B)*, mRNA levels for IL-1β and TNF-α by quantitative RT-PCR *(C&D)* and Tc18SrDNA signal by qPCR *(E&F)*.

shown by *$^{,\#}p<0.05$, **$^{,\#\#}p<0.01$, ***$^{,\#\#\#}p<0.001$ (*normal-versus-infected; $^{\#}$infected versus infected/treated).

## Results

### Mφs elicit subdued IL-1β response to *T. cruzi* infection

THP-1 mφs incubated with *T. cruzi* trypomastigotes (1:3, cell: parasite ratio) or *Tc*-lysate exhibited a ~2-fold increase in IL-1β release at 3 h pi that was consistently increased at 6 h and 12 h pi (data not shown) and maximized to a >9.8-fold increase by 18 h pi (Fig.1A&B). Incubation with higher number of parasites (1:4, 1:5 or 1:6, cell: parasite ratio) did not result in a further increase in IL-1β release at 3 h and 18 h pi (data not shown). LPS treatment (100-ng/ml) triggered a substantially higher level of IL-1β release in THP-1 mφs than was observed with *T. cruzi* infection, the maximal difference being noted at 3 h (Fig.1A&B). Exogenous addition of ATP, the K$^{+}$ flux agent that can trigger caspase-1 cleavage and inflammasome activation in response to PAMPs, elicited a 2-fold and no increase in IL-1β release in *Tc*-infected

and LPS-treated cells at 3 h (Fig.1A&C). No significant effect of exogenous ATP on IL-1β release was observed at 18 h post-incubation (Fig.1B&D)

To determine if IL-1β is required for parasite control, mφs were infected with SYTO$^{®}$11-labeled *T. cruzi* and incubated in the presence or absence of anti-IL-1β antibody for 18 h (Fig.1E). The intracellular SYTO$^{®}$11 fluorescence (indicates parasite presence) was significantly increased in anti-IL-1β antibody - treated THP-1 mφs (Fig.1E.d&f). The qPCR estimation of parasite burden confirmed the microscopic findings and showed a 2-fold increase in *Tc*18SrDNA signal in infected cells treated with anti-IL-1β antibody (Fig.1F). Antibody efficacy in depleting secreted IL-1β is shown in Fig.1G. Together, these data suggested that *a*) mφs respond to *T. cruzi* infection, in comparison to LPS treatment, by a subdued IL-1β release; b) IL-1β release can be enhanced by ATP at an early time-point pi; and c) IL-1β is required for controlling intracellular *T. cruzi*.

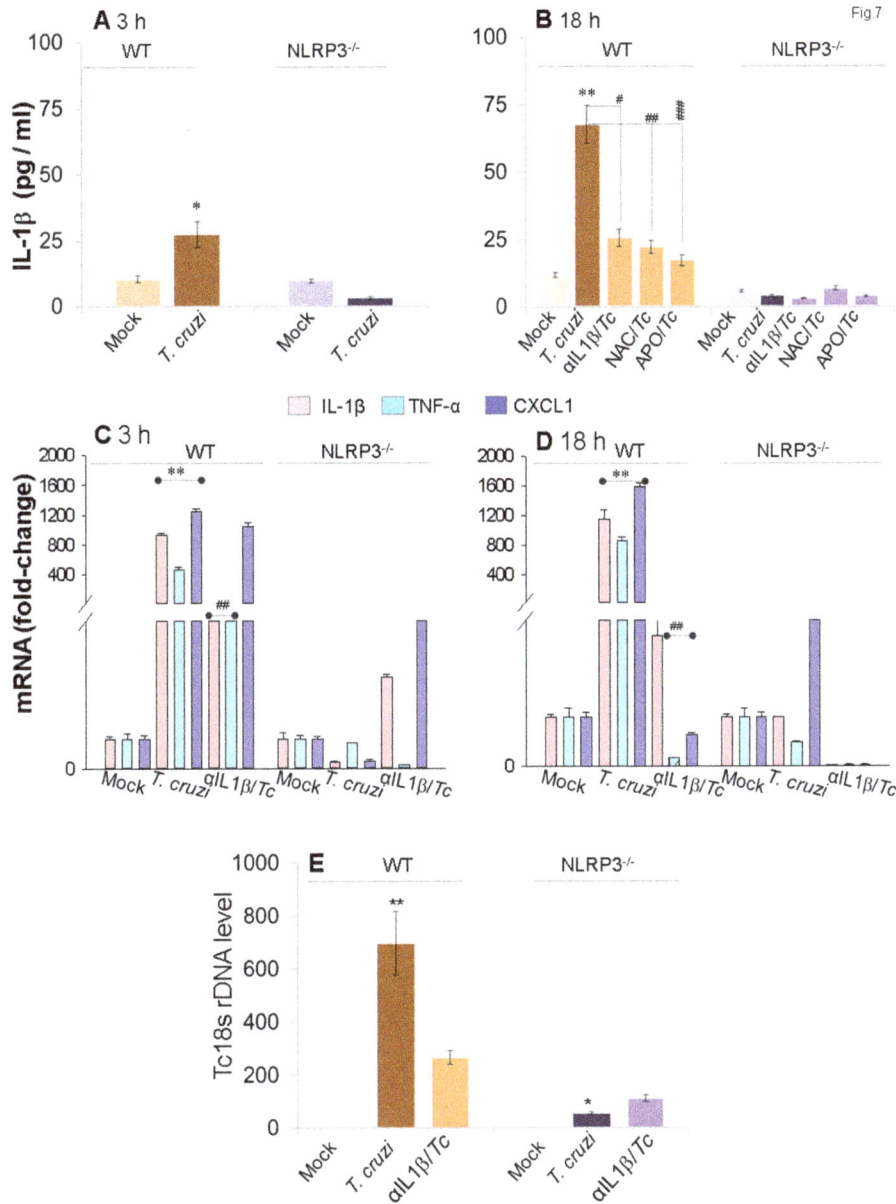

**Figure 7. NLRP3^-/- mφs are compromised in IL-1β activation and inflammatory cytokine gene expression, but equipped to control T. cruzi.** Bone marrow-derived macrophages were isolated from matched WT and NLRP3^-/- mice. Primary mφs were infected with T. cruzi and incubated for 3 h or 18 h in the presence or absence of anti-IL-1β antibody or ROS scavengers (as in Figs.1&4). Shown are IL-1β release by ELISA (A&B), mRNA level for IL-1β, TNF-α and CXCL1 by quantitative RT-PCR (C&D) and Tc18SrDNA signal by qPCR (E).

## Inflammasome-related gene expression in mφs infected by T. cruzi

We investigated the mRNA expression levels of various genes involved in the function of inflammasomes and NOD-like receptor (NLR) signaling. THP-1 mφs were incubated with T. cruzi or LPS for 3 h and 18 h (± ATP treatment) and probed for the expression of 95 genes (including house-keeping genes) utilizing the RT^2 Profiler Inflammasome PCR Arrays. The differential mRNA level was captured by qRT-PCR, and HTqPCR software was employed to attain the statistically significant differential expression in Tc-infected versus control mφs (± ATP) and LPS-treated versus control mφs at 3 h and 18 h (Table 1 & Table S2). Venn

diagrams of comparative analysis of gene expression in Tc-infected and LPS-treated mφs at 3 h and 18 h (± ATP) are shown in Fig.2. Of the 95 genes that were examined, 63 genes exhibited differential expression in one of the studied groups (p<0.05). We noted the differential expression of 18 (8 up-regulated and 10 down-regulated) and 29 (9 up-regulated and 20 down-regulated) genes in Tc-infected mφs at 3 h and 18 h, respectively, in comparison to that noted in controls (Fig.2A, Table 1). When Tc-infected mφs were treated with exogenous ATP during the last 30 min of incubation, 22 (15 up-regulated and 7 down-regulated) and 35 (11 up-regulated and 24 down-regulated) of the inflammasome-related genes were differentially expressed at 3 h and 18 h pi, respectively (Fig. 2B & C, Table 1). Only seven and

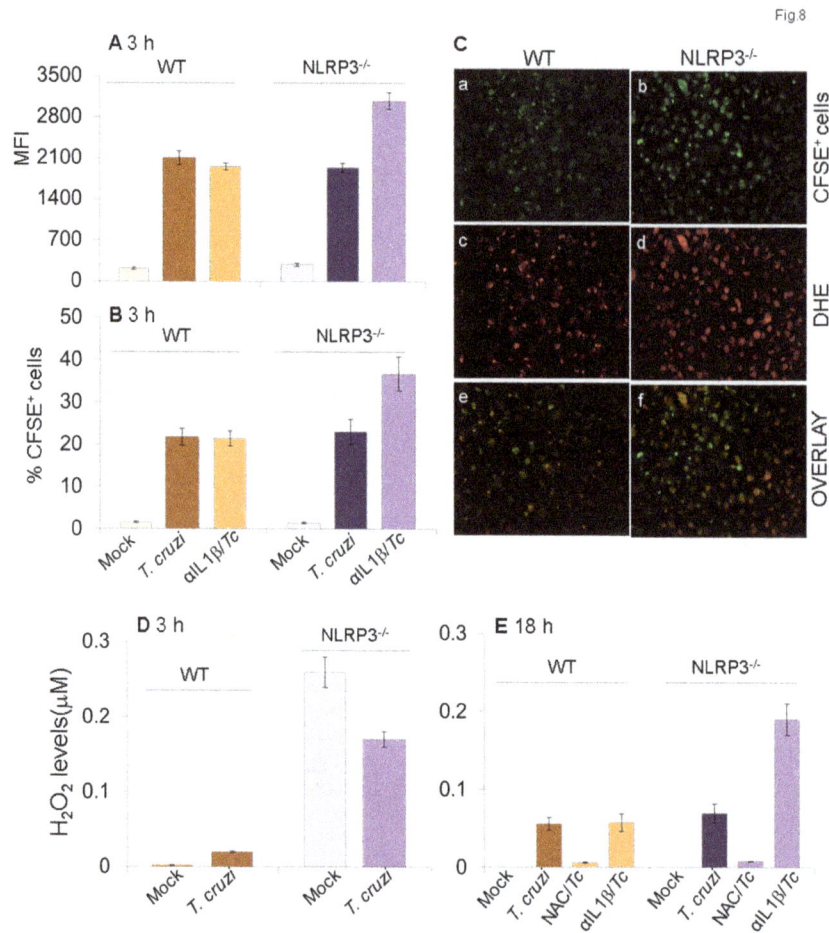

**Figure 8. NLRP3 deficiency is compensated for by increased ROS levels in mφs (± *T. cruzi*).** Bone marrow-derived primary macrophages isolated from matched WT and NLRP3⁻/⁻ mice were infected with CFSE-labeled *T. cruzi* and incubated for 3 h or 18 h in the presence or absence of anti-IL-1β antibody or ROS inhibitors. Shown are the mean fluorescence intensity of CFSE *(A)* as an indicator of # parasites/cell and mean percentage of CFSE⁺ mφs *(B)* as an indicator of parasite uptake efficiency. *(C)* Fluorescence microscopy of NLRP3⁻/⁻ (panels a, c, e) and WT (panels b, d, e) mφs infected with CFSE-labeled *T. cruzi* for 18 h. Shown are representative images of CFSE (green, panels a & b), intracellular ROS-specific DHE fluorescence (panels c & d) and overlay images of a & c and b & d in panels e & f. *(D&E)* Bar graphs show a quantitative measure of ROS release, measured by an Amplex red assay, in NLRP3⁻/⁻ and WT mφs.

eight genes were uniquely expressed in *Tc*/ATP-treated mφs when compared to *Tc*-infected mφs at 3 h and 18 h, respectively. However, other genes, e.g., BCL2, PSTPIP1, and CIITA at 3 h and NAIP1 at 18 h, that were down-regulated by *T. cruzi* infection, were up regulated when ATP was provided exogenously (Fig.2B&C, Table 1). In LPS-treated mφs, we noted differential expression of 31 (12 up-regulated and 19 down-regulated) and 27 (11 up-regulated and 16 down-regulated) of the inflammasome-related genes at 3 h and 18 h pi, respectively, and the LPS-induced gene expression profile was not changed by exogenous addition of ATP (Fig.2D&E, Table S2). These data suggest that very few of the inflammasome-related genes are up regulated in response to *T. cruzi* infection. Exogenous ATP was effective in enhancing the inflammasome-related gene expression at 3 h, but not at 18 h pi. In comparison, LPS served as a potent activator, as evidenced by a significant up regulation of inflammasome-related gene expression within 3 h post-treatment. These data support the results presented in Fig.1 and suggest that *T. cruzi* is a silent invader that elicits low level of inflammatory response in macrophages.

## Functional characterization of differentially expressed genes in *Tc*-infected cells

The top biological and molecular functional analysis of the differentially expressed gene datasets in *Tc*-infected mφs by Ingenuity iReport is presented in Table 2. In infected THP-1 mφs at 3 h pi (± ATP), 13–16 of the differentially expressed genes were indicative of an increase in inflammatory responses and control of macrophages' cell death ($p < 0.001$, z score: $-1.195$ to $-2.333$, $p < 0.001$). Exogenous addition of ATP had no significant effect on the top biological functions altered in response to *T. cruzi* infection. Toxicity analysis indicated that changes in mitoChondrial membrane potential that would likely result in ROS generation would be a key event in inflammatory and cell remodeling/cell death responses ($p < 0.01$). By 18 h pi (± ATP), 14–21 of the differentially expressed genes were involved in inflammation, activation of leukoCytes, and/or migration of phagoCytes (z score: 1.78 to 2.6) and 24–26 of the differentially expressed genes implicated in cell death were down-regulated (z score: $-1.23$ to $-1.25$). The top canonical pathway, toxicity, and upstream regulators analysis suggested the macrophage's attempt to balance inflammation with cellular protection via PPAR-α signaling (p<0.001).

## Caspase-1-mediated IL-1β activation is ROS-dependent, and plays a role in control of *T. cruzi* replication in wild-type mφs

To determine if inflammasomes play a role in IL-1β activation in mφs infected by *T. cruzi*, we pre-treated the THP-1 mφs with selective inhibitors of inflammasome activation for 2 h and continued the inhibition pressure during the infection period. The IL-1β release induced at 3 h and 18 h pi was decreased by 100% and 75–95%, respectively, when mφs were treated with cycloheximide (inhibits protein synthesis and thus the inducible arm of the inflammasome pathway), glibenclamide (bloCks the caspase-1 and pro-IL-1β maturation), Ac-YVAD-CHO (inhibits caspase-1 activity), or KCl (inhibits K$^+$ efflux required for caspase 1 activation) (Fig.3A&B). Importantly, glibenclamide and KCl also bloCked the mφs' ability to control parasite replication, resulting in a 50–70% increase in intracellular Tc18SrDNA, as determined by sensitive qPCR (Fig.3C).

Ingenuity iReport analysis of differentially expressed genes in this study (Table 2) and our previously published reports [22] have led us to suggest that infection by *T. cruzi* would elicit ROS by changes in mitoChondrial MPT or NOX2 activation in mφs, and that ROS may serve as a 2$^{nd}$ signal for inflammasome activation [11,23]. We, therefore, determined whether ROS is induced and plays a direct role in IL-1β production in infected mφs. THP-1 mφs exhibited 3.4-fold and 2.7-fold increases in DCF fluorescence (detects intracellular ROS) at 3 h and 18 h pi, respectively (Fig.4A). When NOX2 inhibitors (DPI or apoCynin) were added during incubation with *T. cruzi*, we noted a 21–44% and 68–80% decline in ROS (Fig.4A&B) and 37% and 45% decline in IL-β release (Fig.4C&D) at 3 h and 18 h pi, respectively. These data suggested that NOX2, at least partially, regulates ROS-dependent IL-1β activation in infected mφs. The observation of a moderate, but a significant (35–45%, p<0.05) decline in DCF fluorescence in infected mφs treated with anti-IL-1β antibody (Fig.4E&F) implied that IL-1β also contributes to activation of ROS production. However, we found no increase in intracellular parasite burden in THP-1 mφs treated with NOX2 inhibitors (Fig.4G). Together, the data presented in Figs.3&4 suggest that a feed-back cycle of caspase-1/IL-1β and ROS activation oCcurs in response to *T. cruzi* infection in THP-1 mφs and is required for control of intracellular parasite replication. While direct inhibition of IL-1β (Fig.1F) or caspase-1 (Fig.3C) affected the WT mφs ability to control *T. cruzi*, inhibition of ROS-dependent IL-1β was compensated for, likely by activation of other immune defenses capable of controlling the intracellular parasite replication and survival (Fig.4G).

## IL-1β signaling of NFκB in *Tc*-infected mφs

Because gene expression analysis has identified several of the signaling (e.g. TRAF6, MYD88, NFKBA) and cytokine (e.g. CCL2, CCl5, CXCL1, TNF) molecules involved in inflammatory responses that were activated in *Tc*-infected mφs (Tables 1 & 2), we determined if IL-1β signals the activation of the nuclear factor κB (NF-κB) pathway of cytokine gene expression in mφs. The mRNA levels for IL-1β, TNF-α and CXCL1 in THP-1 mφs were increased by 1.6 fold, 2-fold, and 2.3-fold, respectively, at 3 h pi. At 18 h pi, IL-1β and CXCL1 were increased by 1.6-fold and 5-fold, respectively, while no increase was observed in TNF-α mRNA level (Fig.5A.a–c). Treatment of infected mφs with anti-IL-1β antibody abolished the increase in IL-1β mRNA in infected

mφs; possibly indicating that the IL-1β engagement of surface receptors induced the IL-1β mRNA expression.

We performed a luciferase reporter assay to verify the role of IL-1β in signaling cytokine gene expression via NFκB in *Tc*-infected mφs. RAW mφs transiently transfected with luciferase reporter plasmid pNF-κB-luc with 3×NF-κB binding site and pREP7-Rluc (expresses renilla luciferase) were infected with *T. cruzi* for 18 h, and NF-κB-dependent luciferase activity (normalized to renilla luciferase) was monitored. We observed a >5-fold and 15-fold increase in luciferase activity, respectively, when mφs were infected with *T. cruzi* or treated with recombinant TNF-α (Fig.5B). Treatment with anti-IL-1β antibody resulted in 70% decline in *Tc*-induced luciferase activity (Fig.5B, p<0.01). Together, the data presented in Fig.5 suggested that IL-1β signals the NF-κB activation of cytokine gene expression in mφs infected by *T. cruzi*.

## ASC$^{-/-}$ mφs were compromised in ROS-dependent IL-1β activation and NF-κB-dependent cytokine gene expression, and exhibited pronounced *T. cruzi* replication and survival

To determine if ASC is involved in caspase-1-dependent inflammasome formation and activation of inflammatory proCesses for parasite control, we utilized primary BM-derived mφs from ASC$^{-/-}$ mice and matched controls. The ASC$^{-/-}$ mφs infected by *T. cruzi* exhibited a significantly compromised IL-1β release (67% and 40% decline at 3 h and 18 h, respectively) when compared to that noted in matched WT controls (Fig.6A&B, p<0.01). Treatment with anti-IL-1β antibody and ROS scavengers (NAC or apoCynin) normalized the IL-1β production in *Tc*-infected ASC$^{-/-}$ and WT mφs to control levels. Further, the expression of NF-κB-inducible pro-IL-1β and TNF-α mRNAs, that were significantly increased in *Tc*-infected WT mφs, were completely abolished in *Tc*-infected ASC$^{-/-}$ mφs (Fig.6C&D). Subsequently, ASC$^{-/-}$ mφs exhibited a 3.2-fold and 2-fold increase in intracellular parasite burden at 3 h and 18 h pi, respectively, when compared to that noted in matched WT infected mφs (Fig.6E&F). As previously observed in THP-1 mφs (Fig.4G), ASC$^{-/-}$ mφs exhibited no further increase in parasite burden upon treatment with ROS scavengers (Fig.6). These data confirmed that the caspase-1/ASC inflammasomes play an important role in parasite control through ROS-dependent IL-1β activation and expression of other inflammatory cytokines in mφs.

## NLRP3$^{-/-}$ mφs exhibited increased ROS-dependent control of *T. cruzi*

To specifically determine if NLRP3 inflammasome activation by ASC/caspase-1 is the primary source of protective IL-1β in the context of *Tc* infection, we conducted further studies in primary bone marrow-derived mφs from NLRP3$^{-/-}$ mice and matched controls. As expected from studies in ASC$^{-/-}$ cells, NLRP3$^{-/-}$ mφs lacked the ability to elicit IL-1β activation in response to *T. cruzi* infection (Fig.7A&B, p<0.01). Treatment with anti-IL-1β antibody and ROS scavengers (NAC or apoCynin) normalized the IL-1β production in *Tc*-infected WT mφs to control levels (Fig.7B). Further, NF-κB-inducible pro-IL-1β, TNF-α and CXCL1 mRNA levels that were increased in *Tc*-infected WT mφs, were almost absent in *Tc*-infected NLRP3$^{-/-}$ mφs (Fig.7C&D). However, to our surprise, NLRP3$^{-/-}$ mφs exhibited a 12-fold decline in intracellular parasite burden at 18 h, when compared to that noted in matched WT infected mφs (Fig.7E).

To determine if NLRP3$^{-/-}$ mφs were simply refractory to *T. cruzi*, we incubated the NLRP3$^{-/-}$ and matched WT mφs with

CFSE-labeled trypomastigotes for 3 h and analyzed for CFSE fluorescence by flow cytometry. We observed no statistically significant difference in intracellular mean fluorescence intensity (MFI) indicative of the number of parasites/cell (Fig.8A) and the percentage of $CFSE^+$ cells indicative of the number of $Tc$-infected $NLRP3^{-/-}$ and WT mφs (Fig.8B). Representative micrographs showing efficient uptake of CFSE-labeled parasites in $NLRP3^{-/-}$ and WT primary mφs are presented in Fig.8C (panels a&b). These data suggested that $NLRP3^{-/-}$ mφs were not refractory to $T. cruzi$ and efficiently phagoCytized parasites in a manner similar to that of the WT mφs. Instead, $NLRP3^{-/-}$ mφs, in comparison to WT mφs, exhibited 8.5-fold higher basal ROS levels (Fig.8D). The ROS production in $NLRP3^{-/-}$ mφs was enhanced upon $T. cruzi$ infection and not inhibited by the addition of anti-IL-1β antibody (Fig.8D&E). The increase in ROS production in $NLRP3^{-/-}$ mφs in response to $T. cruzi$ infection was also evidenced by a significant increase in DHE fluorescence (red, detects intracellular ROS, Fig.8C.c–f). Together, the data presented in Figs.7&8 suggested that NLRP3/ASC/caspase-1 inflammasome mediates IL-1β activation and expression of other inflammatory cytokines in mφs infected by $T. cruzi$; however, NLRP3 deficiency is compensated for by increased ROS levels capable of preventing parasite replication and intracellular survival in $NLRP3^{-/-}$ mφs.

## Discussion

The innate immune response to $T. cruzi$ infection is mediated by diverse PRRs, including receptors of the TLR family [24,25] and ASC-containing inflammasomes (e.g. NOD1 [26], and NLRP3 [27,28]). Genetically modified mice deficient in MYD88 (interacts with TLR-2, -4, -6), TLR-4, NOD1, and ASC exhibited an increased susceptibility to $T. cruzi$, thus pointing to these PRRs as critical determinants of host resistance to $T. cruzi$ infection [24,26,28,29]. In this manuscript, we have utilized cultured and primary mφs and employed inhibitory approaches to investigate the mechanisms of caspase-1/ASC inflammasome activation and their role in the context of $T. cruzi$ infection. We found that mφs respond to $T. cruzi$ infection with suboptimal activation of inflammasome-related gene expression and IL-1β production. Functional analysis of the differentially expressed gene datasets in $Tc$-infected mφs was indicative of an increase in inflammatory responses and control of macrophages' cell death. The IL-1β production in $Tc$-infected mφs was ROS-dependent and could be enhanced by the exogenous addition of ATP. Studies in WT mφs treated with specific inhibitors and $ASC^{-/-}$ mφs suggested that caspase-1/ASC inflammasome played a role in activation of the IL-1β–ROS–NF-κB pathway, that when inhibited, resulted in a compromised inflammatory cytokine response and increase in $T. cruzi$ replication and survival in macrophages. However, $NLRP3^{-/-}$ mφs, despite an inability to elicit IL-1β activation and inflammatory cytokine gene expression, were capable of parasite control. Thus, our data allow us to conclude that caspase-1/ASC inflammasomes play a significant role in the activation of IL-1β/ROS and NF-κB signaling of cytokine gene expression for $T. cruzi$ control in human and mouse mφs. However, NLRP3 balances the mφs' activation of ROS and NFκB/cytokine response, and its deficiency shifted the mφs' responses towards increased ROS-dependent control of $T. cruzi$. To the best of our knowledge, this is the first study demonstrating a double-edged role of NLRP3 in determining macrophage activation of ROS and cytokine response, both of which are required for clearance of $T. cruzi$ infection.

The NLRs are expressed in most cell types of the immune system, but are also reported to be expressed in other tissues. Based upon the expression profile of components of NLRP1, NLRP3, and NLRC1; blood, placenta and thymus are shown to constitutively express inflammasomes. Other tissues (e.g., heart, vascular tissue, bone marrow) require up regulation of one or two components in order to assemble functional inflammasomes [30]. Our data demonstrated that inflammasome-related gene expression is induced in mφs exposed to $T. cruzi$ infection (Table 1) or LPS treatment (Table S2), as well as in non-phagoCytic, human vascular smooth muscle, fibroblast, and microvascular endothelial cells (Table S3). Functional analysis of the gene expression profile indicated that in mφs, an inflammatory response to control $T. cruzi$ infection was assoCiated with significant efforts to prevent cell death (Table 2). Infected mφs also exhibited suppression of several of the genes involved in PPARα signaling that induces apoptosis following activation with TNF-α/IFN-γ [31,32]. In non-phagoCytic cells, TLR/MYD88 signaling of NFκB-dependent cytokines' (IL-6, IFNα/β, IL-12) gene expression and caspase-1/NLRP3-mediated activation of IL-1β gene expression was noted. Canonical analysis, as in mφs, also indicated PPAR-α regulation of gene expression in HMVEC and HFF cells infected by $T. cruzi$ (Table S4). Our observations allow us to surmise that phagoCytic and non-phagoCytic cells responded to $T. cruzi$ infection by induction of diverse inflammasome-related gene expression. Eventually, all cell types appeared to be overwhelmed by the inflammatory signal and tended to switch to signaling events related to prevention of cell death.

Several studies indicate that ROS are an essential secondary messenger for signaling caspase-1/ASC inflammasome activation [11,33]. The use of ROS scavengers controlled IL-1β activation by virtually all agonists of the NLRP3 inflammasome [33,34]. Besides ROS, extracellular ATP, through activation of the P2X7 (purogenic ionotrophic ATP-gated cation channel), triggers $K^+$ efflux, which, in turn, triggers pore formation by pannexin, thereby allowing the delivery of pathogen products into the cytosol, resulting in caspase-1/ASC inflammasome activation [35]. In agreement with the literature, we also observed ROS-dependent IL-1β activation mediated by caspase-1/ASC inflammasome in mφs exposed to $T. cruzi$ (Figs.1–4). The exogenous addition of ATP resulted in a higher level of IL-1β activation in infected mφs at 3 h pi, as also noted by others [28]. The lack of an effect of exogenous ATP on the extent of IL-1β activation in infected mφs at 18 h pi may mean that other signals were generated. We speculate that either $Tc$ kDNA released by dying $T. cruzi$ within mφs or mtDNA released due to bystander damage in infected mφs served as a secondary signal in activating caspase-1/ASC inflammasomes at 18 h pi. This notion is supported by the observation of similar levels of IL-1β activation in mφs incubated with live as well as dead $T. cruzi$ and requires further investigation.

Importantly, we made a novel observation and noted a feedback cycle of IL-1β signaling of ROS activation in infected mφs (Figs.3&4). The molecular mechanism for the ROS production by IL-1β remains to be elucidated. In previous studies, IL-1β was found to stimulate phospholipase $A_2$, promoting release of arachidonic acid. Since arachidonic acid can activate NAPDH oxidase to produce superoxide, it is possible that this fatty acid may serve as an intermediate in the IL-1β-induced activation of enzymes, leading to the production of ROS [36,37]. However, several non-NADPH oxidase-dependent sources, including mito-Chondrial electron transport and arachidonate metabolism, may also be involved in the cytokine-induced ROS generation, to be investigated in future studies.

It is intriguing that both $ASC^{-/-}$ and $NLRP3^{-/-}$ mφs were equally invaded by $T. cruzi$, lacked IL-1β production, and elicited ineffective, NF-κB-mediated cytokine gene expression (Figs.6–8);

yet only ASC$^{-/-}$ mφs were restrictive in their capacity to control *T. cruzi* infection (Fig. 6). ASC$^{-/-}$ and caspase-1$^{-/-}$ mice have been doCumented to exhibit a higher incidence of mortality, cardiac parasitism and heart inflammation, meaning that ASC/caspase-1 inflammasomes are critical determinants of host resistance to infection with *T. cruzi* [28]. Like ASC$^{-/-}$ mice and *in vitro* cultured ASC$^{-/-}$ mφs, NOD1$^{-/-}$ mice and BM-derived derived mφs from NOD1$^{-/-}$ mice also showed an impaired induction of NF-κB-dependent products and failed to restrict *T. cruzi* infection [26]. Our finding that a deficiency of NLRP3 did not affect mφs ability to control parasites allows us to surmise that the caspase-1/ASC requirement for effective control of *T. cruzi* is delivered via formation and activation of inflammasomes with other NLRs, e.g., NLRP1, AIM2 or NLRC4. The ability of NLRP3$^{-/-}$ mφs to efficiently manage parasite killing via enhanced ROS release (Figs.7&8) suggest that NLRP3 suppresses NOX2-dependent ROS production in mφs. Though it is shown that activation of NLRP3 containing inflammasome is not dependent on the function of NOX1-4 [38], our data provide the first evidence that NLRP3 dysregulates the NOX2 function at the superoxide production level and suppresses the mφs' ability to kill *T. cruzi*. Caspase-l, upon activation by NLRP3/ASC, loCalizes to phagosomes and disturbs NOX2 control of pH, thereby triggering acidification and microbicidal activity of phagosomes in mφs infected by *StaphyloCoCcus aureus* [39]. We surmise that caspase-1 loCalization to phagosome and suppression of NOX2/ROS enhances the bactericidal activity of mφs. However, NLRP3 interaction with NOX2 resulting in low level of ROS production likely maintains the alkalization of the phagosomal lumen that prevents the mφs' ability to directly kill *T. cruzi*, but plays a critical role in allowing mφs to function as specialized phagoCytes adapted to proCess antigens for cross presentation and elicitation of adaptive immunity.

Other investigators have shown a severe defect in nitric oxide (NO) production and impairment in mφ-mediated *T. cruzi* killing in NLRP3$^{-/-}$ mice and isolated mφs [27]. In this study, we utilized the BM-derived monoCytes from NLRP3$^{-/-}$ mice that were differentiated to mφs by M-CSF treatment, and infected with *T. cruzi* (SylvioX10) for 3 h and 18 h. Others utilized peritoneal mφs obtained 4 days after intra-peritoneal injection of 1% starch solution in NLRP3$^{-/-}$ mice and infected these mφs with the Y strain of *T. cruzi* for 48 h for all the studies [27]. We propose that NLRP3 deficiency is compensated for by over activation of NOX2/ROS that effectively controlled the early invasion and replication of *T. cruzi* in mφs, as observed in this study. However,

a lack of NF-κB-mediated cytokine response (Fig.7) and iNOS/NO activation [27] prevented the NLRP3$^{-/-}$ mφs from sustained, long-term control of the parasites, resulting in increased susceptibility. It is also plausible that the differences in the source of mφs, parasite isolates, and the time course of infection may explain the observed differences in the ability of NLRP3$^{-/-}$ mφs to control *T. cruzi* infection in our and published studies.

In summary, we have demonstrated that *T. cruzi* interfered with the potent activation of caspase-1/ASC inflammasome-related gene expression and cytokine response in mφs to ensure its survival. We found that caspase-1/ASC inflammasomes played a role in the activation of the IL-1β–ROS–NF-κB pathway, that when inhibited, resulted in an increase in *T. cruzi* replication and survival in mφs. However, NLRP3$^{-/-}$ mφs were compensated for by increased NOX2/ROS activation capable of parasite killing. Our data suggest that the NLRP3/caspase/ASC inflammasome balances the mφs' activation of ROS and NFκB/cytokine response and provide the first evidence for NLRP3 regulation of NOX2 function as an effector mechanism contributing to parasite persistence.

## Supporting Information

**Table S1    Oligonucleotides used in this study.**

**Table S2    Inflammasome-related differential gene expression in THP-1 macrophages in response to LPS treatment (± ATP) in comparison to normal controls.**

**Table S3    Inflammasome-related differential gene expression in non-phagoCytes at 24 h infection by *Trypanosoma cruzi*.**

**Table S4    Ingenuity iReport analysis of inflammasome-related datasets in non-phagoCytes infected by *T. cruzi*.**

## Author Contributions

Conceived and designed the experiments: NJG JE ND. Performed the experiments: ND MS SG MNG. Analyzed the data: NJG MS SG MNG BAL. Contributed reagents/materials/analysis tools: NJG ND BAL RF. Wrote the paper: NJG MS BAL ND.

## References

1. World Health Organization (2010) Chagas disease: control and elimination. UNDP/World Bank/WHO. Available: http://apps.who.int/gb/ebwha/pdf_files/WHA63/A63_17-en.pdf. Accessed 2014 Sept 3
2. Machado FS, Dutra WO, Esper L, Gollob KJ, Teixeira MM, et al. (2012) Current understanding of immunity to *Trypanosoma cruzi* infection and pathogenesis of Chagas disease. Seminars in Immunopathology 34: 753–770.
3. Huang H, Chan J, Wittner M, Jelicks LA, Morris SA, et al. (1999) Expression of cardiac cytokines and inducible form of nitric oxide synthase (NOS2) in *Trypanosoma cruzi*-infected mice. J Mol Cell Cardiol 31: 75–88.
4. Machado FS, Tyler KM, Brant F, Esper L, Teixeira MM, et al. (2012) Pathogenesis of Chagas disease: time to move on. Front Biosci (Elite Ed) 4: 1743–1758.
5. Machado FS, Souto JT, Rossi MA, Esper L, Tanowitz HB, et al. (2008) Nitric oxide synthase-2 modulates chemokine production by *Trypanosoma cruzi*-infected cardiac myocytes. Microbes Infect 10: 1558–1566.
6. Coelho PS, Klein A, Talvani A, Coutinho SF, Takeuchi O, et al. (2002) Glycosylphosphatidylinositol-anchored mucin-like glycoproteins isolated from *Trypanosoma cruzi* trypomastigotes induce in vivo leukocyte recruitment dependent on MCP-1 production by IFN-gamma-primed-macrophages. J Leukoc Biol 71: 837–844.
7. Kayama H, Takeda K (2010) The innate immune response to *Trypanosoma cruzi* infection. Microbes Infect 12: 511–517.
8. Monteiro AC, Schmitz V, Morrot A, de Arruda LB, Nagajyothi F, et al. (2007) Bradykinin B2 Receptors of dendritic cells, acting as sensors of kinins proteolytically released by *Trypanosoma cruzi*, are critical for the development of protective type-1 responses. PLoS Pathog 3: e185.
9. Schmitz V, Svensjo E, Serra RR, Teixeira MM, Scharfstein J (2009) Proteolytic generation of kinins in tissues infected by *Trypanosoma cruzi* depends on CXC chemokine secretion by macrophages activated via Toll-like 2 receptors. J Leukoc Biol 85: 1005–1014.
10. Kanneganti TD (2010) Central roles of NLRs and inflammasomes in viral infection. Nat Rev Immunol 10: 688–698.
11. Garg NJ (2011) Inflammasomes in cardiovascular diseases. Am J Cardiovas Dis 1: 244–254.
12. Ting JP, Lovering RC, Alnemri ES, Bertin J, Boss JM, et al. (2008) The NLR gene family: a standard nomenclature. Immunity 28: 285–287.
13. van de Veerdonk FL, Netea MG, Dinarello CA, Joosten LA (2011) Inflammasome activation and IL-1beta and IL-18 proCessing during infection. Trends Immunol 32: 110–116.
14. Foldes G, Liu A, Badiger R, Paul-Clark M, Moreno L, et al. (2010) Innate immunity in human embryonic stem cells: comparison with adult human endothelial cells. PLoS One 5: e10501.
15. Tousoulis D, Andreou I, Antoniades C, Tentolouris C, Stefanadis C (2008) Role of inflammation and oxidative stress in endothelial progenitor cell function and

mobilization: therapeutic implications for cardiovascular diseases. Atherosclerosis 201: 236–247.

16. Schultz K, Murthy V, Tatro JB, Beasley D (2007) Endogenous interleukin-1 alpha promotes a proliferative and proinflammatory phenotype in human vascular smooth muscle cells. Am J Physiol Heart Circ Physiol 292: H2927–2934.

17. Hirao K, Yumoto H, Takahashi K, Mukai K, Nakanishi T, et al. (2009) Roles of TLR2, TLR4, NOD2, and NOD1 in pulp fibroblasts. J Dent Res 88: 762–767.

18. Boyd JH, Mathur S, Wang Y, Bateman RM, Walley KR (2006) Toll-like receptor stimulation in cardiomyoCtes decreases contractility and initiates an NF-kappaB dependent inflammatory response. Cardiovasc Res 72: 384–393.

19. Dvinge H, Bertone P (2009) HTqPCR: high-throughput analysis and visualization of quantitative real-time PCR data in R. Bioinformatics 25: 3325–3326.

20. Costales JA, Daily JP, Burleigh BA (2009) Cytokine-dependent and-independent gene expression changes and cell cycle bloCk revealed in Trypanosoma cruzi-infected host cells by comparative mRNA profiling. BMC Genomics 10: 252.

21. Wen JJ, Garg NJ (2008) MitoChondrial generation of reactive oxygen species is enhanced at the Q(o) site of the complex III in the myoCardium of Trypanosoma cruzi-infected mice: beneficial effects of an antioxidant. J Bioenerg Biomembr 40: 587–598.

22. Dhiman M, Garg NJ (2011) NADPH oxidase inhibition ameliorates Trypanosoma cruzi-induced myoCarditis during Chagas disease. J Pathol 225: 583–596.

23. Liu W, Yin Y, Zhou Z, He M, Dai Y (2014) OxLDL-induced IL-1 beta secretion promoting foam cells formation was mainly via CD36 mediated ROS production leading to NLRP3 inflammasome activation. Inflamm Res 63: 33–43.

24. Caetano BC, Carmo BB, Melo MB, Cerny A, dos Santos SL, et al. (2011) Requirement of UNC93B1 reveals a critical role for TLR7 in host resistance to primary infection with Trypanosoma cruzi. J Immunol 187: 1903–1911.

25. Bafica A, Santiago HC, Goldszmid R, Ropert C, Gazzinelli RT, et al. (2006) Cutting edge: TLR9 and TLR2 signaling together account for MyD88-dependent control of parasitemia in Trypanosoma cruzi infection. J Immunol 177: 3515–3519.

26. Silva GK, Gutierrez FR, Guedes PM, Horta CV, Cunha LD, et al. (2010) Cutting edge: nucleotide-binding oligomerization domain 1-dependent responses account for murine resistance against Trypanosoma cruzi infection. J Immunol 184: 1148–1152.

27. Goncalves VM, Matteucci KC, Buzzo CL, Miollo BH, Ferrante D, et al. (2013) NLRP3 controls Trypanosoma cruzi infection through a caspase-1-dependent IL-1R-independent NO production. PLoS Negl Trop Dis 7: e2469.

28. Silva GK, Costa RS, Silveira TN, Caetano BC, Horta CV, et al. (2013) Apoptosis-assoCiated speck-like protein containing a caspase recruitment domain inflammasomes mediate IL-1beta response and host resistance to Trypanosoma cruzi infection. J Immunol 191: 3373–3383.

29. Oliveira AC, de Alencar BC, Tzelepis F, Klezewsky W, da Silva RN, et al. (2010) Impaired innate immunity in Tlr4(-/-) mice but preserved CD8+ T cell responses against Trypanosoma cruzi in Tlr4-, Tlr2-, Tlr9- or Myd88-deficient mice. PLoS Pathog 6: e1000870.

30. Yin Y, Yan Y, Jiang X, Mai J, Chen NC, et al. (2009) Inflammasomes are differentially expressed in cardiovascular and other tissues. Int J Immunopathol Pharmacol 22: 311–322.

31. Chawla A (2010) Control of macrophage activation and function by PPARs. Circ Res 106: 1559–1569.

32. Chawla A, Nguyen KD, Goh YP (2011) Macrophage-mediated inflammation in metabolic disease. Nat Rev Immunol 11: 738–749.

33. Bauernfeind F, Bartok E, Rieger A, Franchi L, Nunez G, et al. (2011) Cutting Edge: Reactive Oxygen Species Inhibitors BloCk Priming, but Not Activation, of the NLRP3 Inflammasome. J Immunol 187: 613–617.

34. Pan Q, Mathison J, Fearns C, Kravchenko VV, Da Silva Correia J, et al. (2007) MDP-induced interleukin-1beta proCessing requires Nod2 and CIAS1/NALP3. J LeukoC Biol 82: 177–183.

35. Kanneganti TD, Lamkanfi M, Kim YG, Chen G, Park JH, et al. (2007) Pannexin-1-mediated recognition of bacterial molecules activates the cryopyrin inflammasome independent of Toll-like receptor signaling. Immunity 26: 433–443.

36. Hwang YS, Jeong M, Park JS, Kim MH, Lee DB, et al. (2004) Interleukin-1beta stimulates IL-8 expression through MAP kinase and ROS signaling in human gastric carcinoma cells. Oncogene 23: 6603–6611.

37. Lo YY, Conquer JA, Grinstein S, Cruz TF (1998) Interleukin-1 beta induction of c-fos and collagenase expression in articular chondroCytes: involvement of reactive oxygen species. J Cell BioChem 69: 19–29.

38. van Bruggen R, Koker MY, Jansen M, van Houdt M, Roos D, et al. (2010) Human NLRP3 inflammasome activation is Nox1-4 independent. Blood 115: 5398–5400.

39. Sokolovska A, Becker CE, Ip WK, Rathinam VA, Brudner M, et al. (2013) Activation of caspase-1 by the NLRP3 inflammasome regulates the NADPH oxidase NOX2 to control phagosome function. Nat Immunol 14: 543–553.

# 19

# *Plasmodium falciparum* Transfected with Ultra Bright NanoLuc Luciferase Offers High Sensitivity Detection for the Screening of Growth and Cellular Trafficking Inhibitors

Mauro F. Azevedo[1,9], Catherine Q. Nie[1,9], Brendan Elsworth[1,2], Sarah C. Charnaud[1,2], Paul R. Sanders[1], Brendan S. Crabb[1,2,3], Paul R. Gilson[1,2]*

1 Macfarlane Burnet Institute of Medical Research and Public Health, Melbourne, Victoria, Australia, 2 Monash University, Melbourne, Australia, 3 University of Melbourne, Melbourne, Australia

## Abstract

Drug discovery is a key part of malaria control and eradication strategies, and could benefit from sensitive and affordable assays to quantify parasite growth and to help identify the targets of potential anti-malarial compounds. Bioluminescence, achieved through expression of exogenous luciferases, is a powerful tool that has been applied in studies of several aspects of parasite biology and high throughput growth assays. We have expressed the new reporter NanoLuc (Nluc) luciferase in *Plasmodium falciparum* and showed it is at least 100 times brighter than the commonly used firefly luciferase. Nluc brightness was explored as a means to achieve a growth assay with higher sensitivity and lower cost. In addition we attempted to develop other screening assays that may help interrogate libraries of inhibitory compounds for their mechanism of action. To this end parasites were engineered to express Nluc in the cytoplasm, the parasitophorous vacuole that surrounds the intraerythrocytic parasite or exported to the red blood cell cytosol. As proof-of-concept, these parasites were used to develop functional screening assays for quantifying the effects of Brefeldin A, an inhibitor of protein secretion, and Furosemide, an inhibitor of new permeation pathways used by parasites to acquire plasma nutrients.

Editor: Tobias Spielmann, Bernhard Nocht Institute for Tropical Medicine, Germany

**Funding:** This work was supported by the National Health and Medical Research Council Grants 637406, APP1068287 and APP1021560 awarded to BSC and PRG (URL: www.nhmrc.gov.au). The authors gratefully acknowledge the contribution to this work of the Victorian Operational Infrastructure Support Program received by the Burnet Institute. The funders had no role in study design, data collection and analysis, decision to publish, or preparation of the manuscript.

**Competing Interests:** The authors have declared that no competing interests exist.

* Email: gilson@burnet.edu.au

9 These authors contributed equally to this work.

## Introduction

Malaria is a life-threatening disease caused by five species of parasites belonging to the genus *Plasmodium*, and includes *P. falciparum*, *P. vivax*, *P. ovale*, *P. malariae* and *P. knowlesi*. Despite successful efforts to reduce mortality through transmission control and patient treatment, there are still over 200 million clinical cases per year and tragically about 600,000 deaths [1]. The appearance and spread of multidrug resistant strains of the most deadly species, *P. falciparum* threatens to revert the mortality and morbidity reductions achieved so far and future control and eradication strategies must include the discovery of new anti-malarial compounds.

Recently, extensive high throughput screens of proprietary and publicly available libraries have identified thousands of potent compounds that inhibit the growth of *P. falciparum* blood stages in *in vitro* culture [2–5]. There is now a pressing need to identify the targets of these compounds and the aspects of parasite biology that are inhibited by them. Although the structural family to which

some compounds belong may indicate their likely targets, for most compounds the targets are completely unknown. This is further exacerbated by the fact that even if the targets could be identified by complex and costly laboratory investigation, the functions of more than half of the parasite's proteins are still unknown. This therefore makes it very difficult to deduce what a compound's mechanism of action might be, making it problematic to evaluate if the compound is worth further development.

In the bloodstream, red blood cells (RBCs) are infected by the short-lived extracellular merozoite form of the malaria parasite. During invasion of the RBC, the malaria parasite envelops itself in the parasitophorous vacuole membrane (PVM). In order to grow rapidly and replicate, *Plasmodium* parasites extensively modify the RBC host by exporting hundreds of proteins across the PVM into the RBC cytosol and some then further out to the RBC plasma membrane [6–8]. While the function of most of these exported proteins is unknown, those that have been investigated appear to be associated with virulence related functions such as cytoadherence to vascular endothelium and strengthening of the host cell cytoskeleton [9,10]. Parasites also cause their host erythrocytes to become more porous so blood plasma nutrients can be acquired for rapid growth. At least one secreted parasite protein called RhopH1, has been shown to help establish these new permeation pathways (NPPs) but it is not clear if other parasite proteins, potentially exported into the host, also play a role [11]. Protein export is a unique aspect of the parasite's biology and is essential for their survival and contributes greatly to disease pathology. Recently a novel protein export machine called PTEX, has been identified that resides in the PVM [12] and appears to selectively transport proteins across the membrane into the RBC cytoplasm [13,14]. PTEX appears to constitute a single portal through which hundreds of exported proteins must pass. Therefore, inhibition of PTEX or of any complex required for protein export is likely to be highly effective against the parasite, as this will simultaneously disrupt the functions of 100 s of essential or virulence related exported proteins given they will not be able to reach their functional destinations.

So far, assays used to screen molecules active against *P. falciparum* parasites have mostly been based on measuring inhibition of overall growth of asexual blood stages, and recently of sexual stages [15,16]. Growth inhibition assays have the advantage of being able to test in a high throughput manner large chemical libraries to identify potential hits, but have the disadvantage of not allowing biological validation of the actual targets of candidate drugs. Conversely, specific assays such as the detection of protein export in *P. falciparum* have been established by expressing fluorescent reporters such as the green fluorescent protein (GFP) fused at its N-terminal with a previously character-ized protein export element (PEXEL) [17]. To use such approaches to screen for export inhibitors would be slow and laborious requiring sophisticated microscopes and highly trained personnel.

Luciferase (Luc) is the enzyme required for bioluminescence in a variety of organisms, and it has been adapted as a reporter in several cell types. The firefly and Renilla luciferases, from *Photinus pyralis* and *Renilla reniformes* respectively, have been expressed in *Plasmodium spp.* without any apparent toxicity [18,19]. This reporter has been used to study several aspects of parasite biology [20–22] and for growth/drug inhibition assays [4,16,23–27]. In addition, being an exogenous protein allows Luc to be expressed in fusion with other proteins, including trafficking signal motifs. Protein trafficking in *P. falciparum* has been investigated using Luc, but the inefficient fractionation technology available impaired conclusive results regarding protein export [28].

Here NanoLuc (Nluc), a new smaller and brighter luciferase from deep-sea shrimp [29], is evaluated as a reporter in *P. falciparum*. Parasites engineered to express cytosolic, secreted and exported forms of Nluc have been developed, and new affordable and simple assays to quantify protein secretion and export, NPP functions as well as parasite growth are described.

## Materials and Methods

### Plasmid Construction

Nluc coding sequence was retrieved from pNL1.1 (Promega) digested with *NcoI* and *XbaI* and cloned in the same restriction sites of pPf86 [19] to generate pPfNluc. The stable expression vector pEF-Nluc was made cloning Nluc in the *NcoI* and *SpeI* sites of pEF-Luc [30]. To express a version of Nluc that could be exported into the erythrocyte compartment, a synthetic gene (Genscript) encoding the first 113 amino acids of the *P. falciparum* exported protein hyp1 (PF3D7_0113300), was cloned in pEF-Nluc

via *XhoI* and *NcoI* sites, generating pEF-PEXEL-Nluc. The secreted Nluc was made cloning a synthetic gene encoding the first 23 amino acids of *P. falciparum* MSP1 (PF3D7_0930300) in *XhoI* and *NcoI* sites of pEF-Nluc, generating pEF-SP-Nluc.

### Parasite Culturing and Transfection

*P. falciparum* 3D7 parasites were cultured as per [31] in RPMI-HEPES media supplemented with L-glutamine (Sigma) and Albumax II (Invitrogen). Trophozoite stage parasites were transfected by feeding with erythrocytes electroporated with 100 μg of plasmid DNA [32,33]. For transient transfections, parasites were harvested 4 days later. For stable transfections, parasites were cultured on 2.5 nM WR99210 until a transfected population was established.

### Luciferase Assay

Unless otherwise stated all bioluminescence reagents are from Promega. Transiently transfected parasites were lysed with 0.1% saponin and washed twice in PBS to remove the hemoglobin. For firefly luciferase, parasite pellets were re-suspended in 20 μL of 1x Luciferase Cell Culture Lysis Reagent, mixed with 100 μL of Luciferase Assay Reagent, containing the Luciferase Assay Substrate Luciferin previously dissolved in Luciferase Assay Buffer, and immediately measured for 30 seconds in a FLUOstar Omega Luminometer (BMG Labtech) with the gain adjusted to maxi-mum. For Nluc, parasite pellets were re-suspended in 100 μL of PBS, mixed with 100 μL of 2x Nano-Glo Luciferase Assay Reagent and measured in the luminometer with the same gain and for the same time used for firefly luciferase. Nano-Glo Luciferase Assay Reagent was made by adding one volume of Nano-Glo Luciferase Assay Substrate to 50 volumes of Nano-Glo Luciferase Assay Buffer as recommended by the manufacturer. Cultures of stably transfected Nluc parasites, were diluted in RPMI to 0.1% hematocrit, mixed with 1 volume of Nano-Glo Luciferase Assay Reagent and measured for 2–10 seconds as indicated in each experiment in a luminometer with the gain reduced by 10% for the sample with the highest signal on the plate to prevent saturation.

### Chloroquine Sensitivity

Nluc ring stage parasites were incubated with chloroquine (CQ) concentrations varying from 52 pM to 4 mM in 5-fold increments. After 3 days, cultures were frozen and stored at −80°C. Frozen cultures were thawed immediately before Nluc or lactate dehydrogenase (LDH) quantification [34]. Three experiments were done in duplicate with Nluc independently transfected lines. $IC_{50}$ was determined by non-linear regression analyzed by GraphPad Prism software.

For transient transfections, parasites at trophozoite stage were incubated with 400 μL of erythrocytes previously electroporated with 150 μg of pPfNluc and left to invade for one day in a culture volume of 10 mL (4% hematocrit). Culture volume was increased to 40 mL (1% hematocrit), cultures split in 5 mL volumes and CQ added at each concentration. Cultures were maintained for 3 days and Nluc quantified after hemoglobin removal by saponin lysis as described in the Luciferase Assay section. Results correspond to 3 independent experiments. For each assay 100% growth corre-sponds to parasites grown in absence of CQ and 0% growth to parasites grown in supra lethal dose of the drug (4 mM). The robustness of the assays were evaluated by calculating the statistical parameters Z' value, signal to background ratio (S/B), signal to noise ratio (S/N), %CVmax and %CVmin as previously described [35]. The Z' value was calculated according to the formula: $Z' = 1 - (3 \sigma_{c(+)} + 3 \sigma_{c(-)})/|\mu_{c(+)} - \mu_{c(-)}|$, where $\mu_{c(+)}$ and

$\sigma_{c(+)}$ correspond the mean and standard deviation of the positive control (100% growth) and $\mu_{c(-)}$ and $\sigma_{c(-)}$ correspond the mean and standard deviation of the negative control (0% growth) respectively. The S/B was calculated as $\mu_{c(+)}/\mu_{c(-)}$ and S/N as $(\mu_{c(+)}-\mu_{c(-)})/\sigma_{c(-)}$. CVmax and CVmin were calculated as $100\times(\sigma_{c(+)}/\mu_{c(+)})$ and $100\times(\sigma_{c(-)}/\mu_{c(-)})$ respectively.

## Immunofluorescence Microscopy

Nluc, SP-Nluc or PEXEL-Nluc parasites were settled on poly-L lysine coated wells of a Lab-Tek chamber slide and fixed in 4% paraformaldehyde and 0.0075% gluteraldehyde in PBS [36]. Parasites were probed with an anti-EXP2 monoclonal [37] and rabbit anti-NanoLuc IgG (a kind gift from Lance Encell, Promega) all at 20 µg/mL. Images were acquired with a Zeiss AxioObserver inverted microscope and processed with ImageJ.

## Sub-Cellular Fractionation

Equinatoxin II was expressed in *E. coli* and purified as previously described [38] and stock aliquots at [1 mg/mL] were kept at $-80°C$, thawed and diluted 10-fold in PBS just prior to use. 50 µL of total parasite culture at 1% hematocrit was transferred to 1.5 mL tube and washed with 0.5 mL of RPMI to remove Nluc from the media. After centrifuging at 3,000 g for 3 min, 0.5 mL of the supernatant was discarded. To the 50 µL of culture left in the tube, 1 volume of equinatoxin in PBS was added. After gently mixing for 5 min, 400 µL of RPMI was added and the samples centrifuged at 6,000 g for 3 min. 50 µL of the supernatants were recovered and transferred to a 96 well plate, corresponding to the RBC fraction. Most of the remaining supernatant was then removed and discarded leaving about 30 µL in the tube. This represents 6% contamination, necessary to avoid accidental removal of the cells on the bottom of the tube. 470 µL of 0.01% saponin diluted in RPMI was added, and gently mixed for 2 min. The sample was centrifuged at 6,000 g for 3 min and 50 µL of the supernatant transferred to a 96 well plate, which corresponds to the PV fraction. Most of the solution was discarded, leaving about 30 µL in the tube. 470 µL of RPMI was added to dilute the sample, which was gently mixed and 50 µL transferred to a 96 well plate to become the parasite fraction. Just prior to measuring in the luminometer, 50 µL of Nano-Glo Luciferase Assay Reagent was added to each sample.

## Brefeldin A (BFA) incubation

Trophozoite stage cultures were incubated for 6 hours with 5 mg/mL BFA (Sigma) and harvested for RBC fractionation and Nluc quantification. DMSO was used to resuspend BFA and served as a negative control.

## New Permeation Pathway assay

Parasite culture expressing PEXEL-Nluc were washed to remove excess Nluc from the culture media. The culture containing 1–5% trophozoite stage parasites was diluted to 1% hematocrit. The culture was incubated with 5-fold the final concentration of furosemide for 20 min at RT. 10 µl of culture was then added in triplicate to new wells of an opaque 96-well plate (Nunc). 40 µl of sorbitol lysis buffer (280 mM sorbitol, 20 mM Na-HEPES, 0.1 mg/ml BSA, pH 7.4) or PBS was then added. The cells were then pelleted and the supernatant was collected at each time point and frozen. The supernatant was then mixed with 2x Nano-Glo assay reagent buffer and luminescence was measured. Alternatively, a 1/2000 dilution of Nano-Glo was added directly to the sorbitol lysis buffer or PBS and luminescence was measured at each time point.

The statistical parameters Z' value, S/B, S/N, %CVmax and %CVmin were calculated in order to determine the robustness of the assay at each time point where reporter activity was measured. Parasites incubated with 0 µM or 25 µM furosemide served as positive and negative (background) controls respectively. Experiments were performed in triplicates and repeated once.

## Results

### NanoLuc expression in *Plasmodium falciparum*

Nluc is an ATP independent luciferase nearly two orders of magnitude brighter than firefly and *Renilla* luciferases when expressed in mammalian cells [29]. To evaluate if Nluc would perform similarly in *P. falciparum*, the reporter plasmid pPfNluc was made by replacing the firefly luciferase gene in pPf86 [19] with that encoding Nluc (Fig. 1A). With both firefly luciferase and Nluc under control of the same Hsp86 promoter, the plasmids were electroporated into uninfected RBCs and then fed to *P. falciparum* trophozoite stage cultures so that the parasites would take up the plasmid when they invaded the electroporated RBCs. Four days later, the so-called transient transfectants were assayed for luciferase activity after saponin lysis which selectively permeabilises the RBC to facilitate the removal of haemoglobin, a powerful quencher of emitted light. Extracts of PfNluc parasites when provided with its optimised furimazine substrate (Nano-Glo Luciferase Assay Reagent) produced a strong signal (Fig. 1B), but extracts from the pPf86 parasites did not (data not shown). Firefly luciferase activity in the pPf86 transfectants could only be detected when provided with its D-luciferin substrate diluted in the Luciferase Assay Buffer (Fig. 1B). Extracts of mock-transfected parasites were used as negative controls and to determine the background luminescence for each substrate. After subtracting the background activities, Nluc was approximately 117-fold brighter then firefly luciferase, demonstrating Nluc can be expressed in *P. falciparum* and similar to mammalian cells is considerably brighter than firefly luciferase (Fig. 1B).

### Stable expression in *P. falciparum*

*P. falciparum* tolerance for Nluc expression in continuous culturing was next investigated. The stable transfection vector pEF-Nluc was made by replacing the firefly luciferase of pEF-Luc [30] with Nluc (Fig. 2A) and this vector was stably transfected in *P. falciparum*. To confirm that the transfected parasites were expressing Nluc, 100 µL of parasite culture at 1% hematocrit and 5% parasitemia, corresponding to 500,000 infected red blood cells (iRBC), were washed and resuspended in the original volume of PBS. The cells were then lysed in 100 µL of Nano-Glo Luciferase Assay Reagent (referred to as 1:1 in Fig. 2B) and this, measured for 10 seconds in the luminometer with its gain adjusted to 10% below saturation, yielded almost $10^7$ RLU, which is the value where the signal saturates. The high signal strength prompted us to determine if the Nano-Glo Luciferase Assay Reagent could be more economically used and so it was diluted in 10-fold increments in Luciferase Cell Culture Lysis Reagent, part of the Luciferase Assay System (Promega), and tested with the same number of Nluc expressing parasites in the same conditions. A 10-fold dilution of Nano-Glo decreased the signal by less than 10%, suggesting the substrate was in excess for the amount of enzyme present in the extract (1:10, Fig. 2B). The 100 and 1,000-fold dilutions resulted in signal decreasing about 3 and 24-fold, respectively (1:100 and 1:1000, Fig. 2B). In spite of the lower signal detected, 1,000-fold dilution of Nano-Glo still permitted detection at a S/B ratio of approximately 125 (1:1000 versus 1:1 *wt*, Fig. 2B).

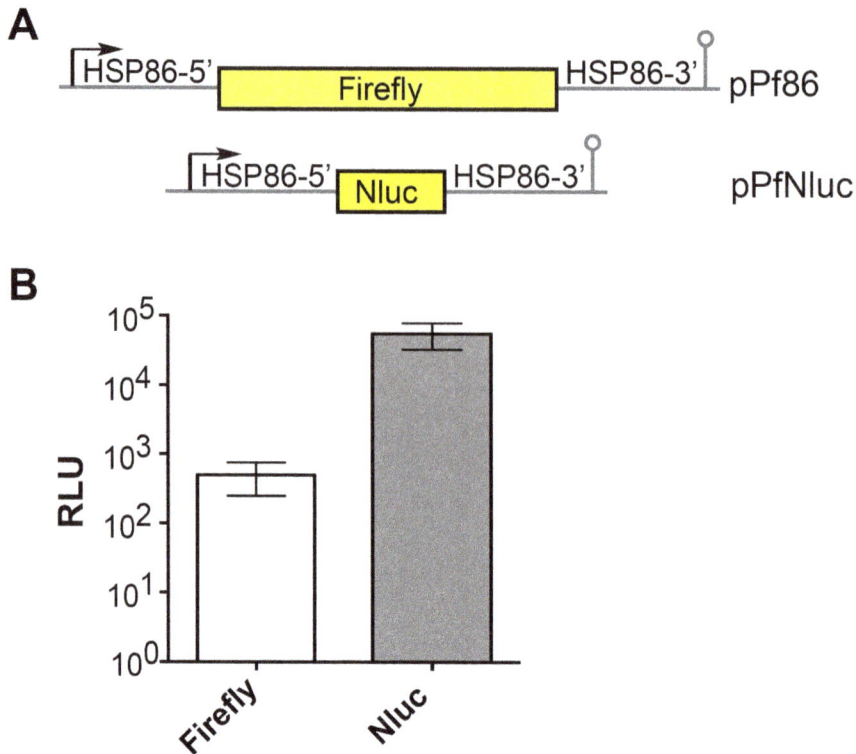

**Figure 1. NanoLuc (Nluc) luminescence in *Plasmodium falciparum*.** (A) Diagrams of firefly and Nluc reporter vectors. (B) pPf86 and pPfNluc were transfected in trophozoite stage parasites and luciferase activity in relative light units (RLU) determined 4 days later. Note that the RLU of each luciferase was measured in its own optimal substrate ie, Nluc with Nano-Glo and Firefly with D-luciferin. Mock-transfected parasites were used as negative control and to determine background luminescence, which was then subtracted from firefly and Nluc activities. The result represents the mean of 3 independent transfections ± standard deviation.

High throughput applications require the reporter detection to correlate linearly with cell number and the reagent to be stable for the duration of the experiment. We determined that 1:1000 dilution is stable for up to 20 minutes (data not shown) and so that this dilution was only used for assays that lasted less than that. A slightly lower dilution (1:500) was stable for up to 40 minutes (data not shown) and this was chosen for most of the assays. In order to determine if detection of reporter activity using the 1:500 dilution would correlates linearly with parasite number, Nluc activity was measured in samples containing a range of 5–42,500 iRBC at the same hematocrit (Fig. 2C). The luminometer's gain was set to 10% below saturation for the brightest sample and measurements performed for 2 s. Linear correlation of RLU and parasite number could be achieved from 78–42,500 iRBC, corresponding to a parasitemia of 0.015–8.5% for that specific experiment (Figure S1). Setting the gain to maximum did not improve the detection threshold (Figure S2A), but saturated the signal at about 10,000 iRBC. Increasing both the gain to max, the measurement time to 10 s, and using the standard Nano-Glo dilution (1:1) allowed detection in the linear range down to approximately 10 iRBC, but the signal saturated at just over 600 iRBC, providing a very narrow window of detection in the linear range (Figure S2B). In order to assure detection in the linear range, all subsequent experiments contained <50,000 iRBC, with the gain set to 10% below saturation for the brightest sample and thus the absolute RLU values cannot be compared among different experiments. Using another cell line with exported Nluc (described next), detection in the linear range was achieved in the range of 71–73,500 iRBC (Figure S3).

The possibility of working at such a high Nano-Glo dilution and keeping a high S/B ratio raised the question whether Nluc could be used as an inexpensive high throughput assay for quantification of *P. falciparum* growth in the presence of inhibitors. To test this, the Nluc parasites, which were transfected forms of the chloroquine-sensitive 3D7 strain, were grown for 3 days in the presence of varying concentrations of chloroquine (CQ). Initial parasitemias were 0.5–1% ring stage parasites, at 1% heamatocrit in a total culture volume of 100 µL. After three days the parasitemias were about 2.5–5% of late trophozoite/schizont stage parasites and the cells present in 10 µL of the culture were harvested and reporter activity detected using the Nano-Glo Luciferase Assay Reagent in its standard dilution or diluted 1,000-fold as in Figure 2B. As a control, the remainder of the cultures were used to quantify growth by the benchmark lactate dehydrogenase (LDH) assay where the activity of endogenous LDH enzyme was measured [34]. The growth curves and $IC_{50}$ were similar for all the methods applied, suggesting Nluc performs similarly to LDH and that 1,000-fold Nano-Glo Luciferase Assay Reagent dilution can be used (Fig. 2D–E).

A caveat of using luciferase to quantify cell growth is that the parasites have to be already stably transfected, which can take several weeks for *P. falciparum*. The RLU levels detected in the transient transfections with pPfNluc (Fig. 1B) suggested the signal might be strong enough to be used for a growth assay. To test that, erythrocytes electroporated with pPfNluc were fed to wild type parasites at trophozoite stage and left to invade for 1 day. After that, varying concentrations of CQ were added and parasites cultured for 3 more days when Nluc and LDH activities were

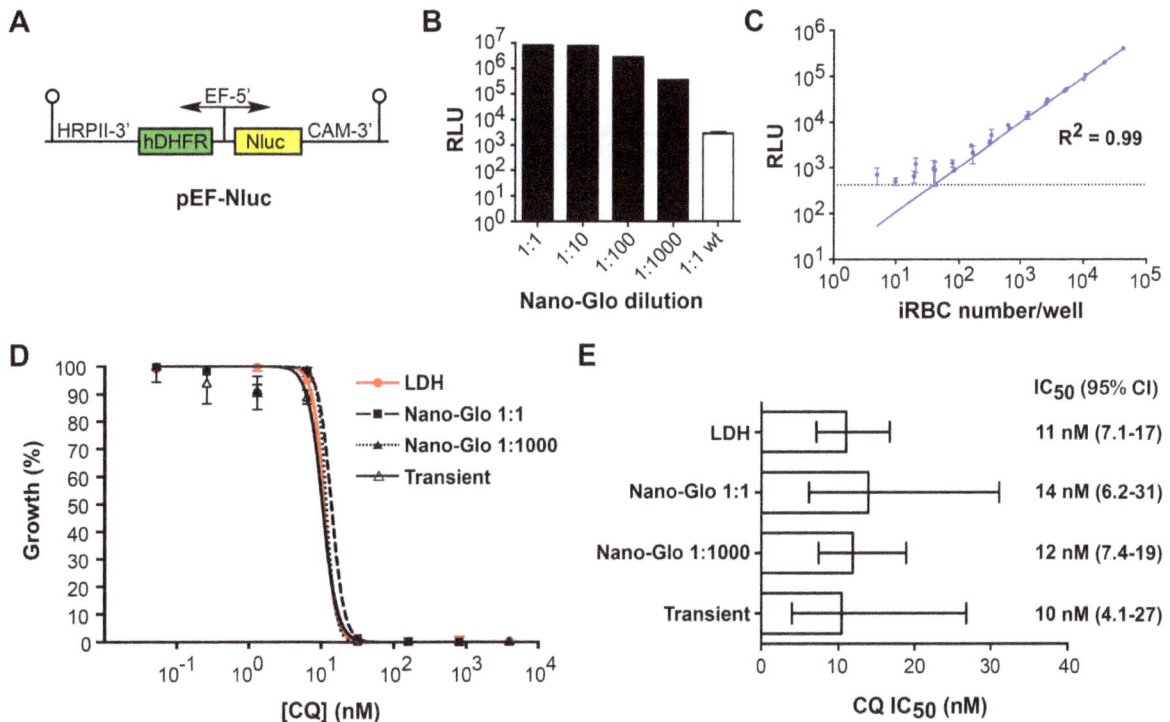

**Figure 2. *Plasmodium falciparum* stably expressing Nluc.** (A) The Nluc gene was cloned in the pEF vector for stable expression in *P. falciparum*. (B) Aliquots (100 µL) of cultures at 1% hematocrit and 5% parasitemia corresponding to 500,000 *P. falciparum* infected RBCs transfected with pEF-Nluc were mixed with 1 volume of Nano-Glo Luciferase Assay Reagent and reporter activity measured (1:1). Nano-Glo Luciferase Assay Reagent was further diluted in 10-fold increments in Luciferase Cell Culture Lysis Reagent and used to determine reporter activity of the same culture. As a negative control, wild type parasites (wt) were mixed 1:1 with Nano-Glo Luciferase Assay Reagent. (C) Parasites stably transfected with pEF-Nluc were diluted in 2-fold increments in RPMI + RBC maintaining 0.5% hematocrit. For each sample, 10 µL of the culture dilutions were mixed with 40 µL of Nano-Glo diluted 1:400 in water and measured in the luminometer for 2 s with the gain adjusted 10% below saturation for the brightest sample. The solid line represents the mean RLU after linear regression. The dashed line represents the background +3 standard deviations. (D) Nluc expressing parasites were cultured in varying concentrations of chloroquine (CQ) and their growth determined by the LDH standard method or by measuring reporter activity using Nano-Glo Luciferase Assay Reagent at its standard dilution (1:1) or diluted 1:1000 as described in (C). Similarly, wild type parasites were transiently transfected with pPfNluc and their growth determined using Nano-Glo Luciferase Assay Reagent (Transient). $IC_{50}$ was calculated by non-linear regression and represents the mean of 3 experiments. (E) The $IC_{50}$s determined in D were plotted with 95% confidence intervals (CI).

determined. Both the growth curve (Fig. 2D) and the $IC_{50}$ were similar to what had been determined for the stable transfected parasites measured by Nluc or by LDH activities (Fig. 2E).

Statistical parameters were calculated to evaluate the robustness of the assay and whether it could be suitable for high throughput drug screening (Table 1). Z' values for CQ sensitivity assay using the stable transfected line and the standard or 1:1000 Nano-Glo dilutions were >0.9 and similar to the LDH assay. The assay using transiently transfected parasites produced a significantly lower, but still suitable (>0.6) Z' value, which was caused mainly by a lower S/B ration and higher variation of both positive and negative controls.

## Targeting Nluc beyond the parasite membrane

Much of the virulence of *Plasmodium* parasites can be attributed to the export of effector proteins into the host compartment, which extensively modifies the iRBC [10,17,39–43]. Consequently, there has been much interest in understanding how parasites export proteins beyond the PVM that envelops them within the RBC cytoplasm and whether the export system can be targeted with drug inhibitors [6,12,39,44]. Methods such as tagging exported proteins with fluorescent markers like green fluorescent protein or by immuno-labeling with protein-specific

antibodies have been extensively applied for detection of exported proteins in *Plasmodium spp* [17,41,45–47]. Microscopy of individual parasites has then been used to follow the tagged proteins but this is time consuming and not amenable to quantification. In order to create a more quantitative approach we attempted to export Nluc into the RBC compartment so we could quantitatively follow its passage. To do this the first 113 residues of the *P. falciparum* exported protein Hyp1 (PF3D7_0113300), containing the PEXEL cleavage site RLLTE, was fused to Nluc at its N-terminus (Fig. 3A). PEXEL-Nluc parasites were analyzed by immunofluorescence (IFA) with antibodies for Nluc and Exp2, a PVM marker used to delimit the parasite boundaries. PEXEL-Nluc did not concentrate and co-localize with Exp2 rather occupying the entire RBC cytosolic region that lies beyond the PV, suggesting it is correctly exported (Fig. 3A). There was also an increased Nluc signal surrounding the parasites' nuclei that could represent newly synthesized Nluc transiting through the endoplasmic reticulum (ER).

Once trafficked from the ER to the parasite surface probably by vesicular transport, exported proteins must transit across the PV before engaging the PVM-spanning export machine PTEX, to cross the vacuole membrane [12]. Any inhibitor of protein export is likely to trap cargo within the parasite or its vacuole and so a control parasite line was created that secreted its cargo into the

**Table 1.** Summary of the assay parameters from the growth assays.

| | Assay and reagent dilution | | | |
| | Nano-Glo 1:1 | Nano-Glo 1:1000 | Transient[1] | LDH |
|---|---|---|---|---|
| Z'-value | 0.96±0.02 | 0.95±0.02 | 0.69±0.11 | 0.93±0.03 |
| S/B ratio | 114±52 | 70±26 | 21±13 | 5±0.7 |
| S/N ratio | 1397±84 | 687±221 | 182±48 | 163±75 |
| %CV$_{max}$ | 1.36±0.60 | 1.56±0.47 | 9.20±3.99 | 1.23±0.97 |
| %CV$_{min}$ | 8.01±3.28 | 9.92±0.66 | 10.50±4.59 | 2.84±1.31 |

[1]refers to luminescence measured from transiently transfected parasites using the standard Nano-Glo dilution.

PV. To do this the first 23 residues of merozoite surface protein 1 (MSP1) was fused to the N-terminus of Nluc (Fig. 3B). This region comprising the signal peptide region (SP) of MSP1 has been reported to efficiently target GFP to the secretory pathway [48]. Parasites expressing SP-Nluc were analyzed by immunofluorescence with anti-Nluc and anti-Exp2 and both proteins seem to mostly co-localize in the PV with some minor labeling in the RBC compartment (Fig. 3B).

In order to demonstrate that Nluc trafficking was dependent on the signals fused to its N-terminal region, the sub-cellular localization was also analyzed in parasites expressing the original Nluc and as expected, the reporter alone has a cytosolic localization (Fig. 3C).

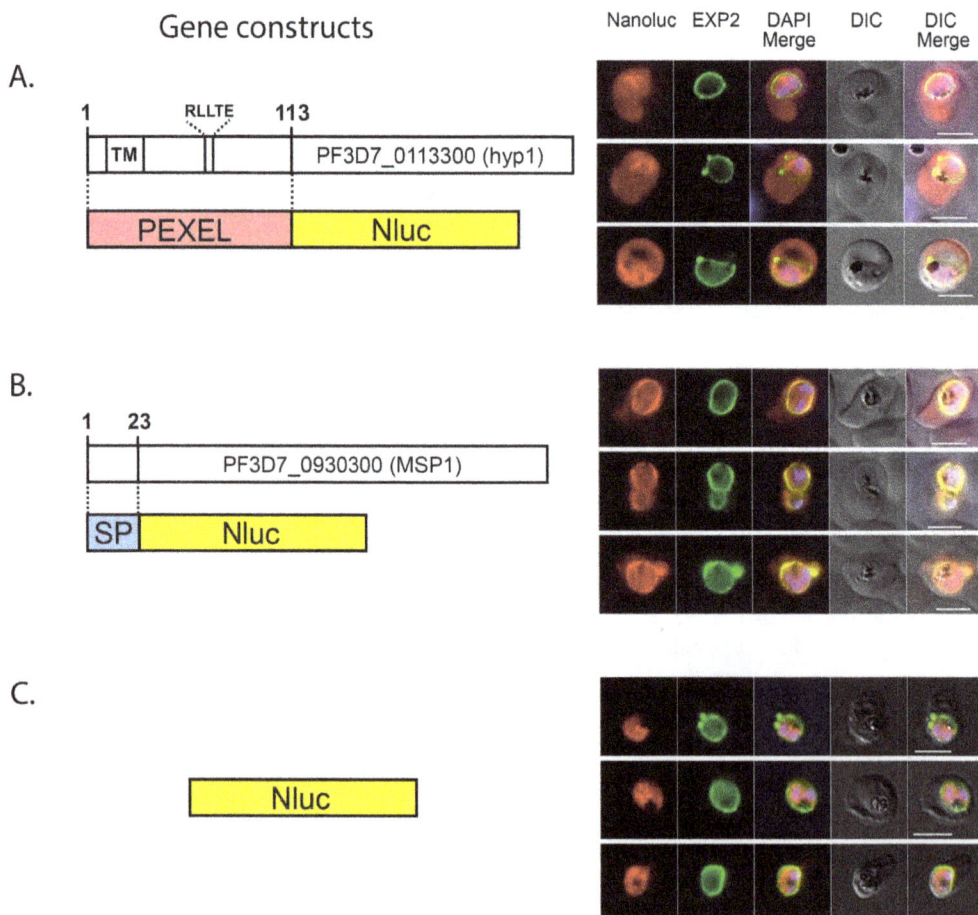

**Figure 3. NanoLuc is targeted to the PV and to the RBC.** Diagrams of gene constructs and the IFA images are shown on the left and right respectively. Nluc fused at its N-terminus to (A) the N-terminal region of an exported protein (PEXEL), (B) to a secretion signal peptide (SP) or (C) original (cytosolic). The gene encoding each fusion protein was cloned in the pEF vector. Transfected parasites were analysed by IFA using antibodies to detect Nluc and the PVM marker Exp2. DAPI was used for nuclear staining. Size bar = 5 μm.

## Quantification of Nluc fusions on iRBC sub-cellular compartments

Before quantifying the export of Nluc, we needed to ensure that hemoglobin a powerful quencher of luminescence detection would not mislead interpretations of the reporter activity quantified in different compartments of the infected RBCs. In order to find conditions where hemoglobin did not substantially block luminescence, PEXEL-Nluc parasites at 1% parasitemia and 10% hematocrit (ie, 10% RBCs in RPMI v/v) were sequentially diluted in RPMI. The infected RBCs were then lysed in an equal volume of Nano-Glo Luciferase Assay Reagent and the luminescence measured. After normalizing the RLU for parasite number that we have shown are directly proportional (Fig. 2B), reporter activity was similar from hematocrits of 0.04% to 0.12% (Figure S4). Hematocrits of $\geq 0.37\%$ showed progressively reduced luminescence, suggesting a hemoglobin quenching effect at higher hematocrits (Figure S4). Thus for all subsequent experiments, cells extracts were always diluted so that the equivalent hematocrit was never higher than 0.1%.

Microscopy of the secreted and exported Nluc parasites indicated the fusion proteins were most strongly detected in trophozoite and schizont stage parasites ($\geq 24$ hours post invasion, [hpi]), because the EF (elongation factor 1$\alpha$) promoter is weaker in younger ring stage parasites (0–24 hpi). This was confirmed by a time course luciferase assay over a single cell cycle (Figure S5) and for this reason, fractionation of the lines expressing the cytosolic Nluc, SP-Nluc and PEXEL-Nluc were first carried out on trophozoites. The subcellular compartments were first fractionated by treating the cultures with equinatoxin which forms pores in the RBC membrane, releasing the contents of the RBC cytosol where soluble exported proteins should reside [38]. The cells were then incubated with the detergent saponin, which forms pores in the PVM, releasing the PV fraction, leaving RBC and PV free parasites, which corresponds to the remaining parasite fraction (Fig. 4A). Equivalent volumes of the RBC, PV and parasite fractions from equal number of cells were then mixed with Nano-Glo Luciferase Assay Reagent and their RLU was measured.

In parasites expressing the cytosolic Nluc, more than 80% of reporter activity was detected in the parasite fraction (Fig. 4B). The remaining activity measured in the PV and RBC fractions may be caused by membrane damage during fractionation. Importantly, the percentage of activity detected in the parasite fraction of SP-Nluc and PEXEL-Nluc lines were significantly lower than in the Nluc line.

In SP-Nluc parasites just over half of the reporter activity was in the PV fraction, which was significantly higher than the PV fractions of the Nluc and PEXEL-Nluc parasites (Fig. 4B). Unexpectedly, about a third of the SP-Nluc luciferase activity was detected in the parasite fraction, which could possibly be newly synthesized enzyme en route to the parasite surface via the ER. About 10% of SP-Nluc activity was detected in the RBC fraction, which may be due to membrane damage or leakage out of the vacuole prior to fractionation. Alternatively, the small amount of NLuc activity detected in the RBC fraction may reflect the microscopy results from earlier indicating that in some SP-Nluc expressing parasites, the RBC cytosol was also labeled with anti-NLuc in addition to the PV (Fig. 3B).

About 70% of luciferase activity was detected in the RBC fraction of PEXEL-Nluc parasites (Fig. 4B), which is substantially higher than in Nluc and SP-Nluc lines. The higher reporter activity in the Parasite fraction in the PEXEL-Nluc parasites compared to the PV fraction could indicate Nluc takes longer to be secreted from the parasite than it does to transit the PV. Taken together these results indicate that Nluc fusion proteins are

targeted correctly to different cell compartments and can be used to investigate and to quantify cellular trafficking.

While the secretory pathway is clearly functioning during schizogony, little is know about the ability of parasites to export proteins during this stage, bringing into question whether PEXEL-Nluc could be exported or would remain trapped in the PV. Cellular fractionation was therefore carried out with segmented schizonts and the pattern was very similar to trophozoites (Fig. 4C). The fact that total luciferase activity in these lines increases until at least 32 hpi (Figure S5) and that a substantial part of the RBC is engulfed by the parasites strongly suggest PEXEL-Nluc is exported in schizonts.

Despite the weakness of the EF-1$\alpha$ promoter in rings (Figure S5), we attempted RBC fractionation and found the cytosolic and exported lines had a pattern very similar to trophozoites and schizonts (Fig. 4D). In rings however, the secreted line had higher proportion of reporter activity in the RBC fraction than in older trophozoites and schizonts (Fig. 4B–D). The proportion of signal in the SP-Nluc Parasite fraction was not altered, but in the PV fraction it had been reduced. It is possible that the ring-stage PVM is more susceptible to some degree of equinatoxin permeabilisation than in mature stages, but given the very low expression of SP-Nluc in rings, there could be an experimental artifact such as some leakage of SP-Nluc into the RBC during invasion when the PVM forms. Despite a higher proportion of reporter activity in the RBC fraction in SP-Nluc ring stages, this line still has the signature of a secreted reporter. The luciferase activity in the PV fraction is still substantially higher than in the equivalent fraction in the cytosolic and exported lines (Fig. 4D). In addition, reporter activity in the RBC fraction of SP-Nluc rings is considerably lower than in PEXEL-Nluc rings (Fig. 4D).

## Inhibition by Brefeldin A

With the Nluc, SP-Nluc and PEXEL-Nluc parasites having such distinct and characteristic reporter fraction patterns, we decided to test known trafficking inhibitors on the lines to assess their potential usefulness for screening for novel trafficking inhibitor compounds. To our knowledge, however there are no known *Plasmodium*-specific protein export inhibitors, so we instead used the antibiotic brefeldin A (BFA), which blocks the translocation of proteins from the ER to the Golgi. Since this pathway is common for both protein secretion and export, SP-Nluc and PEXEL-Nluc lines were treated with BFA followed by RBC fractionation (Fig. 5A). Treatment of trophozoites with 5 mg/mL BFA for 6 hrs in both lines caused an increase in reporter activity in the parasite suggesting luciferase is accumulating in the parasite as expected. In the SP-Nluc line, activity decreased in the PV fraction but not in the RBC, since transit from the PV into the RBC was not expected to be affected by BFA. In the PEXEL-Nluc line, activity was reduced in the RBC, but no change was seen in the PV fraction suggesting that BFA does not block the PTEX protein transporter. The general BFA toxicity to the cells was evaluated by measuring total reporter activity and it dropped by 40 and 35% for the SP-Nluc and PEXEL-Nluc lines respectively, during the assay period (Fig. 5B).

## Inhibition of New Permeation Pathways

As the parasite matures from a ring into a trophozoite it progressively increases the permeability of the iRBC to acquire plasma nutrients and remove waste products that are vital for rapid growth. The increased permeability appears to be due to the formation of anion selective channels in the iRBC plasma membrane. While its clear that parasites play a role in generating the NPPs, it is not known if they directly form the membrane

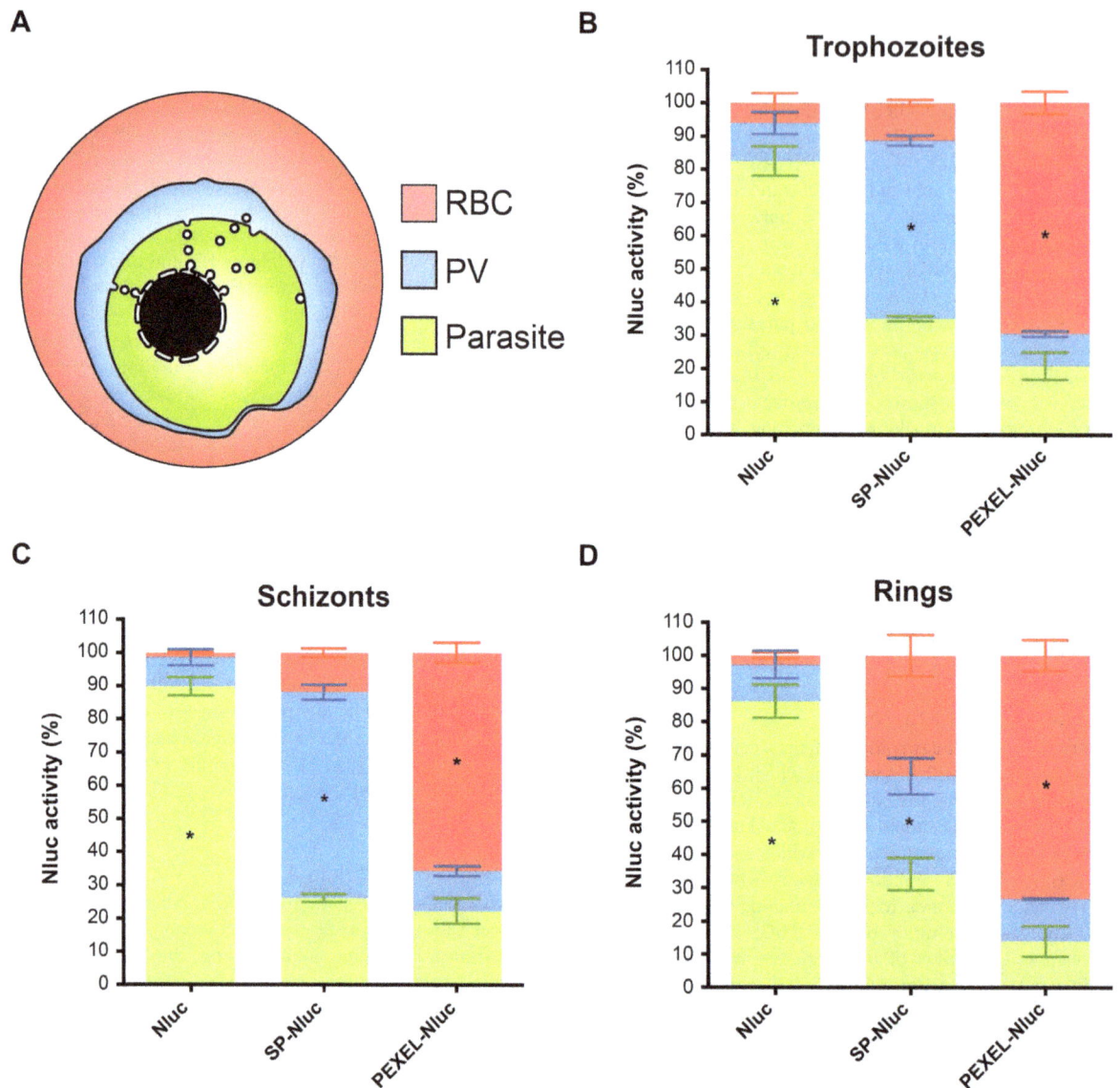

**Figure 4. Quantification of Nluc in cellular compartments of infected RBCs.** (A) Schematic of the iRBC's compartments where RBC represents the exported fraction that is released after Equinatoxin treatment. The PV compartment was then released following treatment with 0.01% saponin and finally, the Parasite fraction was lysed by Nano-Glo Luciferase Assay Reagent. (B) Trophozoite stage parasites transfected with either the original Nluc, secreted SP-Nluc or the exported PEXEL-Nluc fusions were fractionated as shown in (A) and luciferase activity measured. Nluc activities as a percentage of the total for each parasite line are shown and represent the mean of 3 experiments +/−SEM. Similar to (B), (C) Schizonts and (D) Ring stage parasites were also fractionated. Statistical significance (* p<0.05) was determined by 2 way ANOVA test comparing the percentage of reporter activity of each sub-cellular fraction among the 3 cell lines.

channels *de novo* or if they increase the conductance of pre-existing human channels [11,49]. It has long been recognised that if NPPs could be blocked with drug inhibitors then this may starve the parasites severely curtailing their growth. Screens for such inhibitors have been based on the susceptibility of iRBCs to an iso-osmotic sorbitol solution that lyses mature parasites with functional NPPs [50,51]. Compounds that block NPPs render the parasites resistant to the sorbitol and the resultant reduction in haemoglobin from lysed iRBCs is measured. To obtain sufficient sensitivity, large numbers of purified iRBC have to be assayed and we therefore wished to determine whether exported PEXEL-Nluc from lysed iRBCs might be more sensitively detected, and could be quantified without prior isolation of the iRBCs from uninfected RBCs. To test this, PEXEL-Nluc trophozoite-stage parasites at 3–

5% parasitemia were incubated with the benchmark NPP inhibitor furosemide [50] prior to addition of the sorbitol solution containing Nano-Glo. Bioluminescence produced by the released PEXEL-Nluc was measured as a function of time and it was observed that without furosemide pre-treatment (0 μM), sorbitol triggered high Nluc activity that peaked 20 minutes after the addition of sorbitol (Fig. 6). Pre-treatment with furosemide greatly reduced the release of PEXEL-Nluc in a concentration dependent manner (Fig. 6). In order to determine the sorbitol incubation period that would allow optimal detection of NPP inhibition, robustness parameters were calculated for 10–60 min times (Table 2). While the shortest time (10 min) produces an acceptable Z' value, ≥0.5, 20–40 min appeared to be the most suitable incubation period, where the Z' value was the highest. At 60 min,

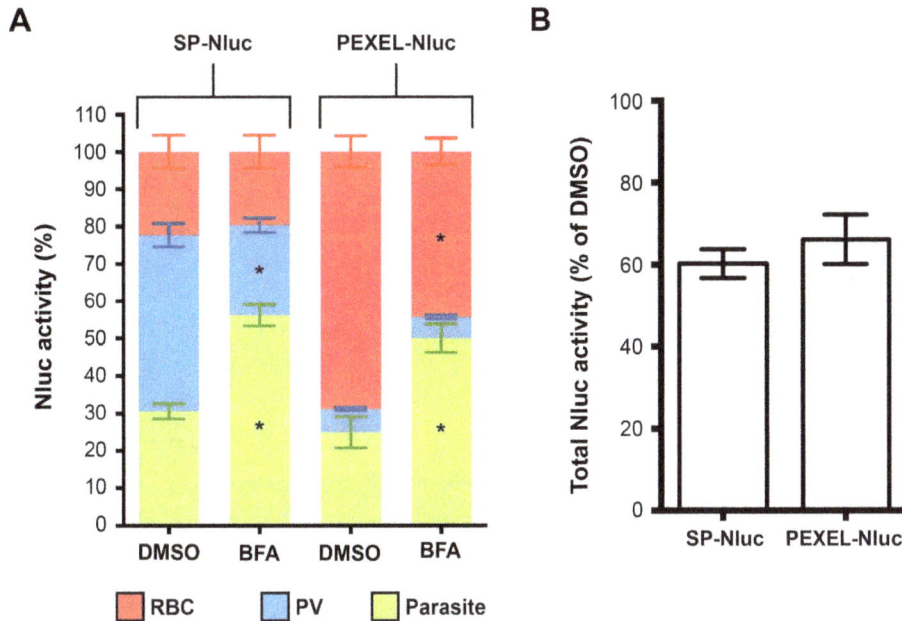

**Figure 5. Detection of secretion and export inhibition.** (A) Parasites expressing the secreted (SP) or the exported (PEXEL) Nluc were treated with Brefeldin A (BFA) for 6 hours and luciferase activity determined in each sub-cellular fraction. Results are the mean of 4 experiments and the error bars represent standard deviations. Statistical significance (* $p < 0.05$) was determined by 2 way ANOVA test comparing the activity detected in the parasite fraction of treated and DMSO control parasites of each line and of PV and RBC fractions of DMSO against treated SP-Nluc and PEXEL-Nluc lines respectively. (B) The toxicity of each treatment was determined by measuring total luciferase activity as relative to the DMSO control.

iRBC tended to burst even in the presence of the highest concentration of furosemide, suggesting the NPP inhibition is partial, resulting in lower S/B and S/N ratios and a highly variable Z' value. The benefit of PEXEL-Nluc parasites is the sensitivity of the assay which requires no parasite enrichment and hence fewer parasites and manipulations compared to previous methods rendering the assay more suitable for high throughput screening.

## Discussion

*P. falciparum* bioluminescent lines have been widely used both in studies of parasite biology as well as in assays where detection and quantification of parasite development or maturation is required [20–22,25,27]. However, this technology has some caveats such as the low transfection efficiency and consequently the long periods required to generate stable transfectants. Here we have shown that the significantly brighter Nluc reporter can overcome some of these difficulties. To begin with we have shown that transiently

transfected parasites can be used for drug assays, opening the possibility of assaying drug sensitivity/resistance in many parasite strains and field isolates without having to generate stable lines. The absence of species barrier for promoter recognition in *Plasmodium* [52,53] suggests this assay could be applied for parasites of other species, including *P. vivax*, a parasite for which *in vitro* culturing is more complex and less affordable. Our experimental design involved two cycles of RBC invasion; the first when RBCs electroporated with plasmid were added to mature stage parasites; and the second when parasites have grown in the presence of chloroquine for 3 days. To overcome poor invasion efficiency in newly cultured *P. falciparum* clinical isolates and of *P. vivax*, transient transfection could be performed by electroporating ring stage parasites and measuring reporter activity as parasites mature. Previous results have shown that luciferase based assays are suitable for monitoring parasite maturation and the effect of fast acting drugs [26,27].

Although funds for malaria control and eradication have increased in the last decade, they are still less than half of the

**Table 2.** Summary of the assay parameters from the NPP assay.

| | Incubation time | | | |
|---|---|---|---|---|
| | **10 min** | **20 min** | **40 min** | **60 min** |
| **Z'-value** | 0.65±0.19 | 0.80±0.12 | 0.81±0.02 | −0.20±1.24 |
| **S/B ratio** | 2.7±0.19 | 4.83±1.96 | 2.72±0.49 | 1.47±0.11 |
| **S/N ratio** | 50.6±13.3 | 54.84±27.54 | 37.28±10.91 | 10.49±11.09 |
| **%CV$_{max}$** | 5.94±3.33 | 3.76±2.79 | 2.33±1.22 | 5.33±4.92 |
| **%CV$_{min}$** | 3.43±0.53 | 9.01±8.09 | 4.61±0.04 | 8.87±8.28 |

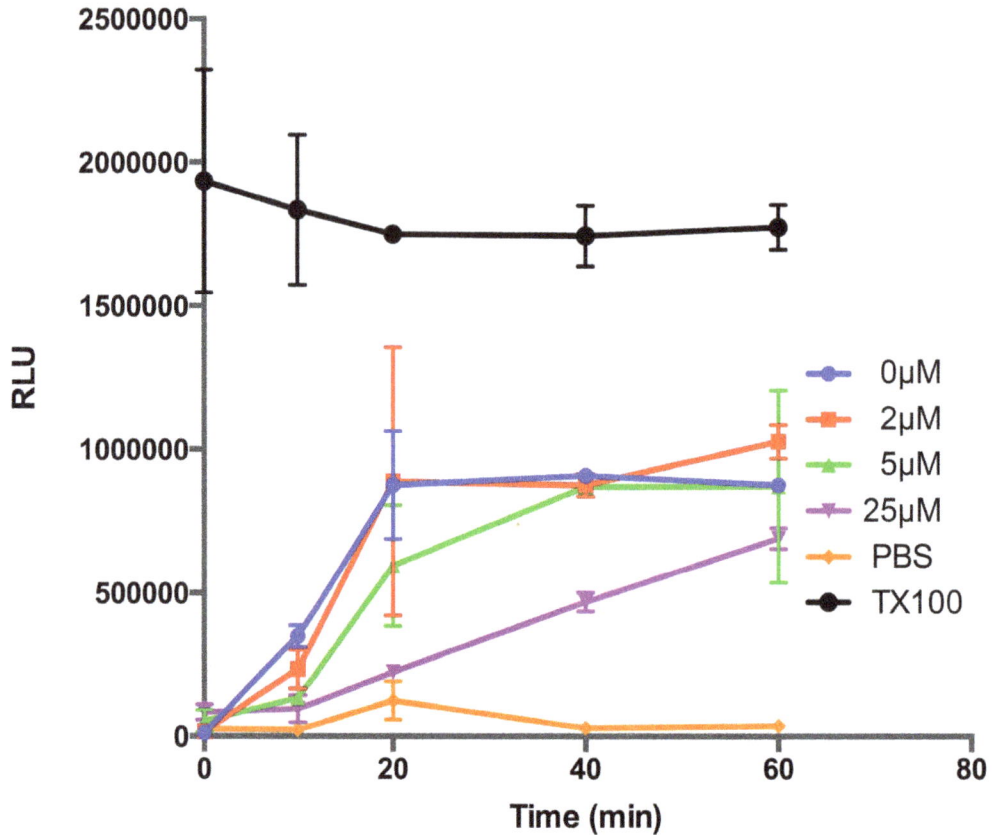

**Figure 6. Quantification of NPP activity and inhibition in PEXEL-NLuc parasites.** Trophozoite stage parasites (3% parasitemia) expressing exported PEXEL-NLuc were incubated with different concentrations of the new permeability pathway (NPP) inhibitor furosemide (µM), and then treated with sorbitol lysis buffer at time zero to induce the release of PEXEL-Nluc which was measured over time. A 1% Triton-X100 in PBS (TX100) treatment was included to simulate rapid lysis and a PBS control for background lysis. Data points represent the mean +/− SD of three replicates.

U$5 billion that is estimated to be required [1]. Since drug discovery is an essential part of control/eradication strategies, reducing the cost of screening cultured parasites to discover new inhibitors is highly desirable. When used at its recommended concentration the cost of the Nano-Glo reagent for screening a 96 well plate of Nluc expressing *P. falciparum* parasites is several dollars. However, we have shown that the Nano-Glo reagent can be diluted a 1,000-fold while still maintaining sufficient sensitivity. This could be suitable for high throughput screening and represent an attractive option to researchers with limited resources. In addition, the stability of Nluc in frozen samples makes it practical for storage and transport to where the Nluc can be measured. In our current drug assays, parasite cultures were kept frozen at −80°C for several weeks or at −20°C for several days without any loss of reporter activity that would affect the results (data not shown).

Protein export plays an essential role for *P. falciparum* infection where it is strongly associated with virulence and severity of disease. Investigation of exported protein trafficking has relied on examination by fluorescence microscopy of protein fusions with fluorescent markers or labelling with fluorescent antibodies. Such techniques require highly trained personnel as well as expensive and sophisticated microscopes. Microscopy is also poorly quantitative since only a relatively small number of cells can be practically examined. Bioluminescence offers a simpler and more quantifiable method to measure protein export, requiring less training and significantly more affordable equipment. We have

established a protocol for the small-scale quantification of protein secretion and export in stable Nluc transfected parasites that can be used in tandem with general drug screening protocols to evaluate the effect on export of compounds previously shown to inhibit parasite growth. Since there are no known specific protein export inhibitors, we could not test whether PEXEL-Nluc expressing parasites could be used to screen for export inhibition. Instead, BFA treatment that blocks transit from ER to Golgi was used to confirm that inhibition of protein export and secretion can be quantified. The fact that export is similarly BFA-inhibited as secretion confirms both trafficking systems share common steps in the secretory pathway [46,47,54,55].

The Nluc fusion proteins generally appear to localize where they were expected with data from both microscopy and reporter activity of sub-cellular fractionation suggesting a small leakage of SP-Nluc from the PV to RBC. It is known that the PVM looses integrity and becomes leaky in late schizonts [56], which could have allowed for the small soluble SP-Nluc (19 kDa) to be detected in the RBC fraction. Strong leakage in rings indicates that the PVM could be more permeable than previously thought or just more susceptible to equinatoxin damage. Despite this, the cytoplasmic, vacuolar and exported parasite Nluc lines once fractionated, have unique signatures that would enable screening for inhibitors that prevent growth, secretion and export. Our microtube cell fractionation protocol is currently not suitable for the high throughput screening of export inhibitors in microplate format due to the large number of steps needed to fractionate the cell compartments. We are however continuing to improve the protocol including its adaption for quantifying export in transiently transfected parasites.

In contrast to our assay for protein export inhibition, the NPP inhibition assay seems considerably more robust and ready to be adapted for high throughput screening. The PEXEL-Nluc NPP assay has the advantage over the existing protocol, which measures haemoglobin release, in that the iRBCs do not need to be purified. Moreover, due it its higher sensitivity, at least 50 times fewer parasites are needed which is better suited for large scale screening. In conclusion, we have shown that the use of the new Nluc reporter in *P. falciparum* has significant advantages when compared to the most commonly used firefly luciferase, offering simple and affordable assays to quantify growth and cellular trafficking to iRBC compartments.

## Supporting Information

**Figure S1   Graph of raw RLU vs parasitemia of same data presented in Fig. 2C.**

**Figure S2   Graph of raw RLU vs iRBC number.** (A) Samples and conditions are the same as in Fig. 2C, except that gain was adjusted to maximum. (B) Samples were measured for 10 sec, gain was set to maximum and Nano-Glo was used in its standard dilution.

**Figure S3   Graph of raw RLU vs iRBC number. PEXEL-Nluc parasites were serially diluted and samples measured as in Fig. 2C.**

**Figure S4   Hemoglobin quenches bioluminescence detection.** Nluc expressing parasites were diluted in RPMI in 3-fold increments and luciferase activity determined. Values were

normalized by parasite number and represented relative to the lowest hematocrit (0.04%). The bars represent the mean ± standard deviation of 4 experiments.

**Figure S5   Nluc stably expressed from the pEF vector shows maximal expression during the last half of the cell cycle.** To tightly synchronise the PEXEL-Nluc parasites so that the bioluminescence could be accurately measured during the cell cycle, heparin was added to schizonts prior to them rupturing. The heparin inhibits invasion and was removed once the schizonts began rupturing and then added back 2 hours later to produce an invasion window of 2 hours. The remaining unruptured schizonts were lysed in 5% sorbitol. The synchronized parasites were harvested every 4 hours until the 32–34 hour time point. Cultures were then diluted to 0.1% hematocrit and added to 1 volume of Nano-Glo Luciferase Assay Reagent and Nluc activity measured. Nluc values are represented as % of the maximum activity. Hours post invasion values are the mid-point of the 2 hour window.

## Acknowledgments

B.E. is recipient of an Australian Post Graduate Awards and S.C.C. recipient of a Monash Graduate Scholarship. We thank the Australian Red Cross Blood Bank for the provision of human blood and Jacobus Pharmaceuticals for providing WR99210. We are grateful to L. Encell and R. Parry from Promega for the Nanoluc antibody.

## Author Contributions

Conceived and designed the experiments: MFA CQN SCC BE PRS PRG. Performed the experiments: MFA CQN SCC BE PRS PRG. Analyzed the data: MFA CQN BE PRG. Contributed reagents/materials/analysis tools: MFA PRS BE SCC BSC. Contributed to the writing of the manuscript: MFA CQN PRG BE SCC.

## References

1. WHO (2013) World malaria report 2013.
2. Gamo FJ, Sanz LM, Vidal J, de Cozar C, Alvarez E, et al. (2010) Thousands of chemical starting points for antimalarial lead identification. Nature 465: 305–310.
3. Guiguemde WA, Shelat AA, Bouck D, Duffy S, Crowther GJ, et al. (2010) Chemical genetics of Plasmodium falciparum. Nature 465: 311–315.
4. Lucumi E, Darling C, Jo H, Napper AD, Chandramohanadas R, et al. (2010) Discovery of potent small-molecule inhibitors of multidrug-resistant Plasmodium falciparum using a novel miniaturized high-throughput luciferase-based assay. Antimicrob Agents Chemother 54: 3597–3604.
5. Spangenberg T, Burrows JN, Kowalczyk P, McDonald S, Wells TN, et al. (2013) The open access malaria box: a drug discovery catalyst for neglected diseases. PLoS One 8: e62906.
6. Boddey JA, Hodder AN, Gunther S, Gilson PR, Patsiouras H, et al. (2010) An aspartyl protease directs malaria effector proteins to the host cell. Nature 463: 627–631.
7. Marti M, Spielmann T (2013) Protein export in malaria parasites: many membranes to cross. Curr Opin Microbiol 16: 445–451.
8. Pasini EM, Braks JA, Fonager J, Klop O, Aime E, et al. (2013) Proteomic and genetic analyses demonstrate that Plasmodium berghei blood stages export a large and diverse repertoire of proteins. Mol Cell Proteomics 12: 426–448.
9. Maier AG, Cooke BM, Cowman AF, Tilley L (2009) Malaria parasite proteins that remodel the host erythrocyte. Nat Rev Microbiol 7: 341–354.
10. Maier AG, Rug M, O'Neill MT, Brown M, Chakravorty S, et al. (2008) Exported proteins required for virulence and rigidity of Plasmodium falciparum-infected human erythrocytes. Cell 134: 48–61.
11. Nguitragool W, Bokhari AA, Pillai AD, Rayavara K, Sharma P, et al. (2011) Malaria parasite clag3 genes determine channel-mediated nutrient uptake by infected red blood cells. Cell 145: 665–677.
12. de Koning-Ward TF, Gilson PR, Boddey JA, Rug M, Smith BJ, et al. (2009) A newly discovered protein export machine in malaria parasites. Nature 459: 945–949.
13. Beck JR, Muralidharan V, Oksman A, Goldberg DE (2014) PTEX component HSP101 mediates export of diverse malaria effectors into host erythrocytes. Nature 511: 592–595.

14. Elsworth B, Matthews K, Nie CQ, Kalanon M, Charnaud SC, et al. (2014) PTEX is an essential nexus for protein export in malaria parasites. Nature 511: 587–591.
15. Adjalley SH, Johnston GL, Li T, Eastman RT, Ekland EH, et al. (2011) Quantitative assessment of Plasmodium falciparum sexual development reveals potent transmission-blocking activity by methylene blue. Proc Natl Acad Sci U S A 108: E1214–1223.
16. Lucantoni L, Duffy S, Adjalley SH, Fidock DA, Avery VM (2013) Identification of MMV malaria box inhibitors of plasmodium falciparum early-stage gametocytes using a luciferase-based high-throughput assay. Antimicrob Agents Chemother 57: 6050–6062.
17. Marti M, Good RT, Rug M, Knuepfer E, Cowman AF (2004) Targeting malaria virulence and remodeling proteins to the host erythrocyte. Science 306: 1930–1933.
18. Goonewardene R, Daily J, Kaslow D, Sullivan TJ, Duffy P, et al. (1993) Transfection of the malaria parasite and expression of firefly luciferase. Proc Natl Acad Sci U S A 90: 5234–5236.
19. Militello KT, Wirth DF (2003) A new reporter gene for transient transfection of Plasmodium falciparum. Parasitol Res 89: 154–157.
20. Azevedo MF, del Portillo HA (2007) Promoter regions of Plasmodium vivax are poorly or not recognized by Plasmodium falciparum. Malar J 6: 20.
21. Frank M, Dzikowski R, Costantini D, Amulic B, Berdougo E, et al. (2006) Strict pairing of var promoters and introns is required for var gene silencing in the malaria parasite Plasmodium falciparum. J Biol Chem 281: 9942–9952.
22. Horrocks P, Lanzer M (1999) Mutational analysis identifies a five base pair cis-acting sequence essential for GBP130 promoter activity in Plasmodium falciparum. Mol Biochem Parasitol 99: 77–87.
23. Che P, Cui L, Kutsch O, Cui L, Li Q (2012) Validating a firefly luciferase-based high-throughput screening assay for antimalarial drug discovery. Assay Drug Dev Technol 10: 61–68.
24. Cui L, Miao J, Wang J, Li Q, Cui L (2008) Plasmodium falciparum: development of a transgenic line for screening antimalarials using firefly luciferase as the reporter. Exp Parasitol 120: 80–87.

25. Ekland EH, Schneider J, Fidock DA (2011) Identifying apicoplast-targeting antimalarials using high-throughput compatible approaches. FASEB J 25: 3583–3593.

26. Hasenkamp S, Sidaway A, Devine O, Roye R, Horrocks P (2013) Evaluation of bioluminescence-based assays of anti-malarial drug activity. Malar J 12: 58.

27. Khan T, van Brummelen AC, Parkinson CJ, Hoppe HC (2012) ATP and luciferase assays to determine the rate of drug action in in vitro cultures of Plasmodium falciparum. Malar J 11: 369.

28. Burghaus PA, Lingelbach K (2001) Luciferase, when fused to an N-terminal signal peptide, is secreted from transfected Plasmodium falciparum and transported to the cytosol of infected erythrocytes. J Biol Chem 276: 26838–26845.

29. Hall MP, Unch J, Binkowski BF, Valley MP, Butler BL, et al. (2012) Engineered luciferase reporter from a deep sea shrimp utilizing a novel imidazopyrazinone substrate. ACS Chem Biol 7: 1848–1857.

30. de Azevedo MF, Gilson PR, Gabriel HB, Simoes RF, Angrisano F, et al. (2012) Systematic analysis of FKBP inducible degradation domain tagging strategies for the human malaria parasite Plasmodium falciparum. PLoS One 7: e40981.

31. Trager W, Jensen JB (1976) Human malaria parasites in continuous culture. Science 193: 673–675.

32. Deitsch K, Driskill C, Wellems T (2001) Transformation of malaria parasites by the spontaneous uptake and expression of DNA from human erythrocytes. Nucleic Acids Res 29: 850–853.

33. Hasenkamp S, Russell KT, Horrocks P (2012) Comparison of the absolute and relative efficiencies of electroporation-based transfection protocols for Plasmodium falciparum. Malar J 11: 210.

34. Makler MT, Hinrichs DJ (1993) Measurement of the lactate dehydrogenase activity of Plasmodium falciparum as an assessment of parasitemia. Am J Trop Med Hyg 48: 205–210.

35. Zhang JH, Chung TD, Oldenburg KR (1999) A Simple Statistical Parameter for Use in Evaluation and Validation of High Throughput Screening Assays. J Biomol Screen 4: 67–73.

36. Tonkin CJ, van Dooren GG, Spurck TP, Struck NS, Good RT, et al. (2004) Localization of organellar proteins in Plasmodium falciparum using a novel set of transfection vectors and a new immunofluorescence fixation method. Mol Biochem Parasitol 137: 13–21.

37. Bullen HE, Charnaud SC, Kalanon M, Riglar DT, Dekiwadia C, et al. (2012) Biosynthesis, localization, and macromolecular arrangement of the Plasmodium falciparum translocon of exported proteins (PTEX). J Biol Chem 287: 7871–7884.

38. Jackson KE, Spielmann T, Hanssen E, Adisa A, Separovic F, et al. (2007) Selective permeabilization of the host cell membrane of Plasmodium falciparum-infected red blood cells with streptolysin O and equinatoxin II. Biochem J 403: 167–175.

39. Boddey JA, Carvalho TG, Hodder AN, Sargeant TJ, Sleebs BE, et al. (2013) Role of plasmepsin V in export of diverse protein families from the Plasmodium falciparum exportome. Traffic 14: 532–550.

40. Heiber A, Kruse F, Pick C, Gruring C, Flemming S, et al. (2013) Identification of new PNEPs indicates a substantial non-PEXEL exportome and underpins common features in Plasmodium falciparum protein export. PLoS Pathog 9: e1003546.

41. Hiller NL, Bhattacharjee S, van Ooij C, Liolios K, Harrison T, et al. (2004) A host-targeting signal in virulence proteins reveals a secretome in malarial infection. Science 306: 1934–1937.

42. Sargeant TJ, Marti M, Caler E, Carlton JM, Simpson K, et al. (2006) Lineage-specific expansion of proteins exported to erythrocytes in malaria parasites. Genome Biol 7: R12.

43. van Ooij C, Tamez P, Bhattacharjee S, Hiller NL, Harrison T, et al. (2008) The malaria secretome: from algorithms to essential function in blood stage infection. PLoS Pathog 4: e1000084.

44. Elsworth B, Crabb BS, Gilson PR (2014) Protein export in malaria parasites: an update. Cell Microbiol 16: 355–363.

45. Boddey JA, Moritz RL, Simpson RJ, Cowman AF (2009) Role of the Plasmodium export element in trafficking parasite proteins to the infected erythrocyte. Traffic 10: 285–299.

46. Knuepfer E, Rug M, Klonis N, Tilley L, Cowman AF (2005) Trafficking of the major virulence factor to the surface of transfected P. falciparum-infected erythrocytes. Blood 105: 4078–4087.

47. Wickham ME, Rug M, Ralph SA, Klonis N, McFadden GI, et al. (2001) Trafficking and assembly of the cytoadherence complex in Plasmodium falciparum-infected human erythrocytes. EMBO J 20: 5636–5649.

48. Gilson PR, O'Donnell RA, Nebl T, Sanders PR, Wickham ME, et al. (2008) MSP1(19) miniproteins can serve as targets for invasion inhibitory antibodies in Plasmodium falciparum provided they contain the correct domains for cell surface trafficking. Mol Microbiol 68: 124–138.

49. Bouyer G, Cueff A, Egee S, Kmiecik J, Maksimova Y, et al. (2011) Erythrocyte peripheral type benzodiazepine receptor/voltage-dependent anion channels are upregulated by Plasmodium falciparum. Blood 118: 2305–2312.

50. Duranton C, Huber SM, Tanneur V, Brand VB, Akkaya C, et al. (2004) Organic osmolyte permeabilities of the malaria-induced anion conductances in human erythrocytes. J Gen Physiol 123: 417–426.

51. Huber SM, Uhlemann AC, Gamper NL, Duranton C, Kremsner PG, et al. (2002) Plasmodium falciparum activates endogenous Cl(−) channels of human erythrocytes by membrane oxidation. EMBO J 21: 22–30.

52. Crabb BS, Cowman AF (1996) Characterization of promoters and stable transfection by homologous and nonhomologous recombination in Plasmodium falciparum. Proc Natl Acad Sci U S A 93: 7289–7294.

53. Fernandez-Becerra C, de Azevedo MF, Yamamoto MM, del Portillo HA (2003) Plasmodium falciparum: new vector with bi-directional promoter activity to stably express transgenes. Exp Parasitol 103: 88–91.

54. Ansorge I, Benting J, Bhakdi S, Lingelbach K (1996) Protein sorting in Plasmodium falciparum-infected red blood cells permeabilized with the pore-forming protein streptolysin O. Biochem J 315 (Pt 1): 307–314.

55. Chang HH, Falick AM, Carlton PM, Sedat JW, DeRisi JL, et al. (2008) N-terminal processing of proteins exported by malaria parasites. Mol Biochem Parasitol 160: 107–115.

56. Collins CR, Hackett F, Strath M, Penzo M, Withers-Martinez C, et al. (2013) Malaria parasite cGMP-dependent protein kinase regulates blood stage merozoite secretory organelle discharge and egress. PLoS Pathog 9: e1003344.

# Permissions

# List of Contributors

Hai-Long Wang, Tie-E Zhang, Li Guan, Hong-Li Liu, Jian-Hong Zhang, Xiao-Li Meng and Guo-Rong Yin
Research Institute of Medical Parasitology, Shanxi Medical University, Taiyuan, Shanxi, PR China

Li-Tian Yin
Department of Physiology, Key Laboratory of Cellular Physiology Coconstructed by Province and Ministry of Education, Shanxi Medical University, Taiyuan, Shanxi, PR China

Min Pang
Department of Respiratory, the First Affiliated Hospital, Shanxi Medical University, Taiyuan, Shanxi, PR China

Ji-Zhong Bai
Department of Physiology, Faculty of Medical and Health Sciences, University of Auckland, Auckland, New Zealand

Guo-Ping Zheng
Department of Biochemistry and Molecular Biology, Shanxi Medical University, Taiyuan, Shanxi, PR China

I-Ping Lee, Andrew K. Evans, Cissy Yang, Melissa G. Works and Zurine De Miguel
Department of Biology, Stanford University, Stanford, California, United States of America

Vineet Kumar
School of Biological Sciences, Nanyang Technological University, Singapore, Republic of Singapore

Nathan C. Manley
Department of Biology, Stanford University, Stanford, California, United States of America
Department of Neurosurgery, Stanford University School of Medicine, Stanford, California, United States of America
Stanford Stroke Center and Stanford Institute for Neuro-Innovation and Translational Neurosciences, Stanford University School of Medicine, Stanford, California, United States of America

Bashir A. Akhoon and Krishna P. Singh
Department of Bioinformatics, Systems Toxicology Group, CSIR-Indian Institute of Toxicology Research, Lucknow, India

Megha Varshney
Interdisciplinary Biotechnology Unit, Aligarh Muslim University, Aligarh, India

Shishir K. Gupta
Department of Bioinformatics, Biocenter, Am Hubland, University of Würzburg, Würzburg, Germany

Yogeshwar Shukla
Department of Proteomics, CSIRIndian Institute of Toxicology Research, Lucknow, India
Academy of Scientific and Innovative Research (AcSIR), New Delhi, India

Shailendra K. Gupta
Department of Bioinformatics, Systems Toxicology Group, CSIR-Indian Institute of Toxicology Research, Lucknow, India
Academy of Scientific and Innovative Research (AcSIR), New Delhi, India

Angela K. Talley, John Whisler, Ruobing Wang, Stefan H. Kappe, Olivia C. Finney and Susan Lundebjerg
Malaria Clinical Trials Center, Seattle Biomedical Research Institute, Seattle, Washington, United States of America

Sara A. Healy and Patrick E. Duffy
Laboratory for Malaria Immunology and Vaccinology, National Institute of Allergy and Infectious Diseases, National Institutes of Health, Bethesda, Maryland, United States of America

Sean C. Murphy
Department of Laboratory Medicine, University of Washington Medical Center, Seattle, Washington, United States of America

James Kublin, Peter Gilbert
Fred Hutchinson Cancer Research Center, Seattle, Washington, United States of America

**Carola J. Salas**
United States Naval Medical Research Unit Number 6, Lima, Peru

**Wesley C. Van Voorhis**
Department of Medicine, University of Washington Medical Center, Seattle, Washington, United

**States of America**
Chris F. Ockenhouse and D. Gray Heppner
United States Military Malaria Vaccine Program, Walter Reed Army Institute of Research, Silver Spring, Maryland, United States of America

**Kit-Ying Choy and Larry M. C. Chow**
Department of Applied Biology and Chemical Technology, The State Key Laboratory of Chirosciences, The Hong Kong Polytechnic University, Hung Hom, Kowloon, Hong Kong SAR, PR China

**Robert M. Sapolsky**
Department of Biology, Stanford University, Stanford, California, United States of America, Department of Neurosurgery, Stanford University School of Medicine, Stanford, California, United States of America
Stanford Stroke Center and Stanford Institute for Neuro-Innovation and Translational Neurosciences, Stanford University School of Medicine, Stanford, California, United States of America
Department of Neurology and Neurological Sciences, Stanford University School of Medicine, Stanford, California, United States of America

**Cheryl C. Y. Loh, Kitti W. K. Chan and Bruce Russell**
Department of Microbiology, Yong Loo Lin School of Medicine, National University of Singapore, Singapore, Singapore

**Rossarin Suwanarusk and Laurent Rénia**
Singapore Immunology Network, Agency for Science Technology and Research, Biopolis, Singapore, Singapore

**Martin J. Lear**
Department of Chemistry, Graduate School of Science, Tohoku University, Aza Aramaki, Aoba-ku, Sendai, Japan

**François H. Nosten**
Centre for Tropical Medicine, Nuffield Department of Medicine, University of Oxford, Oxford, United Kingdom
Shoklo Malaria Research Unit, Mahidol-Oxford Tropical Medicine Research Unit, Faculty of Tropical Medicine, Mahidol University, Mae Sot, Thailand
Centre for Tropical Medicine, University of Oxford, Churchill Hospital, Oxford, United Kingdom

**Kevin S. W. Tan and Yan Quan Lee**
Department of Microbiology, Yong Loo Lin School of Medicine, National University of Singapore, Singapore, Singapore
NUS Graduate School for Integrative Sciences and Engineering, National University of Singapore, Singapore, Singapore

**Rodolfo Thomé, André Luis Bombeiro, Luidy Kazuo Issayama, Thiago Alves da Costa, Rosária Di Gangi, Isadora Tassinari Ferreira, Ana Leda Liana Verinaud and Alexandre Leite Rodrigues Oliveira**
Department of Structural and Functional Biology, Institute of Biology, University of Campinas, Campinas, Brazil

**Catarina Rapôso and Maria Alice da Cruz Höfling**
Department of Histology and Embryology, Institute of Biology, University of Campinas, Campinas, Brazil

**Stefanie Costa Pinto Lopes and Fábio Trindade Maranhão Costa**
Department of Genetics, Evolution and Bioagents, Institute of Biology, University of Campinas, Campinas, Brazil

**Figueiredo Longhini**
Department of Hematology, Faculdade de Ciências Médicas, University of Campinas, Campinas, Brazil

**Philipp Stahl**
Institut für Virologie, AG Parasitologie, Philipps-Universität Marburg, Marburg, Germany

**Volker Ruppert**
Klinik für Kardiologie, Philipps-Universität Marburg, Marburg, Germany

**Ralph T. Schwarz**
Institut für Virologie, AG Parasitologie, Philipps-Universität Marburg, Marburg, Germany
Unité de Glycobiologie Structurale et Fonctionnelle, UMR CNRS/USTL nu 8576, Université de Lille1 Sciences et Technologies, Villeneuve d'Ascq, France

**Thomas Meyer**
Klinik für Psychosomatische Medizin und Psychotherapie, Georg-August-Universität Göttingen, Göttingen, Germany

**Rattiporn Kosuwin, Chaturong Putaporntip and Somchai Jongwutiwes**
Molecular Biology of Malaria and Opportunistic Parasites Research Unit, Department of Parasitology, Faculty of Medicine, Chulalongkorn University, Bangkok, Thailand

**Hiroshi Tachibana**
Department of Infectious Diseases, Tokai University School of Medicine, Kanagawa, Japan

**Yesmalie Alemán Resto**
Research Experiences for Undergraduates (REU) NSF Program - 2013 – Bigelow Laboratory for Ocean Sciences, Boothbay, Maine, United States of America

**José A. Fernández Robledo**
Bigelow Laboratory for Ocean Sciences, Boothbay, Maine, United States of America

**Ryusei Tanaka, Akina Hino, Juan Emilio Palomares-Rius, Ayako Yoshida, Haruhiko Maruyama and Taisei Kikuchi**
Division of Parasitology, Faculty of Medicine, University of Miyazaki, Miyazaki, Japan

**Isheng J. Tsai**
Biodiversity Research Center, Academia Sinica, Taipei, Taiwan

**Yoshitoshi Ogura and Tetsuya Hayashi**
Division of Microbial Genomics, Department of Genomics and Bioenviromental Science, Frontier Science Research Center, University of Miyazaki, Miyazaki, Japan

**Sebastian Emd and Thomas Kuhn**
Institute for Ecology, Evolution and Diversity, Goethe-University, Frankfurt am Main, Hesse, Germany, Judith Kochmann
Senckenberg Gesellschaft für Naturforschung, Biodiversity and Climate Research Centre, Frankfurt am Main, Hesse, Germany

**Martin Plath**
College of Animal Science and Technology, Northwest Agriculture & Forestry University, Yangling, Shaanxi Province, P. R. China

**Sven Klimpel**
Institute for Ecology, Evolution and Diversity, Goethe-University, Frankfurt am Main, Hesse, Germany
Senckenberg Gesellschaft für Naturforschung, Biodiversity and Climate Research Centre, Frankfurt am Main, Hesse, Germany

**Hui Wang, Tao Lei, Jing Liu, Muzi Li, Huizhu Nan and Qun Liu**
Key Laboratory of Animal Epidemiology and Zoonosis, Ministry of Agriculture, National Animal Protozoa Laboratory, College of Veterinary Medicine, China Agricultural University, Beijing, China

**Vladimír Skála and Petr Horák**
Charles University in Prague, Faculty of Science, Department of Parasitology, Prague, Czech Republic

**Alena Černíková**
Charles University in Prague, Faculty of Science, Institute of Applied Mathematics and Information Technologies, Prague, Czech Republic

**Zuzana Jindrová**
Charles University in Prague, Faculty of Science, Department of Parasitology, Prague, Czech Republic
Charles University in Prague, 1st Faculty of Medicine, Institute of Immunology and Microbiology, Prague, Czech Republic

**Martin Kašný**
Charles University in Prague, Faculty of Science, Department of Parasitology, Prague, Czech Republic
Masaryk University, Faculty of Science, Department of Botany and Zoology, Brno, Czech Republic

**Mala Sinha and Bruce A. Luxon**
Department of BioChemistry & Molecular Biology, UTMB, Galveston, Texas, United States of America

**Matthew McKnight Croken**
Department of Microbiology and Immunology, Albert Einstein College of Medicine, Bronx, New York, United States of America

**Yanfen Ma**
Department of Pathology, Albert Einstein College of Medicine, Bronx, New York, United States of America

**Lye Meng Markillie and Galya Orr**
Environmental Molecular Sciences Laboratory, Pacific Northwest National Laboratory, Richland, Washington, United States of America

**Ronald C. Taylor**
Computational Biology and Bioinformatics Group, Biological Sciences Division, Pacific Northwest National Laboratory, Richland, Washington, United States of America

**Louis M. Weiss**
Department of Pathology, Albert Einstein College of Medicine, Bronx, New York, United States of America

**Kami Kim**
Department of Microbiology and Immunology, Albert Einstein College of Medicine, Bronx, New York, United States of America
Department of Pathology, Albert Einstein College of Medicine, Bronx, New York, United States of America
Department of Medicine, Albert Einstein College of Medicine, Bronx, New York, United States of America

**Martin Vostrý**
Charles University in Prague, Faculty of Science, Department of Parasitology, Prague, Czech Republic
Institute of Haematology and Blood Transfusion, Prague, Czech Republic

**Anthony J. Walker**
Molecular Parasitology Laboratory, School of Life Sciences, Kingston University, Kingston upon Thames, Surrey, United Kingdom

**Ji Zhang and Jouni Laakso**
Centre of Excellence in Biological Interactions, Department of Biological and Environmental Science, University of Jyväskylä, Jyväskylä, Finland
Department of Biological and Environmental Science, University of Helsinki, Helsinki, Finland

**Tarmo Ketola and Johanna Mappes**
Centre of Excellence in Biological Interactions, Department of Biological and Environmental Science, University of Jyväskylä, Jyväskylä, Finland

**Anni-Maria Örmälä-Odegrip**
Department of Biological and Environmental Science, University of Helsinki, Helsinki, Finland

**Giacomo Bastianelli and Michael Nilges**
Institut Pasteur, Unité de Bioinformatique Structurale, Département de Biologie Structurale et Chimie, Paris, France

**CNRS UMR 3528, Paris, France**
Anthony Bouillon and Jean-Christophe Barale
Institut Pasteur, Unité d'Immunologie Moléculaires des Parasites, Département de Parasitologie et de Mycologie & CNRS URA 2581, Paris, France
CNRS, URA2581, Paris, France

**Christophe Nguyen and Dung Le-Nguyen**
SYSDIAG, CNRS UMR3145 CNRS-BioRad, Montpellier, France

**Else M. Bijker, Guido J. H. Bastiaens, Karina Teelen, Cornelus C. Hermsen, Anja Scholzen, Robert W. Sauerwein, Joshua M. Obiero, Marije C. Behet, Geert-Jan van Gemert, Marga van de Vegte-Bolmer and Wouter Graumans**
Radboud university medical center, Department of Medical Microbiology, PO Box 9101, 6500 HB Nijmegen, The Netherlands

**Remko Schats and Leo G. Visser**
Leiden University Medical Center, Department of Infectious Diseases, PO Box 9600, 2300 RC Leiden, The Netherlands

**Lisette van Lieshout**
Leiden University Medical Center, Department of Medical Microbiology, PO Box 9600, 2300 RC Leiden, The Netherlands
Leiden University Medical Center, Department of Parasitology, PO Box 9600, 2300 RC Leiden, The Netherlands

**Nilay Dey, Shivali Gupta and Janice J. Endsley**
Department of Microbiology and Immunology, University of Texas Medical Branch (UTMB), Galveston, Texas, United States of America

**Mauro F. Azevedo, Catherine Q. Nie and Paul R. Sanders**
Macfarlane Burnet Institute of Medical Research and Public Health, Melbourne, Victoria, Australia

**Mariela Natacha Gonzalez**
Instituto Nacional de Parasitología "Dr. Mario Fatala Chaben", Ciudad Autónoma de Buenos Aires, Argentina

**Rong Fang**
Department of Pathology, UTMB, Galveston, Texas, United States of America

**Nisha Jain Garg**
Department of Microbiology and Immunology, University of Texas Medical Branch (UTMB), Galveston, Texas, United States of America

Department of Pathology, UTMB, Galveston, Texas, United States of America
Faculty of the Institute for Human Infections and Immunity and the Center for Tropical Diseases, UTMB, Galveston, Texas, United States of America

**Brendan Elsworth, Sarah C. Charnaud and Paul R. Gilson**
Macfarlane Burnet Institute of Medical Research and Public Health, Melbourne, Victoria, Australia
Monash University, Melbourne, Australia

**Brendan S. Crabb**
Macfarlane Burnet Institute of Medical Research and Public Health, Melbourne, Victoria, Australia
Monash University, Melbourne, Australia
University of Melbourne, Melbourne, Australia

# Index

www.ingramcontent.com/pod-product-compliance
Lightning Source LLC
Chambersburg PA
CBHW061251190326

41458CB00011B/3637